How the Earth Turned Green

HOW THE EARTH
TURNED GREEN

A Brief 3.8-Billion-Year History of Plants

Joseph E. Armstrong

The University of Chicago Press
Chicago and London

Joseph E. Armstrong is professor of botany, head curator of the Vasey Herbarium, and director of the Organismal Biology and Public Outreach Sequence for Biological Sciences Majors, all at Illinois State University.

The University of Chicago Press, Chicago 60637
The University of Chicago Press, Ltd., London
© 2014 by The University of Chicago
All rights reserved. Published 2014.
Printed in the United States of America

23 22 21 20 19 18 17 16 15 14   1 2 3 4 5

ISBN-13: 978-0-226-06963-0   (cloth)
ISBN-13: 978-0-226-06977-7   (paper)
ISBN-13: 978-0-226-06980-7   (e-book)
DOI: 10.7208/chicago/9780226069807.001.0001

Library of Congress Cataloging-in-Publication Data

Armstrong, Joseph E. (Joseph Everett), author.
  How the Earth turned green : a brief 3.8-billion-year history of plants /
Joseph E. Armstrong.
      pages cm
  Includes bibliographical references and index.
  ISBN 978-0-226-06963-0 (cloth : alkaline paper) —
ISBN 978-0-226-06977-7 (paperback : alkaline paper) —
ISBN 978-0-226-06980-7 (e-book) 1. Botany. 2. Plants—History.   I. Title.
  QK45.2.A76 2014
  580—dc23
                                                        2013047074

♾ This paper meets the requirements of ANSI/NISO Z39.48-1992 (Permanence of Paper).

CONTENTS

Preface: A Botanist at Large   *ix*

1 A Green World   *1*

Wherein a discussion of "plant" and "plant kingdom" introduces the science of taxonomy and classification, and the nature of science is illustrated by explaining how we know the age of the Earth and why biologists care about elements from stars and molecules from space.

2 Small Green Beginnings   *31*

Wherein the discovery of microorganisms, their amazing numbers, and the places they live are explored; evidence of ancient life is examined; and the nature of metabolisms and the origin of their complexity is explained.

3 Cellular Collaborations   *75*

Wherein the diversity of unicellular organisms is explored, chloroplasts and mitochondria are obtained via symbiotic interactions between cells, and other features such as nuclei and sex are examined to determine what can be learned of their origins and functions.

4 A Big Blue Marble   *131*

Wherein algae are introduced, ocean ecology is explained, and phytoplankton diversity is explored.

5 Down by the Sea (-weeds)   *143*

Wherein coastal environments are contrasted to oceanic environments, and organisms adapt to the new challenges presented by living on coasts by becoming anchored, larger, and multicellular, which in the case of green organisms results in those algae called seaweeds.

6 The Great Invasion   *163*

Wherein the challenges and colonization of the terrestrial environment are examined so as to understand the adaptations of land plants, especially their life cycle.

7 The Pioneer Spirit   *197*

Wherein liverworts, hornworts, and mosses are examined to demonstrate their adaptations to
terrestrial life and their relationships to each other and vascular plants.

8 Back to the Devonian   *217*

Wherein a field trip to the Devonian introduces early vascular plants and examines how,
from such small beginnings, xylem and new ways of branching helped plants produce leaves and
roots and grow into trees, Earth's first forests.

9 Seeds to Success   *265*

Wherein the nature of seeds, their impact on the land plant life cycle, their history, and
the diversity of seed plants is investigated.

10 A Cretaceous Takeover   *295*

Wherein the quite singular ecological dominance and species diversity of the flowering plants are
examined in light of their novel features and their gymnospermous ancestry.

11 All Flesh Is Grass   *337*

Wherein the development of modern vegetation and recent interactions, like agriculture, and
their impact on both plants and humans are examined.

Postscript   *363*

Appendix   *365*

    Brown Algae and Tribophyceans   *367*

    Clubmosses and Fossil Stem Groups   *373*

    Conifers and Ginkgoes   *387*

    Coniferophytes: Cordaitales and Voltziales   *399*

    Cycads   *401*

    Ferns   *409*

    Gnetophytes   *431*

    Green Algae   *439*

    Green Bacteria   *451*

    Hornworts   *455*

    Horsetails   *457*

    Liverworts   *465*

    Mosses   *473*

    Phytoplankton   *481*

    Red Algae   *489*

    Rhyniophytes and Trimerophytes   *493*

Seed Ferns   495

Whisk Ferns   501

Notes   *507*

Glossary   *523*

References   *533*

Index   *553*

A Botanist at Large

It must be said at once that this book is better than having a sausage stuck to the end of your nose.

—Mark Golden, describing a book by S. Pomeroy in *Classical Review*

Laptop computers are a wonderful invention. My computer companion and I, both a little worse for wear, are sitting on a veranda overlooking some Central American rain forest. Far away from office, phone, and university (although the Internet lurks nearby), I am conducting field research on floral biology. Although it is still fairly early in the morning, my flowers opened at first light, so a couple of hours of field work and some breakfast are already out of the way. The stormy season is beginning and the weather here is changeable. For the last three days, hard, straight-down tropical deluges of warm rain have thoroughly soaked the forest. Today dawned pleasantly fair and breezy, rather unusual for the wet tropics where the air is often still and heavy with humidity. A front moved in off the Caribbean and I cannot tell what it will bring, but for the moment, this is tropical weather at its absolute finest. Perhaps this excellent tropical morning has produced the urge to be a bit reflective, an appropriate state of mind for writing a book preface, which is by no means the first thing written. Without all the distractions of modern life there is time to think about why I bothered writing a history of green organisms.

All evidences indicate we live on an ancient Earth, one with a long history. During this time green organisms arose early on from small, simple beginnings, developed, and diversified, culminating in something as interesting, as complex, and as diverse as a rain forest. But this history is a subject well known by only a fairly small cadre of biologists who call themselves botanists, and then only some of them.

So let me ask. Are you a botanist? If so, great! But now I must beg your indulgence. Please understand that this book was not written for you, and therefore is not like the books you usually read. This book was written for everyone else because the challenge I have set for myself is to relate the history of green organisms to nonbotanists in an interesting and understandable manner. As such, my primary purpose will be to function as both an interpreter and a filter. Science must be translated to be understandable to people who do not speak it, read it, or write it. Science is complex and detailed. If science is not winnowed of its many terms and details, people cannot see the forest for the trees. In fact, this is a common problem in teaching science. Translating real science into a more generally understandable narrative always risks allowing some errors in understanding to creep in. Science uses technical language, jargon, in order to be as precise as possible, so attempts to explain what we know in less technical language will mean some loss of precision; it is unavoidable. And of course this author's understandings may not be perfect either, alas. So any botanist still reading this book will have to allow me some latitude, a degree of flexibility, and some understanding because this book is not written wholly in the language of botany. And this plays to my primary expertise, which is explaining science to nonscientists.

This history is about green organisms for several reasons. First, they are what I study. As a botanist I have come to appreciate green organisms' subtlety, quiet resilience, sophistication, and diversity, and I enjoy telling people about them. Second, even though this book deals just with green organisms, so much knowledge exists that it poses a major challenge to assimilate, organize, and relate. My colleagues are industrious and the literature resulting from their scholarly activities accumulates more quickly than anyone can read. Fortunately, many others have helped by summarizing and explaining various parts of the whole story that they know well, and by necessity I must rely on their expertise and knowledge. On too many occasions, portions have been revised to take into account newly published studies, but you have to stop doing that or you never finish a book!

Another reason for writing this book is that we are surrounded by green organisms, but many people, and this includes more than a few nonbotanical biologists, are so plant blind that they perceive only a static green background, a passive scenery acting as a stage upon which the real actors, animals, move (Wandersee and Schussler 1999). Years ago, when still young, I asked a wise older colleague why so many nonbotanical biologists were so disparaging about plants, and he concluded that "it takes a certain mental and emotional maturity to appreciate something as subtle as a plant, which goes a long way toward explaining their behavior."

In our familiar terrestrial environment, flowering plants dominate most landscapes, and they also feed us, clothe us, and make us happy. Thus flowering plants occupy our attentions and as a result they are the primary subjects of most botany and horticulture courses, most books on plants, and most botanical research. But of all the many groups of green organisms, flowering plants were the most recent to appear, although even this event was at least one hundred million years ago. Trying to imagine the Earth without flowering plants is very difficult, yet for the vast majority of life's history on Earth, no flowering plants, not even land plants, existed. So suffice it to say at this point, far from being mere "pond scum," blue-green algae (cyanobacteria) is arguably the most important, successful, and influential group of organisms in Earth history.

Many people fail to understand how we can know anything at all about such ancient historical events. Such events are not subject to an experimental approach, but nonetheless we can construct hypotheses,[1] which can be evaluated. Although no one has ever observed a bacterial-type cell becoming a nucleated cell (better labels will be forthcoming), everything we know about the biology of cells and the diversity of microorganisms suggests such an event took place and many more events like it as well. "Just as in much of theory [hypothesis] formation, the scientist starts with a conjecture and thoroughly tests it for its validity, so in evolutionary biology the scientist constructs a historical narrative, which is then tested for its explanatory value" (Mayr 2004). Such narratives should be plausible biologically, compatible with observations and other data, and internally consistent, and they should have an explanatory power that makes sense out of previously unexplained or puzzling phenomena. Several such narratives will be presented and evaluated during the course of this history. Such hypotheses can be tested to see how well they account for things known. Often these scientific explanations attempt to determine patterns of descent with modification, one of Darwin's two great ideas.[2] To some people these narratives sound like just-so stories, and while they may start that way, ultimately these explanations get shaped by what is known and changed by new findings. So while not experimental, this is science nonetheless.

The task I have undertaken is to write a brief, understandable history of green organisms from their earliest beginnings to present day. Many biology students have found this subject interesting, but many more are interested in pursuing medical careers and avoiding anything sounding even remotely botanical. Still I operate on the premise that lots of people (okay, maybe not lots, but at least some really bright, curious people) will be interested in reading about green organisms, in finding out how many different ones there are, how we have come to have the ones we have, and why they look, grow, and reproduce the way they do. Along the way some basics of biology will be explained, fundamental stuff really, but seldom well explained in textbooks, if explained at all, which points out a major problem in science education.

An hour has passed since I began writing this. Beyond the veranda the lush, dense greenery of a well-watered rain forest is warming up. The day is noticeably hotter now, suggesting that a typical hot, muggy, tropical afternoon will follow. The history of green organisms, like other scientific histories, is composed of a series of events, and like players on a stage, the cast of organisms changes through time; groups of organisms appear and disappear or give rise to modified descendants. What we know of ancient organisms comes in part from the fossil record and in part from the patterns of relatedness observed among living organisms. Most often plant diversity is presented and taught as a series of groups, classification categories, thus putting the cart directly in front of the horse. But the real science is found in the explanations, the hypotheses, both supported by and accounting for the data that resulted in these classifications, and these are my primary subjects. Information about each specific group of organisms is relegated to appendices.[3] Thus in my educational approach I greatly deviate from the approach of botany textbooks by concentrating on what most books omit, and relegating the usual textbook material to appendices. So let me say how thankful I am that a publisher allowed me the latitude to use this approach.

Lastly, although far from perfect or complete, a coherent history of life and its diversity is one of the greatest of human intellectual achievements constructed by logically combining facts and inferences, and sadly, too few people know much about it. Many scientists have spent their careers accumulating these data and producing these explanations, so credit is given where credit is due. If you are so inclined, the references provided will connect you to many scientists and to their work, but no attempt has been made to present an exhaustive or comprehensive review of the literature. Such a list of references alone would be as big as this book. Further, no matter what, some of this book's contents will be out of date because science is an ongoing process; my colleagues continue to collect data, test hypotheses, and generate new explanations. By the time this book is published, some explanations included will have changed and new studies will have been published, but this rapid outdating is a measure of research activity and scientific progress.

In several instances I purposely digress from the main narrative to explain and demonstrate how science operates. This is, or should be, a motive for all teachers of science. The idea is to illustrate not just what we know but also how we know it and why we have confidence in it. Science is more than a body of knowledge; science is the process by which this knowledge was gained. At the present time, far too few people understand this. This lack of understanding allows many people to treat scientific explanations as if they were merely opinions to which, in fairness, their undocumented, unsupported opinions can be equated. This special form of ignorance is growing in influence, a type of know-nothingism that demonstrates the darker side of relying too much on faith and not enough on evidence and reason. Although the nitty-gritty details are omitted in this book, whenever feasible the observations that must be explained are presented to illustrate the tyranny of data. When you do science you cannot ignore the data. Anyone can pose an alternative explanation by ignoring what is known; thinking of a plausible alternative explanation that accounts for everything known is a real intellectual challenge not to be taken lightly. Critics of science routinely and universally fail this challenge.

I hope this approach will improve your understanding and knowledge of green organisms and, along the way, improve your appreciation for the scientific process that figured out all these things. After all, is it possible "for anyone to study plants without an elevation of thought, a refinement of taste, and an increased love of nature" (B. S. Williams, 1868)? I think not, so let us get started.

ONE

A Green World

Wherein a discussion of "plant" and "plant kingdom" introduces the science of taxonomy and classification, and the nature of science is illustrated by explaining how we know the age of the Earth and why biologists care about elements from stars and molecules from space.

In tropical forests, when quietly walking along the shady pathways, and admiring each successive view, I wished to find language to express my ideas. Epithet after epithet was found too weak to convey to those who have not visited the intertropical regions the sensation of delight which the mind experiences.

—Charles Darwin, 1839

IN THE RAIN FOREST

Biologists are fascinated by rain forest, and that is where this book about green organisms begins and ends. When you thought "green organism" did you envision a tree—or a tree frog? You may be disappointed, but the only green organisms considered are those that are green with chlorophyll, the pigment of photosynthesis. As you will see, the chlorophyll green of plants explains why tree frogs are green too, but that is as far as this book goes with animals. The lush and diverse vegetation of the wet tropics is as green as it gets, but these forests haven't always existed. And the same goes for other forests and grasslands, tundra and deserts too. All these types of communities and the organisms in them were different in the past, illustrating a fundamental principle of biology: things change. Organisms alter their environment, and then they must change to adapt

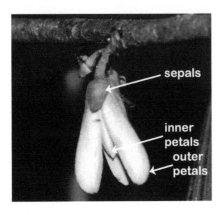

sepals

inner
petals

outer
petals

Fig. 1–1 Flower of *Anaxagorea crassipetala* showing three perianth whorls, the outermost sepals, and two whorls of petals, the outermost of which are very thick and fleshy (approx. life size). (Image source: The Author)

to the new conditions they produced; organisms[1] must evolve or go extinct. Evolution is a necessary aspect of life. This back-and-forth interaction between changing environments and adaptation means that life has a history of change resulting in organisms that possess a myriad of ways to succeed at making a living in the natural world. Biologists make a living explaining both.

The time is not quite 9 in the morning, the locale is tropical northeastern Costa Rica, and I have been up for over four hours because the flowers I am studying open at first light, just after 5 a.m. The flowers belong to a small, understory, rain-forest tree in the custard apple family, *Anaxagorea crassipetala*, and like many tropical organisms it lacks a common name. Rather than having the broad, flat, thin, and colorful petals of familiar flowers, this tree's outer whorl of three petals are a dull creamy color and nearly circular in cross section, looking like three little peeled bananas (Fig. 1–1). I am attempting to determine why this tree invests so much energy, 64% of the flower's dry weight, in making thick petals (*crassipetala* = with thick petals) that provide no visual display to speak of. Such research is evolutionary and yet another test of Darwin's hypothesis of natural selection, which allows us to predict that such a big energetic investment in petals should function to enhance the plant's production of offspring, so I need to figure out how.

Charles Darwin was not the first person to propose evolutionary change. His primary contribution was to propose a mechanism by which species of organisms could change, and that was natural selection. Within any species, organisms display heritable variations. Darwin's idea was that under any particular set of environmental circumstances some variants would be more successful in producing offspring than others. This would result in those variants becoming more common in the next generation. Simply put, nature "selects" certain variants as measured by the number of offspring they produce (the number of times those genes are passed on to the next generation). Natural selection is about differential reproduction, the exact opposite of random reproduction. And of course while "selects" sounds like a conscious decision, nothing of the sort takes place. Although lots of species of *Anaxagorea* and other custard apples have fairly thick, somewhat fleshy petals, none are as thick as those of *A. crassipetala*. So here I presume that the ancestors of this particular species were those trees that produced the fleshiest petals, and as a result, produced the most offspring by setting the most seed. How that might be the case is a question of interest and a worthy intellectual project for a botanical scientist. And before you get too excited, I have yet to find the answer although I have learned many other interesting things along the way.

Flowers and fruits present displays, either colorful (visual) or fragrant (olfactory) or both, and certain animals react to these displays to obtain a reward, which is often, but not always, food. Flowering plants do not provide such rewards to be nice to animals. These payoffs attract animals for the plant's purposes, the dispersal of pollen and seeds. Almost all of the many different rain forest plants in the surrounding community engage in such "cooperative" plant-animal interactions; a couple of ferns provide exceptions. Such cooperative interactions between animals and plants are one hallmark of flowering plants, the angiosperms. Prior to the appearance of flowering plants, the interaction between animals and plants could hardly be called "cooperative"; animals fed upon plants, a rather one-sided interaction. Of course, animals, including ourselves, still feed upon plants, but flowering plants found a way to benefit from some of this animal feeding by using animals as agents of dispersal. These cooperative interactions are one of the reasons for the extraordinary evolutionary success of flowering plants.

Can you envision how natural selection does this? Any individual plant whose reproductive structures were even slightly more attractive and more rewarding got better pollination and better seed dispersal, which resulted in more offspring carrying those genes that resulted in more attractive and more rewarding flowers and fruits. Animals that responded most efficiently to these displays got the most reward, which resulted in them having more offspring, who had similar behaviors. And pretty soon it's hard to tell who invited whom to the dance. But every study of any biological interaction is in one way or another fundamentally about evolution.[2]

Another dimension exists to such studies; they can be extended into time and space. *Anaxagorea* is found in the tropical forests of both Southeast Asia and the New World tropics of Central and South America, different species separated by thousands of miles of ocean. How do we account for such a distribution in plants lacking any means of long-distance dispersal? If you examine a world map you cannot help but notice that South America and Africa look like they could fit together like pieces of a jigsaw puzzle. Antarctica, Australia, New Zealand, New Guinea, and India also fit together with South America and Africa to form the former supercontinent Gondwana (see Fig. 9–5 for a map). Africa and South America rifted apart forming the South Atlantic. More rifting and rafting fragmented the rest of Gondwana, a process that continues to this day. Each land mass "rafted" to its present position with a complement of organisms. The genus *Anaxagorea* is estimated to be 44 million years old (Scharaschkin and Doyle 2006), but even this great age is not old enough for it to have been on opposite sides of the Atlantic Ocean before there was an Atlantic Ocean. The custard apple family is estimated to be over twice as old, so some 80 to 90 million years ago ancestors of this tree were flowering and fruiting while dinosaurs were still the dominant land animals. This still does not explain the distribution of *Anaxagorea*, which remains an unanswered question. Flowering plants appeared well over 100 million years ago, but of all the major groups of green organisms, flowering plants appeared most recently. Does that boggle your mind? Most of us are not used to dealing with such time frames. Humans tend to talk of ancient history in the thousands of years, but biologists toss off millions of years like dollar bills at a cake raffle. If recent events in the history of green organisms took place during the age of dinosaurs, then early events in this history are going to be really, really old, and in fact, the history of green organisms begins when the Earth itself was still quite young.

In our familiar terrestrial habitats, flowering plants dominate most areas. Flowering plants also occupy our human attentions because they are of utmost importance to us. We depend on flowering plants for most of our material needs, and because of this importance and their ecological dominance of Earth's terrestrial landscapes, flowering plants are the primary focus of botany, plant science, horticulture, and agriculture. But rather than just take flowering plants for granted, some of us ask questions. Where did flowers and fruits, and the plants that bear them, come from? How do flowering plants differ from their ancestors, and which differences account for their extraordinary success? What plants dominated the Earth before flowering plants, and how did they reproduce and disperse? To be curious is human, and lucky ones get to be botanists and satisfy some of that curiosity.

The fossil record tells us that other groups of plants appeared and flourished long before the flowering plants appeared. Other seed plants flourished, diversified, and dominated terrestrial landscapes, and while some, like the conifers, are still common and important, many other groups of seed plants have become extinct. Even earlier, over 360 million years ago, clubmosses, horsetails, and ferns were common and diverse components of ancient coal-forming forests. Today only a few descendants of clubmosses and horsetails remain as mere relics of their former glory. A few descendants of these ancient ferns still exist, but new groups of ferns have appeared and become common. While small and often overlooked, mosses, liverworts, and hornworts appear to be still older. All of these afore-mentioned green organisms share a life cycle that produces an embryo, an indication that they all share a common ancestry. All are quite correctly called plants, or even more specifically, land plants or embryophytes (EM-bree-oh-fights),[3] embryo-producing plants.

This brings up Charles Darwin again because he explained evolution as descent with modification. The characteristics of a species change through time because of natural selection (and other mechanisms that have been discovered since), but life is connected through time via common ancestry. The evidence of this is shared characters inherited from ancestors, the basis of classification, although at its inception the science of taxonomy included no evolutionary concept. Darwin fundamentally changed that. Now classification is about common ancestries and that in turn tells us about the histories of organisms and the characters they inherited from those common ancestors. The two, evolution and classification, are interwoven such that biologists talk about lineages and phylogenies (fye-LAH-jen-eez) more than groups. Still, classification provides useful labels.

As hard as it is to imagine, the fossil record tells us that the land was not always green with plants. Even the appearance of land plants and the greening of the land are relatively recent events occurring in the last one-eighth of Earth history. Prior to this, green organisms were found only in aquatic habitats, and most such organisms are called algae. Some algae are quite large, seaweeds that live anchored in coastal regions. Many more are microscopic organisms that drift along in the oceans. Still other green organisms exist for which even a general label, like algae, does not seem appropriate. Some bacteria are green, and one of these groups, the cyanobacteria or blue-green algae, is among the oldest, most common, most successful, and most influential groups of green organisms in Earth history. You may wonder how that can be. Whether any or all of these algae and green bacteria are correctly called plants remains a matter for further discussion. Yet even if technically not plants, all are part of this story, and as a botanist I make my living by learning and teaching about all these green organisms.

WHAT ARE THE CONSEQUENCES OF BEING GREEN?

Some important consequences of being green must be understood for this history to make any sense. Green organisms are green because they possess a pigment called chlorophyll, which captures solar energy. But green organisms cannot use light energy directly. The captured energy is used to synthesize molecules of sugar, so those crystals in your sugar bowl represent sunlight captured, concentrated, and transformed by a plant, probably sugar cane, into a molecular form, which is what photosynthesis refers to. Sugar is crystallized sunlight. No wonder a little candy can brighten your day! Sugars and their polymers (starches and celluloses) are part of a class of organic molecules called carbohydrates, an appropriate name for molecules made from very simple raw materials like carbon dioxide and water. From carbohydrates and their metabolic intermediates, green organisms synthesize every other molecule they need to grow and reproduce. So when I say green, I mean green with chlorophyll. Other green organisms, like some lizards, frogs, and katydids, are green because they have non-photosynthetic pigments that provide camouflage among foliage green with chlorophyll.

Green photosynthetic organisms are autotrophs ("self feeders") and in their abundance they provide for us all, so we refer to them as producers. Organisms that require pre-made organic molecules for energy and raw materials are called heterotrophs ("other-feeders" or consumers). As such we and other consumers must "eat" other organisms (either wholly or in part), their secretions, or their metabolic waste products as food. Eating and being eaten is a fact of life, a process by which the light energy captured by green organisms is passed through a series of consumers, a food chain, before eventually being lost as heat, which dissipates.[4] Everything else is recycled with the able assistance of decomposers, primarily fungi and microorganisms, heterotrophs who obtain their food from dead organisms or their metabolic wastes. A large part of ecology concerns such trophic (TROW-fic) or feeding interactions, the energy transfers that result, and the cycling of biogeochemicals, the elements of life.

Corner (1964) described a plant as "a living thing that absorbs in microscopic amounts over its surface what it needs for growth." Of course, this definition is so broad it would include fungi and bacteria too. But his point was that in one way or another, all the raw materials that green organisms need (light, carbon dioxide, water, and mineral nutrients) are dilute or diffuse, and thus plants must spread a tremendous surface area into their environment. As any solar engineer can explain, the problem with solar energy is that it takes a tremendous surface area to absorb any significant amount. Our familiar plants generate tremendous surface area in both the air and soil. Their roots and stems branch again and again, ramifying until both leafy stems and roots are a network filling the space around a plant, making the plant an environmental obstruction for capturing diffuse and dilute resources. Branches end in leaves, flat arrays of tissue for absorbing sunlight and carbon dioxide. As a consequence of hanging lots of broad, thin leaves in the air, plants constantly lose water, which needs to be replaced. So a network of roots is needed to absorb water and the dilute mineral nutrients dissolved in it, and a conducting tissue is needed to carry water from the network of roots to the crown of leaves. A weighty crown of leaves and branches also requires considerable structural support. In most familiar plants of forest and field, both support and conduction are performed by a vascular tissue called xylem (ZEYE-lem). Xylem cells are dead at maturity, but their thick cell walls form tubes, which are both strong and a convenient shape for conducting water. Whether support or conduction was the initial function will be discussed later. Trees and shrubs produce a new layer of xylem in their stems and roots each year, and the accumulated layers of xylem are called wood.

Plants rooted in the soil, stiff and massive with thick-walled xylem cells, are not motile (MOH-til) and free to move about seeking needed resources and mates, but yet they must acquire both. The form needed to obtain diffuse resources results in immobility, which explains why flowering plants use rewards to entice animals to act as pollen and seed dispersers. Other plants must disperse too, but they largely rely upon movement by wind or water. No costly rewards are needed for wind or water dispersal, but such abiotic dispersal agents generate another cost in the production of vast numbers of dispersal units needed to compensate for the randomness of the physical elements. The physical and biological constraints for acquiring the basic necessities of plant life and the costs of reproduction and dispersal very much shape most of the recent chapters in the history of the green organisms. As major obstructions, plants also greatly influence and change their environment, a lesson many people have not yet learned. Plants are the food and habitats for many other organisms, and they all exert influences on each other. As things change, all these organisms must adapt, reacting to both the environment and the influences of other organisms, so all of this pushing and pulling gets played out in the nonrandom reproductive success of many individuals, with the result that the entire evolutionary system generates biological diversity.[5] To provide you with an understanding of what is and what is not needed, and what was and was not possible, these biological basics must be explored, along with a very ancient history.

WHAT IS A PLANT? A TAXONOMIC PRIMER

You may have grasped from the preceding paragraphs that not all green organisms are correctly called plants. So what is a plant? If all plants are green photosynthetic autotrophs, why is it incorrect to call all green autotrophic organisms plants? The ecological role of a "plant" as a green photosynthetic autotroph tells us nothing about how the diversity of such organisms should be organized. As a term, *plant* is used as a label for one discrete group of organisms. Although presently no consensus exists of this group's membership, one thing is clear: it does not and cannot include all green photosynthetic organisms. In day-to-day usage "plant" and "green photosynthetic producer" are one and the same because of the prevalence of flowering plants, but obtaining a more sophisticated perspective is what education is all about.

Among all green autotrophic organisms you find only three different kinds of photosynthesis,[6] and if you organize all green organisms on the basis of their type of photosynthesis it produces an interesting result. Each unique type of photosynthesis is found in a different group of bacteria: the green sulfur bacteria, the green nonsulfur bacteria, and the cyanobacteria. The metabolic details of the photosynthesis do not matter to illustrate this idea. The type of photosynthesis illustrated in all the textbooks, the kind found in the chloroplasts of all other green organisms including plants, is the photosynthesis possessed by the cyanobacteria (blue-green algae). But this does not make them plants. To further compound the problem, some organisms that are clearly flowering plants on the basis of their structure and reproduction have lost their green color and become parasites on other plants (e.g., beech drops and witch weeds). Without question they are heterotrophs ecologically, but yet they clearly have green ancestors. So being green and photosynthetic cannot be used to group these organisms. As Darwin (1859) understood so clearly, common ancestry provides the key concept for organizing biological diversity and making sense of all of this. Organizing groups on the basis of common ancestry actually produces classifications that can be treated as hypotheses that yield testable predictions. And of course, testable predictions are both necessary and sufficient for doing science, both as a process by which we learn about the natural world, and for the explanations, the knowledge, this process generates.

As an organizing principle common ancestry reflects a very unique pattern in the natural world. Plants that share the features of flowers and fruits are the flowering plants. Flowering plants share the feature of seeds with other seed-producing plants, thus forming a broader inclusive group of seed plants. Seed plants are part of several successively broader and more inclusive groups, until finally they are included in a group of all land plants, which share a life cycle producing an embryo. Diverse evidence indicates that land plants have a common ancestry with green algae. And so it continues. The green algae and land plants are part of a very broad group of organisms with nucleated cells, which includes all large, familiar organisms and a good many microscopic ones as well. Organisms with nucleated cells share even more-basic features—for example, DNA, a genetic code, amino acids, and ribosomes—with bacteria whose cells lack nuclei. Thus all living organisms demonstrate an overall unity, leading biologists to conclude there was a universal common ancestor (a hypothesis). All of this diversity forms a pattern of nested sets based on common ancestry, groups nested within groups nested within groups, and so on. The nested sets outline a sequence of common ancestries and a sequence in which novel shared features appear. This is the fodder upon which many biologists feed.

The traditional classification hierarchy groups species in genera, genera in families, families in orders, orders in classes, classes in phyla, and phyla in kingdoms, producing nested sets. This is how humans organize everything from food to furniture to mailing addresses.[7] Our classifications of human artifacts are totally arbitrary, but to be useful scientifically our classification of life must accurately reflect groupings that resulted from real historical events, common ancestries. Classification as a science predates the theory of evolution by over one hundred years, so biological classification began with no preconceived notions about relatedness. Pre-Darwinian classifiers, called Linnaeans (after Carl von Linné, also known as Linnaeus, the father of taxonomy), made observations of similarities and then organized groups accordingly, and this resulted in nested sets because organisms possess more specific and more general similarities. Darwin grasped that only descent with modification would result in a classification that formed nested sets. Darwin's insights led to a paradigm shift, a change in how the taxonomic data were explained. After Darwin, shared features were interpreted to be the result of common ancestry, unless the similarities prove to be from convergent evolution, where unlike and unrelated ancestors came to have similar-looking descendants because of adaptations to similar environmental conditions.[8]

With Darwinian principles guiding the way, biological classification has continued to grow as a science; there are new classification hypotheses almost daily. Science continues to learn things about living organisms, and new data result in changes in our understandings and hypotheses. Indeed, in the past 20 to 25 years the growth of our knowledge and understanding of common ancestries has changed faster than ever, which results in lots of new hypotheses. Anyone who thinks classification is a dead, do-nothing, going-nowhere science could not be more wrong. All of this research has now caused biologists to question the continued usefulness of traditional classification categories.

HOW CAN KINGDOMS CHANGE?

A plant kingdom consisting of green algae and land plants was proposed by Copeland (1956), but a broader plant kingdom concept prevailed, one that included all algal organisms, as well as fungi, basically putting all the subjects of botanical study into the plant kingdom. This was neat, efficient, and very wrong. This was pretty much the state of affairs when I began studying biology over 40 years ago. My freshman biology textbook, *Life* (Simpson and Beck 1965), used three kingdoms. Members of the plant kingdom included land plants, all the algae, fungi,[9] and fungal-like organisms such as the slime molds and water molds. Of course animals were a kingdom.[10] The third kingdom consisted of protists, diverse unicellular organisms such as amoebae and paramecia. Bacteria were a phylum[11] within the kingdom of protists, but cyanobacteria were placed in a different phylum from the rest of bacteria. Unicellular organisms like euglenozoans and dinoflagellates formed still other phyla. Both euglenozoans and dinoflagellates have green and nongreen species—that is, some possess chloroplasts and some do not. So even when the plant kingdom was defined very broadly, it did not include all green organisms; some were protists.

By the time I was done with my college education, which took nine years from BA to PhD, the classification I had learned as a freshman had been replaced by a five-kingdom classification: animals, fungi, plants, protists, and monerans (Whittaker 1969, Margulis 1974, Margulis and Schwartz 1988) (Fig. 1–2). In this classification all bacteria and bacteria-like organisms were placed in Kingdom Monera. Fungi, long recognized

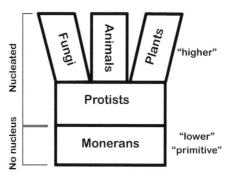

Fig. 1–2 Five kingdoms of living organisms. Monerans (bacteria and all similar organisms) occupy the basal position as lower or primitive life. Protists (composed of largely unicellular and small, simple multicellular organisms) occupies a position intermediate between the monerans and the kingdoms of higher, multicellular organisms: plants, fungi, and animals.

as a distinct group of organisms, were removed from the plant kingdom and treated as a separate kingdom. A number of animal-like, fungal-like, and plant-like organisms, mostly motile and unicellular, remained classified as protists. Four of these kingdoms (animals, fungi, plants, and monerans) were well but narrowly defined, and any organism thus excluded ended up classified as a protist. Perhaps the most surprising change was the removal of all algae from the plant kingdom, leaving it restricted to just land plants!

Traditionally the term protist refers to unicellular and small, simple multicellular organisms—for example, filaments and colonies. While some algae are motile and unicellular or of simple multicellular organization, like other protists, other algae are large, complex, multicellular seaweeds like the giant kelps. Classifying seaweeds as protists did not just enlarge the concept of a protist a little, it bent it completely out of shape. One creative solution was to not change the composition of this "catch-all" kingdom, but rather to change the name of the kingdom from Protista[12] to Protoctista, meaning "first to establish" (Margulis and Schwartz 1988), thus eliminating the conceptual problem of calling great big seaweeds protists via creative relabeling.

Not much came of this proposal. Scientists actually vote on such proposals, not in a formal ballot or in any organized manner, but by using or not using the proposed term or concept in their teaching and publications. If protoctista had been used by enough biologists, then this term would have found its way into mainstream biology. Sometimes the decision to use a new term or classification is partly a matter of opinion or preference; however, if a proposal is accompanied by some new and compelling data, then the likelihood it will be adopted improves greatly because it has the backing of evidence, the scientific trump card. For example, biologists readily accepted a new classification that placed brown algae within a diverse assemblage of algal and fungal-like organisms because they share a novel cellular organization, and for a while this grouping formed its own kingdom, but more on this later.

While this restrictive classification simplified the plant kingdom, and greatly shortened the syllabi for plant diversity courses, this hypothesis produced some huge problems. Even at that time no one ever doubted that green algae and land plants shared a common ancestry even though better evidence would be forthcoming. In this classification, two groups of organisms with a common ancestry are in different kingdoms (protists and plants), and similarly other protists were considered to have a common ancestry with either animals or fungi. This classification only makes logical sense if you view the relationships among these five kingdoms in terms of increasing complexity, as opposed to common ancestry (Fig. 1–2). The three kingdoms (animals, fungi, and plants) that contained all the large, complex multicellular organisms (except for the seaweeds) were viewed as "advanced" or "higher." The protists were unicellular or simple multicellular organisms, so they were viewed as intermediate between the "primitive" or "lower" organisms (the monerans) and the three "higher" kingdoms. The fundamental flaw in this type of thinking would be exposed later, but it resulted in goofy statements like "there is no such thing as a unicellular plant" (Kaveski et al. 1983).

In spite of its flaws, the five-kingdom classification held sway for almost 25 years. By this time, several new lineages of organisms had been identified, leading to a total of seven, eight, or more kingdoms. Clearly the whole question of kingdoms, and the traditional classification hierarchy, had changed dramatically. Many biologists were no longer trying to decide how many kingdoms there were, but rather trying to decide whether the concept of a kingdom made sense any longer at all.

Why does it matter? To those of us who use classifications to explain things, it matters very much. Chlorophyll is found in only two places: in certain bacteria and in chloroplasts, the photosynthetic organelle of all other green organisms. As it turns out, the two places are exactly the same, but for the longest time we did not know that. Evidence strongly supports the hypothesis that chloroplasts were free-living photosynthetic bacteria that became cellular slaves within a host cell. While this hypothesis explains many unusual features of chloroplasts, biologists wonder how many times this evolutionary event has happened. If red algae, green algae, and land plants share a common ancestor, then chloroplasts need to have been acquired only once. Otherwise chloroplasts must have had two or more independent origins. Evidence suggests the chloroplasts of red algae and green algae were passed around quite a bit, and the actual number of times chloroplasts have been borrowed or stolen all depends upon common ancestries.

Historically, this discussion about kingdoms is all backwards. Classification did not begin by deciding on kingdoms, the most general of classification groupings, and working down, it began with individual species and worked its way up. Several species concepts exist, but for our purposes a species is a group of similar organisms. Some definitions emphasize that the members of a species have the potential to interbreed—that is, members of a species share a common pool of genes, and thus species are real biological entities in a reproductive sense. But taxonomists seldom apply the interbreeding rule to decide upon species, and sometimes keep species separate even if they readily hybridize with others or reproduce only asexually. In practice the limits of species are determined by observing patterns of variation among a sampling of individuals that have been collected. When a significant discontinuity in the pattern of variation is encountered, someone decides whether these two "groups" constitute different species, and then each gets its own species name. The underlying assumption is that the discontinuity in variation results from a genetic isolation, and thus the two species concepts are connected in theory if not in practice.

Of course, classification greatly predates science, and many traditional and useful classifications of various portions of the natural world exist and persist. We all refer to weeds and wildflowers, toadstools and mushrooms, trees and shrubs, and so on, collective terms for certain sets of organisms. Such classifications are very arbitrary, and they can vary from time to time, from place to place, and from person to person. Some aspects of scientific classification are also somewhat arbitrary—for example, the amount of specific variation that constitutes different forms, different varieties, and different species can vary from classifier to classifier. Classification criteria are not universal, even in biology— for example, in comparison with botany, the genera and families in ornithology (avian biology) seem extremely narrow and based upon the slightest of differences. There are only about 9,000 species of birds in total, but they are organized into more than 200 families, which make Class Aves about the same size as a big flowering plant family (e.g., the grass family has about 8,000 species). In contrast, the 220,000 or so species of flowering plants are organized into not quite 400 families. Clearly botanists and avian biologists have different family concepts. Scientific classification differs from the many "folk" classifications in its scope and its intent because it attempts to be universal, classifying all living organisms, and in the process, biologists try to document how the natural world is organized, a necessary prelude to seeking underlying mechanisms.

The science of classification, taxonomy, began with the purpose of naming each and every species. In 1753 Linnaeus proposed using only binomial (= two names) species names consisting of a genus and a specific epithet to replace long descriptive names. Most of the 5,900 plant species names coined by Linnaeus still exist and they can be recognized by the initial L. following the species name—for example, *Liriodendron tulipifera* L., tulip tree (Table 1–1). Linnaeus gave catnip, one species of the cat mints, the species name *Nepeta cataria* L. Previously catmint was called *Nepeta floribus interrupte spicatus pedunculatis* (cat mint with stalked flowers in an interrupted spike). Clearly the latter name conveys more information, but such names quickly become a challenge to the memory. Remember *Liriodendron tulipifera* is the binomial, the species name. One of the most common of biological mistakes is to call *tulipifera* the species name; this is the specific epithet, which functions as an adjective, so it cannot stand alone. Specific epithets can be used more than once by combining them with different genera—for example, *Phaseolus vulgaris*, common bean, and *Thymus vulgaris*, common thyme. Each genus must be a unique name.

The genus was the first collective classification category grouping two or more similar species together based on more general shared features. The genus *Magnolia* has about 80 species including *Magnolia virginiana* L., *Magnolia grandiflora* L., *Magnolia tripetala* L., and *Magnolia acuminata* L. Observations of shared features throughout the biological world made it clear that bigger and bigger sets of organisms shared more and more general features, and thus higher classification categories were constructed (Table 1–1). *Magnolia* and *Liriodendron* share numerous general features and so these genera were placed in the magnolia family (Magnoliaceae) along with several other genera. The magnolia family in turn shares features with other families like the nutmegs and custard apples, and so this set of families was placed in the order Magnoliales. This order shares with a great many other orders the feature of embryos bearing two leaves (i.e., two cotyledons) and so formed a class of dicotyledons[13] (Magnoliopsida). Dicotyledonous plants (dicots) together with the monocotyledonous (monocots = embryos with one leaf) plants (grasses, palms, orchids, and gingers) constituted the flowering plants, the angiosperms. In traditional classification, angiosperms are usually treated as a phylum, the Magnoliophyta.[14] Angiosperms, along with all the other phyla of land plants,

Table 1–1. Classification of tulip tree

Species: *Liriodendron tulipifera* L.	Common names: tulip tree, tulip poplar
Genus: *Liriodendron*	

Family: Magnoli**aceae** (Magnolia Family)

Order: Magnoli**ales**

Class: Magnoli**opsida** (Dicotyledonous Plants)

Phylum: Magnoli**ophyta** (Flowering Plants)

Kingdom: Plantae

Genera and binomial species names are written in italics because they are Latin in form. Genera are capitalized; the specific epithet is not. A name, abbreviated name, or initial(s) following a species indicates who named the species, which in this example is L. for Linnaeus. Boldface suffixes are standardized endings denoting that taxonomic rank when added to a generic root. Irregular taxonomic names not based on generic roots can be found in older literature, e.g., Angiospermae, Dicotyledonae. Botanical taxonomy formally uses Division instead of Phylum, but I opted to use the well-known term phylum to eliminate some unneeded jargon.

either alone or with one or more algal groups, constitute the plant kingdom. In a similar manner all biological diversity was organized into a classification hierarchy from species to kingdoms.

Discovering and giving a scientific name to all organisms was a large task, but in the 1770s it seemed an attainable goal, especially from the perspective of European biologists. This is because only a small fraction of the Earth's biological diversity resides in Europe, and little did they know what exploration of the world beyond Europe, especially the tropics, would yield. Today the quest to identify and name all species continues with no end in sight, but modern taxonomists are faced with the rather gloomy prospect that many species will become extinct before they are known to science. This situation is so chronic some of my colleagues are even proposing that we give up trying to name newly discovered organisms, which takes too much time and study, and just try to find out how many species exist. But the number of species gets hard to count, and new species get hard to identify if previously discovered species have not been accurately described and named.

Another problem is that fewer and fewer people are studying taxonomy at the species level. In fact it is getting hard to find experts who can identify organisms for those of us who are not experts. One such taxonomic expert is the youngish fellow who occupies the office next door. He is without question the world taxonomic expert on an order of insects, the Psocoptera. Having accumulated some experience and knowledge, he is at the peak of his game at the age of nearly 80. Who will replace him? It is not that there is no need, but such biological study is no longer fashionable or fundable, and without some major research groups at botanical gardens and museums, this level of taxonomy would just about have ceased to exist. Taxonomic experts are themselves becoming endangered species.

Nonetheless this shift in emphasis away from taxonomy and toward higher-level phylogenetic studies during the past thirty years has changed our understanding of evolutionary relationships more than at any time since Darwin. Our knowledge has grown primarily from molecular studies that use shared characters in the genetic material itself, DNA. DNA is a long linear molecule composed of two strands of alternating phosphates and sugars.[15] Each sugar carries one of four different nucleic acids, ATGC,[16] each of which pairs with a nucleic acid on the other strand, which runs in the opposite direction. Because each nucleic acid always pairs with the same counterpart (A-T, G-C), the two strands have complementary sequences of nucleic acids (CATTAGG—GTAATCC). The nucleic acids are the letters of the genetic alphabet and they form three-letter words, the genetic code. Four letters taken three at a time yield 64 possible codes that with some redundancy stand for one of 20 amino acids plus a code for start and stop. For example, GGG, GGA, GGC, GGT all code for the amino acid glycine. So a particular sequence of nucleic acids can be read like an amino acid book to construct a particular protein with either an enzymatic or structural function.

DNA is copied with great fidelity, but not perfectly, so mistakes, mutations, occur. Mutations can vary from substituting one nucleic acid for another, which happens on the order of one mutation per 100,000 copies, to major rearrangements of genetic material like gene duplications or inverted nucleic acid sequences, which occur far less frequently. Shared sets of changes in the nucleotide base sequences are more likely to have been acquired via a common ancestor than via numerous independent chance events. Confidence increases as the size of the data set and the number of shared features increases because the odds of the same changes happening two or more times independently by chance alone go down as the number of similar changes goes up. Mutations are the ultimate source of genetic variation, although they are often thought of only as harmful changes. Some mutations cause lethal changes—that is, they render a gene nonfunctional; some alter a gene function. The new version might work better or worse, and that may depend upon the environment too. Many mutations are known to be neutral, not altering gene function—for example, a mutation in the third position of the glycine code (above) has no effect on the protein even though there was a mutation. Phylogenetic studies find and compare shared changes in DNA sequences, mutations that have been successfully incorporated into a gene. The more changes two organisms share, the more likely it is that they were acquired via a common ancestry.

In particular, molecular studies have shed considerable light upon the phylogenetic relationships among microorganisms, which in turn has led to considerable discussion of kingdom-level classifications among all living organisms (Blackwell 2004). New kingdoms have been proposed to accommodate newly defined groups with two consequences. First, the five-kingdom classification has been falsified. Second, sooner or later a decision will have to be made about whether each distinct lineage defined by common ancestry should be called a kingdom, or whether the concept of a kingdom remains useful at all. This problem will be examined in a couple of chapters. Even if everyone could agree, a taxonomic problem remains because the current phylogenetic pattern of life requires many more nested sets than provided by the traditional classification hierarchy.

The classification of tulip tree presented above does not show the problem (Table 1–1). But we now know that flowering plants are nested among seed plants, which are nested among woody plants, which are nested among megaphyllous plants, which are nested among vascular plants, which are nested among land plants, which are nested among a certain group of green algae, which are nested among the rest of green algae. And this is not everything! Quite a few new taxonomic levels would have to be added between phylum and kingdom to accommodate what is known. So this is why traditional taxonomy will be largely ignored from this point on, both in this book and in biology.

No consensus exists about what to do. A lot of new nonhierarchical names have been proposed for newly identified or redefined groupings, and some are both informative and useful. However, as much as possible familiar and common place names for groups of organisms will be employed because using all these new names becomes counterproductive when trying to communicate with the more general reader.[17] However, some familiar names refer to groups of organisms that have been found to be artificial assemblages—that is, those classification hypotheses have been falsified, and it would be incorrect or misleading to continue using those names. In some instances, these names can be redefined and used with care; in others, the name must be changed.

This common portrayal of the evolutionary relationships among the traditional five kingdoms of organisms (Fig. 1–2) demonstrates a popular misconception that requires a phylogenetic perspective to correct. The concept implied is simple: monerans are the smallest, simplest forms of life so they gave rise to more complex unicellular and colonial protists, which in turn gave rise to multicellular organisms. While in some general

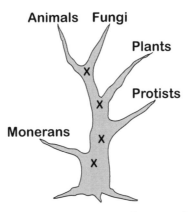

Fig. 1–3 Major branches (lineages) of the tree of life whose branching pattern shows the hypothesized ancestral relationships. Generally members of the Protista are considered intermediate between the monera and the "higher" or multicellular organisms: plants, fungi, and animals. However, this diagram depicts a very different relationship than Figure 1–2. Here all five kingdoms occupy the ends of branches. Xs indicate places where divergent lineages above had a common ancestry.

sense this is true historically, it fails to correctly represent relationships among living organisms. Present-day protists are not ancestral to fungi, plants, or animals. And present-day bacteria are in no way ancestors to living protists. More correctly, plants, fungi, and animals had a common ancestry with organisms that if alive today would be classified as protists, but modern protists have the same ancestors. Similarly, if bacteria-like organisms were ancestors of protists, then modern monerans have the same ancient organisms as ancestors. To correct this diagram, it should be redrawn as a tree where the end of each branch represents living organisms and the joining of the branches to the trunk represents the common ancestries (Fig. 1–3).

The branching pattern of this tree of life only generally hints at hypothesized common ancestries as presently understood. Its shape implies several relationships that have not been explored yet. Fungi were placed closer to animals purposely even though traditionally fungi have been considered botanical organisms. Plants were placed closer to the protists. The animal-fungi branch and the plant-protist branch join further down, suggesting a more ancient common ancestry of all organisms with nucleated cells, and finally at a much lower branch, the monerans have a common ancestry with all the other kingdoms.

While sketches such as this hint at relationships, new classification criteria have made phylogenies and phylogenetic diagrams less subjective. About one hundred years after Darwin, the introduction of cladistics, which uses an explicitly Darwinian approach, began changing classification. Hennig (1966) reasoned that since evolution proceeds by speciation, where various mechanisms partition an ancestral species into two genetically isolated but related species, then every taxonomic group must be derived from such a speciation event and must represent a single lineage of descent. In Hennig's terminology each evolutionary lineage represents a clade. Thus each taxonomic group should be a clade, and a classification hierarchy would represent a series of nested clades. Nested sets within nested sets is fine, but it has proven difficult in many cases to use the traditional taxonomic hierarchy to label a cladistic phylogeny because so much branching has occurred in so many lineages that the traditional classification hierarchy does not provide enough labels. Subsequent chapters will provide plenty of examples.

A cladistic representation of this simple tree of life (Fig. 1–4; compare with Fig. 1–3) would show a series of clades: five clades, one for each of the five kingdom lineages (monera, protists, plants, fungi, and animals), plus a fungi-animal clade; a clade of plants, animals, and fungi; a clade of organisms with nucleated cells (protists, plants, animals, and fungi); and one all-inclusive clade of life (Doolittle 1999). Green organisms are found in three of these clades: monera, protists, and plants. This phylogeny was presented for purposes of illustration, and while it approximates some aspects of current phylogenetic hypotheses, it is overly simplistic and lacking some critical diversity that will be presented later.

One big advantage of Hennig's cladistic approach to classification is that it provides a sequence of common ancestries, which represent a relative temporal sequence of biological events. In other words, cladistic phylogenies predict a sequence of common ancestry, one before another, but without any time scale. But matching a cladistic phylogeny to a sequence

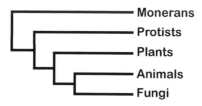

Fig. 1–4 Cladistic diagram (cladogram) of the five kingdoms of living organisms showing a sequence of common ancestries. Animals and fungi have a common ancestry more recent than they have with plants. Then the fungi-animal-plant clade has a common ancestry with protists. And lastly the protists-plants-animals-fungi clade has a more ancient common ancestry with monerans. Note the number of clades: one for each "kingdom" (five), plus an animal-fungi clade, a plants-animals-fungi clade, a protists-plants-animals-fungi clade (all the organisms with nucleated cells), and lastly a clade of all living organism, nine clades in all.

of events in the fossil record allows the phylogeny to be tested, and then the geological history produces a real chronology. For example, if we hypothesize that flowering plants had a common ancestry among nonflowering seed plants, then nonflowering seed plants should appear in the fossil record before flowering plants. In this manner the fossil record has the potential to either refute or provide support for various phylogenetic hypotheses, in so much as the critical characters can be observed in fossils. When the event in question deals with the appearance of nucleated cells, the fossil record is of limited value.

SO WHEN DID THINGS HAPPEN?

Absolute dates for ancient events are provided by geological data. The geological timescale begins with the formation of Earth, an event that took place 4.5 billion years ago.[18] Earth history is composed of four eons, Hadean, Archean, Proterozoic, and Phanerozoic (Nisbet and Sleep 2001) (Fig. 1–5). Dates will be presented as bya (billion years ago) and mya (million years ago). The Hadean eon is pre-life and represents the time from the accretion of Earth during the formation of the solar system, 4.5 bya, to the origin of life estimated to be at 4.0 bya (± 0.2 bya). The Archean eon represents the time from the beginning of life to the origin of protist-like organisms, 4 bya to 2.5 bya. The Proterozoic eon represents the period from the origin of protist-like organisms to the emergence of multicellular organisms visible to the naked eye, 2.5 bya to 0.54 bya (540 mya). Finally the Phanerozoic spans the time from 540 mya to present. The Phanerozoic is broken into many subunits of geological time called periods (Cambrian to Tertiary), and most familiar groups of organisms make their appearance during this eon (Fig. 1–5). Big discontinuities in the geological record make easy markers and thus they define the boundaries of geological periods. The subdivisions of the Phanerozoic often involve the first appearance, or disappearance, of certain diagnostic fossil organisms. On the scale of this geological time table, the point marking the appearance of *Homo sapiens* is so recent it is covered by the thickness of the end line. Sort of puts us Johnny-come-lately organisms in perspective, doesn't it?

These dates are much more than just educated guesses. An enormous amount of scientific effort was expended to generate the geological time scale, and lots of evidence supports it. By the early 1800s geologists understood that the many rock layers were produced by slow and ongoing geological processes, which suggested the Earth was of considerable age and had changed a lot. Such geological ideas greatly influenced Charles

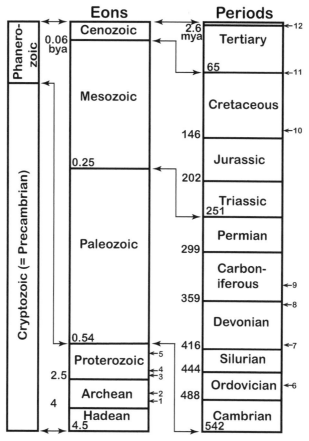

Fig. 1–5 Geological timetable. The first two columns represent all of Earth history. The Cryptozoic (formally called the Precambrian) consists of three eons (Hadean, Archean, and Proterozoic) that account for 88% of Earth history. The Phanerozoic accounts for the remaining 12% and is expanded in the second column and subdivided into the Paleozoic, Mesozoic, and Cenozoic eons, which are further expanded and subdivided in the third column into Periods. Time is shown in mya (millions of years ago) and bya (billions of year ago). Numbered arrows indicate major events in the history of green organisms: 1. oldest rocks, 2. oldest known fossils, 3. estimated age of nucleated cells (eukaryotes), 4. first fossil seaweed, 5. oldest animal fossils, 6. first land plants, 7. first vascular plants, 8. first seeds and ferns, 9. forests of clubmosses, horsetails, and ferns, 10. first flowering plants, 11. K-T boundary extinction, 12. Quaternary begins 2.6 mya—Pleistocene ice ages occur and humans appear. (Image source: After Geological Society of America's 2009 time scale, http://www .geosociety.org/science/timescale/timescl.pdf).

Darwin; if the Earth were old and had changed, he reasoned that organisms must have changed too.

Sedimentary rocks are deposited in layers, strata, one upon top of another. Nowhere on Earth is there a complete sequence of rock strata representing all of Earth history. Partial sequences exist in many places, and by correlating overlapping portions from different locations, a complete sequence was reconstructed by careful mapping and comparisons. The basic principle is simple: within any undisturbed sequence of strata, the oldest layer is on the bottom and the youngest on the top. Old, older, oldest, provides useful relative ages but no precise dates. Rock strata could be thicker or thinner, but when comparing strata, thicker might mean either a longer period of deposition or a faster rate of deposition, so age could only be estimated. Because the age estimates depended upon so many assumptions, the dates were problematic. Nonetheless geologists were still thinking in terms of millions of years rather than thousands of years.[19]

In addition to sedimentary rock strata, there are igneous rock strata that are formed by volcanic events. Some igneous rocks contain radioactive elements and their absolute ages can be determined by radioactive decay. Radioactive elements have known half-lives, a period of time during which half of the atoms will decay spontaneously into atoms of a stable element. The ratio of radioactive parent element to decay element (potassium-argon, uranium-lead, rubidium-strontium) tells you how many half-lives have transpired since the igneous rock was formed. Different radioactive elements have different half-lives, and several radioactive elements with very long half-lives are used to date very old rocks—for example, the half-life of potassium-40 is 1.3 billion years.

If the ratio of potassium-40 atoms to argon atoms in an igneous rock is 1:3 (two half-lives), the rock is 2.6 billion years old. Uranium 238 has a half-life of 4.5 billion years, so if you examine a sample that has a ratio of U-238 nuclei and lead 206 nuclei of 1:1, then the sample is 4.5 billion years old. Ideally, ratios are calculated for several different radioactive elements with different half-lives, which should provide very similar independent estimates of age. Radioactive decay is unaffected by environmental conditions and can be calculated with great precision, including corrections for several different sources of error. Once an igneous rock stratum is dated, then the sedimentary rocks above are known to be somewhat younger, and those below somewhat older. Many such comparisons have generated the dates on our present geological timetable.

Geological forces, weathering, erosion, and plate tectonics destroy rocks, and the older the rock the greater the opportunity for its destruction. Really old rocks are rare, and no rocks derived from the Earth's formation still exist. The oldest known Earth rocks were found in Northern Quebec and date to 4.28 billion years old (O'Neil et al. 2008), significantly older than the previously dated oldest Earth rocks, which were 3.7 to 3.8 billion years old. However, in comparison with the geologically active Earth, the Moon has been geologically inactive. Rocks left over from its molten formation are just lying around for the occasional passer-by to gather, as a few adventurous humans have done. The age of the Earth is based on Moon rocks and the assumption that both were formed at the same time.

A COSMIC CONTEXT

Earth's geological history can be placed into a larger cosmological chronology. Some understanding[20] of the cosmos is needed to understand the phenomenon of life on Earth. Life deals with energy and matter, so it helps to have some idea of where all this stuff came from, and why we have so much of this and so much of that. The universe is estimated to be 13.75 billion years old (± 0.11 by), about three times older than Earth. According to the most popular current theory, everything began when a quantum fluctuation in a void produced an unimaginably hot, unstable bubble of energy that expanded explosively in a Big Bang that ultimately produced all matter and energy. If the universe started out as an infinitesimally small mote, you naturally may wonder where everything, all the stuff, came from. As strange as it may seem, the remarkable expansion was possible because in sum the positives, matter and energy, and the negatives, gravity and dark energy, equal zero, nothing, in total. Everything equals zero, so there was no net change, a very creative accounting! This and many other aspects of modern physics may strike you as counterintuitive or just downright crazy thinking. However, these hypotheses were constructed to explain the available data, so no matter how weird, these explanations account for what is known.

As the universe's bubble expanded, it cooled. When it had cooled to about 10 billion K (degrees Calvin), units of energy began forming pairs of particles, one each of matter and antimatter. Most of these matter and antimatter particles annihilated each other, producing a photon, a packet of light,[21] with each annihilation. With complete matter:antimatter symmetry there would be nothing left but light, a bright but very empty universe. One extra particle of matter out of every three billion or so particle pairs accounts for all the matter in the universe. And in a classic of understatement, all of this took place rather quickly.

Much, much later in the history of the universe, about two minutes after the Big Bang, the universe had cooled to 1 billion K. For the next 300,000 years or so, matter in the expanding universe existed as hot plasma, a liquid gas of sorts, which as it expanded continued to cool. When cooled to about 4,000 K, the opaque plasma became transparent, and light, the photons from all those matter-antimatter annihilations, streamed forth. Let there be light, indeed! The Big Bang hypothesis predicts this event, and the resulting cosmic afterglow, the light from the Big Bang's fireball, has been observed by the COBE satellite, just one of many Big Bang predictions so far confirmed.

A certain frothiness existed within the plasma, and this resulted in an uneven distribution of matter. After a billion years, the matter that would form countless galaxies was distributed in swirling masses of helium and hydrogen. The observed proportions of helium (He) (28% of everything) and hydrogen (H) (71% of everything), the two simplest elements in the universe (check your handy periodic table), are a direct "fossil" of the Big Bang, which theoretically accounts for this ratio. H and He together total 99% of all matter, so the 1% that remains accounts for everything else, all the other elements on the periodic table.

Further condensation formed stars, whose fusion furnaces ignited from the heat generated by gravitational forces pressing their matter together. Light began radiating out from stars at its constant speed (C), but because of the universe's expansion, the stars and galaxies are simultaneously receding from each other at a great rate. When a light source and an observer are moving rapidly away from each other, it causes an apparent slowing of the speed of light, a Doppler effect that shifts the spectrum toward red wavelengths. The greater the Doppler effect, the greater the redshift, and the further away the star. Redshift was discovered by the astronomer V. M. Slipher of the Lowell Observatory; if distance and redshift remain proportionate (Hubble's law), redshift is what provides us with the estimates of the size, and therefore, the age, of the universe. Really distant celestial objects have the biggest redshifts, and the light observed from these objects started moving across the universe a very long time ago. Such light is the fossil record of the cosmos.

Stellar fusion of atomic nuclei produced, and continues to produce, all of the elements heavier than helium, and these elements are flung across the cosmos by the explosive supernova deaths of stars. A swirling cloud of hydrogen, helium, and the heavier elements made in the atomic cauldrons of ancient supernovas coalesced to form the solar system. The third innermost planet is a rocky globe called Earth, and because it orbited its star within a "just right" zone, upon this planet life appeared. Life may have appeared on the fourth planet as well, but we remain uncertain, and presently Mars seems lifeless, although it almost certainly was wetter and warmer in its planetary infancy.

An examination of the elements produced by stellar fusion shows that the basic chemistry of the universe turns out to be very compatible with the biochemistry of living organisms. Life is composed of a relatively simple recipe consisting primarily of just six elemental ingredients, carbon, hydrogen, oxygen, nitrogen, sulfur, and phosphorus (CHONSP) (Oro 1983; De Duve 1995). Together with helium, an inert gas, the CHONSP elements compose 99.9% of the matter in the entire visible universe. The CHONSP elements compose over 99% of living matter, and in proportions very similar to those found in the universe. Living organisms are composed of 62% hydrogen, 27% oxygen, 8.5% carbon, 1.9% nitrogen, 0.1% sulfur, and 0.1% phosphorus. This is remarkably similar to the proportion of these same elements found in the volatile fraction of comets, which are left over from the formation of the solar system (H = 56%, O = 31%, C = 10%, N = 2.7%, S = 0.3%, P = 0.08%) (Delsemme 1998, 2001). Life is composed of hydrogen and cosmic star-stuff, the most common elemental products of stellar fusion in the normal proportions of the universe. Helium, the second most common element, is inert and does not form combinations with other elements. As far as anyone can tell, helium exists to fill balloons.

Numerous origin-of-life hypotheses exist, and it is not my purpose to review them. All involve plausible explanations by which inorganic elements combine to produce organic molecules, whether on Earth or in space, or both, since one does not preclude the other. Under the conditions that formed the solar system, the CHONSP elements will combine spontaneously in a number of ways to form simple molecules that in turn can be used to form the four basic classes of organic molecules: nucleic acids, proteins, sugars, phospholipids. Molecules of life-generating significance, all composed of CHONSP elements, have been identified in interstellar space: hydrogen (H_2), ammonia (NH_3), water (H_2O), carbon monoxide (CO), hydrogen sulfide (H_2S), hydrogen cyanide (HCN), formaldehyde (CH_2O), cyanamide (H_2NCN), phosphorus nitrile ($P[CN]_5$), and many nitrogen-containing polycyclic aromatic hydrocarbons. The presence of these molecules in space has been determined by examining the light of distant stars. As starlight passes through such interstellar molecules, some characteristic wavelengths are absorbed by each type of molecule, producing spectral signatures that indicate the identity of the absorbing molecule. The building blocks of life form and exist in deep space (Cloud 1988).

To test the idea that organic molecules occur in comets, experimental syntheses have been attempted in laboratory environments simulating deep space. Using the raw materials known to be in comets, experiments have produced the three most common amino acids found in proteins of living organisms: glycine, alanine, and serine. In addition, organic molecules, including sugars and amino acids, have been found in meteorites, and the amino acids are the same ones in the same general proportions as those synthesized experimentally (Sephton 2001). Diverse organic molecules including the basic building blocks of life can be synthesized under conditions similar to those found on Hadean Earth (Miller and Lazcano 2002). The conclusion drawn from such research is quite clear: organic molecules from inorganic sources are much more common in the cosmos than anyone ever imagined, and the most easily synthesized amino acids are the most common ones found in living organisms.

Comets could well have been the primary means by which either organic molecules produced in deep space or their biogenic molecular precursors were delivered to Earth's surface. Even at today's low rate of bombardment, comets are estimated to deliver about 300,000 kg, over 68 tons, of organic material to Earth every year (Whittet 1997). During the heavy bombardment of Hadean Earth, the same bombardment that left the Moon heavily cratered, the delivery rate of organic molecules has been estimated at 50,000 tons per year, a rate of delivery high enough to account for Earth's current total organic biomass, the mass of all living organisms, in just 10 million years (Whittet 1997; Delsemme 2001). However, a comet's impact generates heat, which could destroy organic molecules, and this casts some doubt about the role comets played in delivering organic molecules to Earth (Lumine 2001).

Spontaneous synthesis of both life-generating molecules and the organic building blocks of life could also have taken place on ancient Earth. Different combinations of the ten simple biogenic molecules, under plausible pre-life conditions on Earth, will readily form amino acids, nucleic acid bases, sugars, and phospholipids (Oro 2002). Ancient Earth differed from Earth today in two major ways that explain why presently no such abiotic syntheses take place. Ancient Earth lacked living organisms and its atmosphere lacked free oxygen. Presently Earth's atmosphere is 20% oxygen and very oxidative. Without oxygen, the atmosphere of Hadean Earth was either a reducing environment or a neutral mixture of carbon dioxide and nitrogen (Davies 1999; Fry 2000). Even if neutral, volcanic conditions and geothermal vents would have provided places with reducing environments where syntheses of organic building blocks could have taken place easily (Miller and Lazcano 2002). Oxidative versus a neutral-to-reducing environment makes a huge difference in what type of organic chemistry will occur spontaneously. In a reducing environment, spontaneous syntheses of small organic molecules including amino acids and nucleic acids are readily accomplished, even though such reactions are virtually impossible in an oxidative environment. Laboratory experiments mimicking the hypothesized conditions of Hadean Earth have produced a variety of organic molecules spontaneously from very simple materials.

The most famous took place in 1953, the Miller-Urey experiments.[22] A sealed apparatus was constructed containing water, ammonia, methane, and hydrogen, a clear, stable mixture, but with neither of the two major gases forming Earth's atmosphere today, nitrogen and oxygen. The system circulated water from liquid to vapor and back to liquid again. Electrodes provided a spark, miniature lightening, like that seen in a volcanic eruption. Within days the mixture darkened. Chemical analysis revealed a mixture of organic molecules had spontaneously formed, including five common amino acids.[23] The results of such experiments allow us to conclude that the essential molecular building blocks of life are easily synthesized from inorganic materials, and under certain circumstances, organic molecules may have been common and readily available.

Various hypotheses have been proposed to account for how the three basic components of a living system—an environmental interface, a metabolism, and an inheritable information system—can be formed from an array of basic organic molecules (e.g., Cairns-Smith 1985). Unfortunately, or fortunately, depending upon your interest in this subject, it would take another whole book to even outline all these ideas. But what we do know is that life appeared pretty early in the history of Earth.

If life is composed of the most common star-stuff and the basic building blocks of life appear by simple, spontaneous self-assembly, does this mean life in the universe is as common as science fiction writers have always assumed? On a cosmic scale, perhaps it is quite likely that life is common, but that still means life occurs on infinitesimally small specks, planets like Earth spread ludicrously far apart in a universe of mind-boggling dimensions. Words like *big* and *vast* just do not assist the human mind to comprehend what "common" means in a cosmic sense. However, enough planets around other stars have now been detected to suggest planetary systems are relatively common in the same cosmic sense. In fact, at this writing over 3,000 planets have been detected by the orbiting Kepler observatory, some subject to verification, and at the rate they are being found, at least 17 billion Earth-sized planets (smaller, rockier, closer to their star) should exist just in the disk of the Milky Way galaxy alone. As we increase our ability to detect planets of this size, more will be found in the "Goldilocks zone," that distance from the star they orbit that is just right for life, and many will be within 200 light-years of the Sun, a distance which is considered to be the immediate cosmic neighborhood (Borucki et al. 2010).

Two hundred light-years means that a spaceship traveling at the speed of light will take 400 years to do the round trip. Consider what a trip of such duration means to organisms such as ourselves. In 1606 the sailing ship *The Discovery* left London bound for Virginia, where it arrived 5 months later. If *The Discovery* were an interstellar ship capable of light-speed and it was launched in 1606 to travel to a new planet orbiting a star some 200 light-years away, it would just be returning to port in present-day London. Imagine everyone's surprise. A lot has changed in London, and elsewhere, between 1606 and now. This is why even the most optimistic scientists among us remain very skeptical about the possibility of Earthly visits by intelligent extraterrestrials or their mechanical devices. First, the time and energy needed to transit cosmic distances are staggering, and presently we know of no means of circumventing these limitations. Second, if what we know of Earth history is at all typical of life elsewhere in the universe, then life may be quite common, but the vast majority of life-bearing planets would be expected to have organisms no bigger or structurally more complex than bacteria. With just one example to study (Earth) we cannot determine how likely is the appearance of large, complex organisms, let alone ones intelligent enough to contemplate this problem or to solve the technological problems of interstellar travel. Everything else is science fiction.

Nonetheless one close neighbor, Mars, offers an intriguing possibility. Our study of Mars provides evidence that early in its history Mars was warm enough to have running water. When young, Mars also was geothermally active, so places with a reducing environment existed and organic syntheses could occur. And if life is, as we think, pretty much a done deal wherever appropriate conditions occur, then living organisms could well have existed on Mars during its youth.

This brings us to a meteorite named AHL 84001, which was found in Antarctica's deep freeze. Elemental analysis indicates it is nearly 4 pounds of Mars. This meteorite has the notoriety of being the oldest physical object (4.6 billion years) that humans have ever found and examined.[24] This piece of Mars was flung into space by a big meteor impact about 25 million years ago. Some 10 to 20 thousand years ago it wandered into Earth's gravitational field and smacked into Antarctica. Great pains were taken to make certain of no Earthly contamination. Analysis of the interior of this ancient meteorite has found some tantalizing hints of life, what appear to be biochemical signatures of living organisms. However, not everyone is convinced that some as-yet-unknown abiotic process didn't produce these molecules. Stephen Jay Gould's remarks about the evidence indicating life on Mars can be summed up as "so what?" After all, based on what we know about Earth's living organisms and the universe they live in, life of a simple sort was expected. Confirmation would be nice, and if life is 2 for 2 in this planetary system, then we would gain great confidence in our surmise that life is common, in a cosmological context.

The ultimate fate of our universe remains unknown. Scientists have long thought that there are two possible fates (hypotheses). If there is enough matter, then gravitational forces will slow and reverse the expansion, and the universe will end in a Big Crunch. If not enough matter exists, then the universe will continue to expand although at a slower and slower rate, and as entropy increases, the universe will become a cold, dark hulk, ending in a Big Chill. So far not enough matter has been found to suggest a Big Crunch, unless a lot of unobservable "dark matter" exists. Recent observations suggest the universe's expansion actually seems to be accelerating rather than slowing, which requires a third hypothesis. An accelerating expansion requires a new force, a "dark energy," great enough to counteract gravity and push things apart, and this hypothesis predicts a rather shocking cosmic doomsday. As dark energy overcomes the pull of gravity, the universe ends with a runaway expansion at an age of about 35 billion years, and ends with a really Big Rip, when galaxies, stars, planets, and then molecules and atoms break up as the forces holding them together are overwhelmed by dark energy. Thirty years before the idea occurred to scientists, the Big Rip was very accurately depicted in the floor show at Milliways, *The Restaurant at the End of the Universe* (Adams 1980). But such distant futures have no present consequences for life on Earth. Long before the Big Crunch, Big Chill, or Big Rip, the Sun's fusion furnaces will have burned out. *C'est la vie.*

All the events described above have happened in a sequence: universe, first-generation stars, atomic fusion leading to CHONSP elements, supernova deaths of stars, second-generation stars and solar systems, Earth, life on Earth. As this narrative of green organisms unfolds, you will learn that some groups of organisms appeared before others, and some share a common ancestry with later-appearing groups. Since organisms inherit some features from their ancestors, ancestral organisms must be introduced before later-appearing organisms, their descendants. Various innovations of life appear along the way. So to understand biological diversity and green organisms, this narrative must be presented using something of a historical approach, a very basic organizing principle. Since flowering plants appear last, a lot must have happened before most of our most common and familiar plants appeared. This also means that most botanical instruction, which emphasizes flowering plants, starts at the end of the history. As with all books, you may turn directly to the last chapter, but you are likely to find that some important things do not make sense, like seeds, flowers, fruits, and the life cycle that uses them. To get to flowering plants a lot of ground must be covered first.

To condense this narrative into a fairly concise history, this story will concentrate on events that have had a profound impact in shaping the subsequent course of life's history. Eight such major events in the history of green organisms have been identified: the origin of life itself, the development of chlorophyll and photosynthesis, the advent of the eukaryotic (nucleated) cell, the development of multicellular organisms, the invasion of land, the development of vascular tissues, the development of seeds, and the development of flowers and fruit (Niklas 1997). Five of these eight events result in the appearance of new groups of organisms: organisms with nucleated cells, land plants, vascular plants, seed plants, and flowering plants. Most significantly these events and these groups of organisms, which are described in the appendix, appear in a known chronological sequence, from most ancient to most recent. Having spent a good portion of an academic career learning to understand, however imperfectly, the diversity of green organisms, I shall attempt to explain to you why the green organisms that surround us are what they are and the way they are. Organisms green with chlorophyll appeared pretty early in Earth history, diversified, and adapted to oceanic, coastal, and finally terrestrial environments. As this took place, the Earth turned green. To the best of our present knowledge, this is how it happened.

TWO

Small Green Beginnings

Wherein the discovery of microorganisms, their amazing numbers, and the places they live are explored; evidence of ancient life is examined; and the nature of metabolisms and the origin of their complexity is explained.

In wine there is wisdom, in beer there is strength, in water there is bacteria.
—David Auerbach, 2002

About 7,000 years ago, in the southern region of the Tigris and Euphrates river valleys (an area destined to become Sumer and much later Iraq), a clay pot containing a gruel made from sprouted barley grains was forgotten for a day or two, and by the time the owner of the pot discovered his or her absent-mindedness, the mixture was topped by a prominent bubbly froth. Rather than being spoiled beyond human consumption, the mixture had a particular odor and tang to it that was not altogether unpleasant. After the first mug or so, the taste, and the taster's attitude, actually seemed to improve, and it was an elementary matter of logic to think that adding the foamy froth from this pot to the barley gruel in another pot would cause a similar conversion. And when the foam was mixed with wheat flour and water, the resulting puffed-up sour dough could be baked into a quite tasty golden crustiness with a lighter bready texture. What made these magical transformations happen was a mystery, but the long association of leaven (the foamy froth) with beer and bread making had begun. The oldest such sour dough yeast culture is thought to reside in a bakery in Samarkand that claims to have been in continuous business for over 4,000 years.

Without understanding it, humans had begun interacting with, and in the process domesticating, a microorganism. The "miraculous" transformation of barley and water into beer could not be explained, and certainly no one thought that this and many other such mystical transformations should be attributed to the metabolisms of countless inconceivably small organisms. This idea would not become fixed even in the minds of biologists until after the famous investigations of Louis Pasteur who, just some 200 years ago, demonstrated that wine would not spoil unless microorganisms were present. Human ignorance of microorganisms did not prevent us from using microorganisms, and even manipulating their metabolisms, to make such important things as beer, bread, cheese, pickles, wine, vinegar, and yogurt. No one associated these marvelous transformations, or nasty things like infections and diseases, with tiny organisms in countless numbers. Even today when the presence of "germs" on everything is taken for granted,[1] rather few people actually grasp that we live in and are adapted to a world teeming with microorganisms. So naturally if this idea is difficult for people to grasp, then the idea of a world without all of our large, familiar organisms is simply unimaginable. Yet not only can such a world exist, one did, on Earth, and it worked quite well for eons.

A planetary biology composed solely of microorganisms sounds very alien, not at all like our familiar habitats, but actually Earth's biology is still fundamentally a microbiological ecology. Virtually all of the really important ecological processes are based on the metabolisms of microorganisms, and as you will see, this includes photosynthesis. A biology based solely upon microorganisms has all the same parts and involves all the same processes as our familiar large-organism biology, but microorganisms operate at a molecular level and communities of microorganisms are organized on a very small scale, sometimes occupying quite unusual places. This idea is important because the earliest beginnings of green organisms must be sought on a young Earth, a planet occupied solely by microorganisms. Understanding such a world is difficult because the evidence of early life is fragmentary and rare, and it must be interpreted carefully if we are to understand the biology correctly (Knoll 2003).

The study of microorganisms is fairly new because for the longest time, scientists did not even know microorganisms existed even though some of their activities were well known. The study of microbial communities is even newer because for the longest time, microbiologists did not know ecology existed even though its basic principles were well known. Our understandings of ancient microorganisms, early Earth ecology, and microbial ecology have developed only within the past 50 years.

Their very small size means a very large number of microorganisms can live in a very small space, but still the numbers have a way of staggering our imagination. Every cubic centimeter of soil, a volume about the size of a sugar cube, is estimated to contain between 9 million and 21 million microorganisms. Seawater may look empty, but an identical volume (1 ml) can contain a million cells, some of which will be green. Even when green organisms are so diffuse that they fail to noticeably tint the water, such tiny green organisms may conduct more photosynthesis on a worldwide basis than all the big green organisms, like trees and grasses, combined. Usually we think about microorganisms in our bodies only when we have some sort of disease or infection, but our bodies normally, regularly provide a home to so many microorganisms that they actually outnumber the millions and millions of cells that compose our bodies. A little over one pound of your weight can be attributed to these tiny organisms. Big organisms like ourselves live in a symbiotic (sim-bye-OH-tic, literally "living together") relationship with these microorganisms; they are part of our normal biology, and our health depends on them, just as they depend on us for their food and lodging.

The bottom line is simple but not widely recognized because we have a large-organism bias. Large organisms would not and could not exist without the ecological recycling performed by microorganisms, without the symbiotic interactions, without their metabolisms. Life on Earth is dominated by microscopic organisms in terms of numbers, diversity, and importance.

Even though biologists are well aware of this, the diversity and the surprising places microorganisms are being found have the potential to amaze even us. The identities and relationships of many microorganisms still remain unknown to science, and no one knows how many different species of microorganisms exist or how many have yet to be discovered. A great deal of the history of biology is the gradual realization that microorganisms and large organisms are connected not only ecologically but metabolically and ancestrally. For at least a century biologists have thought all organisms had a common ancestry with those microorganisms commonly called bacteria, but only in the past few decades has this been demonstrated beyond all reasonable doubt. In fact, large organisms are all part bacterial in a much more integral sense. Every single cell in your body carries direct evidence of your bacterial heritage.

WHO PUT THE CELL IN CELL BIOLOGY?

In the 1670s Anton van Leeuwenhoek used homemade microscopes to observe "very many little animalcules," which both were numerous and were found just about everywhere he looked. This was the first time any-one had observed microorganisms. No one could guess how many such tiny organisms might inhabit the world around us, but tiny drops of water teeming with tinier organisms suggested the total number was astound-ing. And while many of these tiny organisms are motile, which among large organisms is a feature generally associated with animals, some of these "animalcules" racing around in the water were green.

The biological connection between microorganisms and large organ-isms began when Robert Hooke (1635–1703) observed the empty cell walls of cork, the bark tissue of the cork oak, through his microscope, and they reminded him of many small rooms, cells. Such observations be-gan the field of cell biology and led to the concept that the cell was the fundamental unit of life. Large organisms are large because they are com-posed of many tiny subunits, cells. And of course large, multicellular or-ganisms also develop by cell division from a single cell, a fertilized egg, a zygote (ZEYE-goat). Logically, if large organisms can develop and grow from a single cell, then large, multicellular organisms could have evolved from unicellular ancestors. For a long time, biologists thought that such development (ontogeny) reiterated the evolutionary history (phylogeny) giving rise to the much quoted, much misused, and now discredited state-ment "Ontogeny recapitulates phylogeny."[2] Nonetheless, the general concept of cell biology was correct. All life was cellular, and big, multicel-lular organisms were derived, both ancestrally and developmentally, from small, unicellular beginnings.

But unlike dead, empty cork cells, whose waxy cell walls provide pro-tection to living tissues within the tree's trunk, and to the contents of wine bottles, living cells are not empty. A clear "sap" called protoplasm[3] (PRO-toe-plaz-ehm) fills living cells, and it was considered the ultimate constituent of life until improved microscopes and new staining tech-niques revealed that a cell's protoplasm contained even tinier structures, organelles. Cells of all fungi, plants, and animals, as well as many uni-cellular organisms, contain a nucleus, which we now know houses the genetic material, DNA, which is organized into chromosomes. Cells have mitochondria, organelles where much of a cell's respiration takes place. Plants are green because their cells contain numerous green-pigmented organelles, chloroplasts. Plants appear uniformly green only because with-out a microscope our vision cannot resolve individual chloroplasts.

However, no matter how good the microscope, no matter how care-ful the observations, no organelles, no nucleus, no mitochondria, and no chloroplasts were ever observed within the really small cells of bacteria.

Green bacterial cells are uniformly pigmented. These observations led to the recognition that a primary dichotomy existed between those organisms whose cells had a nucleus and other organelles versus those whose cells lacked a nucleus and other discrete organelles. Ernst Haeckel (1834–1919) was among the first to recognize the significance of microorganisms whose cells lacked a nucleus. Although commonly called bacteria, Haeckel designated all such organisms Monera (moe-NEH-rah). The now familiar term prokaryote (pro-CARRY-oat), meaning "before the nucleus" (*karyon* = kernel, meaning nucleus), was introduced by E. Chatton in 1937 to describe the cells of bacteria. By contrast, eukaryote (you-CARRY-oat), *eu-* meaning true, referred to all organisms with nucleated cells, which includes many unicellular organisms and all the large organisms. Unfortunately this label seems to define prokaryotes by what they lack (nuclei) rather than by what they share (circular, naked DNA attached to the cell membrane). Nonetheless, the term prokaryotic carries with it the implication, the generally held assumption, that bacteria were ancestors of and gave rise to eukaryotic organisms.

Until very recently the differences between prokaryotes and eukaryotes represented the most significant and basic division in biological diversity. In spite of such differences, from a biologist's perspective the most notable aspect of biological diversity is the underlying sameness among all living organisms. All organisms share very general characters: a membrane-bound, cellular organization; a common genetic code; ribosomes[4] composed of RNA (ribonucleic acid[5]); proteins constructed from the same set of twenty amino acids; and the same hereditary material, DNA, composed of the same four nucleotide bases, A-adenine, T-thymine, C-cytosine, and G-guanine. Since all observations indicate that life is a continuous process, and living organisms arise only as offspring of other living organisms, what we do not see is also significant. Biologists have known since Pasteur's time that living organisms do not arise from inanimate matter and that presently the spontaneous generation of life is not occurring. Observations of a fundamental sameness, very basic characters shared among all living organisms, have generated the hypothesis that all living organisms share a common ancestry. Darwin's descent with modification accounts for both their fundamental sameness and organismal diversity. However, if large organisms have a small, simple, unicellular heritage, then it suggests that life on Earth must be ancient for there to be enough time for so many changes to have taken place.

HOW DO WE KNOW WHAT WE KNOW ABOUT THE HISTORY AND ANTIQUITY OF LIFE?

This is a very fair question. The idea of knowing anything about ancient life on Earth strikes many people as ridiculous, and in some places, doubtful students have been trained to ask biology teachers, "Were you there?"[6] Science has two sources of information about historical events: phylogenetic relationships and the fossil record. Phylogenetic hypotheses, patterns of common ancestry, have the potential of allowing us to make inferences and predictions about the nature of ancient life even though no one was there to observe it. Of particular interest to this history will be the origin of photosynthesis and the first green organisms (Hartman 1998). Biologists are pretty excited about this because phylogenetic hypotheses with sufficient detail to allow such inferences are all relatively new so there are a lot of new ideas to consider and evaluate. The basic task here is to determine if what is already known makes more sense in light of, is consistent with, or actually refutes these new phylogenetic hypotheses. But for the longest time, the fossil record seemed hopeless when it came to ancient life.

An ancient history of life on Earth once seemed glaringly at odds with the fossil record. The Cambrian period marked the "sudden" appearance of fossil organisms, and the most ancient fossils clearly belonged to known animal phyla: sponges, comb jellies, lamp shells, mud dragons, velvet worms, annelids, arthropods, echinoderms, and chordates (Briggs et al. 1994; Tudge 2000). Fossil seaweeds similar to both green and red algae were present too. Geological history from the beginning of the Cambrian (540 mya[7]) to the present day was called the Phanerozoic (fan-air-oh-ZOH-ick), meaning evident life, because of the numerous fossil-bearing rock strata (Fig. 1–5). All Precambrian strata were termed the Cryptozoic, meaning hidden life, because these older rocks contained no evident fossils. The name was derived from fairly common Precambrian geological structures, minutely layered hummocks of mostly limestone named cryptozoa, a name suggesting they were associated with living organisms. These structures are now called stromatolites (stroh-MAT-oh-lights). However, the names cryptozoa and Cryptozoic reflect the assumption that evidence of ancient life was just hidden, and life had an older, longer history than revealed by the fossil record. Although it probably was not a serious suggestion, Seward (1933) proposed that the Agnostozoic would have been a better term for the Cryptozoic because biologists believed life existed; there just was no evidence for it.

Critics of evolution have long claimed the fossil record does not support the theory of evolution, but as you will see, to make this claim requires them to ignore the fossil data. As it actually turns out, even fossils of large organisms do not "suddenly appear" in the Cambrian (Schopf 2000). Although less common, conspicuous fossils extend back far before the Cambrian, but most people still do not know this, so the false claim that living organisms suddenly appeared on Earth continues to be made.

Fossils are either direct or indirect evidence of living organisms. A fossil can represent the actual organism in various degrees of detail depending upon the mode and quality of preservation. Impressions and compressions may show only shape, size, and surface details, while petrifactions may preserve cellular details with great clarity. Bigger, harder organisms or parts thereof are more likely to produce fossils than smaller organisms or softer parts. The so-called Cambrian "explosion" has been commonly misinterpreted, and the suddenness with which fossils appear can be attributed to the appearance of not just larger organisms, but organisms with more hard parts. Soft-bodied organisms left an earlier fossil record, one much more difficult to detect, but have no doubts about it, a long Precambrian fossil record exists, and in its general pattern it supports evolutionary hypotheses quite well.

Some fossils represent indirect evidence of an organism's existence, usually by recording their activity—for example, tracks, burrows, or metabolic secretions. Precambrian "burrows" are a testament to the presence of soft-bodied, worm-like organisms even in the absence of the organisms' fossils. Fossil evidence also can be "hidden" if the fossil is too small to be seen by the human eye, or if all that exists is some biochemical artifact of living organisms. So if ancient life is small and simple in form, and if the smallest and simplest organisms known are bacteria, whose cells are far too small to be seen with the naked eye, then how large are fossils of ancient life expected to be? Clearly, they must be microscopic.

HOW GOOD IS THE EVIDENCE OF EARLY LIFE?

Paleontologists searching for evidence of early life must look for fossils in ancient rocks. This is not easy for several reasons. When it comes to rocks, the older the rock, the rarer the rock is. Most really old rocks have been destroyed by active geological processes. As it turns out the Earth's crustal material gets recycled, returned to the Earth's molten interior in subduction zones, as new crustal material gets formed at spreading zones, like the mid-Atlantic ridge. Most really old rocks also have been greatly altered by heat and pressure—that is, they are metamorphic, which can destroy evidence of life. And of course only rocks of sedimentary origin are expected to have fossils. Fossils of ancient organisms must be sought with a microscope, and you cannot do this in the field. You may wonder how rocks are observed using a microscope, and while the process is laborious,[8] the technique is simple enough. Once you find rocks of the right type and age, ancient rocks of sedimentary origin, you gather specimens and haul them back to your laboratory. This is not made easier because such rocks seem to be found in inconvenient locations such as Greenland, the Australian outback, and Siberia. In the laboratory, thin slices are cut from the rocks using diamond saws and the slices are glued to glass slides. Then the rock slices are polished away until only a very thin, translucent layer of rock is left on the slide. Then, and only then, do you get to look and see if your field work and all that effort has been successful or not.

In this manner, Barghoorn, Tyler, and Schopf (Tyler and Barghoorn 1954; Barghoorn and Schopf 1966; Schopf and Barghoorn 1967; Schopf 1994) began examining some of the Gunflint formation of Western Ontario, rocks of some 2 billion years old, only about halfway along in Earth history, but relatively convenient to obtain. And much to their delight, they found abundant evidence of fossil organisms, some very familiar looking and some of an unknown nature. The oldest really convincing fossils of microscopic organisms consist of chains of cells called filaments (FILL-a-mints) dated to 3.2 bya in mineral deposits typical of hot springs (Rasmussen 2000), places where thermophilic (heat-loving) organisms would live. Fossils have been reported from the Apex Chert of Western Australia, dated to almost 3.5 bya, but the nature of these simple "filaments" remains open to debate (Schopf 1994; Knoll 2003). Now more fossil bacteria from a very nearby site, the Strelley Pool Formation, have been reported at an age of 3.4 bya and interpreted as having a sulfur-based metabolism (Wacey et al. 2011). By 2.7 bya, fossils show enough diversity to demonstrate that all the major lineages of cyanobacteria (blue-green algae) observed today had appeared (Fig. 2–1) (Schopf 2002). The wishful thinking of geologists was justified: while "hidden," Cryptozoic rocks contained ample evidence of early life. These discoveries have lengthened the known history of life by more than five times. The fossil record clearly demonstrates that living organisms, some of which appear very similar to modern cyanobacteria, have inhabited Earth for at least 2.7 billion years, and maybe for as long as 3.5 billion years!

Evidence of early metabolisms yields similar conclusions about the antiquity of life. The use of carbon isotopes to find evidence of early life is a great example of scientific creativity. Carbon-12 and carbon-13 are stable, naturally occurring isotopes[9] of carbon. Both carbon isotopes can

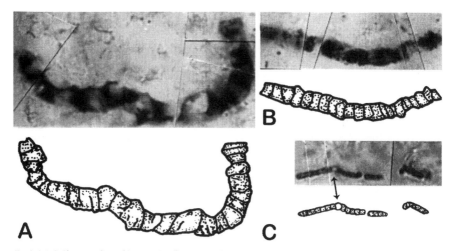

Fig. 2–1 A-C. Photographs and interpretive illustrations of ancient microfossils similar to cyanobacteria in size and form. The upper images are the actual images of the fossil as observed microscopically in a very thin slice of the rock. The photograph is pieced together because numerous photos were taken at slightly different focal planes and the areas in best focus pieced together. The line drawing is the interpretation based on these observations. Arrow in C indicates a larger, empty-appearing cell similar to a heterocyst (see Appendix: Green Bacteria, Fig. A40). (Image source: After Schopf 1993)

be found in the atmosphere or dissolved in the oceans as carbon dioxide, but the isotopes are not equally abundant and photosynthetic organisms do not use them in direct proportion to their abundance. Carbon-12, the lighter isotope, is 90 times more common and is preferentially used in photosynthesis, thus the ratio of carbon-12 to carbon-13 in organic molecules is higher than 90 to 1. Modern photosynthetic organisms skew the carbon isotope ratio in organic molecules by about 18 parts per million (Nisbet and Sleep 2001; Schopf 2002). Since living organisms preferentially use more carbon-12, a little bit less is left in the atmosphere for composing carbon-containing inorganic materials, so the ratio found in things like limestone can be lower in carbon-12 by some 7 parts per million, which adds up to a total organic-inorganic isotopic skew of 25 parts per million, a biochemical signature of life. A skewed carbon isotope ratio indicating life has been found in carbon-containing materials dating back to an age of 3.5 to 3.8 bya (Schidlowski 1988; Schopf 2002). The oldest graphites from ancient seafloor sediments, dated to 3.8 bya, are somewhat less convincing because the skew is less than 25 parts per million. Other biochemical signatures—for example, sulfur isotopes, equally indicative of living organisms, provide similar dates and add confidence by being data of a different sort.

The underlying assumption is that ancient photosynthetic organisms have the same carbon isotope preference as modern green organisms. Presently there is no reason to think otherwise, so the assumption remains reasonable, particularly since, as it appears, the photosynthetic apparatus of modern green organisms, including the same key carbon-fixing enzyme (rubisco), has been inherited more or less intact from ancient organisms.

If you think this assumption unreasonable, then you are free to offer another plausible explanation of these skewed carbon isotope ratios; however, the tyranny of data is a critical component of science. In science you are free to disagree with an explanation only if you can show that the data gathered were the result of some type of error, or if you can offer another plausible explanation of the observed results. When very critical and skeptical scientists, and that is the way we are, are finally convinced by the data and explanations, then people who continue to dissent begin to sound like cranks. To say, "I just do not believe it" and dismiss data and the corresponding explanations out of hand is against the rules of science. What you believe privately remains up to you, but such personal opinions do not get published in peer-reviewed scientific journals. However, in this day and age science critics and denialists who cannot meet the challenge of science still manage to convince many gullible nonscientists that their opinions or beliefs have validity.

Stromatolites provide data of a very different sort. Stromatolites are one of the most common of ancient geological formations associated with living organisms. Modern stromatolites are formed by microorganisms growing in thin mat-like communities upon substrates submerged in shallow water. Such microbial mat communities are thin, often no more than a millimeter or two thick, but on the size scale of bacterial cells these mats are towering forests. Like forests these microbial mats are layered communities where the photosynthetic producers dominate the upper layers, the canopy of this small-scale forest, if you will. The photosynthetic producers of modern microbial mat communities are predominately cyanobacteria. Microbial mat communities dominated by cyanobacteria often form on the soil surface at the bottom of puddles that persist for a few weeks. The geological structure of stromatolites forms as these mats of microorganisms capture layer upon layer of fine tidal sediments as the water moves in and out. Stromatolites continue to grow under the mat, building up microscopically layered hummocks of sediments cemented together by secreted carbonates (Fig. 2–2). Ancient stromatolites, dated to 3.5 bya, are constructed identically, but the possibility always remains that they are of inorganic origin. Somewhat younger stromatolites of about 2.7 bya bear

Fig. 2–2 Modern stromatolites in Shark Bay, Australia. A bacterial mat community constructs these structures by capturing and building up minute layers of sediment. (Image source: Courtesy of P. Harrison, Wikimedia Creative Commons. See entry in References for WCC license by attribution.)

fossil evidence confirming the presence of cyanobacteria. So these geological formations, if understood correctly, provide evidence that is completely consistent with the early fossils and metabolic signatures of life.

Presently stromatolite formation is rare and occurs in only a few locations, but during the Precambrian these microbial mat communities were common and widespread. Stromatolite abundance declines in the late Precambrian, which is when fossils of large multicellular organisms become fairly common, suggesting that one event was involved with the other. How could large organisms cause microbial mat communities to become rare? The usual explanation is that the mat communities were eaten by multicellular grazing animals, probably invertebrate grazers like slugs, which formerly did not exist. An equally likely possibility was that stromatolite communities were newly subjected to competition for light by the appearance of larger algae, seaweeds. This latter possibility has not been considered previously, and seaweeds appear in the fossil record just prior to the beginning of the decline in stromatolites. In chapter 4 you will learns that seaweeds are an evolutionary response to competition for limited space and light in shallow water and coastal areas.

All three independent types of evidence—microfossils, metabolic signatures of life, and stromatolites—yield very similar conclusions that bacteria-like organisms and photosynthesis were present by 2.7 bya, probably present by 3.5 bya, and maybe even present as early as 3.7 to 3.8 bya. When diverse, independent evidences agree, they are conciliatory and compatible with each other, and it increases our confidence in the general conclusion. Each independent study constituted a test of evolutionary hypotheses based on previous observations. Greatly different results have the potential to falsify such hypotheses. And indeed, independent reexamination of the data can increase or decrease confidence in new hypotheses. For example, cell-like microfossils were reported from the 3.5-billion-year-old Apex Chert of Western Australia (Schopf 1994), but the nature of the rocks and the low abundance of the putative fossils have failed to convince all biologists that these structures are of a biological nature (Knoll 2003).

Rocky continents formed about 4.2 bya, yet scientists studying the origin of the solar system estimate that 3.8 bya is about as early as life could have existed on Earth's surface because prior to that time the meteor bombardment, which left the moon heavily cratered, was too intense for life to survive. Impacts were big enough and frequent enough that the heat they generated could boil any oceans present, sterilizing the Earth's surface. If life could not exist on Earth's surface before 3.8 bya, perhaps it could exist earlier in deep, dark, safe places, a possibility that will be explored. But organisms in deep, dark places would not be photosynthetic, so even if older life occurred in deep, dark places, sometime between 3.8 and 2.7 bya life adapted to Earth's surface and some organisms turned green.

SOME LIKE IT HOT: ORGANISMS ADAPTED TO EXTREME HABITATS

Other recent discoveries have proven significant to our understanding of early life on Earth. Familiar living organisms occupy a thin surface layer of the Earth, a skin-like biosphere where conditions are within normal ranges of environmental parameters—for example, temperatures between 0° C and 40° C, a pH that is near neutral,[10] and a salt concentration not much higher than sea water. When we venture beyond this thin layer, or into some geothermally active places, we encounter habitats that seem too hot, too cold, too acidic, too salty, or too deep within the Earth's crust to harbor life because such environments would cook, denature, or crush virtually all familiar life forms. So, what a surprise for biologists to discover that many such habitats teem with life!

Organisms adapted to such conditions are called extremophiles, "lovers of extreme environments." Without worrying about whether these organisms feel affection for their environments or not, from our perspective, extremophiles are adapted to conditions well outside a "normal" environmental range for living organisms. Several different types of extremophiles are known: those adapted to extremely hot and just very hot habitats (hyperthermophiles and thermophiles), very salty habitats (halophiles), very acidic habitats (acidophiles), and very high pressure habitats (barophiles) (Rothschild and Mancinelli 2001). However, just knowledge of their existence does not make research on extremophiles easy. Organisms from such habitats are not easily collected and cultured, a necessary step for laboratory study. Extremophile habitats are hard to duplicate and maintain in the laboratory, so our knowledge of extremophiles remains rather limited.

Prokaryotic organisms have been found growing in the superheated water of undersea volcanic vents and deep in Earth's crust where water temperatures exceed 100° C but where pressure prevents the water from boiling (Tunnicliffe 1992). One such organism, *Pyrolobus fumarii,* literally "fire lobe of the chimney," a lobed cell found in a deep sea vent, grows at temperatures up to 113° C. *Pyrolobus* cannot grow at temperatures below 90° C because it "freezes," solidifying into inactivity. A personal favorite, the aptly named *Sulfolobus acidocaldarius*, grows optimally at pH 3 and 80° C, conditions found in hot, acidic springs, basically hot sulfuric acid, conditions that can easily dissolve your pocket change. Organisms have been found in very acidic (pH 1) mine wastes, conditions very like battery acid. Boiling sulfurous mud baths in Iceland contain up to 10^8 (100,000,000) living cells per milliliter (= 1 cm^3). At lower temperatures, 50° C–60° C, these thermophilic microorganisms become inactive and dormant, but such "frozen" cells are in a suspended animation and can remain viable for years. Dormant thermophilic organisms have been found drifting in seawater samples, on the order of one viable cell per each cubic meter of water, demonstrating how such organisms can disperse by drifting from one hotspot, an "island" of geothermal habitat, to another.

Life also abounds in deep, dark, hot places within the Earth's crust. Production fluids from 3,500 meters deep (over two miles deep) in the geothermally heated oil fields of the North Sea and Alaska contain up to 10^4–10^7 living cells per milliliter. These extremophile organisms do not merely tolerate such conditions; they grow optimally under these conditions. Optimal growth is a relative concept because in some of these environments cells grow so slowly they may divide only once every 100 years, but slow growth is not a problem in very stable, long-lived environments such as these. All extremophiles from deep, dark, hot places are anaerobic (living without free oxygen), thermophilic prokaryotes; most are members of a group called archaebacteria, meaning ancient bacteria. As more and more new organisms are being discovered in "unlikely places," microbiologists now estimate that less than 10%–20% of known prokaryotes can be isolated and cultured, a necessary prelude to scientific study (Madigan and Marrs 1997).

Biologists' initial reaction to the discovery of extremophiles was to ask, "How does life adapt to such extreme environments?" Although logical enough, it was the wrong question. Diverse geological evidences suggest the environment of ancient Earth was hot, was very geothermally active, and had an atmosphere lacking free oxygen. In other words, extremophiles are quite well adapted to ancient Earth environments, conditions that remain in geothermally heated mineral springs and many deep, dark, hot places. Thus biologists began to realize that what was "extreme" from *our* perspective was perhaps just normal for organisms whose ancestors were adapted to the environmental conditions of a young Earth.

Because deep, dark, hot places were safe long before the surface of Earth was habitable, many biologists now think life may have begun in such places. If this proves true, then living organisms adapted to Earth's frigid, bright surface, drenched in ultraviolet radiation, shortly after it became habitable at about 3.8 bya. Thus the question biologists should have been asking was, "How did organisms adapt to the extremely cold, bright, oxidative conditions that we now think of as normal?" No one thought that photosynthesis could evolve in deep, dark places, but our thinking on this may have to change because bacteria have recently been discovered that can use the extremely low "light" energy of infrared (heat) wavelengths produced by deep sea geothermal activity. How photosynthesis is possible in such dim (pitch-black to us) "light" remains unknown. Most extremophiles capture energy from geochemical reactions rather than light, so at this point most biologists think such ancient metabolisms, which can function in deep, dark places, preceded photosynthesis (Martin and Russell 2003). However, one thing is clear, green organisms appear fairly early in Earth history.

WHICH CAME FIRST, PRODUCERS OR CONSUMERS?

All modern ecosystems rely upon autotrophic producers to capture energy and form the first step of a food chain because heterotrophs require premade organic molecules for energy and raw materials. If what is true at present was also true on ancient Earth, then autotrophs had to appear before heterotrophs. This is very logical, but we must consider the possibility that very ancient Earth communities did not operate like present-day communities. Heterotrophic organisms could have appeared first if an abiotic source of food molecules existed. This is not a new idea and was first proposed in the 1920s by the famous Russian biologist A. I. Oparin (Miller and Lazcano 2002). A spontaneous organization of life is a hypothesis based upon self-assembly from premade abiotic organic molecules, and if true, then by its very nature the first life was heterotrophic, relying upon an abiotic source of organic molecules. As it turns out, abiotically produced organic molecules are easy enough to find in the absence of life (chapter 1), so the first living organisms may well have been heterotrophic scavengers of abiotic organic molecules, breaking them down to gain energy. Such simple beginnings are the raw materials, the component parts, from which more complex metabolic pathways were assembled piece by piece (Lazcano and Miller 1999). Fermentations are the simplest known metabolisms, and biologists like explanations where simple precedes more complex. Photosynthesis appears to be a metabolic add-on because all photoautotrophs have basic heterotrophic metabolisms too, and many prokaryotic autotrophs can function quite well as heterotrophs if provided food molecules, which is expected if heterotrophs became autotrophic secondarily.

This suggests two possible means by which photosynthesis began: either a heterotroph becomes a green photosynthetic autotroph by the addition of components to its metabolism, or a chemoautotroph became a photoautotroph by shifting the source of captured energy from chemical reactions to light. The biogeochemistry of Hadean Earth centers on sulfur and iron compounds, which are very reactive (Liu et al. 1997). One possibility for an early, simple autotrophy would be capturing energy from reactions involving hydrogen and iron-sulfur compounds. Such fairly simple chemoautotrophic metabolisms are currently found in a number of prokaryotic organisms inhabiting geothermally heated hot places (Liu et al. 1997). Like the isotopic evidence for biological carbon, there is isotopic evidence of a sulphate-reducing metabolism dating to 3.47 bya. Presently the only organism known to reduce sulfur in a similar manner is a hyperthermophilic prokaryote (Shen, Buick, and Canfield 2001). A brief examination of respiration and photosynthesis and their component parts will assist in developing scenarios and assessing the possibilities.

HOW DOES LIFE WORK? AN INTRODUCTION TO METABOLISMS

Most biologists think it is easier to explain how life works than to define what life is, but perhaps the two things are related. The universe is composed of just two things, matter and energy. Living organisms consist of highly ordered matter in the form of organic molecules. Living organisms store their energy in the form of organic molecules, and metabolisms are how this energy is captured, stored, and used. Life is an interesting phenomenon because if our understanding is correct, the general nature of the universe is for energy to dissipate, for disorder or entropy to increase. In simple terms the universe is working against living organisms. Life cannot and does not violate the laws of the universe, and the second law of thermodynamics does not make life impossible.[11] The basic trick of living organisms is that they delay rather than defy entropy, the dissipation of energy. Ultimately energy will dissipate, moving from hot places to cold places, until no differences remain. Autotrophs have the ability to capture energy, convert it into a molecular form, and thus delay its dissipation for a time, and that is all it takes for life to operate.

Biochemistry and cell biology can be intimidating subjects. All those integrated pathways, all the arrows, all the molecules and enzymes make for a complex memory challenge, but memorization does not usually bring understanding. The construction of such complex systems by natural processes seems quite unlikely when we first confront them as a whole. However, our ability to recognize and understand the components of metabolisms and to interpret them from a phylogenetic perspective has slowly made it apparent that rather than being elegantly designed assemblies, metabolisms are nature's "most wonderfully contrived Rube Goldberg machines" (Knoll 2003). All metabolisms are constructed of components, smaller, simpler sets of reactions, some of which are so versatile that they can function in two or more different ways. These components have been cobbled together in various ways in different groups of organisms, and the contraptions work. But metabolisms are not the wonderfully elegant things they may seem on first or superficial examination. Fortunately for you, metabolisms need to be examined only in a very conceptual way to gain an essential understanding. The details can be looked up in a biology book, if ever the occasion calls for them.[12]

The two basic things about such pathways you need to understand are pretty straightforward. First, metabolic pathways are a stepwise synthesis or deconstruction of organic molecules. Cell chemistry is never done all at once. This is like climbing stairs. The distance from one floor to the next is the same, but trying to jump up to or down from one floor to the next requires or releases too much energy, too much force, all at once, so we climb the distance in little steps, each of which takes or releases much less energy, although in the end, the same total amount is needed or released based upon your mass and the height between floors. Second, enzymes are proteins that make the stepwise reactions more energetically favorable because they act as a catalyst. This narrative does not require any understanding of the specifics beyond being aware of the stepwise nature of metabolic pathways and that enzymes are used to facilitate the process.

All metabolisms are either aerobic or anaerobic (requiring oxygen or operating in the absence of oxygen) and either autotrophic or heterotrophic, four combinations in total. All four metabolic combinations are found among prokaryotes with considerable variation in each category. All organisms respire[13]; they break down a substrate, some organic molecule, step by step to obtain the energy locked up in the molecule's chemical bonds. In addition to respirations, autotrophs have a synthetic component to their metabolism, one that first captures energy from an external source in the environment, and then uses this captured energy to synthesize organic molecules from simpler raw materials. The synthesized organic molecules are both building blocks (raw materials for the organisms' growth and reproduction) and the stored chemical energy that can be respired to provide energy for cell functions. Autotrophs do not use captured energy directly; they must first synthesize molecules and then respire them just like heterotrophs. So in this sense, green organisms are just like us consumers.

Many bacteria respire anaerobically, by fermentation. Some bacteria respire aerobically. All large organisms respire aerobically, both heterotrophs and autotrophs, and this is the metabolism that gets taught in most biology classes. Some unicellular organisms like yeast, which is a fungus, and some cell types within complex multicellular organisms, like your muscle cells, may be facultative, capable of changing between aerobic and anaerobic respiration depending upon whether oxygen is available or not. Aerobic organisms, unless facultative, cannot live without oxygen, and oxygen inhibits the growth of most anaerobic metabolisms, so unless they are facultative, organisms do not switch between aerobic and anaerobic habitats.

Oxygen is much misunderstood biologically. Most people think of oxygen as a life-sustaining, beneficial gas. In reality, oxygen is toxic, especially to anaerobic organisms and all the anaerobic components of aerobic metabolisms, which are considerable. In fact aerobic respiration appears to be the result of an add-on to standard anaerobic respiration, which makes sense of what we know about the history of Earth's atmosphere. Earth's present atmosphere of 21% oxygen (O_2) and 78% nitrogen (N_2) is highly oxidative. Other gases, mostly argon (0.9%), plus much smaller amounts of carbon dioxide (0.036% and increasing), plus much smaller amounts of neon, helium, krypton, xenon, methane, hydrogen, and ozone, account for the remaining 1%. But such an atmosphere did not occur on ancient Earth. Diverse evidence suggests the atmosphere on ancient Earth had no free oxygen, so only anaerobic respirations were possible. Significant amounts of oxygen did not accumulate until about halfway through Earth history, so aerobic respiration could evolve only after a lot of free oxygen came from somewhere. Since photosynthesis evolved in an ancient, anaerobic world, it comes as no surprise that photosynthesis is an anaerobic metabolism, able to operate without free oxygen. So it is somewhat ironic that one form of photosynthesis is responsible for oxygenating Earth's atmosphere.

Earth's ancient anaerobic atmosphere helps explain something a bit strange about the primary carbon-fixing enzyme of photosynthesis, ribulose 1, 5 bisphosphate carboxylase, nick-named "RuBisCO." Two factors combine to make rubisco the most abundant enzyme on Earth. First, rubisco is an inefficient enzyme because of a rather slow turnover rate, so green organisms must have lots of it. Rubisco uses captured light energy to couple ("fix") a carbon dioxide onto a five-carbon sugar, ribulose, producing a six-carbon sugar, but then rubisco is slow to release the product and to take up more reactants. Second, photosynthetic producers occupy the bottom rung of communities, so there are lots of them, they have a high biomass, and they all have lots of rubisco. Rubisco is the enzyme that prefers carbon-12, and the 3.5 to 3.8 billion year history of a skewed carbon isotope ratio highlights not only the antiquity of life but the very real possibility that this enzyme is equally as old.

Like many enzymes, rubisco can also catalyze the reverse reaction because oxygen competes with carbon dioxide for the active site on the enzyme. When working in reverse, rubisco respires an organism's sugar molecules, its stored energy, releasing carbon dioxide. Some aerobic bacteria actually use rubisco as an oxidative enzyme in respiration. When oxygen is available rubisco makes it harder for autotrophs to store up energy. Far from being a well-designed enzyme, rubisco, the most common enzyme on Earth, is both inefficient and wasteful. But remember, according to our hypotheses about the atmosphere of ancient Earth, rubisco evolved in an anaerobic world where there was no free oxygen to compete with carbon dioxide. Under its original anaerobic conditions rubisco would not have wasted fixed carbon, nor could it be used in aerobic respiration. So what happened?

An enzyme like rubisco makes sense only when viewed from a historical perspective. One group of green organisms, cyanobacteria, changed Earth's environment because they liberate oxygen as a by-product of their photosynthesis. In a changed environment, an oxygenated atmosphere, rubisco is a less well-adapted enzyme, but unless a more efficient variant of the enzyme appears, natural selection has no means of generating any change. Since rubisco has a virtual monopoly on carbon fixation among green organisms, no evolutionary mechanism can change it. An engine that runs poorly will still go further than an engine that does not run at all, so all green organisms are stuck with the same inefficient enzyme. Evolutionary processes cannot manufacture a more efficient enzyme just because the organism needs it or would be better off with one. The existence of such inefficiency directly contradicts the primary premise of supporters of intelligent design who think the perceived "perfection" and efficiency of organisms, especially their biochemistry, support the idea of creation by an intelligent designer.

The situation with rubisco is like QWERTY (Nisbet and Sleep 2001). Do you recognize this word? Well, QWERTY is not a word but rather the six letters typed by reaching upward with the four fingers of the left hand on a standard keyboard. The QWERTY key arrangement was invented for typewriters over 120 years ago with the supposed purpose of slowing the typist down so that the manual arms carrying each type character wouldn't jam. Balls replaced jamming arms and PCs with word processing software replaced typewriters, and now internet phones and tablets are replacing PCs, but we still have the inefficiently arranged QWERTY keyboard. Other arrangements of the keys have been shown to generate faster typing of characters,[14] so why do we keep QWERTY? Once you have learned to unconsciously hit the correct keys, no one wants to learn a new key arrangement. So because QWERTY keyboards are a monopoly,[15] we continue to use this inefficient interface with our electronic world and QWERTY probably will remain until voice-, eye-, or thought-activated input replaces keyboards.[16]

Some green algae like *Chlorella*, a unicellular organism much used to study photosynthesis, lose as much as 30% of their fixed carbon because of rubisco's inefficiency (Graham 1993). But rubisco's monopoly does not mean green organisms cannot or have not adapted to an oxygenized atmosphere. An adaptation to recapture fixed carbon appeared in one small group of green algae. A new enzyme called glycolate oxidase uses oxygen to reverse the loss of fixed carbon, and interestingly enough, this group of algae shares a common ancestry with land plants, who live in a habitat where oxygen is readily available. Glycolate oxidase for glycolate recovery can certainly be considered a preadaptation for conquest of terrestrial environments (Graham 1993), and this may be one of the reasons land plants are green rather than red or brown—that is, related to green rather than to other algae. Other green algae have other adaptations for reducing the loss of fixed carbon.

OXYGEN, THE TOXIC BYPRODUCT

Why does one particular type of photosynthesis liberate oxygen? And then how do anaerobic metabolisms become aerobic? Photosynthesis requires hydrogen to combine with carbon dioxide to make carbohydrates, and this is important to note. In spite of the name, carbo-hydrate, sugars are not synthesized by adding carbon to water; they are synthesized by adding hydrogen to carbon dioxide. The most common form of photosynthesis uses water (H_2O) as its hydrogen source (H) and thus ends up releasing oxygen as a byproduct. Why oxygenic photosynthesis became the predominate autotrophic metabolism of green organisms will be discussed presently, but one of the reasons is that water is a very common raw material in comparison with other sources of hydrogen. The 21% oxygen in our present atmosphere came about from a slow accumulation of oxygen that gradually changed the Earth's surface environment from anaerobic to aerobic over a couple of billion years (Riding 1992; Kasting 2001).

The release and accumulation of oxygen as a photosynthetic byproduct has been geochemically documented. Iron is quite soluble in ocean water containing no dissolved oxygen, but dissolved iron and oxygen will combine readily producing iron oxide, rust, a precipitate. Ancient banded iron oxide deposits are evidence of oxygenic photosynthesis. Only after precipitating all the dissolved iron from the oceans would the atmospheric concentration of oxygen slowly begin to rise. Today we mine these ancient deposits of iron oxide. Smelting reduces iron oxide to elemental iron, its familiar metallic form, from which we construct various artifacts that immediately begin to rust in an oxygen-rich atmosphere.

About 2.4–2.8 bya, anaerobic organisms faced an oxygen crisis when the atmospheric oxygen concentration reached about 0.5%. This is only 2% of the present day oxygen concentration, but 0.5% oxygen is enough to inhibit most anaerobic metabolisms (Bilinski 1991; Knoll 1999). The oxygen crisis produced several different reactions by organisms. Anaerobic organisms survived either by retreating into environments that remained anaerobic or by adapting to the increasingly aerobic conditions. Ultimately, many anaerobic organisms found and adapted to new anaerobic environments found inside the bodies of large multicellular organisms, but such large organisms would not appear for nearly 2 billion more years. Some organisms adapted to oxygen by respiring aerobically (using oxygen metabolically), which has the additional benefit of detoxifying oxygen into harmless water from whence it came.

The oxygen crisis provides an example of how life both changes its environment and then adapts to these changes, which in turn change the environment demanding new adaptations. An evolutionary process is a biological necessity because organisms alter their environment. Life can have no history without an evolutionary process.[17]

The appearance of eukaryotic (nucleated cell) organisms (chapter 3) occurs at about the same time as the oxygen crisis, suggesting that one had something to do with the other. Indeed, the nature of the eukaryotic cell suggests that a lineage of anaerobic archaebacteria became eukaryotes by adapting to aerobic conditions in an ecological manner, via symbiosis. If this is true, a eukaryotic cell represents an association between at least two different organisms, a fundamentally anaerobic host cell with an ancestry among archaebacteria and an aerobic bacterial cell, which became a mitochondrion, the organelle within which aerobic respiration takes place.

Although this sounds as if organisms had a choice about becoming aerobic, this is not the case. Just because an organism would be better off with or needs aerobic respiration does not mean it will happen. Evolution just does not work that way. Natural selection operates only by tinkering around with variations of preexisting adaptations. This does not seem a likely avenue for big changes or real innovation, but our perceptions and preconceived notions fool us into misjudging the amount of innovation involved. For example, aerobic and anaerobic respiration seem very different, but the change is not really so great. I learned a similar lesson while building automobile carburetors in summers to earn my college tuition. Only a couple of small components changed and you were constructing carburetors for Oldsmobiles instead of Buicks or Chevys.[18] If you think of these metabolic pathways as assembly lines, the only change needed would be found at the very last stage of assembly; all the rest remains the same. Keep this in mind.

WHERE DID THE COMPLEX RESPIRATION METABOLISMS COME FROM?

When viewed as a whole, especially by students, respiration and photosynthesis have an aura of great complexity, but complex metabolisms are actually composed of smaller, simpler components all of which are found either together or separately in various groups of bacteria. One of the biggest and most common errors in teaching science is to provide too many details, too soon, and in the process to obscure concepts. This helps perpetuate a common perceptual error; you observe complexity and assume it must have arisen as a whole. Consider the invention of the motor vehicle. My first motor vehicle was a 1949 Ford pickup truck that I acquired in the early 1960s. By today's standards my truck was utterly primitive, very much simpler than modern vehicles, yet it still managed to do almost everything you would want a truck to do (except go over 50 mph). Not many things were under the hood: a coil, a distributor, a fuel pump, a carburetor, a radiator and water pump, and a big, flat-head six-cylinder engine. In today's vehicles I now find it hard to recognize many of these parts, or their functional counterparts. Many more gadgets exist under the hood whose functions remain unknown to me. But the complexity of today's vehicles did not come about all at once; they were changed little by little over generations since their invention more than one hundred years ago. The originals were very, very much simpler, and all were composed of components already on hand.

No matter how complex, a vehicle is composed of many simpler components, nuts, bolts, levers, springs, pistons, cams, gears, axles, wheels, and so on, all of which existed long before motor vehicles. The invention of the automobile was largely the result of combining preexisting components in a new manner. A new combination of preexisting components was the real innovation. Another type of innovation is finding new uses for old components. Wheels had been used for rolling vehicles across the terrain for centuries before one was used for steering the vehicle. Like wheels, some metabolic components had a previous function or have more than one present function. A better understanding of where biological complexity arises can be gained by identifying all the major components of respiration and photosynthesis and then figuring out their origins.

Fermentations are pretty simple metabolisms. These anaerobic respirations use various substrate molecules and produce different products, but they all take a substrate, some particular organic molecule, and break it down stepwise into one or more products, smaller molecules, to extract the energy stored in the substrate molecule's chemical bonds. The most familiar fermentation is alcoholic, where a 6-carbon sugar is broken down into two 2-carbon molecules of ethanol and two molecules of carbon dioxide. Even this simple fermentation consists of two components. The core of alcoholic fermentation is a nearly universal 10-step pathway called glycolysis (gleye-COLL-ee-sis) that "splits" a 6-carbon molecule of glucose into two 3-carbon molecules of pyruvate (pie-RUE-vate).[19] Glycolysis yields energy, in the form of four a̲denosine tri̲phosphate (ATP) molecules per glucose. The net yield is 2 ATP because the process requires two ATP as input at the beginning, and that has to be paid back before a gain is realized. Then a second metabolic component breaks down each pyruvate into a 2-carbon molecule of ethanol and a carbon dioxide, a total of 2 ethanols and 2 carbon dioxides per 6-carbon sugar. No additional energy is gained from the alcoholic respiration of pyruvate, but ethanol and carbon dioxide are molecules easily released from the cell as metabolic waste products to diffuse away unless trapped in a convenient container, like a wine bottle.

ATP is an energy-linking/energy-carrying molecule of cells. ATP consists of the nucleic base adenine, ribose (a 5-carbon sugar), and a chain of three phosphate groups. Even this simple metabolic component, a nucleotide, has two very different functions. Carrying a single phosphate this nucleotide is one of four subunits that compose RNA (ribonucleic acid), a molecule that carries genetic information around a cell and synthesizes proteins. ATP cycles back and forth with ADP (D = diphosphate) in coupled reactions by accepting or delivering chemical energy along with a third phosphate group. ATP to ADP delivers energy along with a phosphate; ADP to ATP picks up energy along with a phosphate. However, ATP is not a good molecule for energy storage, which is why carbohydrates or lipids are synthesized and used for energy storage.

Waste products or end products of one metabolism can be used as substrates for other respirations. Pyruvate, the end product of glycolysis, still contains considerable energy, so it too can function as a substrate of other respirations. Even ethanol can be aerobically respired into acetic acid by the bacterium *Acetobacter*, transforming wine into vinegar. Thus it comes as no surprise to find that glycolysis is a component of many different respirations, even aerobic respiration. As a result of such component couplings, pyruvate acts as a molecular junction of sorts, a starting point for other pathways. Any organism that acquires the means of coupling another metabolic component to glycolysis gains more energy from the original sugar molecule, so any such variants will be highly favored by natural selection. As a result, glycolysis, which takes place in a cell's cytoplasm rather than in mitochondria, is a nearly universal piece of metabolic machinery. Even our own aerobic respiration has a glycolysis "backbone." Such a fundamental piece of cell metabolism suggests glycolysis is an ancient metabolic component inherited from the last common ancestor of all organisms.

Yeast cells convert pyruvate into ethanol and carbon dioxide only under anaerobic conditions; otherwise yeast respire aerobically. So to brew or bake the environment must be manipulated to control the metabolism. The bacterium *Lactobacillus* takes lactose (milk sugar), converts it to glucose, and after glycolysis further respires pyruvate into lactic acid, which is released into the environment to diffuse away. If enough lactic acid accumulates, it lowers the pH, and the acidic conditions cross-link (denature) the milk proteins producing yogurt. Each teaspoon of yogurt contains zillions of *Lactobacillus* cells. These organisms obtain no additional energy by making alcohol or lactic acid, but they transform pyruvate into more easily disposable molecules to keep pyruvate from accumulating and having a repressive feedback on glycolysis, which causes the cell to stop making ATP until the excess pyruvate is removed.

Muscle cells normally respire aerobically, but they are facultative. When you are exercising heavily, oxygen is used faster than it can be provided, and muscle cells switch to anaerobic respiration. When respiring anaerobically, your muscle cells convert pyruvate into lactic acid, a 3-carbon molecule, so that they can continue producing ATP via glycolysis, which is needed to drive the muscle motor proteins. But muscle cells cannot dispose of lactic acid, and eventually enough of this toxic waste product accumulates causing muscle fatigue. When aerobic conditions return, muscle cells can reverse the pathway, converting lactic acid back to pyruvate for further aerobic respiration. Aerobic conditioning shortens the payback time of the oxygen debt, which relieves the muscle fatigue and soreness cause by the lactic acid accumulation.

Aerobic respiration, including your own, consists of glycolysis, plus two more components: the citric acid cycle, also called Krebs cycle, and an electron transport system (Fig. 2–3). The citric acid cycle uses the pyruvate produced by glycolysis as a substrate. A carbon dioxide is removed from pyruvate, and the resulting 2-carbon molecule enters the citric acid cycle by combining with a four-carbon molecule to make a six-carbon molecule of citric acid. With several steps, two carbon dioxides are removed, reducing the six-carbon citric acid back to a four-carbon molecule, and in the process generating 1 ATP per pyruvate. The four-carbon molecule is then ready to pick up another 2-carbon subunit from pyruvate for another trip around the cycle. By coupling the citric acid cycle's respiration of pyruvate to glycolysis, the ATP produced from the original glucose molecule was doubled from two to four.

Here it is worth noting that by shifting the citric acid cycle into reverse, it acts as a carbon-fixing synthetic pathway by using ATP to add CO_2 to a four-carbon and then a five-carbon molecule. Such a photosyn-

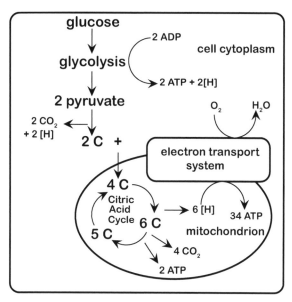

Fig. 2–3 The major components of aerobic respiration: glycolysis, citric acid cycle, and electron transport. Glycolysis takes place in the cell (outer box) cytoplasm; the citric acid cycle and electron transport take place within a mitochondrium (oval shape). See text for explanation.

thetic pathway is found in the green nonsulfur bacteria, and this metabolic pathway has been characterized as what a simple, primitive, carbon-fixing metabolism would be like (Bada and Lazcano 2002). In some organisms, such metabolic components are used in both directions.

So far nothing about aerobic respiration is aerobic at all because both glycolysis and the citric acid cycle are completely anaerobic, which comes as no surprise if these metabolic components both had ancient origins on a young Earth with no free oxygen in the atmosphere.

Carbohydrates are composed not only of carbon and oxygen, which are released as carbon dioxide, but also of hydrogen, which has to end up somewhere. Hydrogen could be released as a gas, but hydrogen is highly reactive, so a safer approach is to combine the hydrogen with another element. Combining the reactive hydrogen with another element also is energetically favorable and will yield more energy that can be captured as ATP. In aerobic respiration, the electron transport system, the third component, generates more ATP out of the original six-carbon glucose molecule by the way the hydrogen protons, [H], are handled. [H] is generated by glycolysis and the citric acid cycle every time carbon dioxide is released. A carrier molecule picks up the [H] and passes it to an electron transport system, a series of cytochromes, pigment molecules. In aerobic respiration, ultimately two electrons and two [H] are combined with an oxygen to make a molecule of water, a molecule that is easily disposed of. A series of coupled reactions generates a lot more ATP.

Oxygen is not the only possible [H] acceptor of electron transport, so electron transport works fine in anaerobic respirations too. Electron transport is used by many anaerobic bacteria, where either sulfur or carbon acts as the [H] acceptor, releasing H_2S, hydrogen sulfide (rotten egg smell), or CH_4, methane (swamp gas), as metabolic waste products. This also shows that aerobic respiration is anaerobic until the last step, where oxygen is used as the ultimate [H] acceptor. Respiration changes from anaerobic to aerobic just by changing the final [H] acceptor.

Each of the three components of aerobic respiration can be found alone or with different metabolic components in different groups of bacteria. Among bacteria, electron transport has been linked to different fermentations a number of times. But what we do not yet understand is how an organism could have acquired a new metabolic component from a non-ancestral organism. Biologists have proposed the mechanism of lateral gene transfer, but no one knows how it works, only that it appears to have happened. The advantage of linking such metabolic components together is obvious. Metabolically coupling the citric acid cycle (2 ATP) and electron transport (34 ATP) to glycolysis (2 ATP) yields 38 ATP per glucose molecule, nineteen times as much ATP as glycolysis alone. By linking metabolic components, the maximum amount of energy is extracted from each substrate molecule, and such efficiency is one reason aerobic respiration is nearly universal among large, multicellular organisms. The other reason is that the citric acid cycle and electron transport take place only inside the mitochondria of eukaryotic cells, the cell's powerhouses. The next chapter will consider how cells acquired mitochondria and these metabolisms. Suffice it to say, it was a package deal.

The understanding that emerges is quite basic. Complex metabolic "machinery" did not arise all at once by outrageous chance but came about by the assembly of smaller, simpler, preexisting metabolic bits and pieces (Morowitz 1992). Metabolisms are constructed of two or more separate components, and each component is functional by itself, although in some cases a component's function is reversed to yield a new function. Any organism happening upon a linkage of metabolic components resulting in a gain in ATP production would be greatly favored by natural selection. Using electron transport with oxygen as the final [H] acceptor is a double winner because it couples an efficient ATP-generating mechanism with the detoxification of oxygen to safe water.

Like respiration, photosynthesis can be best understood as a series of separate components, pieces of metabolic machinery of different origins and different prior functions. Photosynthesis consists of three basic components: a light-absorbing pigment functioning as an energy-capturing antenna of sorts; an electron transport system/ATP pump, now called a photosystem, where the captured light energy produces ATP; and a synthetic pathway where the ATP is used to incorporate ("fix") carbon from an environmental source into organic molecules for energy storage (Fig. 2–4).

Instead of being photoautotrophs, many prokaryotes are chemoautotrophs. Chemoautotrophy is similar to photosynthesis except that the energy that drives the production of ATP and synthesis of organic molecules comes from the oxidation of inorganic substances—for example, hydrogen, ammonia, and hydrogen sulfide. Chemoautotrophic organisms mostly live in deep, dark, and often hot places because a reducing environment and the needed iron-sulfur geochemistry are common in such places. In addition to carbon dioxide, some chemoautotrophs use carbon monoxide (CO) or methane (CH_4) as a source of carbon. The specifics of their energy and carbon sources often restrict chemoautotrophs to very specialized and localized habitats. Since chemoautotrophy is simpler than photoautotrophy, and perhaps considerably older, it raises the possibility that the ancestors of photoautotrophs lived in deep, dark, hot places, and that photoautotrophy was derived from chemoautotrophy. Once chemoautotrophic organisms could live on the surface of Earth, widespread and abundant light would represent a new alternative energy source that is immensely greater and more widely available than that available from geological sources.

Almost all biology textbooks provide details of these autotrophic metabolisms; almost none tell you anything at all about when, where, and how the various components of these metabolisms came together. In seeking to explain how photosynthesis may have arisen in heterotrophic or chemoautotrophic organisms, a circular thinking problem is encountered, and many people fail to see a logical way out of this conundrum. Photosynthetic organisms need a light-absorbing pigment, but if an organism is not photosynthetic it does not need a photosynthetic pigment, so how did it ever get one? Did a complex pigment just suddenly appear in an unlikely event, making the organism the lucky winner of a crazy biological lottery? No luck and no outrageous chance were involved, just preexisting metabolic parts that can function in new ways. Any functional organism already had pigment molecules, like the cytochromes used in electron transport and respiration pathways (Dismukes et al. 2001). After this discussion of metabolisms, the easiest approach is to start with components that are already familiar: respirations and electron transport. Photoautotrophs have such metabolisms because they respire like all other organisms. But, both of these components have the potential to function in reverse.

ROLE REVERSAL YIELDS THE BASIC MACHINERY OF PHOTOSYNTHESIS

The simplest way to obtain a synthetic pathway is to reverse a respiration pathway. Biochemically pathway reversal is not a big deal. All those metabolic pathways are drawn with two-headed arrows (↔) because under proper conditions the reactions can go either direction.[20] Of the 10 enzymes used in glycolysis to break 6-carbon glucose down into two 3-carbon pyruvates, seven of these enzymes also catalyze the reverse reaction. Thus pyruvate can be used as a raw material to synthesize a 6-carbon sugar by an ATP-driven reversal of glycolysis with only three enzyme modifications. The sugar would then be polymerized into starch for storage. Any respiration pathway is fair game for reversal. The citric acid cycle is found functioning as a synthetic pathway in green sulfur bacteria, but because of their phylogenetic position, it remains uncertain whether this pathway started out in respiration and got co-opted for photosynthesis or vice versa. So obtaining a synthetic pathway seems pretty simple.

Electron transport appears to have begun as a mechanism for maintaining a proper cellular pH in an acidic environment (Dyer and Obar 1994), which may have been produced by the waste products of fermentations themselves—for example, lactic acid and pyruvic acid. To avoid feedback, end products must be either further respired or excreted from the cell, which acidifies the cell's immediate surroundings. In acidic environments hydrogen ions [H] tend to leak back into cells, so they must be pumped out continuously to maintain a proper cell pH. Some organisms use ATP to drive a molecular pump to remove excess hydrogen ions from within the cell. Others use a membrane-bound electron transport system that efficiently shuttles hydrogen ions outside cells without requiring any ATP input. If an organism had both systems, the electron transport system could generate acidic conditions around the cell by pumping out excess hydrogen ions. Then the locally elevated pH would allow hydrogen ions to "leak" back in via a reversal of the ATP pumping mechanism. However, when working backwards, the pump mechanism generates ATP rather than using it. This is the basic trick of autotrophy, and it demonstrates what a small step really exists between autotrophy and heterotrophy, an energy-driven reversal of an ATP pump. All that is needed is a means of capturing energy from an environmental source to drive an ATP pump, and a heterotroph can be converted to an autotroph. When light energy is absorbed by a pigment molecule, electrons are "excited," and if enough energy is absorbed, an electron is displaced and channeled through an ATP pump to generate ATP from ADP. And there is the energy needed to drive a synthesis pathway.

Cyanobacteria and chloroplasts actually have two different but interconnected photosystems (ATP-generating electron pumps). All other photosynthetic organisms, prokaryotes all, have but a single photosystem.[21] Photosystem I is based upon iron-sulfur compounds, and it is found in the green sulfur bacteria and photosynthetic gram-positive heliobacteria (*Heliobacterium*) (see Appendix: Green Bacteria). Photosystem II is based upon pigments and proteins and is possessed by purple bacteria and the green nonsulfur bacteria. How cyanobacteria came to have both photosystems is an interesting question, but a relatively new phylogenetic study of bacteria places the gram-negative cyanobacteria nested among the gram-positive bacteria, not among the gram-negative bacteria (Fig 2–8). This means cyanobacteria share a common ancestry with green nonsulfur bacteria and the heliobacteria in the gram-positive lineage, groups possessing both photosystems. On the basis of this phylogeny, green nonsulfur bacteria may equally make claim to being the oldest green organisms. As a result of having both photosystems interconnected via a modification of photosystem II, cyanobacteria are the only green organisms among prokaryotes that use water as a hydrogen source in photosynthesis. One of the other implications of this bacterial phylogeny is that cyanobacteria must have become gram-negative separately from the rest of the gram-negative bacteria, and perhaps studies will demonstrate whether they both produce the outer gram-negative cell envelop the same way or not.

Using water as a hydrogen source has an advantage and a disadvantage. Water is an abundant and readily obtained molecule, but the water molecule is more difficult to split than hydrogen sulfide. Photosystem I functions at a slightly lower energy wavelength than photosystem II, so water molecules can be split only when photosystem I interacts with photosystem II (Margulis 1982). The interaction of the two photosystems also improves the efficiency of photosystem II under aerobic conditions, which is interesting because their interaction is what splits water, releasing oxygen and producing the aerobic conditions. However, the initial advantage to linking and using two photosystems in concert was that

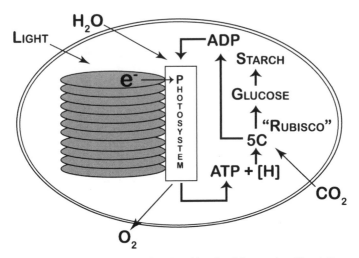

Fig. 2–4 Metabolic components of photosynthesis in a chloroplast: light-capturing chlorophylls on stacked membranes (chlorophylls *a* and *b*) function as an "antenna," an electron transport system functions as a photosystem, and a synthetic pathway "fixes" carbon. The captured light energy excites an electron, which when passed through the photosystem uses this energy to decouple [H] from water, releasing oxygen and converting ADP to ATP. The ATP and [H] are used in a synthetic pathway wherein the carbon fixing enzyme "rubisco" adds a carbon dioxide to a 5-carbon molecule to produce glucose and then starch, thereby storing the energy in the chemical bonds of a carbohydrate.

it made a previously unused and abundant resource, water, available to cyanobacteria.

Some cyanobacteria, like *Oscillatoria limnetica*, provide us with an example of a metabolic missing link[22] between organisms that use hydrogen or hydrogen sulfide as a hydrogen source (e.g., green nonsulfur bacteria) and organisms that use water for photosynthesis (all other cyanobacteria) (Margulis 1982). Under anaerobic conditions and when hydrogen sulfide is abundant, *Oscillatoria limnetica* uses hydrogen sulfide to obtain hydrogen for its photosynthesis rather than water. Facultative use of hydrogen sulfide and water in photosynthesis demonstrates that the two metabolisms are compatible, and when biologists figure out how this works, it may explain how the switch from using H_2S to H_2O was accomplished.

Okay, you may be thinking, so what? The significance and consequences of using water as a raw material in photosynthesis are probably still unclear. Using something really common like water as a hydrogen source turned photoautotrophic organisms loose upon Earth's surface. Since all three necessary materials (light, water, and carbon dioxide) are widespread and common, organisms are no longer restricted to small, narrow, specialized habitats, like hot springs, where something like hydrogen sulfide is locally abundant.

The switch from hydrogen sulfide to water for photosynthesis has had an even more dramatic impact on the history of life. When hydrogen is removed from water, which is H_2O, the leftover oxygen is released, so this metabolism is called oxygenic (oxygen-generating) photosynthesis. This metabolism is solely responsible for all atmospheric oxygen, which now makes up 21% of the atmosphere. As you now know, most metabolisms are fundamentally anaerobic; you probably do not think of oxygen as toxic, but in an anaerobic environment and to anaerobic metabolisms, it is. The aerobic respiration of all large organisms was impossible before a significant amount of oxygen accumulated in the atmosphere, and this took quite a while. This means the evolution of oxygenic photosynthesis is one of the most significant events in Earth history because this new metabolism changed the Earth's surface environment, a change that affected virtually all surface-dwelling organisms and perhaps made the evolution of large organisms possible. We humans breathe an oxygen-rich atmosphere and respire aerobically all because cyanobacteria somehow came to possess two different photosystems from two different sources.

An attempt to figure out how this happened requires an understanding of phylogenetic relationships among prokaryotes. As mentioned above, presumably cyanobacteria obtained the two photosystems from common ancestry with green nonsulfur bacteria and gram-positive heliobacteria. In just the time it has taken to write this book, a mere decade or two by now, new hypotheses have altered our explanations. How chloroplasts came to have the same photosynthetic mechanism as cyanobacteria is well known and this will be explained in the next chapter. Now back to metabolism.

Thus two of the three components needed for photosynthesis, a photosystem and a synthetic pathway, were obtained by conversion of metabolic components already on hand for respiration. No major innovations were involved, just tinkering with preexisting components. If such functional shifts occurred, then we might expect that some metabolic components would continue to play dual roles (a prediction). This is absolutely true for cyanobacteria, which only respire at night because their electron-transport machinery is used in the light, working in reverse, for photosynthesis. In the cells of plants this is not a problem because separate respiration pathways function in the cell's protoplasm and in the mitochondria, while the chloroplast has a dedicated photosystem. This ability to respire and photosynthesize simultaneously may explain one of the primary advantages of having chloroplasts.

The components of metabolism can be diagrammed to show where the components of oxygen-producing photosynthesis came from starting with fermenting thermophilic bacteria as universal common ancestors (Fig. 2–5). Arrows link similar components and indicate the general type of change involved. To keep the diagram as simple as possible, not all of the possibilities were shown. The idea that complex metabolisms were constructed from components of simpler metabolisms is still a relatively new concept, and now that the tools exist to find them, examples abound.

Although not directly germane to this discussion, we can consider the diverse metabolisms of the purple bacteria. Purple bacteria derive their name from the reddish-brown to purplish-brown color of their photosynthetic members that possess bacteriochlorophylls. However, the purple bacteria display a considerable range of metabolisms in addition to photoautotrophy: heterotrophy involving nitrogen reduction and oxidation as well as heterotrophy involving sulfur reduction and oxidation. All of these metabolisms are based upon the same electron transport chain. The nature of the specific inputs and outputs, and the direction the electron transport chain is running—forward when metabolizing substrates to generate ATP, and backward when used autotrophically to synthesize organic molecules and fix carbon dioxide—determine the specific metabolism. Hydrogen sulfide, hydrogen, methane, ammonia, and light-activated bacteriochlorophylls can all act as hydrogen sources, and oxygen, sulfates, nitrates, nitrites, and ferric ions can all be used as [H] acceptors. This metabolic versatility represents the purple bacteria's diverse adaptations to numerous different habitats where they play many significant ecological roles in elemental cycling. Within this one lineage, albeit a large and diverse one, so much metabolic diversity demonstrates how readily one basic piece of metabolic machinery can be used in many different ways. Again nature demonstrates remarkably little real innovation and a great deal of "recycling."

This leaves only one component of photosynthesis to account for: a light-capturing pigment. Green photosynthetic organisms employ the light-capturing pigments called chlorophylls, and clearly their origin is quite central to this narrative. Chlorophylls[23] are green because they absorb wavelengths at the red-orange and blue-indigo-violet ends of the visible spectrum (ROYGBIV[24]), which leaves mostly green wavelengths left over. Chlorophylls are the most common light-capturing pigments of living organisms (green bacteria and chloroplasts). Other photosynthetic bacteria are "purple" because they use bacteriochlorophylls, which appear brownish red because they absorb energy in the near-infrared and blue-indigo-violet wavelengths, leaving ROYG wavelengths unabsorbed. Bacteriochlorophylls are present in the purple bacteria and the heliobac-

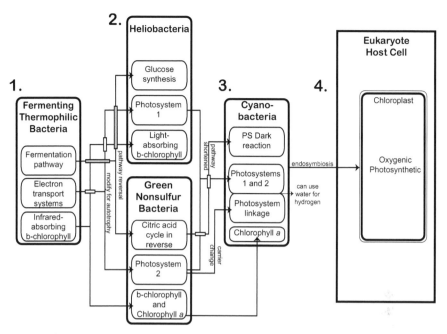

Fig. 2–5 Evolutionary relationships among the metabolic components of photosynthesis. 1. Thermophilic bacteria sense heat via bacteriochlorophyll (b-chlorophyll) that absorbs in the infrared wavelengths. Their metabolism is a fermentation pathway. Electron transport helps regulate cellular pH in the acidic environment caused by the metabolic wastes of fermentation. 2. Both heliobacteria and green nonsulfur bacteria use light-absorbing bacteriochlorophylls obtained from thermophilic ancestors. Green nonsulfur bacteria may be the first organisms to have chlorophyll *a*, which is a derivative of bacteriochlorophyll biosynthesis. Synthesis pathways are derived from fermentation pathways running in reverse, one of which is similar to the citric acid cycle of aerobic respiration. Photosystems are electron transport systems coopted and modified for photosynthesis. 3. Cyanobacteria obtained their two photosystems from both green nonsulfur bacteria and heliobacteria, with which they share a common ancestry according to recent phylogenetic studies (Fig. 2–8). Among photosynthetic prokaryotes, only cyanobacteria have two photosystems operating together, linked by a modification of photosystem 2, which allows them to use water as a source of hydrogen in photosynthesis. 4. Eukaryotes obtained oxygenic photosynthesis intact by transforming a cyanobacterium into a chloroplast via endosymbiosis. The origin of chloroplasts is discussed in chapter 3. Types of modifications to components are shown on connecting arrows. (Image source: After Schopf 1999).

teria, and they co-occur with chlorophyll *a* in the green sulfur bacteria and green nonsulfur bacteria.

The molecular structures of chlorophylls and bacteriochlorophylls are based upon porphyrin (POUR-fur-in) rings, a highly conserved molecular arrangement that also forms the molecular backbone of cytochromes, which are the pigments composing all electron transport systems and [H] pumps (Fig. 2–6). Absorbing light energy causes an electron to move around the ring-like molecular structure of these pigments, and when coupled to a photosystem, this energy generates ATP. Porphyrin-based pigments are universal in known organisms and can be assembled from

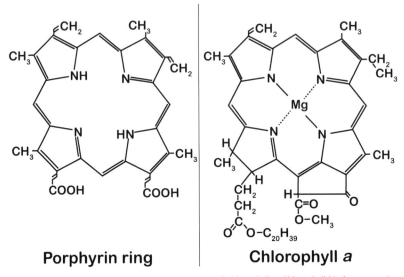

Porphyrin ring | **Chlorophyll *a***

Fig. 2–6 Comparison of basic porphyrin ring structure and chlorophyll *a*. Chlorophyll binds a magnesium (Mg) into the center of the ring structure and has an additional long side chain (CH_2-CH_2-etc. 17 times ending with CH_3; 20 C and 39 H in all).

the same basic organic building block molecules as the rest of life, so here our inferences of their origin must stop.

Chlorophyll may have originated as a derivative of a cytochrome converted from ion pumping to photosynthesis (Dyer and Obar 1994), but more likely chlorophyll arose as a derivative of bacteriochlorophyll. Chlorophyll can be synthesized from intermediate molecules used in the biosynthesis of bacteriochlorophylls, and this would predict that organisms with bacteriochlorophyll are ancestors of organisms with chlorophyll, and phylogenetically, this is true, mostly. If bacteriochlorophylls were found to have a function other than photosynthesis, this would support and strengthen the concept that bacteriochlorophylls preceded chlorophyll and maybe even photosynthesis. Since chlorophylls take better advantage of the visible spectrum by absorbing more light energy than bacteriochlorophylls, any organism chancing upon chlorophyll via a mistake (mutation) in their bacteriochlorophyll biosynthesis pathway would have an advantage. If chlorophyll is derived from a precursor in the synthesis of bacteriochlorophyll, the pathway need only be truncated to shift pigments (Lockhart et al. 1996) or, as some bacteria do, to use both pigments together. If bacteria had not happened upon chlorophyll, the title of this book would have to be *How the Earth Turned Brownish-Red*.

A non-photosynthetic function for bacteriochlorophylls has recently been discovered. Motile deep sea thermophilic bacteria were discovered to be capable of moving toward sources of heat by sensing infrared energy. Infrared energy is composed of wavelengths beyond the red end of the visible spectrum, and bacteriochlorophylls absorb these wavelengths. Thermophilic bacteria from different ocean depths had different bacteriochlorophylls, each of which absorbed infrared at a different peak wavelength that matched the wavelength of their environmental infrared energy source (Nisbet et al. 1995), a finding that helps confirm their heat-sensing function. Bacteriochlorophylls, adapted for use in the "dark" as thermo-sensing pigments, were preadapted for light absorption if the descendants of such bacteria invaded the bright surface environment. Although heat energy can be abundant in certain hot habitats, infrared wavelengths are not energetic enough to be used in autotrophy, so there are no known thermoautotrophs.

The steady stream of new discoveries that cause you to rethink your understandings of things is one of the niftiest things about science. You constantly must learn new things, often changing your way of thinking, but it also means frequently revising your statements. An obligate photosynthetic green sulfur bacterium has been discovered living in a deep-sea hydrothermal vent (Beatty et al. 2005). This organism captures the extremely dim infrared "light" given off by the hot magma, pitch-black to us, and uses it to split hydrogen from hydrogen sulfide. The peak absorbance of this organism's photosynthetic pigments is actually just beyond the visible spectrum in infrared wavelengths, so in effect it is a "red-hot" thermoautotroph. This wavelength is not energetic enough to split a water molecule, but hydrogen sulfide is easier to split. The existence of such a photosynthetic organism suggests photosynthesis could have originated in a deep, dark, hot environment after all. Isn't science wonderful?

Perhaps this discussion has caused you to wonder why, out of the whole electromagnetic spectrum, the narrow band of wavelengths called the visible spectrum is also the band of wavelengths primarily used in photosynthesis. Is this some kind of coincidence, or is there a scientific explanation? The narrow range of electromagnetic wavelengths called light is both energetic enough and common enough that sight-sensory organs are adapted to detect them; thus, from our human perspective, these electromagnetic wavelengths are called the visible spectrum, but this label is quite species specific. Not all organisms see the exact same spectrum—for example, bees and lots of other organisms see a visible spectrum that extends into the ultraviolet. Beyond the red end of the visible spectrum is infrared (heat), and these wavelengths are not energetic enough to use in normal photosynthesis. Beyond the violet end of the visible spectrum are ultraviolet (UV) wavelengths, which are energetic enough to harm many biological molecules. Thus the abundance and energy range of wavelengths in the visible spectrum, energetic enough without being too energetic, explain why photosynthetic organisms absorb these wavelengths, but this does not explain why so many photoautotrophs are green.

Green organisms seem to be wasting a lot of light energy by not absorbing green wavelengths, and the greenness of our world is a testament to how much green light is being wasted. A black photosynthetic pigment would absorb all the visible wavelengths, capturing more energy, something any sixth grade solar engineer can tell you. Why has not a more efficient light-capturing pigment, a "nigerophyll," evolved?

The answer to this question demonstrates something fundamental to understand about evolution. Organisms are constrained by their evolutionary history, and they simply cannot manufacture a "desired" adaptation. A more useful variant of a preexisting feature is the only tool-kit organisms have available to produce new adaptations. Chlorophylls provide an advantage over bacteriochlorophylls because the former makes better use of visible light. Green pigments absorb more of the visible spectrum than brownish-red (purple) pigments. As will be demonstrated in subsequent chapters, for the vast majority of Earth history, photosynthetic organisms were limited to aquatic environments, and a green pigment works very well in aquatic environments because blue and red wavelengths penetrate water better than green. Land plants are green because their aquatic ancestors were green, and why else would land plants have an aquatic-adapted photosynthetic pigment? To put it another way, chlorophylls are poorly designed for a terrestrial environment, but no alternatives exist.

So we arrive at an important realization in explaining why life is the way it is, a basic environmental parameter: the abundance of energy available in the visible portion of the electromagnetic spectrum *under water* is the reason photosynthetic organisms are predominately green. Good luck finding that idea in biology textbooks, and yet this is a very fundamental understanding.[25]

Biological organisms possess many such "imperfections" and "inefficiencies," and they make sense only from an evolutionary perspective. Land plants using chlorophyll and the carbon-fixing enzyme rubisco are botanical versions of the panda's thumb, one of S. J. Gould's (1980) favorite examples of evolution. Pandas have five digits, but lack an opposable thumb. They strip leaves from bamboo shoots by grasping them between their 5-digit paw and an enlarged wrist bone, which is not an elegant design, just a useful contraption for feeding itself. Pandas lack opposable thumbs because the ancestors of pandas are carnivores whose paws are good at running and clawing, but not good at grasping tree limbs or pens. On the other hand, we have inherited an opposable thumb from tree-dwelling primate ancestors, and it has proven useful for grasping coffee cups and tapping a space bar, among other things.

PROKARYOTE PHYLOGENY

To develop a full understanding requires a synthesis. Our current knowledge of ancestral relationships among living organisms must be melded to what we know of ancient life from geological and fossil evidence and combined with what we know about the components of metabolisms to gain an understanding of photosynthesis and green organisms. Determining ancestral relationships among prokaryotes has not been easy, or even possible, until recently. Large organisms display many readily observed shared characters that can be used to group them into categories presumed to represent a lineage descended from a common ancestry.

Bacteria do not display a lot of diversity in form, just little bitty spherical cells (cocci), little bitty rod-like cells (bacilli), and cute little curlicue cells (spirilli), and cell envelopes that stain either purple (called gram-positive) or pink (gram-negative). (This staining process was named for its discoverer, Hans Christian Gram.) The traditional *Bergey's Manual* classification (Buchanan and Gibbons 1974) grouped bacteria on the basis of similar metabolisms and/or similar cellular organizations—for example, fermenting bacteria, endospore-producing bacteria, cyanobacteria, spirochaetes, enteric bacteria, and gram-positive bacteria. The real diversity of bacteria is expressed in their metabolisms, and bacteria have been classified by how they capture energy, by what substrate molecules they respire, and by what metabolic waste products they produce. While similar metabolisms and similar cellular characteristics are shared features suggesting that all the members of these groups have a common ancestry, such characters provide little basis for evaluating ancestral or phylogenetic relationships among the diverse groups. This was such a hopeless situation that for the longest time many microbiologists did not think evolutionary relationships among prokaryotes could ever be studied, or that such studies even mattered very much. Of course, they were wrong on both counts because often in science it is hard to predict where the next breakthrough will occur.

Research that began about 40 years ago has totally changed our understanding of life's diversity. Macromolecular sequence data from the ribosome, a universal cellular organelle that functions in protein synthesis, have been used to generate a picture of relatedness among all organisms. These data have demonstrated a surprising evolutionary pattern among prokaryotic and eukaryotic organisms (Fig. 2–7). Carl Woese (Woese and Fox 1977) discovered that two distinct lineages existed among prokaryotic organisms: one lineage includes most of the familiar groups of bacteria, and a second lineage includes bacteria that inhabit hot, acidic, or salty environments, or whose metabolism produces methane. Members of this latter group are the archaebacteria, and as you may have guessed from the preceding sentence, the vast majority of them are extremophiles.[26] Very significantly, and much to the surprise of biologists, archaebacteria have a more recent common ancestry with eukaryotic organisms (Keeling and Doolittle 1995). These three lineages form the most basic phylogenetic pattern in biological diversity, and generated the proposal to recognize three lineages, bacteria (or eubacteria), archaea (or archaebacteria), and eucarya (all eukaryotes), as a super-kingdom classification category called Domains (Woese et al. 1990). They argued that designating three domains emphasizes that the differences between bacteria and archaea were as great as or greater than the differences between prokaryotic and eukaryotic organisms, which were long thought to be the most fundamental differences in biological diversity. Domains were never accepted as an official classification category; nonetheless, Woese's research opened the door to the study of prokaryote phylogeny.

Like all shared novel features, these shared ribosomal RNA sequences are interpreted as evidence of common ancestry. Such data can be explained and understood without actually examining the nucleotide base sequences. Eukaryotic organisms have 80S (read as "eighty-S") ribosomes in their cytoplasm, which are about 50% bigger than the 70S (seventy-S) ribosomes characteristic of bacteria. 70S and 80S refer to sedimentation rates in a centrifugation density gradient, and thus are logarithmic. 80S ribosomes are bigger because they possess two additional nucleic acid sequences, big insertions, one in each of the two ribosomal subunits. Archaea possess one of the two additional sequences, meaning their ribosomes are larger than 70S, but smaller than 80S. Archaea and eukaryotes share one of the two inserted sequences, a change which occurred in their last common ancestor after the archaea-eukaryote lineage had diverged from the rest of bacteria. The common ancestor of all eukaryotes acquired the second additional sequence, so all eukaryotes share the character of 80S ribosomes in addition to having cells with a nucleus. All molecular phylogenetic hypotheses are constructed using similar data, although not necessarily big insertions, and similar reasoning.

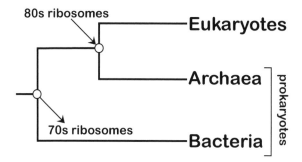

Fig. 2–7 Phylogeny of life on Earth. All living organisms are members of these three lineages, sometimes called domains: bacteria, archaea, and eukaryotes. Eukaryotes have a more recent common ancestry with archaea (upper arrow), formerly known as the archaebacteria. The eukaryote-archaea clade has a common ancestry with bacteria (lower arrow). Archaea and bacteria have a prokaryotic cell organization, but since they do not form a single lineage, this phylogenetic relationship falsifies the taxonomic hypothesis of a prokaryotic group or lineage (e.g., Kingdom Monera).

The phylogenetic relationships among bacteria, eukaryotes, and archaea (Fig. 2–7) form a total of five clades or lineages: (1) archaea, (2) bacteria, (3) eukaryotes, (4) eukaryotes-archaea, and (5) all living organisms. Monera does not exist as a clade including all prokaryotic organisms (bacteria and archaea); therefore, as a classification group and a taxonomic hypothesis, the Kingdom Monera has been falsified and must be abandoned. Prokaryote is no longer synonymous with a taxonomic group, and now designates what is called a grade, a level of organization shared by more than one lineage.

Similar molecular studies have produced a more detailed picture of phylogenetic relationships among groups of prokaryotes. New data are constantly updating this picture in recent years, so a great deal may well change between now and the time this book is published. The archaebacteria and the eubacteria still form two clades, but new data have greatly changed branching patterns among bacteria (Fig. 2–8).

The earliest branching lineage, thermophilic bacteria, consists of two genera which are chemoautotrophs. In terms of habitat, they are extremophiles. The remaining bacteria fall into one of two major clades, one that is basically gram positive and one that is gram negative. Gram-positive bacteria is the earliest diverging lineage and includes the photosynthetic heliobacteria. Cyanobacteria are shown here with the green nonsulfur bacteria as their sister group, but the placement of this gram-negative group remains questionable. An equally likely alternative places the green nonsulfur bacteria as sister group to the gram-positive bacteria. Some previous studies had placed the green nonsulfur bacteria as sister group to the thermophilic bacteria. The actinobacteria decompose cellulose and

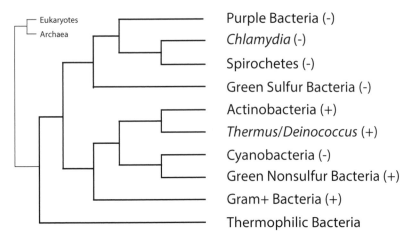

Eukaryotes
Archaea

Purple Bacteria (-)
Chlamydia (-)
Spirochetes (-)
Green Sulfur Bacteria (-)
Actinobacteria (+)
Thermus/Deinococcus (+)
Cyanobacteria (-)
Green Nonsulfur Bacteria (+)
Gram+ Bacteria (+)
Thermophilic Bacteria

Fig. 2–8 Phylogenetic relationships among major groups of bacteria. This diagram shows an expanded but still simplified version of bacterial phylogeny, a consensus of three studies. Diagram upper left shows relationship of bacteria to the other two lineages as shown in Fig. 2–7. (Image source: After Bern, Goldberg, and Lyashenko 2006)

produce the odor of soil, and they are sister group to another group of extremophile bacteria represented by the appropriately named *Thermus*. This group stains gram-positive but they have a second cell envelope, like the gram-negative bacteria. The gram-negative clade consists of the gram-negative bacteria including the parasitic *Chlamydia*, the Green Sulfur Bacteria, the Spirochetes, and the Purple Bacteria (also called Proteobacteria), which consists of five lineages not shown in this simplified diagram.

What does this phylogenetic hypothesis suggest about the origin of photosynthesis? Heliobacteria (gram-positive) have a unique form of bacteriochlorophyll that could be ancestral to all other forms. Heliobacteria also possess a simpler photosystem, which is similar in organization to photosystem I, and their cells lack the infoldings of the cell membrane found in other photosynthetic prokaryotes to provide more surface area for membrane-bound components of photosynthesis. The green nonsulfur bacteria may have been the first organisms to have chlorophyll, a hypothesis supported by their possession of a very simple and unique carbon-fixing metabolism that is basically the citric acid cycle (CAC) running backwards, assuming its original function was in respiration. They also possess photosystem II and are facultative heterotrophs capable of growing in the dark if appropriate substrates are available. They can use hydrogen or hydrogen sulfide for photosynthesis, but they do not release sulfur as a byproduct like the green sulfur bacteria.

Cyanobacteria in this phylogenetic position could have acquired both photosystems and chlorophyll from common ancestry with the gram-positive heliobacteria and the green nonsulfur bacteria. This phylogenetic hypothesis suggests cyanobacteria are a very early diverging lineage of gram-negative bacteria. Earlier studies placed cyanobacteria in the gram-negative lineage, often near the green sulfur bacteria. By placing cyanobacteria nested within a clade of gram-positive organisms, it suggests the gram-negative condition was derived from gram-positive bacteria at least twice (once for cyanobacteria and another time for all the rest). Whether this will be supported by subsequent studies remains to be seen. If cyanobacteria were nested in the gram-negative clade, the only way they could have obtained a photosystem from gram-positive bacteria was by the mysterious and as yet undemonstrated "lateral gene transfer." So perhaps having the gram-negative condition evolve twice is the more probable explanation.

The organisms occupying the deepest branch of the bacteria, the earliest diverging lineage, are anaerobic thermophilic extremophiles. This is also true, although not shown, for the Archaea. This is the same environment, hot mineral springs or deep sea geothermal vents, in which the oldest well-documented fossil bacteria occur. Therefore this phylogenetic hypothesis suggests the common ancestor of all living organisms was anaerobic and thermophilic (Woese 1998; Kerr 1999), but were these ancestors green, purple, or plain? While some members of these basal lineages are chemoautotrophic, none are photoautotrophs, so they are neither green nor purple. In summary, green organisms are not likely to have been the most ancient or earliest appearing organisms, especially if a deep, dark, hot origin proves true. However, nothing in this hypothesis is contrary to the idea that green organisms appeared very early in Earth history, about 3.5 bya. Our present phylogenetic hypotheses do not resolve the question of whether the earliest organisms were autotrophic or heterotrophic because both autotrophs and heterotrophs occupy the earliest diverging lineages of prokaryotes.

On the basis of modern rates of evolution among prokaryotes, biologists estimate that the process of life assembly from abiotic materials to photosynthetic cyanobacteria may have taken no more than 10 million years (Lazcano and Miller 1994), so from our distant perspective, all these events would have occurred almost at once. Remember that environmental circumstances may not have allowed some events—for example, living on the Earth's surface prior to 3.8 bya or having aerobic respiration prior to 2.5 bya when enough oxygen had accumulated.

If simple fermenters using preexisting organic molecules of abiotic origin were the first organisms, they were living on borrowed time. As life began to flourish, at some point the consumption of organic molecules would proceed at a rate faster than they were being abiotically synthesized, and a food crisis would result. This produces a situation where any organism capable of energy capture and synthesis of complex molecules from simpler raw materials would have a tremendous advantage. Natural selection would greatly favor any form of autotrophy. Ultimately the waste materials, consumption, or decomposition of autotrophs rescued heterotrophs from the food crisis, and modern food chains with an autotroph at the base had their origins.

Such hypotheses are logical constructs based upon what is known and they emphasize the interconnectedness of biology. They also allow us to make predictions about other things that should also be true; in other words, they allow us to do science. Phylogenies were constructed from shared changes in macromolecular sequences, and when a phylogeny makes logical and orderly sense out of the distribution of other characters, our confidence in the evolutionary explanation increases. In a very real sense, every new phylogenetic study has the potential to falsify previous hypotheses by making a complete mess out of character distributions and other data, like the sequence in which organisms appear in the fossil record. Similarly, every study of specific characters tests our phylogenetic hypotheses. For example, what if prokaryote phylogeny suggested that aerobic organisms formed the basal lineages of life? If true, it would argue that our hypotheses about an anaerobic Hadean Earth environment based on geological evidences, the origin of life, and the evolution of metabolisms were in error. Biologists continually run this risk of being found in error by new studies, but conversely, our confidence in the truth of such explanations grows with every new study that provides corroborating results and every time some previously anomalous observation is now satisfactorily explained.

Take a deep breath. Exhale slowly. Become one with the metabolisms. That was probably more than you ever wanted to know about the evolution of metabolisms, but too often these basic understandings are overlooked in standard biology classes. Most biology students do not know that prokaryotes "invented" all metabolisms by cobbling together basic metabolic components, often using them in new ways. The rest of this history leading to large organisms, both green and nongreen, will produce no new metabolisms, although metabolisms from bacteria and archaea become combined in a most surprising manner in eukaryotic cells. Chlorophyll became the photosynthetic pigment of choice because it absorbs more of the visible light spectrum under water than does bacteriochlorophyll, which started out functioning as a heat sensor, probably. Photosynthesis in cyanobacteria incorporates two different photosystems that interact allowing only these autotrophs to obtain hydrogen from water. As a consequence oxygen is released as a toxic byproduct in an anaerobic world, and the slow accumulation of oxygen had a major impact on life on Earth. All of this took place while the Earth was still quite young.

The rest of this history will be about the elaboration of cells, multicellular organisms, and changes in form in relation to new habitats and new challenges. In this arena prokaryotes display very limited innovation, producing a couple of specialized cells and nothing more complex or larger than a filament (Appendix: Green Bacteria). However, something about the eukaryotic condition opened the door to many new possibilities.

THREE

Cellular Collaborations

Wherein the diversity of unicellular organisms is explored, chloroplasts and mitochondria are obtained via symbiotic interactions between cells, and other features such as nuclei and sex are examined to determine what can be learned of their origins and functions.

Some Physiological Descriptions of Minute Bodies Made by Magnifying Glasses with Observations and Inquiries Thereupon.
—R. Hooke, 1665

Unicellular organisms are so successful, so numerous, and so diverse that an unbiased description of life on Earth could be summed up with just two words: *mostly unicellular*. This fairly accurate oversimplification echoes the rather brief updated entry for Earth in *The Hitchhiker's Guide to the Universe*: "mostly harmless" (Adams 1992). On a daily basis we humans operate at the large end of the organismal size scale. Although we do not tend to think of ourselves as particularly big or, for that matter, even as animals, humans are on the large end of the organism size spectrum. In terms of volume, from the smallest known prokaryotic organisms (mycoplasmas) to the largest organisms ever to live, sequoia trees, size increases about 23 orders of magnitude (powers of 10). A sequoia tree with a mass of 1,000–2,000 tons is an order of magnitude larger than the largest animal ever to live on Earth, a blue whale (100 tons) (see Fig. 5–9 C-D). And a blue whale is 3–4 orders of magnitude larger than us, but humans are 18–19 orders of magnitude larger than a mycoplasma. This means humans are larger than 99.9% of all animals, yet we seem most enthralled by those animals near our size or larger, the type that dominate most zoos. How many people would visit the world's largest beetle[1] zoo?

Part of our large-organism bias can be attributed to human visual acuity. The resolution limit of our eyes is such that we simply cannot see the vast majority of organisms. The existence of tiny living organisms became apparent only after the microscope was invented and used to study nature. Three hundred and fifty years ago the field of cell biology was in its infancy, its discovery phase, as aptly described in the subtitle of Robert Hooke's *Micrographia* (above), one of the first publications in cell biology. Our knowledge of cells and cell biology has grown exponentially, especially in recent decades, largely because of electron microscopy and molecular biology. But in spite of all that knowledge, with the exception of biologists, most people fail to grasp how common and how diverse tiny organisms are.

Wrapping your brain around cell sizes and cell numbers is not easy because on a day-to-day basis we just don't operate at these scales. According to the best estimates, your body consists of some 10 to 100 trillion (million million) cells, and even more amazingly, your body plays host to so many microscopic organisms that they greatly outnumber the cells that compose your body but they probably don't constitute much more than a thousandth of your mass. The same thing can be said about every other big organism too. The implications of this are clear, but staggering. All the organisms that can be seen with the naked eye, as numerous as they are, are outnumbered by unimaginable orders of magnitude by tiny, unseen organisms. We live in a world teeming with tiny creatures and only seldom do we notice their presence, although as noted in the previous chapter we frequently notice the effects of their metabolisms.

Microorganisms, referring to all microscopic organisms, are both prokaryotic and eukaryotic, but all large organisms are eukaryotic. Having already considered, however briefly and incompletely, prokaryote diversity, the astounding diversity of eukaryotic organisms requires a bit of attention also. On the basis of common patterns of cellular organization, 71 distinct groups of eukaryotic organisms have been identified (Patterson 1999) (Table 3–1), but large organisms are found in only five groups: green algae and land plants, red algae, brown algae, fungi, and animals. The overwhelming majority of eukaryotic organisms, in terms of both sheer numbers and diversity, are unicellular or small, simple multicellular organisms. The five groups with big organisms either include unicellular organisms or have common ancestry with unicellular organisms, which suggests large multicellular organisms evolved independently in each group, and probably more than once in some—for example, green algae and land plants. Another part of our biased thinking is the idea that evolution has a directionality from little to big. In only one sense is this generalization

Table 3–1. Distinct groups of eukaryotes (after Patterson 1999)

All the groups in this list have a distinctive cellular organization, a definite shared characteristic. Furthermore, each is treated as a unique lineage because presently there is no agreement about their sister groups—that is, to whom they are most closely related or with whom they share their most recent common ancestry. Numbers in parentheses indicate genera and species of general groups; names in italics are genera and numbers in parentheses are species. All groups with green photosynthetic members are marked with an asterisk. The brief description is not intended to describe the unique shared character defining the group. Abbreviations: GI, gastrointestinal; syn, synonym; w/o, without.

Acantharea (syn. radiolarians) (50/150) Marine unicells with radiating skeleton of spicules.

Actinophyrids (syn. heliozoa) (2/?) Marine unicells with radiating arms & microtubular skeleton.

*Alveolates (syn. protists, protozoa) (many) Diverse motile unicells with cytoplasmic alveoli.

Ancyromonas (1/3) Small, gliding, flagellated unicells.

Apusozoa (2/?) Biflagellated unicells with dorsal organic sheath.

Biomyxa (1/?) Amoeboid unicells with stiff, branching/anastomosing pseudopodia.

Caecitellus (1/1) Biflagellated unicells with lateral ingestion gullet.

Carpediemonas (1/2) Biflagellated unicells with flagella inserted at end of ventral groove.

*Centroheliozoa (syn. heliozoa) (14/85) Similar to Actinophyrids; some symbiotic with algae.

*Chlorarachniophytes (4/6) Reticulate amoeboid unicells.

Coelosporidium (1/1) Amoeboid parasites of cladocera.

Collodictyon (1/several) Flagellated (4) unicells with broad ventral groove/gullet.

Copromyxids (2/?) Cellular slime molds.

Cryothecomonas (1/several) Biflagellated unicell with gullet & covered with theca.

*Cryptomonads (25/?) Unicells with 2 flagella inserted in an anterior groove.

Desmothoracids (3/?) (syn. Heliozoa) Amoeboid unicells with stiff pseudopodia.

Dimorphids (2/?) Flagellated (2 or 4) unicells with stiff microtubular axopodia.

Diphylleia (1/1) Unicell with 2 flagella inserted apically at end of broad gullet.

Diplomonads (~9/?) Flagellated (2 or 4) unicells lacking mitochondria.

Discocelis (1/several) Asymmetrical biflagellated unicell with organic endoskeleton.

Ebriids (2/3) Biflagellated unicells with siliceous branching skeleton.

Ellobiopsids (5?/?) Multicellular parasites of marine invertebrates.

Entamoebae (3/5) Parasitic amoeboid unicells without mitochondria.

*Euglenozoa (lots) Unicells with 2 flagella inserted in anterior gullet; some heterotrophic.

Table 3–1. (*continued*)

Fonticula (1/?) Cellular slime mold.

Fungi (many) Unicellular and filamentous, yeast, molds, rusts, smuts, mushrooms.

*Glaucophytes (5/20) Diverse unicells with reduced cyanobacteria endosymbionts.

Granuloreticulosa (many/40,000+)(syn. Foraminifera) Amoeboid unicells with pseudopodia.

Gymnophrea (2/?) Amoeboid unicells with fine pseudopodia.

Gymnosphaerida (3/3) (syn. Helizoa) Amoeboid unicells with stiff radiating arms.

Haplosporids (3/35) (syn. Sporozoans) Unicellular parasites of marine invertebrates.

*Haptophytes (several/300+) Biflagellated unicells with haptonema organelle & scales.

Heterolobosea (?) Unicells changeable as amoeboid flagellates or slime molds (syn. acrasids).

Hyperamoeba (1/1) Variable unicell, amoeboid to uniflagellate.

Jakobids (3/?) Biflagellated unicells with ventral feeding groove.

Kathablepharids (3/?) Large biflagellated unicells with fine scales and feeding gullet.

Komokiacea (12/?) Large amoeboid, benthic marine organism with thin pseudopodia.

Luffisphaera (several) Small rounded marine unicells with scales and spines.

Microsporidia (20 families) Minute parasitic unicells without flagella, mitochondria.

Ministeria (1/2) Small marine unicell with 20 stiff symmetrically radiating pseudopodia.

Multicilia (several) Flagellated (up to 30) unicell with rounded body.

Nephridiophagids (5/12) Parasitic unicells of insect kidneys.

Nucleariidae (5/?) Amoeboid organisms with fine, microtubule-supported pseudopodia.

"Opisthokonts" (1,000,000+) (syn. Fungi + chytrids + animals + choanoflagellates)

Oximonads (?/100) Four-flagellated unicells commensal in termite GI tract; no mitochondria.

Parabasalids (several/?) Multiflagellated unicells commensal in wood-eating insect GI tracts.

Paramyxea (several genera/?) Amoeboid (multicellular?) parasites in GI tract of marine invertebrates.

Pelobionts (4/~200) Basal lineage of uniflagellated unicells w/o mitochondria or dictyosomes.

Phaeodarea (100/?) (syn. Radiolaria) Marine unicells with axopodia and silicaceous skeleton.

Phagodinium (1/1) Intracellular parasitic protist of Stramenopila.

Phalansterium (1/several) Uniflagellate colonial organism producing a globular organic matrix.

*Plantae (many) Unicellular, filamentous, complex multicellular; green algae & land plants.

Plasmodiophorids (10/40) Amoeboid parasite/cell predator with unusual cell penetration organ.

Polycystinea (many?) (syn. Radiolaria) Marine unicell with axopodia & silicaceous skeleton.

Pseudodendromonads (3/?) Biflagellated unicells or colonial organisms with gullet.

Pseudospora (1/?) Biflagellated or amoeboid unicell mostly parasitic on plant cells.

Ramicristates (many/many) Naked, lobose amoebas.

*Rhodophytes (?/4,000) (syn. Red Algae) Unicellular, filamentous, pseudoparenchymatous algae & seaweeds.

Reticulomyxa (1/1) Large, naked amoebas with reticulate pseudopodia.

Retortamonads (2/?) Unicells with 2 or 4 flagella at end of ventral feeding groove.

Rosette agent (1/1) Unicellular obligate intracellular parasite of salmon spleen/kidneys.

Spironemidae (3/?) Elongate unicells with many flagella arranged in rows; 2 large body plates.

Spongomonads (2/several) Biflagellated unicells or colonies in iron-rich mucoid globules.

Stephanopogon (1/several) Dorsoventrally flattened marine, multiflagellate unicells.

Sticholonche (1/2) Marine unicells with 4 rows of mobile axopodia inserted in nuclear sockets.

*Stramenopiles (many) Diverse group of unicellular to complex multicellular organisms with unique biflagellated motile cells; autotrophs include brown algae and diatoms, heterotrophs include water molds and slime net amoebas.

Telonema (1/2) Common marine anteriorly biflagellated predatory unicell.

Thaumotomonads (7/?) Biflagellated unicells using pseudopodia for feeding.

Trimastix (1/3) Flagellate (4) unicell with cruciate feeding groove, lacking mitochondria.

Vampyrellids (6/?) Amoebas with fine pseudopodia predatory on contents of algal/fungal cells.

Xenophyophores (15/50) Multinucleate marine amoebas.

true. The unicellular condition by necessity precedes the multicellular because the latter develops from the former by cell division.

Many people think big organisms such as us (especially ourselves) are the inevitable outcome of evolution, but if evolution always leads from small to large organisms, then over 90% of the known groups of eukaryotes failed to read the script. Unicellular organisms are the rule; large, multicellular organisms are the exceptions. And it now seems certain that the ecology of Earth operated on a unicellular basis for at least two-thirds of life's history on Earth. All of this suggests that if Earth is typical, then even if life on other planets is common, it will mostly be microscopic. Even multicellular organization is no guarantee of large size. While a multicellular organization is a necessary prerequisite to becoming big, multicellular organisms in most groups remain microscopic.

Why, you may ask? This question has a fairly simple evolutionary answer. Large size is an advantage only under certain circumstances or conditions. Natural selection will operate to increase the size of organisms only when larger size provides an increase in the organism's reproductive success, which places more genes for larger size into the next generation.[2] Microorganisms continue to exist because their small size is adaptive for their circumstances. Another reason unicellular organisms abound is that large organisms, and their activities and waste products, have produced lots of opportunities for tiny organisms to make a living. No inevitable trend toward larger organisms exists although clearly larger organisms became more common and more diverse during the past 500 million years, the result of something akin to a positive feedback loop. In the case of green organisms, larger size is associated with adaptations that allowed colonization of coastal and terrestrial environments.

Concepts like advanced and primitive have purposefully been avoided. Thinking of big organisms, such as humans, as advanced and of small unicellular organisms, such as an amoeba, as primitive is common enough. But such thinking leads to lots of wrong ideas. Big organisms display more structural and developmental complexity because they are composed of a huge number of cells organized into different types of specialized cells forming tissues and organs, thus forming a complex assemblage. But at the cellular level each type of specialized cell is relatively simple, usually having a simplified form adapted for a single function. On the other hand, a unicellular organism is a general purpose cell, a cell that must deal with all of the same demands of making a living that face a large organism (obtaining needed resources, reproduction, etc.). At the cellular level a general purpose cell is more complex than any individual cell in a large multicellular organism.

Large multicellular organisms have small unicellular organisms as distant ancestors, so large organisms are derived from them. But since contemporary unicellular organisms are also derived from the same unicellular ancestry, they cannot be called primitive or ancestral, although certainly they exhibit less structural and developmental complexity. Rather than thinking of descendants of common ancestry as advanced or primitive, it is more correct to think in terms of organisms that retain more of the ancestral characteristics and organisms with more derived or modified characteristics.

One of the reasons the multicellular condition does not inevitably lead to larger size is that multicellular organization itself exists along a continuum of structural complexity that involves both cellular specialization, ranging from organisms with no specialized cells to organisms composed of virtually all specialized cells, and developmental complexity, from relatively little to a great deal. Colonial organisms show a simple level of multicellularity, having a fixed or limited number of identical, or nearly so, cells, where each cell has an organization similar to its unicellular ancestor. For example, some species of the colonial green algae *Pediastrum* consist of 16-celled organisms forming a flat, snowflake-shaped colony (Fig. 4–1). Each cell in the colony is fully capable of all functions including reproduction. Each cell can produce a new colony by dividing exactly four times to produce 16 daughter cells ($1 \to 2 \to 4 \to 8 \to 16$). Another simple multicellular form is the filament, a linear chain of cells. This differs from colonial organisms in having an indeterminate number of cells. Filaments develop when a single cell divides and the two daughter cells stay attached, which, as will be explained, demonstrates that simple multicellularity arises from a unicellular condition by means of a shift in developmental timing. The developmental instructions for a filament do not require a great deal of information: establish a cellular polarity, divide perpendicular to this axis, repeat, and the filament grows. All the cells in such organisms are essentially identical, and each cell in such simple multicellular organisms displays a complexity similar to related unicellular organisms.

At the next level of complexity, multicellular organisms have at least one type of specialized cell for one particular function. This is the maximum complexity displayed by prokaryotic organisms: a filament with intermittent specialized cells—for example, heterocysts in filamentous cyanobacteria (Appendix: Green Bacteria). Slightly more complex multicellular organisms, but still quite small, may have two or three types of specialized cells (Appendix: Green Algae, *Volvox*). Organisms with more complex organizations and more types of specialized cells can be found among larger eukaryotic organisms, and this increase in structural complexity continues until finally seed plants display three tissue systems and about thirty types of specialized cells, and complex animals have many tissues and about 50–60 types of cells.

The emergence of a eukaryotic cell organization is a prelude to more complex multicellular organizations and ultimately large organisms. However, eukaryotic cell organization alone is not sufficient to explain the emergence of large organisms because most groups of eukaryotic organisms remain unicellular. Before multicellularity and the emergence of large organisms can be explored further (chapter 5), some background knowledge about eukaryote diversity and the organization and origin of the basic unit, the eukaryotic cell, is needed.

Slowly the evolutionary relationships among eukaryotic organisms are becoming clearer. With a few notable exceptions, cellular organizations, which are what Patterson identified, have been of limited use in determining common ancestry. The differences were observed, but that only allowed biologists to group those organisms that shared a unique cellular organization. The fossil record has been of very limited usefulness because most of the features that distinguish eukaryotic cells from each other and from prokaryotic cells do not fossilize. However, the search for derived characters within the genetic material has been most informative. Although many questions remain, a broad pattern is emerging.

HOW DIFFERENT ARE PROKARYOTES AND EUKARYOTES?

Consider the major differences between eukaryotic and prokaryotic (Table 3–2). Eukaryotic cells possess considerably more internal structures and organization than prokaryotic cells. Eukaryotic cells have a nucleus containing molecules of DNA associated with histone proteins. The DNA condenses into chromosomes for nuclear division (mitosis) by winding around the histone proteins, which also have a regulatory function. The nucleus is defined by membranes, sort of double-membrane sacs that are continuous with the endoplasmic reticulum, which is an interconnected network of membranes throughout the cytoplasm that serves to package and deliver materials around the cell. The membranes surrounding the nucleus have many large pores and they do not compose a compartment wall, which is to say, the interior of the nucleus is not isolated from the cytoplasm by a semipermeable membrane of the type that separates all cells from the external environment. Eukaryotes have double-membrane-bound organelles called mitochondria, the site of aerobic respiration, and

Table 3–2. Comparison of bacterial prokaryotes and eukaryotes

Bacterial Prokaryotes	Eukaryotes
1. Unicellular*; cells small (1–10 microns).	1. Unicellular and multicellular; cells 10–100 microns.
2. No nucleus.	2. Membrane-bound nucleus.
3. DNA circular; no histone proteins.	3. DNA in chromosomes with histones.
4. DNA attached to cell membrane.	4. Chromosomes free.*
5. Cell division by binary fission.	5. Cell division by mitosis.*
6. No sexual reproduction.*	6. Sexual reproduction in most groups.
7. Multicellularity limited to a few filamentous forms; few specialized cells.	7. Many multicellular forms; many specialized cells.
8. Some motile cells (gliding or simple flagella).	8. Motile cells common; flagella or cilia, amoeboid movement, gliding.
9. 70s ribosomes.	9. 80s ribosomes in cytoplasm.
10. No microtubules.	10. Microtubules present (tubulin proteins) in flagella, cytoskeleton, mitotic spindles.
11. No ER, Golgi, or other membrane-bound organelles; cell membrane can be infolded for PS & RS.	11. ER, Golgi, other membranes present.
12. No mitochondria or plastids.	12. Mitochondria and plastids present.*
13. Diverse chemo- and photoautotrophy.	13. Green, oxygenic photoautotrophs only.
14. Very diverse metabolisms, many kinds of autotrophy; many strict anaerobes.	14. Aerobic with glycolysis, Krebs cycle & cytochrome system; some facultative anaerobes.
Abbreviations: ER, endoplasmic reticulum; PS, photosynthesis; RS, respiration.	*Exceptions exist.

some have chloroplasts, the site of photosynthesis. And these organelles are separate compartments. The eukaryotic flagellum has a "9 + 2" skeleton of microtubules, a ring of nine plus two in the middle, and a basal body that acts like a rotary motor. Prokaryote flagella lack microtubules and are made by filament-forming proteins. All of these eukaryote features must be derived from a simple prokaryote ancestry if our eukaryotes-have-a-common-ancestry-with-prokaryote hypothesis is correct.

Of particular interest to this narrative is the origin of chloroplasts, the eukaryotic cellular organelle that contains the green pigment chlorophyll. Chloroplasts in land plants have diversified for some new functions in addition to photosynthesis. As fruits mature, chloroplasts become chromoplasts containing bright red-orange-yellow pigments—for example, chili peppers. Indeed, chloroplasts always contain yellow xanthophylls and gold-to-orange carotene pigments in addition to chlorophyll(s), but they are usually masked by the chlorophyll.[3] Some flowers are similarly pigmented. Plants store their food, starch, in still another kind of modified chloroplast, an amyloplast, which generally has no pigments at all. Chlorophyll is found in only three other places: inside the cells of green sulfur bacteria, green nonsulfur bacteria, and cyanobacteria (Appendix: Green Bacteria). After the appearance of chlorophyll-mediated photosynthesis, the origin of eukaryotic cells is the second big event in the history of green organisms. The origin of chloroplasts is part of that event.

WHAT DO PHYLOGENETIC RELATIONSHIPS AMONG EUKARYOTES TELL US?

All eukaryotic organisms share a number of novel characters—for example, a nucleus and DNA in chromosomes and 80s ribosomes, which argues in favor of treating eukaryotes as a single lineage with one common ancestry. Many recent studies have generated phylogenetic hypotheses that have begun to organize what was previously just a grocery list of eukaryotic organisms like that presented in Table 3–1. Presently eukaryotic organisms appear to form two big clades or lineages, the unikonts and the bikonts (a phylogenetic hypothesis) (Fig. 3–1), characterized by motile cells having either one or two flagella (-kont) (Embley and Martin 2006; Cedergren et al. 1989; Cavalier-Smith 1993, 1997; Kumar and Rzhetsky 1996; Van der Peer and De Wachter 1997; Baldauf et al. 2000; Van der Peer et al. 2000; Zhang et al. 2000; Stechmann and Cavalier-Smith 2002). While this sounds so nice and neat, at least one biflagellated organism has been found that according to molecular data belongs to the unikont clade (Kim et al. 2006). This is not totally a surprise because no one is certain whether unikont or bikont is the ancestral condition, and in this case if the common ancestor had two flagella, a member of the unikont clade might still retain the ancestral condition, especially if it occupied a basal position in the unikont clade indicating an ancient common ancestry with the rest of unikonts. However, if too many problems arise, then this bikont-unikont-in-separate-clades hypothesis may find itself in trouble, but do not worry because another hypothesis will have been proposed to explain the new findings.

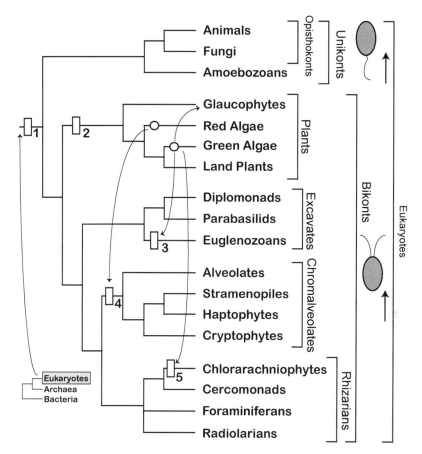

Fig. 3–1 Broad phylogenetic relationships among major lineages of eukaryotes showing two major lineages, unikonts and bikonts, that differ in the type of motile cell they possess (cell diagrams at right; arrows show direction of movement). Mitochondria (1) were acquired by the common ancestor of all eukaryotes. The bikonts consist of two big lineages: the plants and a clade containing the excavates, rhizarians, and chromalveolates. Chloroplasts (2) were first acquired by the common ancestor of the plant lineage. Chloroplasts from the green algae were acquired secondarily by the common ancestor of the euglenozoans (3) and chloroplasts from the red algae were acquired secondarily by the chromalveolates (4) and chlorarachniophytes (5). Nongreen members of these clades would have lost their chloroplasts. See text for further explanations. Cladogram of life (Fig. 2–7, lower left) shows phylogenetic relationship of eukaryote lineage to archaea and bacteria. (Image source: After Stechmann and Cavalier-Smith 2002; Hampl et al. 2009)

Members of the bikont clade sampled so far share a gene fusion, a shared derived character; the two genes remain separate in members of the unikonts and in all prokaryotes, thus unikonts retain the ancestral condition. Members of the unikont clade sampled so far all share another novel genetic rearrangement, a big inverted repeat, which means a big chunk of the genome is duplicated and appears elsewhere in reverse order. Phylogenetic groupings based on such data use the principle of parsimony. It is far more likely that novel shared characters have been inherited from a common ancestor than for such unique and specific changes to have arisen two or more times independently.

The unikont clade is composed of two lineages, the opisthokonts (oh-PEAS-though-konts), motile cells with one posterior—that is, pusher, flagellum; and amoebozoans (ah-me-bow-ZOH-anz), which can switch between a uniflagellated form and a nonflagellated amoeboid form. The opisthokonts include two lineages, the fungi and animals, which share a number of other characteristics: both are heterotrophs, both use glycogen as a storage carbohydrate, and both can make chitin (KEYE-tin), an impermeable material composing the cell walls of fungi and the exoskeletons of insects and other arthropods. Higher fungi possess no flagellated cells, relying instead on wind-dispersed spores, but basal lineages are unicellular flagellated organisms called chitrids (KIT-rids).

No members of the unikont clade possess chlorophyll pigmentation, so no animals are truly photosynthetic either. Green frogs, green katydids, and the like have other green pigments for camouflage among foliage. A few marine invertebrates—for example, corals and mollusks, appear green because they harbor symbiotic algae. Some marine slugs graze on algae and retain the chloroplasts in body cavities, so they can look green for a time. Although no fungus is autotrophic, some fungi form symbiotic (sim-bye-OH-tick, literally "living together") relationships with algae; these composite organisms are called lichens and they function as photosynthetic autotrophs. Lichens are generally considered fungal because the organism has a fungal structure and reproduction.

For nearly one hundred years, fungi and animals have been treated as separate kingdoms, and such a classification hid or overlooked their similarities. The narrowly defined animal and fungi kingdoms of my school days also excluded flagellated unicellular organisms that are clearly related to both and help make their common ancestry more sensible. Like the three domains, this is an example of how recent phylogenetic research has produced results at odds with traditional taxonomic groupings.

The bikont clade is composed of four major lineages, chromalveolates (chrome-al-VEE-oh-latts), plants, rhizarians (rye-ZAIR-ee-ans), and excavates (EX-cah-vates), of which perhaps only the euglenozoans (you-glean-oh-ZOH-anz) may be familiar. The plant lineage includes the red algae (rhodophytes—ROW-doh-fights), the green algae (chlorophytes), and the land plants (embryophytes). The chromalveolate clade is composed of the alveolates (al-VEE-oh-latts) and stramenopiles (strah-MEN-oh-piles).

Alveolates are mostly motile, unicellular organisms and include many core members of the former Kingdom Protista like the ciliates (e.g., *Paramecium*); dinoflagellates, many of which are photosynthetic; and apicomplexans (ah-pea-com-PLEX-ans). The last group are parasitic flagellates that can form a nonmotile resting stage, a group formerly called sporozoans (spore-oh-ZOH-ans), which includes *Plasmodium* (plaz-MOH-dee-uhm), the organism which causes malaria. The name alveolate refers to alveoli (al-VEE-oh-lye), membranous sacs located around the periphery of the cells, a novel shared feature, although no one as yet knows their function.

Stramenopiles include the brown algae and a bunch of other algal- and fungal-like organisms: diatoms, golden algae, yellow-green algae, water molds, slime net amoebas (labyrinthulids [lah-BRIN-thoo-lids]), and a number of other unicellular photosynthetic organisms. The biflagellate motile cells in this clade are asymmetrical, having two flagella inserted in a sub-apical to lateral position (see Appendix: Phytoplankton, stramenopile). The two flagella differ in form with one flagellum long and hairy, and the other shorter and smooth (Fig. A50). The flagellar hairs are tubular and three-parted, and the name stramenopile is derived from this shared, novel feature. Before the fungal-like water molds and amoeboid labyrinthulids were included, the group was basically algal and called Kingdom Chromista or the heterokont ("different flagella") algae. Two groups of unique unicellular "algae," for lack of a better term, the cryptophytes and haptophytes, are sister groups to the stramenopiles. All of the photosynthetic organisms in the chromalveolates clade have the same basic chloroplasts possessing chlorophylls *a* and *c*.

Stramenopile and alveolate are a new type of name for designating lineages based on a novel shared character that defines the clade. This way even if characteristic organisms get swapped to another group (on the basis of new data), the reference name does not have to change. These names are also nonhierarchical; they do not imply a taxonomic level like class, phylum, or kingdom. Both of these are pragmatic actions at a time of so much change. Although still unfamiliar, these new names are being used because they are better labels for these lineages than any of the alternatives.

Rhizarians are mostly free-floating, unicellular marine organisms, zooplankton (zoh-oh-PLANK-tun) like the foraminifera (for-am-in-IF-er-ah) and radiolarians (ray-dee-oh-LAIR-ee-ans). This grouping also include the cercozoans (sir-coh-ZOH-ans) and chloroarachniophytes (KLOR-oh-are-ach-KNEE-oh-fights—meaning green spider plants), both of which are basically amoeboid cells. The remaining bikont lineage is referred to as the excavates, which only recently has been shown to be a single clade (Hampl et al. 2009) although this had been suspected for several years. They include a number of parasitic unicellular organisms and the somewhat more familiar euglenozoans. Most euglenozoans and the chloroarachniophytes are the remaining green photosynthetic organisms. Chloroarachniophytes are chloroplast-possessing marine amoebae with long, thin pseudopodia that make their cells look like microscopic spiders (arachnids), sort of miniature spiderworts.[4] Both euglenozoans and chloroarachniophytes have similar chloroplasts, like those found in the green algae with chlorophylls a and b, but these three groups are not even closely related. How is that possible? One explanation (hypothesis) is that these two groups acquired chloroplasts from green algae.

No surprise here, virtually all of the organisms in the plant lineage are green, although among land plants, more specifically seed plants, a few have become secondarily nongreen by evolving into parasites. Green algae range from unicellular to modestly large seaweeds in both freshwater and marine environments. The red algae are largely marine organisms that range from unicellular to modestly large seaweeds. Although it is convenient to include the red algae in a plant lineage, the phylogenetic position of red algae is by no means certain. Red algae lack motile cells completely and have a fungal-like filamentous organization, so they could belong in the unikont clade. Some earlier studies placed red algae as a sister group to a plant clade and to a chromalveolate clade (Van der Peer and De Wachter 1997), or as a sister group to a plant-opisthokont clade (Kumar and Rzhetsky 1996). If red algae, green algae, and land plants form a single lineage, it means chloroplasts require only one origin (Delwiche 1999), so until this phylogenetic grouping is convincingly falsified and the placement of red algae determined with some confidence, they will remain basal to the plant lineage because it makes for the simplest explanation.

Eukaryote phylogeny is far from resolved. A recent study (Hampl et al. 2009) suggested a relationship between the haptophyte-cryptophyte lineage and either red or green algae. Such a shift in relatedness would greatly complicate the history of chloroplasts and alters the definition of both chromalveolate and plant clades. To keep the chloroplast hypothesis uncomplicated, the haptophyte-cryptophyte clade is left within the chromalveolates for now. But now that this hypothesis has been presented, other researchers will turn their attention to this possibility and seek data that could confirm or refute those results.

A "plant kingdom" that includes the red algae, green algae, and land plants is similar to the more expansive but polyphyletic plant kingdom[5] of my youth that included all photosynthetic organisms. One small lineage, the glaucophytes, a group of motile unicellular organisms, is also treated as part of the plant clade, but its ancestry is uncertain as well. Summing up, at a minimum chloroplast-possessing organisms are found in four lineages of eukaryotic organisms: the plants, the chromalveolates, the euglenozoans, and the chloroarachniophytes, so although several distinct groups of green organisms occur in each lineage, we have to account for chloroplasts appearing in only four places, for now. Over the past 50 years, a robust hypothesis has come together piece by piece from diverse evidences to explain how all these organisms obtained chloroplasts.

Before departing this overview of eukaryote diversity and phylogeny, consider this. How many kingdoms of eukaryotic organisms are there? Two, bikonts and unikonts, or five, or are these superkingdom groupings? If the unikonts are not a kingdom, then do animals, fungi, and amoebozoa all get kingdom status? No attempt will even be made to suggest how many lineages should be recognized among the bikonts. However, some hope of larger, more inclusive groups remains because as fundamental genetic similarities are found, some unique organisms have been placed within larger groups. Now the question arises, does a kingdom-level classification remain useful? Certainly a classification problem exists, and it will not be solved here. Until phylogenetic studies get reconciled with taxonomy, and some will argue this is an impossible task, the only practical approach is to use reasonable labels for lineages and not worry about a taxonomic hierarchy. All available data agree that eukaryotes are a well-defined lineage sharing well-documented cellular features. Several major lineages of eukaryotes have been established as well as some of the relationships among them, but lots of smaller lineages exist too. Nonetheless, compared with the grocery list they replaced, these hypotheses represent significant progress.

Recent progress in our understandings of patterns of common ancestry notwithstanding, some relationships among even major lineages remain uncertain. As this was being written I made many trips to the library seeking the latest studies on the relationships of major lineages of eukaryotic organisms, but you have to stop sometime, so here's my guarantee. This narrative is neither going to present to you with the latest results nor going to be wholly correct. The accumulation of data and new studies even between the time this is sent to press and published will require that we abandon some hypotheses and generate new ones. In an active field of scientific study, you must realize that books are never, ever going to be able to provide you with the latest information. Not only can you never catch up, but different studies based on different data sets have produced different results so several possible hypotheses must be evaluated. Such uncertainty is not a weakness of science; rather it indicates the activity of research presently ongoing and the necessary revisions of thinking and hypotheses that accompany new information and new understandings. Even if all the studies do not completely agree, some earlier hypotheses have been falsified by nearly all recent studies, so we have gained knowledge even while remaining less than certain.[6]

Clearly the discovery phase of biology is not over, and the Linnaean quest to find, name, and classify all organisms continues, but it primarily continues using a microscope and a DNA sequencer. Twenty-two of the groups listed in Table 3–1 consist of a single genus (names in italics) and their relationship to other organisms is so obscure no one can decide if they belong to a larger group. Another 220 genera of eukaryotes have been described, but either none of the seventy-one listed groups is appropriate, or there is insufficient information to place them into any of the seventy-one well-defined groups. These unknown or misplaced genera are grouped broadly into five categories: free-living heterotrophic flagellates, parasites, free-floating green organisms, amoeboid organisms, or, my personal favorite, "organisms of uncertain nature" (Patterson 1999). Sometimes science begins to sound like science fiction.

HIGHLIGHTING THE DIFFERENCES BETWEEN PROKARYOTES AND EUKARYOTES

Phylogenetic studies have confirmed what biologists have thought for well over a century: eukaryotes have a common ancestry with prokaryotes. Somehow all the features we associate with eukaryotic cells arose from a prokaryotic ancestry. How 80s eukaryote ribosomes arose from 70s prokaryote ribosomes in two stages by two big insertions of genetic material has already been discussed (in the "Prokaryote Phylogeny" subsection). Several other features are important for purposes of this history. Prokaryotic cells have a single, circular molecule of DNA attached to the cell membrane. The DNA is "naked," meaning there are no associated histone proteins. When prokaryotic cells divide, first the DNA duplicates along with the point of attachment to the cell membrane. The daughter DNA molecules are physically separated by a shift in their points of attachment, and then the cytoplasm divides by a furrowing of the cell membrane across the middle of the cell. This type of cell division is called binary fission.

Eukaryotic cells have a nucleus containing two or more linear molecules of DNA, chromosomes. The nuclear envelope is composed of porous, flattened membranous vesicles that are continuous with the cell's endoplasmic reticulum. Although the nucleus is described as "membrane bound," the nuclear envelope is very different from the discrete membrane envelopes surrounding mitochondria and chloroplasts. At the time of nuclear division (mitosis), the DNA molecules condense into visible chromosomes[7] by winding around histone proteins. In most eukaryotes the chromosomes are free within the nucleus, but in some groups the chromosomes are attached to the nuclear envelope. There are two different types of nuclear division, open and closed, referring to whether the nuclear envelope disperses (open) or persists (closed) during mitosis. Open mitosis is the type typically depicted in textbooks where a spindle forms; the nuclear envelope disperses (refer to Fig. 3–6). Having been previously duplicated, daughter chromosomes remain attached to each other by their centromeres, an obvious constriction on the chromosome. Centromeres also function as points of attachment for spindle fibers. The mitotic spindle is composed of microtubules, which also function as part of the cell's cytoskeleton. Spatial separation of daughter chromosomes by a mitotic spindle is accomplished by assembly and disassembly of microtubules, which are composed of tubulin proteins. After daughter chromosomes are spatially separated, a new nuclear envelope is produced around each daughter nucleus. Mitosis and cytoplasmic division, called cytokinesis, usually are linked so that mitosis phases directly into cytokinesis, but the two can be separate functions and cytokinesis may not immediately follow mitosis. Prokaryotes have no nuclear membranes, no chromosomes (although microbiologists sometimes call the circular DNA a chromosome), no microtubules, and no cytoskeleton. These features point out the considerable differences between prokaryotes and eukaryotes in just this one aspect of their cells.

Mitochondria and chloroplasts are distinctive features of eukaryotic cells (Fig. 3–2), and these prominent organelles are double-membrane bound. Unlike the nuclear envelope, no connections ever form between these two discrete membranes. Inner membrane is folded into invaginations that provide surface area for pigments (cytochromes and chlorophylls, respectively). Mitochondria are essentially identical in all organisms and a universal feature. At one time biologists thought some unicellular and largely parasitic groups of eukaryotes lacked mitochondria, which could be a primitive condition, but those thought to lack mitochondria all have either hydrogenosomes or mitosomes, which are

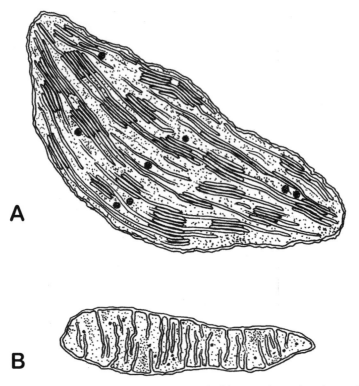

A

B

Fig. 3–2 Chloroplast (upper) and mitochondrion (lower). Each of these prominent eukaryotic organelles is double membrane bound. In both organelles a proliferation of membranes (invaginations) is continuous with the inner membrane. In the chloroplasts of green algae and land plants the double-layered membranes (thylakoids) occur in stacks called grana. Chlorophyll is embedded in these membranes and the light reaction of photosynthesis takes place on their surface. In mitochondria the membrane invaginations are called cristae and electron transport takes place on cytochrome pigments embedded on their surface. Magnification approx. 100,000×. (Image source: Illustration by D. DeWitt, used with permission, Armstrong and Collier 1990)

interpreted as either highly modified or much reduced mitochondria (Embly and Martin 2006) and therefore a derived condition. Their basic sameness throughout eukaryotes suggests mitochondria had one origin, but a single origin is not true for chloroplasts. Chloroplasts differ among groups in their chlorophyll complements: chlorophyll *a* alone or with one or two more molecular forms (*b, c, d,* or *e*) and two, three, or four membrane envelopes. Although components of eukaryotic cells, both mitochondria and chloroplasts possess their own naked, circular DNA attached to the inner membrane and these organelles have 70s ribosomes in their cytoplasm. Both organelles divide by binary fission, and both are similar in size to prokaryotic cells. Somehow this mismatch between these "prokaryotic" organelles and the surrounding eukaryotic cell must be resolved.

Generally eukaryotic cells are bigger than prokaryotic cells, but actually a continuum in size exists such that the largest prokaryotic cells and the smallest eukaryotic cells are about the same size (Table 3–1). This makes judging the age of eukaryotes in the fossil record very problematic because cell size is the primary criterion for judging if a cellular microfossil is prokaryotic or eukaryotic. Cells large enough to be eukaryotic are dated to a bit more than 2 bya just around the beginning of the Proterozoic (Fig. 1–5; Fig. 5–9). Some fossil cells appear to have nuclei, but this is questionable because there is no way to distinguish between a little round artifact and real fossil organelle.

Another approach to determining when a group of organisms appears is to generate estimates of how much time has passed since their last common ancestry with another group of organisms, in this case, the common ancestry between prokaryotes and eukaryotes. This requires constructing a molecular clock, which is done by calculating the rate at which sequence changes in DNA occur, and then dividing that rate into the total number of changes actually observed. The answer is the total time it took to accumulate the total number of changes. This assumes rates of change are more or less constant, a controversial topic, but some confidence comes when different molecular clocks yield similar results. One recent study places the divergence time for eukaryotes and prokaryotes at 3 bya (Nei et al. 2001), which on the basis of fossil cell size is earlier than the age of eukaryotes. Since larger cell size need not have been a prerequisite or an early event in the evolution of eukaryotes, the fossil record produces a very conservative date for the origin of eukaryotes. Together these figures produce a hypothetical age range for the appearance of eukaryotes between 2.3 and 3 bya.

Large multicellular organisms appear about 2 billion years later, so a considerable period of time elapsed before the eukaryotic condition resulted in larger organisms—that is, big enough to be seen with the naked eye. So for at least 2.5 to 3 billion years unicellular and small multicellular organisms were the dominant life forms on Earth. And virtually all inhabited aquatic environments. The oceans still teem with a diversity of free-floating unicellular organisms; in fact, that is where most of the unfamiliar organisms listed in Table 3–1 are found.

No one suggests the eukaryotic cell arose all at once, complete and fully constructed, because that is pretty much impossible. Different features of the eukaryotic cell almost certainly had separate origins, and the sequence or order in which these features appears is rather much in question. So instead of taking a linear approach, we'll start with the most certain and well-documented events and proceed from there to the less certain.

THE ENDOSYMBIOTIC ORIGIN OF CHLOROPLASTS AND MITOCHONDRIA

The endosymbiosis ("en-doh-sim-bee-OH-sis") hypothesis explains the origin of mitochondria and chloroplasts, and accounts for their similarities to prokaryotic organisms (Margulis 1967, 1981, 1996). Mitochondria and chloroplasts are hypothesized to have had their origins as free-living bacteria that became intracellular (endo-) symbionts (sym-, together; -biosis, living) of a host cell, an organism most likely to have had a common ancestry with archaeans. Thus the eukaryotic cell is an organismal composite, a chimera, a symbiotic combination of two, or three, or possibly more prokaryotic organisms. Symbioses are quite common in nature and many prove mutually beneficial to the participating organisms and are referred to as mutualisms.

For example, lichens are a symbiotic "organism" formed by an algae interacting with a fungus. Together they form a tough, successful organism, one capable of living in some difficult and severe habitats, one capable of living on bare rocks, bark, or exposed ground. On the basis of their biology lichens are organisms, yet the fungus and algae can be separated and still can function as independent free-living organisms. However, when living separately neither the algae nor the fungus displays any aspect of the lichen or its biology, and neither organism alone can live where or how the lichen lives. When living together, the resulting lichen is a uniquely different organism. Algal cells do not live within the fungal cells; instead they reside within a densely interwoven filamentous body of the fungus. So while many lichens look like they have a tissue organization similar to land plants, they do not.

If the endosymbiosis hypothesis is true, then mitochondria and chloroplasts should share a common ancestry with their bacterial counterparts, and molecular studies indicate those would be purple bacteria and cyanobacteria, respectively. Further, the endosymbiosis hypothesis means a nongreen, heterotrophic organism would become a green autotrophic organism once the symbiosis was complete. Photosynthesis among eukaryotes is a package deal; the entire metabolism was obtained intact by forming an association with a photosynthetic prokaryote that became a chloroplast. Which other eukaryotic features were acquired by the host cell prior to mitochondria and chloroplasts remains more uncertain, as does the ancestor of the host cell.

Since the mitochondria of all eukaryotes are virtually identical and mitochondria are a universal feature of eukaryotic organisms, although reduced and modified in a few parasitic organisms, the endosymbiotic origin of mitochondria is hypothesized to have taken place once, early in the history of eukaryotes and prior to the acquisition of chloroplasts (Sogin 1997; Embly and Martin 2006). Common ancestors of the plant lineage acquired the original chloroplast. Other photosynthetic eukaryotes acquired their chloroplasts later, secondarily, from either a red algae in the case of chromalveolates or green algae in the case of euglenozoans, chloroarachniophytes, and glaucophytes (Fig. 3–1) (Delwiche 1999; Palmer 2000; Embly and Martin 2006).

The critical initial step in acquiring an endosymbiont, either a mitochondrium or a chloroplast, is to place an appropriate organism, an aerobic purple bacterium or a cyanobacterium, inside the host cell (Carsaro et al. 1999). Unicellular heterotrophs feed either by absorption or by consumption. Most heterotrophic microorganisms (bacteria and unicellular eukaryotes) literally live in or on their food, secreting enzymes outside their cells to digest their food and absorbing the resulting soluble food molecules. Absorptive feeders also scavenge any soluble molecules they encounter (those secreted into the environment by other living organisms as metabolic waste or molecules released after death as decomposition proceeds). Fungi live in or on their food and are absorptive feeders too, having retained this life style from unicellular ancestors.

Heterotrophic consumers obtain food by either partly or wholly ingesting another organism. Multicellular consumers like animals ingest and digest food via an internal digestive tract, which is technically still "outside" the organism because they are organized like a hollow tube, a rather unflattering but accurate depiction. Ingestion also can take place at a cellular level. Unicellular organisms ingest food particles and whole living prey by phagocytosis (faye-goh-seye-TOE-sis), which uses an active invagination of the cell membrane to form a membrane-bound vacuole around the prey cell or food particle (Fig. 3–3). Enzyme-containing vesicles then fuse with the food vacuole and digestion begins, but now completely within the predatory cell's cytoplasm. Ingestion, in comparison with external digestion, reduces the diffusion and loss of enzymes and soluble food molecules.

Phagocytosis not only puts the "prey" cell inside the host cell, it accounts for the double membrane organization of mitochondria and chloroplasts. The inner membrane of the organelle would correspond to the

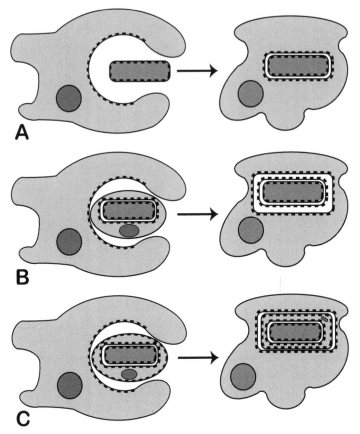

Fig. 3–3 Phagocytosis and capture of a prokaryotic cell (A) or a eukaryotic cell (B-C) (left) and the resulting membrane envelops and their orientation (right). A. Primary endosymbiosis of prokaryotic cell within a food vacuole to become a mitochondrion or chloroplast resulting in a double membrane. Note that the vacuolar membrane is inverted with its inner surface facing out. B-C. Secondary endosymbioses of a photosynthetic eukaryote within a food vacuole to produce either a three-membraned (B) or four-membraned chloroplast (C). In B the prey's cell membrane, nucleus, and cytoplasm are consumed, leaving only the double-membraned organelle with the vacuolar membrane. In C the prey's cell membrane and vestiges of its cytoplasmic contents remain, including a vestigial nucleus (upper right). Dashes on membranes show inner facing surface. See text for explanation of membrane number and orientations. These diagrams show a nucleated amoeboid cell, but the endosymbiotic origin of the mitochondrion may have preceded nuclei. Cytoplasm is shaded light gray; nuclei are shaded a darker gray.

cell membrane of the ancestral prey organism. The outer membrane represents the food vacuolar membrane of the host cell. This makes sense of the observation that the outer membrane of mitochondria and chloroplasts can form connections with the host cell's endoplasmic reticulum, its membrane network, but the inner membrane always remains discrete. The two membrane envelopes never interconnect because, according to this hypothesis, they are of different origins.

The most familiar form of phagocytosis involves cells that possess an amoeboid form, a lobed, irregular cell shape, or a phagocytotic zone on the cell surface—for example a gullet. The diversity of amoeboid organisms among eukaryotes (Table 3–1) suggests such a cell form is very successful and played a significant role in generating the diversity of cellular organizations observed. The advantage conferred upon any organism that developed the ability to wholly consume another cell or a particle of food is obvious. Food would be obtained in a concentrated, energy-rich packet without the waste of absorptive feeding, which wastes some of your enzymes and soluble food molecules. Phagocytotic feeding results in predatory organisms capable of seeking and engulfing prey organisms.

Although some amoeboid cells are quite large, phagocytotic consumption does not require the predatory cell to be much bigger than the prey cell. Predator and prey cells can be relatively similar in volume as long as the predatory cell lacks a cell wall and has a relatively large cell membrane surface area. Expanding cell membrane surface area without increasing the volume of cytoplasm requires a flexible and irregular cell shape, an amoeboid form. If the predatory cell has a volume equal to that of its prey cell, then the predatory cell would require a cell membrane area about 2.5 times larger than that of the prey cell to construct a food vacuole around the prey cell, and still have enough cell membrane left to enclose both its cytoplasm and a food vacuole of similar volume. Smaller prey cells or food particles pose no problem. This means a larger cell size was not needed initially for phagocytotic feeding, although the efficiency of consumptive feeding would have paved the way for larger cells if size was limited by the amount of food acquired.

The process of phagocytosis requires an active shape-changing mechanism, which may not have evolved for the purpose of feeding since the same mechanism also functions in locomotion. Amoeboid movement involves a directional movement of cytoplasm within a flexible cell membrane. This is accomplished by actin, which you may recognize as a muscle motor protein. Actin is a universal component in eukaryotic cells, and it has many functions in addition to moving muscles. Long polymers of actin molecules cross-linked by actin-binding proteins (ABP) produce a flexible network that stabilizes the cytoplasm into a gel-like state (Stossel 1990). Upon receiving a stimulus, the ABP releases the actin network beneath a portion of the cell membrane changing the cytoplasm to a fluid state. Unlinking the actin protein network changes the osmotic equilibrium, and water diffuses into the area to dilute the locally increased cytoplasmic solutes. This increases the cell volume locally and creates a pressure between the stiffer, still-gelled cytoplasm deeper inside the cell and the cell membrane. This increased volume pushes against the cell membrane causing it to protrude outward, producing a lobe-like pseudopodium (false foot). The newly extended pseudopod is then stabilized by ABP-actin linkages, and the process can begin anew. Phagocytosis begins with the amoeboid cell producing pseudopodia that engulf another cell or food particle. When the pseudopodia meet, the membranes fuse with the invaginated portion forming a vacuole. The cilia of non-amoeboid unicellular organisms may direct food particles into a gullet, a sac-like invagination where a vacuole can form. This process can be observed under a microscope by feeding small unicellular algae to amoebae or paramecia.

If this hypothesis is true, then a cytoskeleton composed of actin proteins and the resulting amoeboid cell may well have been the first steps in a series of events that led to eukaryotes (Roger 1999). The endosymbiosis hypothesis suggests that a host cell lineage possessing 80s ribosomes acquired an actin cytoskeleton and diverged from prokaryote ancestors. No prokaryote possesses actin, so no bacterium or archaean is capable of amoeboid movement or phagocytosis. Bacteria with proteins very similar to actins have been identified, and these proteins have a cytoskeletal (structural) function (Egelman 2001, 2003; Van den Ent et al. 2001; Mayer 2003). Similar proteins have not yet been identified in archaeans, but they remain much less studied.

Discovery of a symbiotic prokaryote living within the cell of a non-phagocytotic prokaryote suggests that other alternatives are possible (Embley and Martin 2006). Invasive endoparasitism, which is found among prokaryotes, is an alternative to consumption for getting one cell inside another cell (Guerrero et al. 1986). In this scenario, a smaller endoparasite has the ability to breach the host cell defenses (its cell wall and cell membrane). Several mechanisms of penetration are known, and at least one means of penetration does not disrupt the host cell membrane and generates a parasite-containing vacuole in a nonphagocytotic cell (Corsaro et al. 1999). This scenario might work for mitochondria but not for chloroplasts, because why would a photoautotroph also be an endoparasite?

Evolutionary theory predicts that parasites and disease organisms should become less detrimental to the host with time, assuming the host survives the initial infection (Ewald 1997). Immediately killing your host cell is not in the best interest of an endoparasite because it seeks to grow and reproduce at the expense of the surrounding host cell. An overly harmful endoparasite dies with its host, so natural selection will generate an accord between parasite and host where the former is not overly harmful and the latter not overly harmed, and as a result, both, and that is the key word, host and parasite increase their longevity and reproductive success through a symbiosis of some type. The same is true for disease organisms.

After phagocytosis, digestion of a prey cell within a food vacuole is not an immediate process, so the prey cell remains alive inside the host cell's vacuole for some period following consumption. Rather than dwelling upon the rather gruesome image of being alive within a food vacuole,[8] consider this situation. If a predatory cell obtains some extra benefit as a result of having a live, functional prey cell within a vacuolar membrane, then those consumers who digest their prey most slowly will obtain the greatest amount of extra benefit, so those intracellular associations that last the longest produce the greatest benefit to both parties. Consumer cells whose genes allow a prolonged association will produce the more offspring, and the slow-to-digest-get-the-biggest-benefit type of consumer will be more common among members of the next generation. Whenever the consuming cell obtains some slight benefit over and above nutrition, natural selection favors slower and then indefinitely delayed digestion whereby the two organisms can establish a symbiotic interaction and interdependency. This scenario will operate for either mitochondria or chloroplasts, although they offer the host cell different benefits.

A consumed photosynthetic autotroph could provide a continuing source of food molecules, and the engulfed autotroph would obtain a ready supply of carbon dioxide provided by the respiration of the "host" cell. If the engulfed cell were an aerobic bacterium, it might allow an anaerobic host organism to exist in a broader range of habitats, moving out of strictly anaerobic environments and into those with higher concentrations of oxygen. Such an endosymbiont organism might be similar to *Paracoccus denitrificans*, an aerobic purple bacterium that presently is identified by ribosomal RNA sequences as having the most recent common ancestry with mitochondria. An engulfed bacterium might use a metabolic waste product produced by its "host" as a substrate, and produce a harmless or even useful molecule in return, sort of a molecular swap meet where the exchange leaves both parties benefiting from the result (Morowitz 1992).

Data based on ribosomal RNA sequences place eukaryote host cell ancestry among archaea, and eukaryotic genes for replication, transcription, and translation of DNA (informational genes) have counterparts or homologues among archaea (Ribeiro and Golding 1998), but no prokaryotic organism has yet been identified as most similar to the eukaryotic host cell (Embley and Martin 2006). Within archaea are organisms that possess at least one of the characteristics required of the host cell lineage. Thermophilic archaeans lack cell wall envelopes and have relatively large, flexible cells, a prerequisite of amoeboid cells, but no prokaryotic cells can actively invaginate their cell membrane to form vacuoles. To further complicate things, operational genes of eukaryotes, those that control metabolic and biosynthetic functions, have a common ancestry with bacterial genes. Not only that but these bacterial genes were not acquired from the mitochondrial endosymbiont. No one understands how this has happened, but this clearly demonstrates that eukaryotes are cellular and genetic chimeras of the two prokaryote lineages, archaea and bacteria.

The endosymbiont hypothesis can be tested by predictions logically derived from the explanation. If the outer membrane of mitochondria and chloroplasts originated as a vacuolar membrane, by either phagocytosis or endoparasitism, then the normal inside-outside orientation of the cell membrane will be inverted when invaginated to produce a vacuole (Fig. 3–3 A). An asymmetry in the number and distribution of proteins embedded in inner and outer surfaces of cell membranes allows their topography to be observed using an electron microscope. The predicted inversion is found; the outer membrane of mitochondria and chloroplasts has its "inner surface" facing outward. The endosymbiosis hypothesis also was tested by phylogenetic studies that found that a purple bacterium and the cyanobacteria were the most recent common ancestors of mitochondria and chloroplasts. Mitochondria and chloroplasts are more closely related to free-living bacteria than to the rest of the eukaryotic cell. This provides very strong confirmation of the endosymbiosis hypothesis. Furthermore, mitochondria and chloroplasts are sensitive to antibiotics, and if they are removed from eukaryotic cells, the cells cannot manufacture more organelles. Mitochondria and chloroplasts duplicate themselves via binary fission, like other prokaryotes. Mitochondria and chloroplasts have 70s ribosomes, which is a complete mismatch with the 80s ribosomes in the surrounding cytoplasm. The endosymbiosis hypothesis is successful and accepted as true because it explains so many observations so well. And remember, you cannot reject this hypothesis out of hand without providing an alternative explanation that accounts for these observations.

How fast could such a symbiotic interaction evolve? In a laboratory experiment, amoebae were fed pathogenic bacteria. Bacteria are common prey of amoebas, but feeding upon *pathogenic* bacteria can result in death. As expected the amoebae suffered a high mortality rate; most died after ingesting the pathogenic bacteria, but some few survived. Those surviving amoebae not only gave rise to a population of amoebae resistant to the pathogenic bacteria, but their progeny became dependent on having the formerly pathogenic bacteria in food vacuoles in no more than 300 generations (Jeon 1991). In comparison to geological time scales, this evolutionary shift from lethal pathogen to dependent endosymbiont in 300 generations of a unicellular organism is essentially instantaneous.

Many such cellular associations are known to occur in nature. The giant amoeba *Pelomyxa*, a pelobiont, is a multinucleate unicellular organism that possesses a mitosome, a much reduced organelle derived from a mitochondrium, but no fully functional mitochondria. Presently we do not know why this is the case, but this amoeba has three different kinds of bacterial endosymbionts housed within vacuoles that each perform some mitochondrial functions (Margulis and Swartz 1988). The bacterial endosymbionts are essential for the survival of the amoebae; if a culture of amoebae is treated with antibiotics to kill the bacteria, the amoebae will weaken and die. *Pelomyxa* lives in low-oxygen environments, marginal between anaerobic and aerobic, and one of the endosymbiont bacteria functions to detoxify oxygen, another removes a toxic cellular waste product by metabolizing it. *Pelomyxa* is essentially an anaerobic organism that has adapted to more aerobic environments via bacterial endosymbionts, a situation exactly like the hypothesized early stages of mitochondrial evolution. However, these bacterial endosymbionts are not some new type of mitochondrion because these bacteria are still capable of being free-living organisms. This suggests the bacteria–*Pelomyxa* association is of relatively recent origin, or perhaps there is a regular turnover of bacteria: some are released and new ones acquired, for reasons not yet understood.

Accumulating phylogenetic data and the nature of chloroplasts themselves argue that chloroplasts were acquired by endosymbiosis in the common ancestors of the plant lineage (Fig. 3–1). Chloroplasts in the red algae, green algae, and land plants have only two membrane envelopes, suggesting their chloroplasts were acquired directly from a cyanobacterial ancestor. The chloroplasts of red algae possess only chlorophyll *a* plus biliprotein pigments, the same as most cyanobacteria. Chloroplasts in green algae and plants have chlorophylls *a* and *b*, and they lack the bluish biliproteins. Chlorophyll *b* is biosynthetically and phylogenetically derived from chlorophyll *a*. When three genera of cyanobacteria with chlorophylls *a* and *b* were first discovered, biologists thought, "Bingo! Here are the ancestors of chloroplasts," but chloroplasts with chlorophylls *a* and *b* are not closely related to any of these three genera of cyanobacteria. In fact, these three genera do not even form a single lineage within cyanobacteria, so chlorophyll *b* has arisen from chlorophyll *a* at least four times, three times in the cyanobacteria and once in the chloroplasts of green algae and land plants. Chlorophyll *b* has a slightly different light absorbance spectrum, but the primary importance may be how chlorophyll *b* interacts with the chlorophyll-bearing membranes to produce grana stacks (Fig. 3–2 A). This results in a denser packing of the light-absorbing pigments.

Glaucophytes do not have true chloroplasts, just a slightly reduced cyanobacterial endosymbiont, which can be thought of as an intermediate stage of the endosymbiotic origin of chloroplasts. This suggests a relatively recent endosymbiosis event and that glaucophytes obtained their "chloroplasts" separately rather than as a leftover from the original acquisition of chloroplasts. This also means that their not-quite-a-chloroplast, even though it contains chlorophylls *a* and *b*, cannot be used as evidence for including glaucophytes in the plant clade as the basal lineage. Perhaps the glaucophyte "chloroplast" will be found to be closely related to one of the cyanobacteria that possess chlorophylls *a* and *b*, clearly demonstrating this was a separate acquisition event.

All other green eukaryotic organisms obtained their chloroplasts from other eukaryotes, either red or green algae, via a secondary endosymbiosis. The observations that led to this hypothesis are chloroplasts possessing either one or two additional membrane envelopes, making chloroplasts with a total of three or four membranes. In all cases the outermost membrane is interpreted to be the vacuolar membrane formed when the capture took place. Logically the two innermost membranes represent the vacuolar membrane from the initial endosymbiotic event and the original cyanobacterial cell membrane (Fig. 3–3 B). If four chloroplast membranes are present, the second outermost membrane is interpreted as the cell membrane of the eukaryotic host cell that was ingested along with its chloroplast (Fig. 3–3 C). If only three chloroplast membranes are present, the original host cell and its cell membrane are assumed to have been digested. Even if this explanation seems a bit far-fetched, it makes sense of the available facts and is indeed testable.

The orientations of the additional membrane envelopes can be predicted (Fig. 3–3 B–C). If the outermost membrane envelope is vacuolar, then it should be inverted. The two innermost membrane envelopes would be inverted (vacuolar membrane from the first capture) and normally oriented (the original cyanobacterial cell membrane). If there are four envelopes, they should be inverted, normal (captured eukaryote host cell membrane), inverted, and normal. The observed membrane topologies of these various chloroplasts are as predicted, supporting the second-

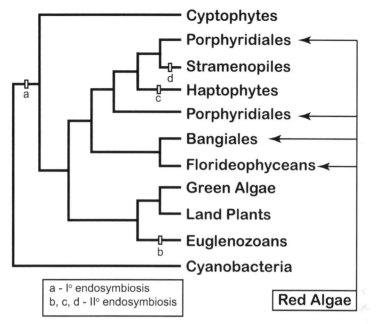

Fig. 3–4 Molecular phylogeny of chloroplasts. This phylogenetic study tests the endosymbiosis hypothesis of chloroplast origins by examining chloroplast DNA to determine common ancestries among them using cyanobacteria, the predicted source of chloroplasts, as an outgroup. Chloroplasts of members of the plant lineage (cyptophytes, red algae, green algae, and land plants) form a single lineage themselves with cyanobacteria as their sister group, which is consistent with their hypothesized primary endosymbiotic origin. Chloroplasts of stramenopiles, haptophytes, and euglenozoans are nested within the plant clade, indicating their common ancestry with either the chloroplasts of green algae (b) or the porphyridiales of the red algae (c, d) via a secondary endosymbiosis. (Image source: After Oliveira and Bhattacharya 2000.)

ary endosymbiosis hypothesis. Still in doubt of this explanation? In those chloroplasts having four membrane envelopes, a vestigial nucleus from the original eukaryote host cell can be found located between the second and third membrane envelopes, which is pretty convincing (Fig. 3–3 C).

The endosymbiosis hypothesis was tested by constructing a molecular phylogeny of chloroplasts based on their own DNA (Fig. 3–4). If members of the plant lineage enslaved cyanobacteria that became chloroplasts, then phylogenetically, plant chloroplasts should have a common ancestry with cyanobacteria. If euglenozoans and brown algae (stramenopiles) and haptophytes acquired their chloroplasts via a secondary endosymbiosis from green algae and from red algae, respectively, then euglenozoan chloroplasts should show a common ancestry with plant chloroplasts, and brown algae and haptophyte chloroplasts should show a common ancestry with red algal chloroplasts (Oliveira and Bhattacharya 2000;

Müller et al. 2001). This study found these predictions to be true, providing strong confirmatory support for the endosymbiotic evolution of chloroplasts.

Endosymbiosis would become irreversible and the endosymbiont a cellular organelle when one or more gene transfers take place from the endosymbiont genome to the host cell nucleus, thus removing some of the genes necessary for independent life from the endosymbiont genome. Genetic studies demonstrate that such shifts have taken place from mitochondria and chloroplasts to the eukaryote cell nucleus, although the mechanism of gene transfer remains uncertain. Such transfers are known to take place because a gene related to photosynthesis is found in the nucleus of a parasitic organism (*Plasmodium*: Alveolates). This organism has a common ancestry with photosynthetic organisms (Fig. 3–1), and while it lost chloroplasts in becoming a parasite, it did not lose a chloroplast gene transferred to the nucleus (a hypothesis). All of this together with many other examples of intracellular symbionts leads us to the conclusion both that such intimate interactions can be highly favorable to the organisms involved and that they evolve fairly easily. Such findings increase our confidence in the plausibility of the endosymbiont hypothesis for the origin of chloroplasts and mitochondria. Since eukaryotic organisms resulted from a composite association between two or more prokaryotic organisms, the tree of life presented earlier must be redrawn with some elaboration (Fig. 3–5; compare with Fig. 1–3).

Is the endosymbiotic origin of chloroplasts and mitochondria a fact? It depends upon your definition of a fact, but in science a very well-supported hypothesis comes very close to the strictest meaning of fact. Presently this is the only explanation that accounts for all the observations, and therefore the endosymbiotic origin of chloroplasts and mitochondria is widely accepted as true. In legal parlance, the endosymbiosis hypothesis is true beyond all reasonable doubt, and in science that is as good as it gets. So biologists are well justified in calling such knowledge a fact.

But to be fair, an alternative hypothesis of mitochondria and chloroplast origin should be considered. Perhaps mitochondria and chloroplasts evolved through a stepwise elaboration of previously existing cellular structures (autogenesis = self generating). The autogenesis hypothesis generates several predictions—for example, the cytoplasm within the organelle should match the surrounding cytoplasm (e.g., same-sized ribosomes in both), the cell should be capable of synthesizing and constructing new organelles, and both membranes of the double-membraned organelles should be capable of forming interconnections with other cellular mem-

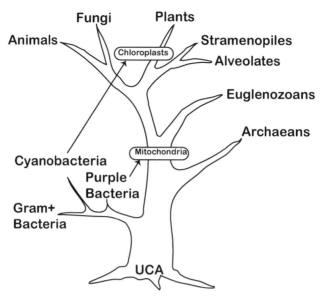

Fig. 3–5 Tree of life diagram revisited to better reflect the composite nature and history of eukaryotic organisms. UCA means universal common ancestor. Compare with Fig. 1–3.

brane systems. But none of these predictions are true. Autogenesis cannot explain the mismatch between cytoplasm's 80s ribosomes and organelle's 70s ribosomes, the presence of prokaryotic DNA with a prokaryotic organization in the organelle, or why their DNA has more shared characters with free-living bacteria then with their own cell. Further, if mitochondria and chloroplasts evolved within a prokaryote cell by a stepwise series of elaborations, then some intermediate organisms might be expected with protochloroplasts or protomitochondria, but we only find fully formed organelles. Those intermediates that we do find are not partially formed organelles, but endosymbionts that are even more like their free-living counterparts—for example, the "cyanelles" of glaucophytes.

To be even fairer, consider still another hypothesis, intelligent design. Proponents of intelligent design creationism (ID) were much in the news while this book was being written. Supposedly ID is a valid alternative to evolution, and proponents of ID want it taught as such in high school biology classes. However, ID proponents have been virtually mute in explaining biological diversity. How all of the observations presented could represent an intelligent design is not at all clear. Nothing in what we know suggests irreducible complexity as ID predicts and requires. Clearly ID is not ready to be an alternative explanation for many biological phenomena. But since creationism deals with diversity, consider the simpler creationist explanation that preceded and gave rise to ID. "If they [prokaryotes and eukaryotes] were, in fact, separately created (as we believe), one needs no difficult theories to account for their existence" (Frair and Davis 1983).[9] Now who can argue with an explanation that covers all possibilities so well? Prokaryotes and eukaryotes share certain features; no problem, they were created that way. Extra membranes around a chloroplast and a vestigial nucleus; no problem, it was created that way; their function just hasn't been figured out yet.[10] 70s ribosomes in organelles surrounded by cytoplasm with 80s ribosome; no problem, it was created that way. Such all-encompassing, all-purpose explanations clearly demonstrate why creationism is not science. Such explanations are not testable, and in terms of guiding further study, they are useless and intellectually bankrupt. Creationist explanations cannot lead to any greater understanding, but they could lead to a new dark age. So far ID has done no better and produced no new insights or advancements in understanding.

Are we biologists dogmatic about our ideas? The creationist literature routinely makes this assertion. There is no doubt that biologists can be more than a little bit stubborn about changing our minds, but at other times, things change so fast it makes our heads spin. Not only are alternative scientific explanations regularly proposed, they just as regularly rejected. Just a few years ago, biologists thought that those few eukaryotic organisms that lack both mitochondria and chloroplasts had their last common ancestry with eukaryotes before these organelles were acquired—that is, their last common ancestor predated the acquisition of mitochondria. This hypothesis explained the observations (eukaryotes whose nucleated cells lack mitochondria and chloroplasts), and the hypothesis was expressed by grouping these organisms into a basal lineage of eukaryotes called archezoans (ancient animals) (Kumar and Rzhetsky 1996).

Most of these organisms are probably unfamiliar: entamoebae, pelobionts (giant amoebae), parabasalids, microsporidians, oxymonads, retortamonads, and diplomonads, which includes *Giardia lamblia*, the water-borne human parasite often responsible for diarrhea[11] (Table 3–1). The pre-mitochondrial-divergence hypothesis does make predictions, and therefore it could be and was tested by seeking evidence of mitochondrial genes in the nuclear genome (Roger 1999). If the last common ancestry between archezoans and all other eukaryotes took place before mitochondria were acquired, then no mitochondrial genes could ever be transferred to their nucleus. So far all of the archezoans tested show some evidence of prior possession of mitochondria, so the archezoan hypothesis has been falsified. Subsequent work has demonstrated that in place of mitochondria these organisms have reduced or modified mitochondria called hydrogenosomes or mitosomes. Many of these organisms are parasitic, and extreme reductions and losses often are associated with a parasitic life style. Further, the phylogenetic positions of these organisms have not conformed to the archezoan hypothesis either; archezoans do not group into single lineage, basal or otherwise. So the archezoan hypothesis is now considered just plain false. Some of the former archezoans are now placed with amoebozoans in the unikont clade and the rest were found to be part of various lineages in the bikont clade. This is another good example of how we gain knowledge of past events by seeking evidence of the logical consequences of the presumed event.[12] That, or maybe they were just created that way.

Endosymbiosis does not account for all eukaryotic cell features. Other features of eukaryotic cells (actin/tubulin proteins and microtubules, nucleation of the genetic material, chromosomes, and dividing by mitosis) are hypothesized to have arisen in the host cell lineage by stepwise elaboration, by autogenesis. And indeed, as expected, organisms with nuclear organizations and division mechanisms that are intermediate between binary fission and mitosis do exist, and some may have ancient common ancestries with the rest of eukaryotic organisms. So we may reasonably conclude that many of the suite of eukaryotic cell characteristics (Table 3–2) arose via autogenesis, a series of stepwise elaborations. Both autogenetic and endosymbiotic hypotheses remain viable for the origin of flagella. Biologists aren't certain whether the host cell lineage acquired nucleation prior to mitochondria or the other way around. Altogether at least seven scenarios have been proposed for different sequences of events; four propose nuclei came first and three propose mitochondria came first (Embley and Martin 2006).

Whether actin proteins, amoeboid cellular form, and phagocytosis appeared before other eukaryotic cellular features like mitochondria, nucleation, and chromosomes remains uncertain too, but it makes an attractive hypothesis. If mitochondria were acquired prior to nucleation (Martin 2000), a mechanism to get the mitochondria inside a host cell vacuole still had to come first. Although a cytoskeleton composed of tubulin proteins is involved in nuclear and cytoplasmic division in eukaryotes, it remains possible for the nucleation of genetic material to precede actins and even chromosomal organization of the DNA, both of which must precede mitosis.

FLAGELLA: CELLS ON THE GO

One of the great thrills and frustrations of biology is trying to follow a tiny unicellular organism on the move using a microscope. Microscopes invert everything so novices often shift the specimen the wrong way, a sure way to lose a fast-moving cell. Microorganisms have three basic types of motility: gliding, swimming, and amoeboid crawling. Many prokaryotic cells have the ability to glide, which is accomplished without any moving parts by the excretion of hydrogels (Hoiczyk and Baumeister 1998). The rapid hydration of the gel produces a push that propels the cell away from the hydrating gel. Perhaps the most famous of bacterial gliders is the cyanobacterium *Oscillatoria*, which as the name suggests, glides or oscillates back and forth (Appendix: Green Bacteria). A few eukaryotes with stiff cell walls, like diatoms, retain this motility mechanism. Amoeboid cells use cytoplasmic streaming and a flexible, irregular cell shape to shift their position, as described previously; this motility involves actins and actin-binding proteins.

Swimming is accomplished by the rhythmic beating of flagella and cilia, long, whip-like extensions of the cell. Some bacteria have simple protein-fiber flagella, but eukaryotes have structurally more complex flagella with an endoskeleton of microtubules attached to a basal body, which acts as a rotary motor. Among eukaryotes, a considerable number of unicellular and small colonial organisms, as well as sperm and zoospores (zoh-oh-spores), are flagellated and capable of swimming through their aqueous environments. Cilia and flagella are constructed similarly and are basically the same organelle, although generally cilia are shorter in length and more numerous. A euphonic general term, undulopodia (uhn-doo-loh-POH-dee-ah), has been proposed for both (Margulis et al. 1989), but the term remains little used.

Cytoskeleton proteins (actins and tubulins) are the key components for both amoeboid movement and flagella, and their origins have implications for the origin of the nucleus, which is the hallmark of the eukaryotic cell. Like flagella, spindle fibers used in mitosis are composed of microtubules, which in turn are constructed of tubulin protein subunits assembled into long tubes. Static microtubules function as structural components of a cytoskeleton, and active assembly or disassembly of the subunits functions to move anything attached to their ends, like chromosomes (Fig. 3–6).

In eukaryotic cells, microtubules are assembled or disassembled at and attached to microtubule organizing centers (MTOCs). Assembly and disassembly makes microtubules grow longer or shorter by polymerizing and

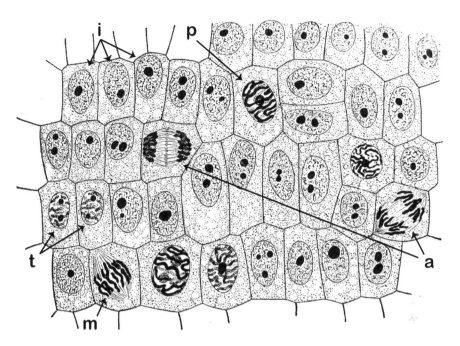

Fig. 3–6 Cell division of land plants. Cell division involves two processes that usually occur together: mitosis (division of the nucleus) and cytoplasmic division. The labeled stages of land plant cell division are inter-phase (i), prophase (p), metaphase (m), anaphase (a), and telophase (t) although the process is continuous. Interphase is the state of the nucleus when the cell is not dividing. In prophase (p) a typical interphase nucleus begins with a granular appearance caused by the condensation of DNA that as it continues looks ropey as the chromosomes thicken. This is followed by a dispersal of the nuclear membrane. As prophase ends, spindle fibers proliferate at each pole and extend either pole to pole, or from pole to each chromo-some attaching at its centromere. At metaphase (m) the chromosomes are lined up across the plane of cell division. At this time the two daughter chromosomes separate. During anaphase (a) a lengthening of the pole-to-pole spindle fibers and a shortening of the pole-to-chromosome spindle fibers pulls the two sets of chromosomes apart. Spindle fibers proliferate centrally at the plane of division and a cell plate grows out-ward, forming new cell membranes and then new cell walls between the two daughter cells. In telophase (t), daughter nuclei re-form by a reversal of prophase events and they are separated by a newly formed cell wall. (Image source: After Wilson 1900).

depolymerizing alpha and beta tubulin proteins, and this assembly/disas-sembly allows them to move anything attached to their ends. MTOCs are associated with the basal bodies of flagella and the poles of mitotic spin-dles. In many bi-flagellated unicellular organisms the flagella are with-drawn into the cell and their microtubules disassembled prior to nuclear/cell division, presumably to provide a needed supply of tubulin subunits. Deactivation of the flagella also frees the two flagellar MTOCs to function in mitosis. MTOCs are too small to be seen with a light microscope, but biologists long ago observed that flagellar basal bodies, called centrioles (SIN-tree-ohlz), migrated to the nucleus before the spindle formed. The centrioles just tag along with the MTOC and play no direct role in mitosis, but for years biology textbooks depicted centrioles in diagrams of animal cell division suggesting they had a function in mitosis. Electron micrographs never show microtubules converging or connecting to centrioles; they connect to an adjacent amorphous area, the real MTOC.

Centromeres on chromosomes also function as MTOCs, and they duplicate when chromosomes duplicate. Thus, like chloroplasts and mitochondria, MTOCs demonstrate a degree of autonomy. Since MTOCs and microtubules are critical to both nuclear division and flagella, MTOCs either arose as part of the motility apparatus and then became associated with the nucleus and cytoskeleton, or vice versa. This question remains central to the problem of accounting for the nucleus as an organelle. And as before, autogenesis and endosymbiosis are competing hypotheses.

Margulis and others (2000) have suggested that microtubules, MTOCs, and flagella as organelles of motility arose from another symbiosis between the eukaryotic host cell and flagellated, motile bacteria, like spirochetes, motile bacteria with a helical (corkscrew) shape that move or swim with a twisting, turning motion. Their flagella are composed of proteins similar to tubulin, but their flagella are attached to the cork-screw shaped cell at both ends. Spinning the flagella spins the corkscrew-shaped cell, producing motion. Spirochetes are known to attach to eukaryotic cells by the millions—for example, flagellated organisms from the gastrointestinal tract of termites. The synchronous motion of the spirochetes provides motility for the host cell, whose own flagella simply steer. No one fully understands this symbiotic association. The advantage to the eukaryotic cell seems obvious (motility), but it already had flagella, so why trade one for the other? No benefit for the spirochetes is obvious, unless like remora fish attached to sharks, they obtain more food via the association. Also, no one understands how the rhythmic motion is coordinated. Regardless, the presence of a spirochete-eukaryote association suggests a spirochete-prokaryote symbiosis is also possible, although none are known. But this alone is not enough to falsify the endosymbiosis hypothesis of flagellar origin.

If flagella and MTOCs are the vestigial remains of a formerly free-living bacterium, then they would be expected to possess their own genetic material (a prediction), but this remains in question. Some studies suggest MTOCs have a small circular or linear genome, perhaps just a few motility function genes; other studies found no such evidence. The MTOC has proven a particularly difficult cellular structure to study, so the results of many MTOC studies remain controversial or contradictory. None have presented irrefutable evidence of genetic material. Rather than allow negative results to ruin a good hypothesis, researchers have incorporated the lack of evidence of DNA associated with MTOCs into a new version of the hypothesis suggesting that flagella and the nucleus arose together from the same endosymbiont. The nucleus just represents the abscised genome of the spirochete-like symbiont, so everyone was just looking for the missing DNA in the wrong place (Margulis et al. 2000). Perhaps this is where the bacterial regulatory genes found in eukaryotes came from.

Good scientists always seek to explain unexpected results such that their hypothesis is still correct, but they have to explain what happened to the host cell genome. Since an endosymbiotic hypothesis of nuclear origin generates testable predictions, it will stimulate research that will either support or ultimately falsify the explanation, but presently the origin of the nucleus by autogenesis remains more plausible.

While chloroplasts and mitochondria exhibit biochemical, physiological, and cytological similarities to their free-living counterparts, the nucleus does not (Martin 2000). While the nucleus is described as "double-membrane bound," suggesting the same membrane topology as chloroplasts and mitochondria, the similarity is only superficial. The nuclear envelope is a single membrane folded double and perforated by numerous large pores. A nuclear pore complex is composed of proteins that act as gatekeepers restricting the ingress and egress of large molecules like proteins and ribonucleic acids (RNAs). The nuclear envelope is continuous with the endoplasmic reticulum, the membrane network responsible for connecting cellular organelles, protein synthesis, and material packaging. Unlike chloroplasts and mitochondria, every time a typical eukaryotic cell divides, the nuclear envelope disperses and is reformed in each daughter cell. While each of these observations is unexplained by endosymbiosis, they are consistent with an autogenetic origin.

If MTOCs, microtubules, the nucleus, and flagella had an autogenic origin, then these structures must have come from other preexisting structures. The homology between chromosomal MTOCs (centromeres) and motility MTOCs raises the possibility that microtubules functioned as cytoskeleton and in genomic division prior to becoming involved in motility. If the nucleus and chromosomes are of autogenic origin, then perhaps MTOCs also began as part of the nuclear-chromosomal division apparatus. This would mean centromeres (MTOCs on chromosomes) are more ancient than motility MTOCs. The tubulin-like proteins found in spirochetes and other bacteria show similarities to bacterial cell division protein ftsZ (van den Ent et al. 2001), which functions to attach the circular prokaryote DNA molecule (the whole genome) to the cell membrane.

Cell division protein ftsZ lacks a critical component that allows self-assembly into microtubules. Microtubules might have arisen from cell division protein ftsZ, and its role in cell division fits the idea of MTOCs functioning in cell division prior to locomotion and the cytoskeleton. This is far from conclusive, but certainly the autogenesis hypothesis remains plausible.

A number of other observations are consistent with the autogenesis hypothesis and its prediction of intermediate conditions in basal lineages. Several lineages of eukaryotes have their chromosomes permanently attached to the nuclear envelope by their centromeres (MTOCs), a structural arrangement that can be considered homologous to the attachment of the prokaryotic genome to the cell membrane by cell division protein ftsZ. Chromosomes are densely packed molecules of DNA tightly wound around histone proteins, and a similar type of DNA packaging is found in the thermophilic archaean *Thermoplasma* (Searcy et al. 1978; Searcy and Stein 1980). Histones function to stabilize the DNA in hot, salty, and acidic environments, conditions that normally denature DNA. The protective function of histones is certainly an adaptation in and of itself, and in a less extreme environment, natural selection could shift the function of histones from protection to regulation.

If centromere MTOCs first functioned in nuclear division, then their involvement with locomotion came later. An intermediate stage can be envisioned where centromeres (chromosomal MTOCs) serve double duty in both mitosis and locomotion. This would require the nucleus to be positioned next to the flagella so that the nuclear MTOCs embedded in the nuclear membrane could produce the basal bodies and flagella. Such double duty is found in the chrysophytes, a group of motile, unicellular algae related to the brown algae in the stramenopile lineage. The membrane-bound centromeric origin of MTOCs also means that until chromosomes become free, the nuclear membrane must persist during mitosis. So the autogenesis hypothesis also explains closed mitosis as an intermediate and ancestral condition retained in many lineages. Unfortunately, if MTOCs first arose in association with flagella and their involvement with the nucleus was secondary, then we might also envision an intermediate condition where the nucleus and flagella shared the same MTOCs. In other words, the chrysophytes may still represent an intermediate condition, but we do not know in which direction evolutionary history moved. Drat!

An autogenetic origin of the nucleus, chromosomes, microtubules, and MTOCs also means that typical mitosis arose from prokaryotic binary fission by a series of elaborations, and therefore, intermediate types

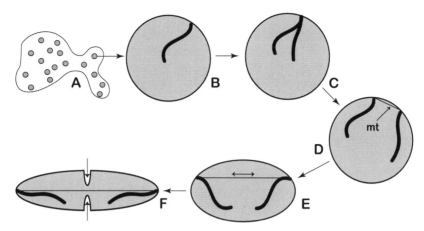

Fig. 3–7 Nuclear division of giant amoeba, *Pelomyxa*, a pelobiont. A. This organism is multinucleate so nuclei divide without cell division following. When this amoeboid cell does divide, each daughter cell gets a complement of the many nuclei. B-F. Nuclear division. B. Unlike most eukaryotes, the chromosomes are permanently attached to the nuclear membrane by their centromeres. For simplicity only one chromosome is shown with a terminal centromere. The nuclear membrane remains intact throughout division and no external spindle is produced. C. After chromosome duplication, the daughter chromosomes are still attached at the same point. D. Microtubules (mt) are assembled between the two centromeres (microtubule organizing center [MTOC]). E. A lengthening (assembly) of the microtubules results in physical separation of the daughter chromosomes and an elongation of the nucleus. F. Nuclear division is complete when the nuclear membrane furrows or blebs to separate the daughter nuclei.

of nuclear division would be expected. This prediction proves true, and these "intermediate" mechanisms offer suggestions of how something as complex as mitosis evolved in a stepwise fashion. Nuclear division in giant amoebae (pelobionts) (Fig. 3–7) is neither mitosis nor binary fission, but incorporates aspects of both. Chromosomes are attached permanently to the nuclear envelope by their centromeres, and the nucleus remains intact throughout division (closed). After DNA duplication each pair of daughter chromosomes is held together by their centromeres at the point of attachment to the nuclear membrane. Microtubules assemble between the centromeres (MTOCs) of each pair of daughter chromosomes to push them apart; no spindle apparatus forms outside the nucleus, so no MTOCs external to the nucleus are needed. This results in something akin to a spindle within the nuclear envelope that elongates until the nucleus pinches itself in half. This type of "mitosis" is like binary fission of the nucleus with the addition of microtubules, which might have come about by a change in cell division protein ftsZ that allowed polymerization. This mitosis is not associated with any type of cytokinesis[13] (seye-toh-ken-EE-sis) because these amoebae are multinucleate. When multinucleate cells divide, each daughter cell gets a portion of the many nuclei, and while a bit haphazard, it works. Whenever a hypothesis makes logical sense out of an otherwise discordant observation (a strange nuclear division in giant amoebae), our confidence in the explanation increases, so an autogenetic origin of nuclei remains a viable hypothesis.

Presently what remains unclear is how or why the genetic material became nucleated, isolated on a membrane island. Clearly this event is related to the origin of an elaborate internal membrane network, the endoplasmic reticulum, another eukaryotic cell characteristic. One hypothesis is based on the idea that mitochondria were acquired before nucleation occurred, which is certainly possible if a cytoskeleton and amoeboid movement were the first full-fledged eukaryote feature to arise (Martin 2000; Embley and Martin 2006). The observation that prompted this hypothesis is that archaebacteria and eubacteria have fundamentally different types of lipids, a key component of membranes, and eukaryotic cells synthesize eubacteria-type lipids, which is both unexpected and unexplained if the eukaryotic host cell has a common ancestry with archaebacteria. This hypothesis suggests the eukaryotic genes for lipids came from mitochondria (eubacteria) that got transferred to the host (archaean) cell genome. As a consequence of the gene transfer, eubacteria lipids were synthesized into a cytoplasm of archaebacteria origin that was "unprepared" to deal with them. As polar molecules, many lipids will spontaneously form membrane-like double layers in aqueous solution. The accumulation of lipid sheaths or vesicles around the site of synthesis (nucleus) could have led to the nuclear envelope (Martin 2000). Unfortunately for this hypothesis, such eukaryotic genes are not similar to those in mitochondria as predicted (Ribeiro and Golding 1998), although their specific origin remains unknown.

A nucleated cell must have some functional advantage over the nonnucleated cells for natural selection to operate in favor of nucleation no matter what their origin. Three suggestions have been made: (1) The nucleus more efficiently packages an increasingly large genome, allowing faster duplication and more accurate separation of copies of the genome. (2) The nucleus protects the genetic material from oxygen damage in an environment with an increasing concentration of free oxygen. (3) Nucleation functionally separates nuclear division and cytoplasmic division.

The first idea is based upon the observation that eukaryote genomes consist of two or more chromosomes, each a DNA molecule. Presently the typical eukaryote genome is much larger than a prokaryote genome, which consists of a single circular molecule of DNA. But would this have been the case early in the history of eukaryotes? Natural selection would not have permitted prokaryotic genomes to grow so large as to be inefficient, so it is difficult to see how an increasingly inefficient genome could exist in the first place. More likely nucleation and packaging of the genetic material into chromosomes allowed the eukaryote genome to increase in size, and the size differences we presently observe in the genomes of eukaryote and prokaryote are an after-the-fact development.

The second reason for nucleation comes from the observation that eukaryotic organisms appeared at about the time scientists think the oxygen crisis was becoming a serious problem, but beyond this coincidence, nothing else supports the hypothesis. No physiological evidence supports the supposed anti-oxidation function of the nuclear membrane, and the cytoplasmic environment of cells remains essentially anaerobic even in aerobically respiring organisms. Indeed, if mitochondria were acquired prior to nucleation, the proto-eukaryotic cell may have already adapted to increasingly aerobic environments via its aerobic endosymbiont. So this explanation has relatively little going for it. The nuclear pores must play some regulatory role that is of benefit.

The third idea, my own contribution, deals with the fact that in prokaryotic cells division of the genome and division of the cytoplasm are completely linked because the mechanism of separating daughter genomes requires cytoplasmic division. But if the genetic material is isolated upon its own membrane "island," a nucleus, it is free to duplicate and divide without an accompanying cytoplasmic division, as described above for the giant amoebae. If nucleation took place before any other changes in the genome, then the initial eukaryotic nucleus would have had a naked circular genome attached to the inner nuclear membrane. Nuclear division would have taken place more or less exactly like binary fission, but the cell would lack a mechanism for assuring that each daughter cell would receive one nucleus when cytoplasmic division eventually took place.

Nuclear division might well trigger cytoplasmic division, and the haphazard location of the daughter nuclei might result in daughter cells with two nuclei or with none. A cell without a nucleus dies, but the surviving cell gets two nuclei. When nuclear division next occurs, both nuclei would divide, perhaps at two different locations within the cell, increasing the odds that after cytoplasmic division, each daughter cell would have at least one of the four daughter nuclei. If the nuclei failed to apportion into daughter cells again, the lucky daughter cell would now have four nuclei. After the next division cycle there would be eight nuclei, and then 16, and at some point, the number of nuclei will virtually assure that some would be apportioned into each daughter cell even without a specific mechanism involved. No question about it, nuclear division without a coordinated cell division mechanism is haphazard and inefficient, resulting in some cell deaths, but it would work. Nucleation without a specific mechanism tying nuclear division to cytokinesis is not a death penalty, particularly if this new condition confers some advantage.

The multinucleate condition may also be an advantage because it produces a new functional relationship between the cytoplasm and genome. Additional copies of the genome tend to increase the cytoplasmic volume, which along with the delay in cell division immediately paves the way for larger cells. So it comes as no surprise that all of the largest unicellular organisms, like the giant amoebae, plasmodial slime molds, some green algae,[14] and the zygomycote (zeye-goh-MY-coat) molds (fungi) are all multinucleate. The advantage of a larger cell size is fairly obvious when considering amoeboid cells that capture and ingest prey organisms. Nucleation may have been the event that produced the *Tyrannosaurus rex* of unicellular organisms, the largest, most ferocious predator of the early Proterozoic world, the multinucleate amoeba, a gargantuan beast that relentlessly pursues and consumes all manner of prokaryotic and eukaryotic prey organisms.

Unless you have raised a slime mold yourself, your imagination may not do justice to the reality of a large, viciously predatory, multinucleate amoeboid cell. One easily grown slime mold, *Physarum polycephalum*, consists of a lemon-yellow plasmodium (plaz-MOH-dee-um), a membrane-bound blob of cytoplasm that will stretch your concept of a single cell. These amoeboid organisms can become 8–10 cm wide, and I have had them crawl out of their Petri dish and invade my desk drawer (they grow best in the dark). You may wonder why we don't find leopard-sized amoebae roaming the Serengeti plains engulfing antelope or a dolphin-sized *Paramecium* zooming through the oceans engulfing herring. For reasons that will be explored later, there are functional limitations to the size of unicellular organisms. These limits can be pushed by reinforcing the cell membrane with a cell wall and adopting a complex form as in some large unicellular seaweeds, but by far the most common and successful means of being large was to become multicellular.

Although the multinucleate condition is fairly common, cells with a single nucleus per cell predominate. In the hypothesis just presented, the multinucleate condition would be ancestral to cells with a single nucleus. Although a large-sized multinucleate cell may be advantageous under some circumstances, manufacture of numerous nuclei would be wasteful under other conditions. If so, natural selection would favor more efficient and precise mechanisms of mitosis and cytokinesis that allow an organism to reduce the number of nuclei to one. More elaborate mechanisms involving a cytoskeleton of microtubules leading to mitosis would follow and would be selected for as the genome became larger and more complex, placing a premium on more efficient packaging and more efficient division of chromosomes.

Such hypotheses may seem like they represent purely wishful thinking on the part of biologists, but such "brainstorming" is a creative part of biology. By considering different possibilities, biologists think about ways to test these ideas, and many will be falsified in the process. Also, the history of biology tends to forget most wrong ideas, so biologists floating such trial balloons risk little, and if correct, they may end up being referred to as "the first to suggest that . . ."

SEX

Now the discussion must turn to sex, and this is not a lame attempt to spice up a botany book. To improve sales, sex would have to be in the title and not a topic buried in the third chapter. Having taught biology for over 35 years, I nominate sex as one of the most misunderstood of biological phenomena. Sex is usually thought of as the same as copulation, reproduction, and gender. My colleagues in psychology get very upset when I contend they are different and that what they study is gender, not sex. One of the primary reasons for this misunderstanding is that sex, reproduction, and gender are inextricably linked together in most large, multicellular organisms like us. Sex is not a universal feature among eukaryotes, but sex becomes increasingly common, to completely obligatory, as organisms get larger and more complex. Among unicellular organisms, both prokaryotic and eukaryotic, sex and reproduction remain very different functions.

By way of definitions, reproduction means producing offspring, an increase in the number of individuals, period. Many organisms reproduce asexually, without sex, resulting in a clone, a population of offspring genetically identical to the parent and each other. Since no time and no energy were used to find mates or engage in any other reproductive activities, generally asexual reproduction is more efficient, resulting in more offspring in a shorter period of time. Clearly if you can reproduce without sex, reproduction and sex are not the same at all.

The simplest type of asexual reproduction is cell division, which is the prevalent mode of reproduction in all unicellular organisms. Some multicellular organisms reproduce asexually by growing branches, buds, plantlets, or other some modular unit that is capable of detaching and growing independently. Plants, fungi, and similar organisms have modular growth (repeated subunits), tailor-made for asexual reproduction via growth. As they grow, they fragment and each fragment continues as an independent organism. Reproduction by growth is sometimes called vegetative reproduction because it is so prevalent in plants. In some cases humans use this ability to our advantage—for example, the vegetative reproduction (cloning) of plants like potato and sugar cane. The potato tubers (modified stems) or cane stems are cut into modules capable of independent growth to increase their numbers. So while common enough in fungi and plants, vegetative reproduction is very uncommon in higher animals like vertebrates. Yet, such is the zoological bias of humans that most people still think reproduction and sex are the same.

Sex functions to produce new genetic combinations in offspring by combining the genotypes of two genetically different parents. Sex involving meiosis (my-OH-sis), a type of nuclear division, has arisen in virtually every lineage of large multicellular organisms, and it has arisen either repeatedly or else once in a common ancestor near the base of the eukaryote clade. The standard evolutionary explanation for sex is that it produces genetically variable offspring, and in a changeable, patchy environment at least some of the offspring from two successful parents, successful in the sense that their genes allowed them to live long enough to reproduce, are a good bet to be successful. Even in an unchanging environment, some of the offspring's genotypes may be better (capable of leaving more offspring) than either parent as the result of new gene combinations, but, of course, many new combinations will be worse too. Since the environment constantly changes, shifting and fluctuating, producing genetically variable offspring increases the odds that some portion of your genome will be successfully passed along to the next generation, which is the Darwinian scorecard of evolutionary success. Darwin argued that when more genetically variable offspring were produced than could possibly survive, natural selection would be inevitable. Now countless examples of natural selection in nature and the laboratory have demonstrated that he was correct. Natural selection is a fact. However, this explanation leaves much about sex unexplained.

Under ideal conditions many unicellular organisms, particularly prokaryotes, can reproduce asexually, growing and dividing so rapidly that huge clonal populations can be generated in short periods of time. With a reproduction rate of more than two generations an hour, a single bacterium growing geometrically by a power of 2 produces a clone of several million individuals in half a day. Even without sexual reproduction, this rate of reproduction allows bacteria to adapt readily to even drastic changes in their environment. When a single bacterial clone[15] encounters a lethal substance, like an antibiotic, the mortality rate is way over 99%, but several resistant cells survive to produce new colonies. Still, this is no big surprise. Each colony of survivors is descended from an individual cell whose genes conferred some sort of resistance to the antibiotic, some particular adaptation to the toxic environment. But since the whole population was a clone descended from one individual by asexual reproduction, why do any genetic variants exist? The answer is mistakes happen. In such large populations mutation becomes a significant mechanism for producing genetic variations.

Mutations, the ultimate source of genetic novelty, occur because of mistakes made during the replication of DNA, which is not a perfect process. Mutations occur at predictable rates depending upon whether a particular sequence of nucleic bases is more or less prone to copy errors. Mutation rates are low, on the order of one mutation per gene for every 1,000,000 replications, but to produce a very large clone there are so many cell divisions that even a low rate of mutation can generate considerable genetic diversity. Mutations produce random changes in genes, so some are inconsequential (neutral to gene function) and many are lethal (producing nonfunctional genes). But some small fraction of mutations may result in new gene forms that may enhance that organism's ability to survive. In such situations, natural selection operates in favor of individuals possessing a beneficial mutation; the rest die. Cells possessing such genes rapidly proliferate. This evolutionary phenomenon in action explains why the overuse of antibiotics has generated numerous strains of antibiotic-resistant bacteria.

If mutations can provide an evolutionarily adequate supply of genetic variation, why should sex have evolved? The answer to this question requires that you understand two facts. One, as organisms get larger and more complex, either they cannot reproduce asexually or they can reproduce asexually but the rate of reproduction is nowhere near as fast as unicellular organisms. As the rate of asexual reproduction slows, the odds of a beneficial gene variant appearing among offspring via random mutations decrease. When mutations can no longer provide enough genetic variation, then natural selection will favor any mechanism linking sex to reproduction. Two, sex originally had a different function, which is survival of adverse environmental conditions. However, sex proved to be a convenient mechanism for generating new gene combinations through the recombination of parental genes. Finally, in many large organisms the coupling of sex to reproduction becomes obligate.

The origin and history of sex cannot be reconstructed with complete certainty because most of the features involved leave no fossil record, but certain events can be inferred. Some prokaryotes engage in conjugation, a one-way transfer of genetic material from a donor individual to a recipient individual resulting in new genetic combinations, but only one individual. The donor, having contributed some or all of their genome to the recipient, dies. Yes, that is correct; two individuals engage in sex, but only one survives! This clearly demonstrates that conjugational sex and reproduction are not the same. Margulis and Sagan (1986) hypothesize that conjugation evolved as a DNA repair mechanism, a means of repairing DNA damaged by ultraviolet (UV) radiation or some other environmental mutagen. This would be an important mechanism when the ancient Earth atmosphere lacked molecular oxygen so there was no ozone (O_3) layer to filter out UV radiation. DNA repair operates by using a donated length of a DNA molecule as a comparison template to repair damaged portions. Because conjugation can transfer some or all of the donor genome to the recipient, some mechanism must exist whereby a segment of the donated DNA can be matched up with the appropriate complementary portion of the recipient's DNA. The two portions would be homologous, carrying the same genes, but the genes would not be identical because the donor was a different individual. And, of course, both molecules could be damaged partially.

Here in a prokaryotic DNA repair mechanism is the molecular origin of the ability of homologous chromosomes to pair, a necessity for sex in higher organisms. The process of DNA repair results in new genetic combinations because some of the repaired genome came from two different individuals. The association of conjugation with DNA repair is very compatible with the idea that sex began as a last-ditch attempt for survival. This hypothesis is supported by the observation that the ability to conjugate and a sensitivity to UV radiation are genetically linked in bacteria like *Escherichia coli*, the common intestinal bacterium and laboratory organism.

The conjugation hypothesis generates a testable prediction. If sex originated as a DNA repair/survival mechanism, then sex should be associated with challenging environmental conditions. Populations of prokaryotes growing under ideal conditions seldom undergo conjugation; and considering the fact that in passing on its DNA, the donor dies, that makes sense. Bacteria conjugate only when environmental conditions become very unfavorable and their survival is in jeopardy, so conjugation can best be understood as an act of biological desperation, an attempt to survive and pass along their genes. Similarly among unicellular organisms, sexual reproduction is almost universally associated with changing or deteriorating environmental conditions. For many organisms environmental deterioration is seasonal and sexual reproduction is associated with the production of a resistant stage for enduring either cold or dry conditions. Conversely, among unicellular organisms living in oceans, a large and rather homogeneous environment, few are known to have sexual reproduction. They experience few environmental changes drastic enough to resort to sex.

STEP TWO: PRODUCTION OF A DIPLOID PHASE

In eukaryotic organisms the production of a tough, resistant survival stage is accomplished by fusion of two gametes[16] (GAHM-eats); these can be either two sex cells or two whole unicellular organisms acting as gametes. Since all gametes and most unicellular organisms are haploid, having a single set of chromosomes bearing a single set of genes, the resulting fusion cell is diploid (n + n = 2n), possessing two sets of chromosomes each bearing a set of genes. By definition the diploid fusion cell is a zygote (ZEYE-goat), the eukaryotic analogue of conjugation. In fact when two nonflagellated gametes of similar form fuse, it is still referred to as conjugation (see Appendix: Green Algae, charophyceans, *Spirogyra*, Fig. A37), whether in algae or fungi.

The two haploid genomes are homologous but not identical, so the zygote is likely to have different gene forms, alleles (ah-LEE-uhls), for any number of genes. Perhaps the most familiar alleles are those for human red blood cell proteins (rh+ and rh-; or A, B, and O). A, B, and O are different forms (alleles) of the same gene, and each allele produces a different protein. As a diploid organism you have two copies of the gene, one from each parent, resulting in four possible blood types, type A (AA, AO), type B (BB, BO), type AB (AB), and type O (OO). Positive and negative blood types come from another gene producing two forms of another protein. If you possess at least one rh protein allele (rh+rh+, rh+rh-) you are rh positive, and if not (rh-rh-) you are rh negative. The medical community often omits the rh and says something like "your blood type is O+," which causes some confusion because they are different genes and different proteins. Here we see one big advantage of the diploid condition; it embodies a lot more genetic diversity with two copies of each gene than the haploid condition. Of course, just as in bacterial conjugation, DNA repair can occur in the diploid condition. Homologous chromosomes must pair with each other to line up similar genes.

To function as a resistant stage, the zygote becomes a zygospore (ZEYE-goh-spor) by producing a thick, protective wall capable of withstanding severe conditions. The zygote of many unicellular organisms incorporates the cytoplasm of both parental cells, so cellular fusion has an energetic advantage over prokaryote conjugational sex, which just transfers a portion of the genetic material. Sex among many unicellular organisms allows two organisms to pool the cellular and genetic resources for survival. The toughness and survivability of the diploid phase is an idea that will be returned to when considering the origin of land plants, because the dominant, well-adapted phase of all vascular land plants is developmentally derived from the zygote, a tough, resistant phase in the life cycles of land plants' aquatic, algal ancestors.

STEP THREE: GETTING BACK TO A HAPLOID CONDITION

One thing just leads to another with sex. Presently we have gotten to third base, the production of a diploid zygote, and have survived until conditions favorable for growth have returned. Now what? Since most unicellular organisms are haploid, possessing only one set of chromosomes, some mechanism is needed to get from the diploid condition back to the haploid condition. In all sexual life cycles the mechanism for changing from diploid to haploid is special nuclear division called meiosis (Fig. 3–8). Do you think this is a logical conundrum, a chicken and egg problem? How can sexual reproduction occur before meiosis, and why would meiosis evolve if you didn't have sexual reproduction? Rather than being one of those irreducibly complex which-came-first problems, a solution lies in the observation that meiosis is quite similar to mitosis. In fact the second part of meiosis is just standard mitosis (Fig. 3–8 D-F), so it seems reasonable to conclude meiosis evolved from and is operationally derived from mitosis, which is what you would expect if sex arose via an evolutionary process. Mitosis evolves from binary fission, and then gets co-opted for sexual reproduction, but modified to reduce the chromosome number from diploid to haploid. Oh, and it just will not do to simply halve the number of chromosomes. Homologous chromosomes are like socks; in a diploid organism they come in pairs, and like socks of different colors and designs, each chromosome carries a different complement of genes. Although it makes no sense to do this with socks, to be functional each resulting cell will need one chromosome from each pair.

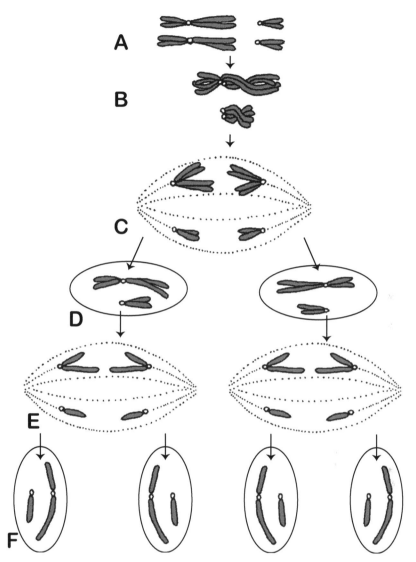

Fig. 3–8 Diagram of meiosis, a nuclear division that changes the diploid chromosome number ($2n = 4$ in this example) to a haploid chromosome number ($n = 2$). Chromosomes occur in homologous pairs and at the time of division consist of two chromatids (copies) each. Meiosis involves two successive divisions. In the first division homologous chromosomes pair at metaphase (A) and reciprocal exchange of some portions (crossing over) (B) can occur. Homologous pairs move to the metaphase plane (B) and pairs of chromosomes are separated in anaphase (C), resulting in two nuclei with just two chromosomes (D), but each still consists of two copies. The second division is operationally identical to mitosis and daughter chromosomes are separated (E), producing four haploid cells with two chromosomes each (F). (Image source: After an illustration by D. DeWitt in Armstrong and Collier 1990)

Prior to regular nuclear division (mitosis), the DNA is duplicated and thus each chromosome consists of two identical sister chromosomes. Each sister chromosome retains half of the original DNA molecule because each half acted as a template for synthesis of a new complementary half. As mitosis begins, each chromosome consists of two daughter chromosomes connected at their centromeres. Mitosis separates each pair of daughter chromosomes resulting in two identical sets, and if cell division follows, then one set ends up in each daughter cell. A zygote possesses two sets of homologous chromosomes (chromosomes bearing the same genes), one set from each parent, and prior to any cell division the DNA duplicates, so it possesses two copies of each chromosome as well. This is why meiosis is a two-step division, first separating each pair of homologous chromosomes and then separating the two copies of each chromosome. In meiosis I, homologous chromosomes pair[17] and the members of each pair are separated so each resulting nucleus has only one set of chromosomes—that is, it is haploid. The chromosomes in each haploid daughter nucleus still consists of two copies, so the second part of meiosis (meiosis II), which is operationally identical to mitosis, separates the two copies of each chromosome. This means that in the normal case, meiosis results in four haploid cells.

The first division of meiosis generates lots of genetic variation because when the homologous pairs of chromosomes are separated, it is without any regard to what parental set they came from. Each homologous pair consists of one maternal and one paternal chromosome, or one plus (+) and one minus (-) chromosome.[18] At the time of separation each pair of chromosomes acts independently, so a daughter nucleus has a 50:50 chance of getting either the maternal or paternal chromosome from each pair. In fact this coin-flip probability for assorting pairs of parental chromosomes is called independent assortment. If a diploid cell with three pairs of chromosomes undergoes meiosis, there is a one-eighth chance of the resulting haploid cell will get all three maternal or all three paternal chromosomes, one-half to the third power (three flipped coins all coming up heads). Of course, when one haploid nucleus gets all three of one parental type, the other nucleus must get the other parental set, so in this instance, two out of eight chances result in haploid nuclei of the parental types. The other three-quarters of the time the resulting haploid cells will be some combination of the parental genotypes. When you get up to 23 pairs of chromosomes like humans, there are only two chances out of more than 8,000,000 they will be of the parental genotype (two chances out of two to the twenty-third power[19]). So meiosis generates lots of new chromosome combinations, and therefore new gene combinations. In this way sex with meiosis operates like shuffling a deck of cards and dealing new genetic hands to offspring.

The shuffled card analogy helps understand how evolution works. First, by shuffling the deck, you generate lots of new card combinations. And each of these hands is in an environment that consists of a population of other genetic "hands." This means any particular combination of cards (genes) has no absolute value; it depends upon their environment. A straight (5-6-7-8-9) wins if no other hand has a higher combination, but it loses to a full house (10-10-10-J-J). This is the same with genes; their value (measured in number of offspring produced) is determined by the environment.

Meiosis serves the same function in all sexual life cycles (Fig. 3–9), reducing the diploid genome to a haploid genome. However, the haploid cells produced by meiosis have different functions depending upon the life cycle. When two diploid organisms, such as us, reproduce sexually, meiosis produces haploid gametes. The fusion of gametes produces a zygote, first cell of the new diploid offspring generation. When two haploid organisms sexually reproduce, since they are already haploid, they can act as gametes if unicellular, or produce gametes by mitosis. Fertilization results in a diploid zygote, which must divide by meiosis to return to the haploid condition. The resulting haploid cells will function as unicellular organisms, or they will be spores (zoospores, if motile) that develop and grow into a new haploid individual. A third sexual life cycle evolved in some seaweeds and in all land plants, an alternation of generations, a life cycle involving both haploid and diploid adults. The haploid organisms produce gametes, as before, but after fertilization, the zygote, rather than dividing by meiosis, grows directly into a diploid organism, one that looks either the same or different from the haploid organism. Cells on the diploid organism divide by meiosis to produce haploid spores that grow into haploid adults. One critical difference, the production of an embryo, distinguishes land plants from seaweeds, but more on this later.

The way out of the sex-meiosis which-came-first conundrum is to assume that the diploid condition is not a complete dead end without meiosis. A few zygotes will produce haploid offspring even if the zygote divides only haphazardly by mitosis. In the simplest circumstance where an organism has a single chromosome in the haploid condition, the diploid condition consists of one pair of homologous chromosomes. If such a diploid cell undergoes pairing of homologous chromosomes for DNA repair, and then undergoes a haphazard mitotic division, one daughter cell might get both chromosomes, and the other daughter cell gets none. Both results lead ultimately to death. But if all chance outcomes are equal,

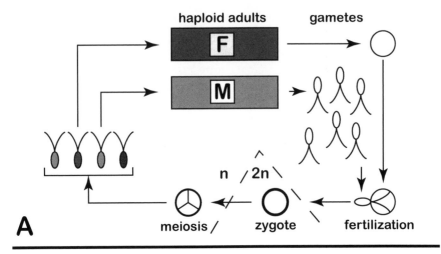

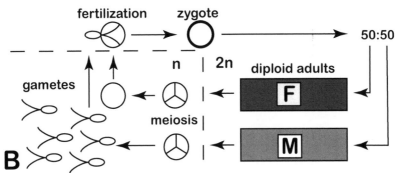

Fig. 3–9 Sexual life cycles with either haploid or diploid adults. A. Haploid adult life cycle. Haploid adults produce gametes (eggs or sperm) directly by cell division. Fertilization produces a zygote (diploid) that then divides by meiosis, producing zoospores and restoring the haploid condition. Spores or zoospores grow directly into adults. B. Diploid adult life cycle. Diploid adults use meiosis to produce gametes, the only haploid stage in this life cycle. Fertilization results in a zygote, restoring the diploid condition. The zygote, either male or female, grows directly into an adult.

then one-third of the time each daughter cell will get one of each pair resulting in two new haploid organisms. Since the sexual fusion and diploid condition was a desperate survival attempt, having one-third of the zygotes produce viable haploid cells is more than enough to reestablish populations of unicellular organisms and make this primitive eukaryote sex successful even without meiosis. Obviously the rate of survival decreases as the number of chromosomes increases, but mitotic divisions of diploid cells would still result in some haploid individuals. Since sex was not a dead end and resulted in survival, meiosis has an opportunity to evolve after the fact to improve the number of surviving haploid cells. The assumption that a small number of chromosomes was involved is reasonable because prokaryotic cells have only a single DNA molecule. Why organisms came to have more than one chromosome is a bit beyond where this discussion needs to go, but when the genome is divided into more segments, sex generates more combinations.

Meiosis does not require a big or drastic modification of mitosis. Following DNA duplication, a pair of sister chromosomes remain attached by their centromeres. In mitosis, the two sister chromosomes are separated efficiently because each centromere (MTOC) is attached to an opposite pole of the mitotic spindle. As the spindle fibers are disassembled at the poles, the identical daughter chromosomes are pulled in opposite directions. The pairing of homologous chromosomes by matching up centromeres may render the two adjacent MTOCs nonfunctional, if only by blocking adjoining connection points for microtubules. If so, then the mitotic apparatus would perform meiosis I and separate the pair of homologous chromosomes without even requiring a change in the mechanism. Then when each haploid cell detects that each chromosome still consists of two daughter chromosomes, this condition triggers another mitotic division, meiosis II, to separate the copies. Natural selection favors any variants that accomplished separation of homologous chromosome pairs more efficiently because they would produce more viable offspring. Any variation that increased the number of viable haploid cells will quickly spread through a population.

Sex becomes a necessity, completely linked to reproduction, when the generation-to-generation reproduction rate slows to the point that mutations occur at too low a rate to provide enough genetic variation for populations to adapt. This means a strict reliance upon sex as a means of reproduction is mostly a feature of larger and relatively slowly reproducing organisms that require sex as a mechanism (in addition to mutation) for generating genetic variants in offspring.

The eukaryotic cell generated a profusion of organismal diversity, so clearly the evolutionary potential of this innovation was tremendous. Perhaps some of the potential was realized because of the composite nature of the eukaryotic cell, resulting from an intimate, and now obligate, interaction between two or more different prokaryotic organisms. Like metabolic complexity, this cellular complexity comes from smaller, simpler components, which like Lego building blocks can be put together in different ways to generate many possibilities. Of particular significance to this narrative is the fact that cyanobacteria, as free-living organisms and as chloroplasts, are responsible for the vast majority of photosynthesis. When we look at all that green world around us, we are looking at the most spectacularly successful and important organism that ever lived on Earth, a cyanobacterium that almost became lunch.

The endosymbiotic origin of chloroplasts and mitochondria is quite well established, quite well confirmed, and at this stage there are no competing hypotheses. This hypothesis makes sense out of lots of otherwise confounding observations, like chloroplast genes in parasitic flagellates. Other features we associate with eukaryotic cells are of less certain origin, although the autogenesis of a cytoskeleton, nucleus, and chromosomes seems likely. As the genetic counterparts of eukaryotic features are identified in prokaryotes, some hypotheses will be falsified and others will be supported. The answers to some of the questions posed, like the origin of actins, actin-binding proteins, and amoeboid movement, may be waiting to be discovered among the as-yet little-studied archaeans.

In no small measure sex contributed to the diversity of eukaryotes. Not only did nucleation of the genetic material separate nuclear division from cytoplasmic division, but also organization of the genetic material into separate chromosomes allows sex to constantly generate new genetic combinations in offspring for the environment to evaluate via natural selection. Sex and dispersal become tied to reproduction, and as will be seen, the demands of the coastal and terrestrial environments make sexual reproduction nearly universal.

If eukaryotic cell organization is such a tremendous advantage, why do a profusion of prokaryotic organisms continue to populate the Earth? The answer to this question is fairly simple. While eukaryotic cell organization opened the door to countless new possibilities, many distinct habitats remain to which prokaryotic organisms are well adapted. In some cases, evolutionary events produced new possibilities for the prokaryotes. Many prokaryotes are obligate anaerobes, and some have found new anaerobic habitats within the bodies of multicellular eukaryotic organisms. The activities of larger organisms produce a more complex environment thus providing numerous new opportunities for tiny organisms, so microorganisms of all types continue to abound.

Had eukaryotic cells not appeared, prokaryotes would have remained Earth's sole inhabitants, as they were for nearly two-thirds of Earth history. No one can guess how likely or unlikely the evolution of the eukaryotic cell is, or if some counterpart of the eukaryotic cell is possible. The fact that it happened once, on Earth, does not argue either way. Symbioses are common, but we cannot know if some other possibility exists whereby prokaryotic cells can give rise to large, complex organisms. For example, cellular slime molds live as amoeboid unicellular organisms until reproduction, and then thousands of individuals mass together to produce a large, slug-like "organism" that can actually crawl a short distance before producing a macroscopic sporangium. What potential does such a cooperative multicellularity have? We have no way of knowing. If the evolution of eukaryotes is rather unlikely, then we would predict that while life may be rather a common feature of our universe—as its common constituents, its simple chemical self-assembly, and its ability to exist in "extreme" environments suggests—then in a cosmic sense many worlds will harbor life. But the organisms will be no bigger or more complex than prokaryotes, a universe of stromatolite worlds, or even more likely, sterile-appearing worlds inhabited by thermophilic chemoautotrophs deep beneath the planetary surface. Counterparts of extremophiles and cyanobacteria are not the kind of extraterrestrials that stir the human imagination, but they are the ETs most likely to exist.

FOUR

A Big Blue Marble

*Wherein algae are introduced, ocean ecology is explained, and
phytoplankton diversity is explored.*

Sponges grow in the ocean. That just kills me. I wonder how much deeper
the ocean would be if that didn't happen.
—Stephen Wright

For the first 21 years of my life, a large, freshwater, inland sea occupied
the northern horizon. Lake Ontario is a beautiful but moody body of water, and in particular I always liked its rocky shores, which have far more
character than its occasional sandy beach. The campus of my undergraduate alma mater sits on its shore, and just behind my freshman dormitory,
the shore consists of large slabs of rock cut by deep watery channels. If
you look into these deep, clear waters, a luxuriant green lawn covers the
rock surfaces undulating to and fro with the watery pulses. In a marine
environment (with salt water instead of fresh water), similar locations
would display much larger seaweeds and considerably more biological diversity. In fact, in terms of major phylogenetic lineages, more "phyla" can
be observed in a rocky marine tidal pool than just about any other single
place on Earth. Not only are there many more animal phyla, but also you
will find a diversity of red, brown, and green seaweeds. Some seaweeds
are small, like Lake Ontario's green lawn of algal filaments, which in spite
of its freshwater habitat has the general adaptations associated with seaweeds. Some seaweeds are quite impressively large and complex, like the

offshore forests of kelp, brown algal seaweeds that wave in the tides and currents, enormous versions of Lake Ontario's green algal lawn.

This leads to one aspect of early life on Earth that may have escaped your attention. Life on land, our familiar terrestrial environment, is a relatively recent event occurring only in the last 10% of life's history on Earth. With the exception of deep, dark, hot environments, all earlier events—the first 90% of life's history—took place in aquatic, marine environments, so no history or study of green organisms is complete without exploring algae.

WHAT ARE ALGAE?

Algae is neither an ecological nor a taxonomic designation, just a general term for a very diverse assemblage of mostly aquatic green organisms ranging from microscopic unicellular organisms to large, complex seaweeds measuring up to 100 meters long. Algae generally are eukaryotic organisms; the term has also been applied to prokaryotic organisms although now that is avoided. The cyanobacteria used to be called the blue-green algae, and this does illustrate that the common names of algal groups mostly refer to their pigmentation: red algae, brown algae, green algae, golden-brown algae, yellow-green algae. All have chlorophyll, but accessory pigments often mask or alter the shade of green.

In a broad ecological sense, two primary sorts of algae exist, phytoplankton (FEYE-toe-plank-ton) and seaweeds, each adapted to a different habitat. Phytoplankton are tiny, free-floating algae adapted to open ocean habitats. Seaweeds are larger, anchored algae adapted to coastal habitats. Both phytoplankton and seaweeds are found in many unrelated lineages, so both have evolved more than once. Lots of lineages have members that are phytoplankton. Seaweeds are found in just three groups: the red algae, brown algae, and green algae (chapter 5; see individual appendices). Freshwater algae, mostly green algae, have the same two categories even though phytoplankton and seaweed imply a marine environment. Although mostly aquatic organisms, algae are common enough in terrestrial habitats where they live on wet rocks, in porous rock, in or on soil, on bark or leaves, and in symbiotic relationships with some land plants—for example, forming coralloid roots of cycads,[1] or with fungi forming lichens.

Phytoplankton consists of unicellular, colonial, and small multicellular organisms including euglena (euglenozoa) and dinoflagellates, which may or may not be called algae. Presently diatoms (Appendix: Phytoplankton, stramenopiles), dinoflagellates (Appendix: Phytoplankton, dinoflagellates), and haptophytes (Appendix: Phytoplankton, haptophytes) are the most common constituents of phytoplankton. Although all three groups produce distinctive microfossils, no Precambrian fossils of these organisms have been found, so these common phytoplankton groups appear to be no older than seaweeds. Zooplankton (ZOH-oh-plank-ton) refers to small, free-floating, heterotrophic organisms adapted to oceanic habitats. Groups like the dinoflagellates include both phytoplankton and zooplankton species.

Seaweeds are anchored and generally larger, multicellular algae, although a few remain unicellular. By this definition Lake Ontario's green algal lawn is composed of seaweeds because while small, they are anchored and multicellular. Herein the usage of the term seaweed is broader than usual because no logical demarcation can be drawn along a continuum of seaweed sizes and complexities.[2] Some small seaweeds grow upon and are attached to larger seaweeds. Although this means seaweeds range from small unicellular and simple filamentous organisms to large complex organisms with well-differentiated organs and cell types, all seaweeds are adapted to the same habitat, the coasts, shallower water at the land-water interface. A few seaweeds have become secondarily free-floating oceanic organisms—for example, *Sargassum*, the brown algal seaweed that dominates the Sargasso Sea.

Fossil seaweeds are the oldest macroscopic[3] organisms to appear in the Proterozoic (Han and Runnegar 1992; Knoll et al. 2006). The earliest fossil seaweed is now dated to 2.1 bya. While it stood just a few centimeters tall, this seaweed would tower over the microbial mat communities. Although relationships are hard to determine with certainty, some of these early seaweeds are very similar to red and green algae. Microfossils clearly indicate an increasing diversity of eukaryotes and the divergence of major lineages during the middle Proterozoic from 1.3 to 0.72 bya (1300–720 mya) (Knoll et al. 2006). Fossil burrows indicate the presence of worm-like organisms by 1.1 bya (Seilacher et al. 1998). These data indicate that the Cambrian (540–490 mya) "explosion" is an artifact and that large organisms and biological diversity did not just "suddenly" appear in the Cambrian; rather, they had much longer, older histories that left scant but increasingly solid evidence. Fossils are rare because the older the rock the rarer the rock and because these small, soft-bodied organisms were not prone to easy fossilization.

Presumably some types of phytoplankton are much older because the eukaryote lineage is estimated to have arisen much earlier, and all seaweeds appear to have had phytoplankton ancestors. Red algae and green algal lineages include both phytoplankton and seaweeds. Brown algae are largely seaweeds, but they have a common ancestry with several phyto-plankton groups. To better understand the difference between phytoplank-ton and seaweeds, their adaptations to very different environments, and how one could have evolved from the other, you must understand the difference between oceans and coasts. The remainder of this chapter will be devoted to oceans and phytoplankton. Coastal habitats and seaweeds will be examined in the next chapter.

UNDERSTANDING THE DEEP BLUE SEA

Other than its fleecy white clouds of water vapor, the great blue oceans are planet Earth's most stunning feature. Together they make Earth appear from outer space like a big blue-white marble. Although prominent and of singular significance to Earth's biosphere, oceans remain Earth's most misunderstood feature. Oceans cover more than two-thirds of Earth's sur-face, and since they receive an abundance of sunlight, oceans would seem a likely place for abundant life, but life does not abound in the oceans. In fact, oceans are actually wet deserts with sparse, thinly spread life com-posed largely of very tiny, mostly unicellular organisms. The oceans ap-pear blue because phytoplankton are too sparse to tint the water green.

In deserts, life is sparse because water is in limited supply. Oceans have sparse life because two other necessities are in limited supply. First, a goodly amount of the sunlight that strikes the ocean's surface is reflected, a fact all sunburned boaters know. Further, water is a very good light fil-ter and quickly absorbs whatever solar energy penetrates its surface. The more water light passes through, the greater the filtering, so a reason-able abundance of light is found only in the upper few meters, a thin, bright surface layer. Light intensity decreases quickly as it passes through water. At a depth of 1 centimeter, just this deep |————|, more than one-quarter of the light intensity at the surface (about 27%) has been filtered out. At a depth of 10 meters (approximately 33 feet), almost 80% of the light intensity at the surface has been absorbed. A biologically significant amount of light penetrates only about 30 meters below the surface, and at this depth, light is pretty dim, about 1/60th the intensity of the light at a depth of 1 meter. So the vast majority of the oceans' volume is dark depths. Photosynthesis takes place only in this thin, 30-meter, light-rich surface layer,[4] but still life doesn't abound in this thin, bright surface layer because another critical resource limits phytoplankton growth.

The second fact that renders oceans largely infertile is that photosyn-thetic organisms need mineral nutrients, fertilizer, to be productive and multiply. Nutrients are endlessly cycled, taken up by producers, passed on through a food chain, and returned to environmental nutrient reservoirs by decomposers. In terrestrial ecosystems, decomposition takes place in the upper regions of the soil and nutrients released there are taken up by waiting roots of plants, starting the cycle over. In oceans, photosynthetic organisms occupy only the thin, bright surface layer, but decomposition and the recycling of mineral nutrients takes place hundreds of meters below in the dark depths. As the dying bodies of organisms sink, the nutrients incorporated in their bodies are constantly being removed from the bright, warm surface layer to the dark depths. Since warmer surface water is less dense than the cold water below, they do not readily mix, and as a result, nutrients tend to remain in the dense cold depths where no photosynthesis is taking place. As a result, the growth and reproduction of photosynthetic organisms in the bright surface layer are limited by a paucity of nutrients, so oceans remain largely barren and unproductive.

Like deserts, oceans are dotted with the occasional small oasis where life is luxuriant and larger organisms are common. In deserts, oases are the result of locally abundant water, but in oceans oases occur where nutrients are locally abundant. In some few places luxuriant oceanic oases of high productivity occur where currents from the depths bring nutrients to the bright surface—for example, the Grand Banks off Newfoundland, or where nutrient-laden fresh water mixes with salt water in estuaries like Chesapeake Bay, or in those few shallow areas, mostly coastal, where light and decomposition occur together. Here the water appears cloudy and murky with a greenish-brownish tint. Productive aquatic communities are not picture-postcard pretty. In spite of its aesthetic appeal, clear blue water is biologically barren.

Deep water and shallow water do mix in the high latitude oceans because the surface water stays colder. As a result the high latitude oceans are relatively productive areas seasonally; the seasonality is caused by having either lots of light (summer) that permits productivity or little to none (winter). The same number of hours of sunlight fall upon every portion of the Earth's entire surface in a year, but the light received differs in quality and pattern. Near the equator the light is intense because it passes through the atmosphere at near to right angles, and the amount is similar throughout the year, about 12 hours daily. Near the poles the light is filtered because it passes obliquely through the atmosphere so it becomes dimmer and redder (the sunset effect), and the daily amount varies from 0 hours per day in midwinter to 24 hours per day in midsummer.

Organisms adapt to severe nutrient limitations by remaining small. A larger organism's growth would quickly become nutrient limited and result in a stunted organism of reduced reproductive ability. Thus, the small size of phytoplankton is actually an adaptation for the oceanic environment. Since the photosynthetic producers of oceanic food chains are extremely tiny, the length of food chains is affected. Phytoplankton are the "grass" of oceans, and its grazers must be very small themselves to find, efficiently capture, and readily consume really small, individual, free-floating photosynthetic organisms. Large consumers could not capture enough energy if they had to feed upon such sparse, tiny individuals. It would be like trying to feed yourself on rice by having to find each grain in a grassy lawn, then pick up and eat each grain individually with chop sticks. These tiny consumers are then preyed upon by somewhat larger organisms and so on up the food chain, and because they start with such tiny organisms, oceanic food chains tend to be very long. A typical fish may be six, seven, or more trophic levels removed from the photosynthetic producers.

OCEANIC FOOD CHAINS AND TRANSFER OF ENERGY

Nature "taxes" each step in a food chain, so the greater the number of steps an organism is removed from producers for feeding, the smaller the amount of energy available. This ecological fact very much affects the community structure of oceans. Phytoplankton, the ocean's producers, capture solar energy and use it to grow and reproduce. Some energy is lost as heat, so in general terms, up to 90% of the total energy captured by phytoplankton is used by them or lost. This means that only 10% of the energy captured is available for organisms that consume phytoplankton. A convenient means of measuring energy is to calculate the biomass, the mass of all the organic molecules composing an organism, basically the dry weight of the organism. The maximum total biomass of consumers of phytoplankton, the grasshoppers of the oceans, can be no more than about 10% of the biomass of the phytoplankton upon which they feed. And the same formula is true for these phytoplankton consumers; most of the energy they consume is used for their growth and reproduction, or is lost, so the next step in the food chain, the consumers of phytoplankton consumers, would have available only 1% of the solar energy the phytoplankton captured (10% of 10%).

Energy transfer up a food chain exerts a heavy tax on the populations of top consumers. Those organisms at the top end of food chains are relatively few in number because they obtain their energy a number of steps along the food chain, and with each step the tax on the energy originally captured grows. 10% of 10% of 10% of 10%, and so on, up a long food chain means that only a very small fraction of the total solar energy originally captured by phytoplankton is available to top consumers. This is why oceanic life, particularly large organisms, is so sparse. Since phytoplankton are thinly spread across a large surface area it takes a tremendous surface area of ocean to support just a few big organisms. An important ecological generality is that all big ferocious animals are relatively rare, either on land or in the oceans, because they are the top consumers at the end of long food chains (Colinvaux 1978). Any good fisherman can tell you that large organisms are not evenly or commonly distributed about the oceans; such large organisms live in or near oceanic oases.

Organisms feeding at a lower trophic level have access to more energy, but this poses other problems resulting in other solutions. Phytoplankton populations are numerically huge, but they are tiny cells spread thinly throughout the upper layer of water, so their consumers must be small themselves but big enough to capture phytoplankton cells. One interesting solution to the problem of feeding at lower trophic levels is to become gigantic, like whale sharks or baleen whales,[5] organisms capable of taking in massive volumes of water and sieving out countless numbers of the tiny organisms with each gulp. Even in regions of relatively high productivity, these huge consumers harvest populations of consumer organisms, like the tiny shrimp-like crustaceans called krill that are still several trophic levels removed from the phytoplankton producers. Baleen whales feed in the high latitude oceans during summer months when these oceans are productive, which forces them to migrate a great distance seasonally, something only very large, very motile organisms with big reserves of stored food can do. Quite large ocean predators, like tuna or marlin, are both relatively rare and very motile because they must move from place to place seeking populations of prey. When prey populations are locally depleted, the predators move on.

Our common and incorrect perception that oceanic life is abundant results from human interest, familiarity, and exploitation of productive places, coastal areas, estuaries, and upwelling zones, but these oases represent only a small fraction of the oceanic habitat. Still many humans perceive the oceans as a vast untapped storehouse of plenty. Anyone who thinks we can "farm" the oceans to harvest food for our ever-growing human population simply doesn't understand ocean ecology. Those few luxuriant oases where nutrient availability allows high productivity are already under intense pressure from humans. Virtually all ocean fisheries have been damaged by overharvesting. Whales have been harvested to the point of near extinction. Many formerly abundant species have been fished to a point where they no longer can be exploited economically, and marine biologists wonder if these populations will ever recover because after depleting one species, human fishermen sometimes turn their attention to species at a lower trophic level, thus reducing the food supply that could gradually restore the depleted species.

Humans also harvest sea creatures without regard for or knowledge of food chains, and so our activities produce unexpected consequences. Harvesting sharks, a top predator, allows skate populations, one of their prey species, to increase in number. Skates feed upon shellfish, like scallops, so more shellfish get eaten by skates because humans are harvesting their predators (sharks) for just their fins, which ironically are processed to make imitation scallops. The bottom line is simple. Oceans are barren places and offer no salvation for hungry humans.

Phytoplankton and zooplankton also occur in freshwater environments. Large bodies of fresh water, like the Great Lakes and Lake Baikal in the central Asian part of Russia just north of Mongolia, have light/nutrient environments very much like oceans, just a much lower concentration of salts. Smaller and shallower bodies of fresh water differ. Light penetrates to the depths of shallower ponds and lakes, and in deeper lakes during the winter the water temperature of the surface layer becomes cold enough that it can mix with the deep nutrient-rich water. Such environments can be very productive; you can judge by their appearance. If clear and blue, they aren't productive, if greenish and murky, life abounds.

HOW TO SUCCEED AS PHYTOPLANKTON

To obtain light, phytoplankton must remain in that thin, light-rich layer at the ocean's surface. Small size is an adaptation not only for dealing with the nutrient limitations explained previously but also for staying at the top. Very small prokaryotic cells may remain suspended in sea water, but larger cells tend to sink, and it gets dark down below. Small size slows the rate of sinking because generally small organisms have a higher surface area–to–mass ratio than larger organisms. Nonetheless, many phytoplankton have densities slightly greater than water, and sinking slowly is still sinking. Shape also contributes to sinking rates. Nonspherical shapes; cells with long, fine projections or protrusions; flat, disk-like cells; and flat arrays of cells can further slow sinking rates or help to optimally space the organisms (Trainor and Egan 1988) (Fig. 4–1). Species of *Pediastrum* are colonial organisms consisting of a flat array of 8, 16, or 32 cells, depending upon the species. Peripheral cells bear one or two spiky projections,

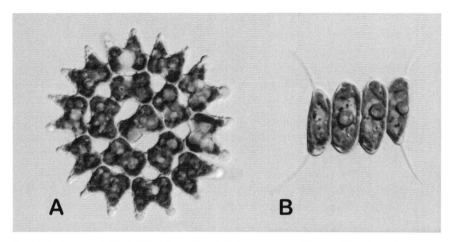

Fig. 4–1 Green algae—small colonial phytoplankton. A. *Pediastrum* shows a flat array of 16 cells, each of which is broadly H-shaped. New colonies form when one of the cells in a mature colony divides four times successively, resulting in 2, 4, 8, then 16 cells. B. *Scenedesmus* (*Desmodesmus*) colonies are commonly composed of four or eight cells. Some species of *Scenedesmus* have two spiny outgrowths on the end cells. Both organisms are magnified over 1,000×. (Image source: A, B—Used with permission of Tim Rockwell)

giving the colonies the appearance of a green snowflake. Such a shape and the projections greatly reduce the rate of sinking, just as a large, flat snowflake will gently float to Earth while an ice pellet of similar mass will continue to accelerate. Increased buoyancy also can slow sinking rates, and this can be accomplished by storing oil rather than starch as energy reserves, or by producing gas-filled vacuoles.

Lastly, many planktonic organisms, both phyto- and zoo-, are motile via flagella and they have eyespots composed of light-sensitive pigments. Since the surface of oceans, seas, and large lakes is constantly being churned, a tiny, light-sensitive organism simply swims constantly toward the brightest horizon, the surface. Such organisms exhibit a positive phototropism, always moving toward light. Even motile plankton drift with the currents; they swim to stay at the top, basically treading water. Although swimming rates vary, they average several times greater than an average sinking rate of 0.5 m per day (Sournia 1982). Some large[6] and quite motile fresh water phytoplankton like *Volvox*[7] (Appendix: Green Algae, Fig. A37) sink several meters at night to more nutrient-rich waters and then migrate back to the bright surface by day. Motility also shifts a unicell linearly through its aqueous medium, which prevents depletion of scarce mineral resources in the immediate vicinity of the cell. Flagellated phytoplankton exhibit a motility and sensory behavior usually associated with animals, which is why designations of plant or animal make little sense among such organisms. And of course, as mentioned above, some phytoplankton organisms have relatives that are zooplankton, so designations of plant or animal are especially meaningless.

Zooplankton must stay at the top, not because of sunlight but because they prey upon phytoplankton, which stay at the top. Ornamentations and multicellularity of phytoplankton, often assumed to be adaptations to slow sinking, may also be adaptations for avoiding predation. Small cells or colonies of cells made wider by cell wall projections can make an organism too big for easy consumption by equally small zooplankton. Experimentally this has been demonstrated. A small nonmotile unicellular green algal organism (*Chlorella vulgaris*) became multicellular in the presence of a predatory organism, or rather the consumer organism selected for those individuals whose asexual reproduction tended to produce colonies rather than single cells (Borass et al. 1998). In evolutionary terms, the organisms that were multicellular and too big to eat had a higher reproductive success, so this trait was passed on to their offspring. Those that remained unicellular were more likely to be consumed, and therefore fewer unicellular offspring were produced. When the selective force (the consumer/predator) was removed, the phytoplankton population shifted back to being largely unicellular. This shows that although small size has a buoyancy advantage, organisms can be subject to more than one selectional force (sinking and predation), resulting in a different optimal size, small enough to remain near the surface but big enough to avoid being eaten.

In a similar experiment, populations of colonial *Scenedesmus* (Fig. 4–1 B) increased the average size of the spines on their two end cells in the presence of predators, suggesting that the spines function to prevent ingestion by small predators without greatly adding to the colony's mass (Hessen and Van Donk 1993). The spines are less prevalent when predators are absent because individuals whose genes cause them to grow the spines anyway are spending some energy on spines that could be used for reproduction, thus putting them at a disadvantage in a predator-free environment. Such experiments demonstrate that evolution is a necessary mechanism because it allows organisms to respond to such changes in their environment, and such adaptive shifts have been documented not only experimentally but also in nature by numerous studies (Endler 1986).

Oceanic habitats cover great homogeneous expanses, large continuous areas having very similar conditions. To be sure there are many gradients, bright to dark, warm to cold, still to dynamic, but oceans are not

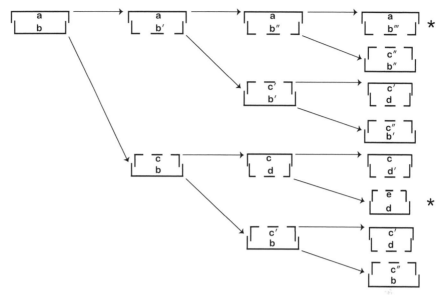

Fig. 4–2 Asexual reproduction of diatoms. The cell wall consists of two halves, a larger half (a) over lapping a smaller half (b). When the cell divides, each daughter cell inherits one of the two halves of the cell wall, and then each synthesizes a new inner half (b', c). So one daughter cell can never grow as large in volume as the original cell*. Continued cell division results in lineages of smaller and smaller cells (*e-d) until they escape this cycle using sexual reproduction and starting the cycle over.

nearly as patchy, as variable, in environmental parameters as the coastal or terrestrial environments. As a result, asexual reproduction via simple cell division predominates among planktonic organisms. Because the oceanic environment changes slowly and is homogeneous over large areas, a clone, a population of genetically identical organisms, is at no disadvantage. Among planktonic organisms, sexual reproduction is usually associated with production of a survival stage for weathering unfavorable environmental conditions or overcoming some functional limitations of their cells.

For example, diatoms have rigid glass walls surrounding their cells (Fig. 4–2). Each daughter cell gets one half of the original cell wall, and then produces a new inner half. So one daughter cell can grow as large as the original cell, but the other daughter, having inherited the inner half of the maternal cell wall, gets limited to a somewhat smaller size because they always produce a new inner half. As this continues generation after generation, diatoms get progressively smaller because they inherit the smaller side of the cell wall. At some minimal size they reproduce sexually to start the cycle over again.

Many planktonic organisms are prokaryotic, and any random bucket of seawater will contain bacteria, including cyanobacteria, not to mention countless virus. Except for parasites, almost all of the recognized groups of eukaryotic organisms are wholly or partly aquatic (Table 3–1), and since most are wholly or partly unicellular, almost all can be found as planktonic organisms. With the exception of those in deep, dark, hot places, most of the unknown organisms still to be discovered will be planktonic, including a newly discovered class of very tiny eukaryotic organisms (Van der Staay et al. 2001).

Eukaryotic phytoplankton (see Appendix) include members of the red algae, euglenozoa, glaucophytes, dinoflagellates (alveolates), haptophytes, cryptophytes, prasinophytes, and green algae (plants) and quite a number of groups in the stramenopile clade (Appendix: Phytoplankton, stramenopiles). Many of these unicellular organisms are motile, have light-sensitive eyespots, and exhibit a positive phototropism (Fig. A50). Adaptations to an aquatic environment have produced many similarities not based on common ancestry, which are called convergent evolution (e.g., the streamlined shapes and fins of penguins, seals, dolphins, and fish). Convergences also occur in phytoplankton because they are adapted to the vast open ocean habitats. Although photosynthetic, many of these unicellular organisms are facultative autotrophs and can live as heterotrophs if an appropriate supply of food molecules to absorb or food materials to consume is available. Even though they have this ability, which was predicted by the endosymbiont hypothesis of chloroplast origin, no one knows how often facultative photosynthetic organisms live as heterotrophs in nature. They probably do whatever conditions allow.

Although life is small and thinly spread in the oceans, the immense area of oceans means these tiny organisms have a tremendous impact on life as a whole. Terrestrial life is neither phylogenetically nor ecologically divorced from oceanic life. Although all large organisms have a common ancestry with unicellular plankton, phytoplankton and zooplankton continue to exist because their habitat continues to exist. While they are simple in organization, it is wrong to think of these tiny organisms as primitive; they are adapted to the open ocean environment, and organisms like these have occupied these habitats for over 2.5 billion years. On several occasions different groups of phytoplankton have invaded and adapted to the coastal environment's diverse habitats, which presents different challenges and requires different adaptations. So let us go down to the coast and sea, and see what these might be.

FIVE

Down by the Sea (-weeds)

Wherein coastal environments are contrasted to oceanic environments, and organisms adapt to the new challenges presented by living on coasts by becoming anchored, larger, and multicellular, which in the case of green organisms results in those algae called seaweeds.

Every time we walk along a beach some ancient urge disturbs us so that we find ourselves shedding shoes and garments or scavenging among seaweed and whitened timbers like the homesick refugees of a long war.
—Loren Eiseley

Our human fascination with and enjoyment of sea coasts is quite evident. As much as we enjoy seaside recreation, we generally fail to recognize the biological significance and uniqueness of coasts as the interface between aquatic and terrestrial habitats. Both coasts and oceans may be considered part of marine environments, meaning they both involve salt water, but the coastal environment is much different than the oceanic environment explored in the previous chapter. In comparison with vast oceanic habitats, the coastal environment is shallow, very small, very long, and extremely narrow, although its width varies inversely with the steepness of the land-water interface. To a biologist, the coastal environment spans the area from the highest place on the shore that water regularly reaches to those portions of the coast reached by the deepest penetration of light in an amount sufficient to allow photosynthesis, about 30 m in depth (Fig. 5–1). From the human perspective, the extremely small size of coastal habitats accounts for the high cost of coastal property; this limited commodity is sought by many, so the supply is low and the demand high. As you will see, things are not so different for coastal organisms; space is

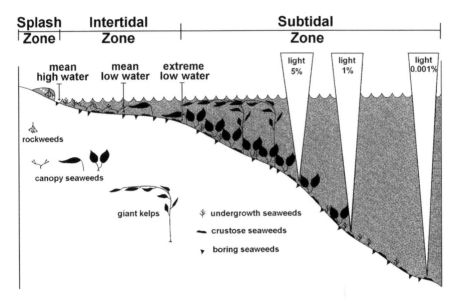

**Splash
Zone** | **Intertidal
Zone** | **Subtidal
Zone**

mean
high water | mean
low water | extreme
low water

light
5%

light
1%

light
0.001%

rockweeds

canopy seaweeds

giant kelps

✤ undergrowth seaweeds

⌐ crustose seaweeds

▾ boring seaweeds

Fig. 5–1 Coastal zonation of seaweeds showing a splash zone above the mean high water mark, an intertidal zone, and the subtidal zone, which is zonated further by decreasing light levels with increasing depth. (Image source: After Lüning 1985).

severely limited, and this has a quite significant impact on the biology of organisms that live there.

The coastal environment is composed of many small and heterogeneous habitats; in other words, coasts are patchy. The patchiness of the coastal environment is derived from several factors. The coastal environment is divided vertically into horizontal zones of various widths depending upon how steeply the land meets the water; the steeper the interface, the narrower the coast and its zones. At the upper end of the coastal environment is the intertidal zone that constitutes an area between the levels of the lowest and highest tides. Intertidal zones are inundated and exposed twice daily with the rising and falling tides, the regular sloshing of water caused by the Earth's rotation and lunar gravity tugging on the oceans. Organisms living in the highest intertidal zone are exposed to air longer and inundated more briefly than those living in the lowest intertidal zone, and so the intertidal zone itself represents a number of horizontal life zones. In some places, because of the land-ocean topography, the magnitude of tides is slight and in other areas, the tides rise and fall rapidly over a distance of several meters.[1] Above the intertidal zone is a variable splash zone wetted only by wave spray. Below the intertidal zone, continually inundated zones of increasing depth exist where light intensity provides a more gradual zonation because it becomes less intense and more limiting as the depth increases. In addition to tides, waves regularly pulse in and out, and currents push or pull water toward or away from the coast. Differences in substrate, ranging from rocky to sandy to muddy, further contribute to the patchiness.

All these factors create numerous small habitats, very close together in a dynamic environment. An environment with so many different habitats presents both new opportunities and new problems for organisms living there. Patchiness means a diversity of habitats exist to which organisms can adapt. Organisms adapted to one particular type of patch must stay put and this is difficult where water is always coming and going, pushing and pulling them. A patchy environment also means space is a truly limited resource. Each patch is only so big, and only so many patches of a particular type exist. Variability among patches means a continuum of habitats exists of varying quality for each particular species, and because the most suitable patches are a very limited commodity, competition for such spaces will be intense. All organisms, it seems, must pay a high price for coastal real estate.

Compared with the open ocean environment, coastal environments have abundant light, carbon dioxide, and mineral nutrients. Nutrients are relatively plentiful because decomposition takes place within these communities and even more nutrients wash in from adjacent terrestrial areas. And of course, it's easier to stay in the well-illuminated portion of the water column because even if you sink to the bottom, light still remains available, meaning that on the coast green organisms are no longer constrained in size by lack of nutrients and the need to float. The dynamic action of waves mixes air and water, so carbon dioxide is readily available. The two biggest challenges are locating a suitable patch, and having located a good patch, staying there. These challenges are met by two adaptations: (1) producing lots of offspring to disperse, increasing the odds that some offspring will find a suitable patch; and (2) anchoring to ensure an organism stays in a favorable patch once a good bit of space is acquired.

Genetically diverse offspring can better deal with a very diverse environment because individuals with slightly different characteristics can find slightly different patches to which they are well adapted. This places an advantage on sexual reproduction. Asexual reproduction remains valuable because once an organisms is anchored in a favorable patch, spreading clonally is a good means of competing with neighbors and securing more of this limited resource.

Coasts were not barren before seaweeds and coastal animals appeared; shallow water lagoons and bays had been home to prokaryotic mat- and

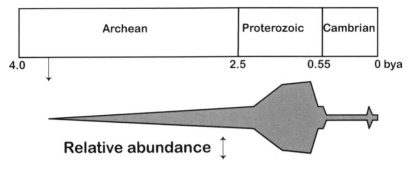

Fig. 5–2 Relative abundance of stromatolites through geological time. From their first appearance some 3.5 bya (billion years ago) the abundance of stromatolites increased until it experienced a sharp decline in the late Proterozoic. (Image source: After Tucker and Wright 1990)

stromatolite-forming communities for nearly 3 billion years, and such communities remained common until late in the Proterozoic (Fig. 5–2). Although eukaryotic organisms appeared more than 2 billion years ago, the appearance of eukaryotic organisms adapted to coastal habitats is a much more recent event, one that is associated with increased size. No one knows why eukaryotic organisms took so long to invade the coast or even when they did. Prokaryotic mat communities are well protected against both drying and UV exposure by gelatinous slime, but eukaryotic organisms do not have such slime; perhaps much of the coastal zone remained uninhabitable to eukaryotic organisms until enough oxygen accumulated in the atmosphere to form an ozone layer to filter out some of the damaging UV radiation. Ozone screening takes a considerably higher oxygen concentration than the metabolic oxygen crisis that favored the endosymbiotic origin of mitochondria. The time required for this additional oxygen accumulation may account for the lag between the appearance of eukaryotes and the invasion of coastal areas by eukaryotes. Although the decline of stromatolite-forming mat communities is commonly attributed to the appearance of mat-grazing invertebrate animals, the earliest large organisms to appear in the fossil record are red algal seaweeds. At 2–5 cm in height, such seaweeds would tower over bacterial mat communities, easily winning a competition for light; thus the decline in stromatolites is just as likely to be due to competition with seaweeds.

HOW TO MAKE A SEAWEED

Seaweeds always were presumed to have a common ancestry with organisms that function ecologically as phytoplankton because small and unicellular to large and multicellular made developmental sense, and phylogenetic studies seem to bear this out in all instances. So how do free-floating or motile, mostly unicellular organisms give rise to organisms that would be called seaweeds? The evolution of seaweeds involves three events: (1) colonization of the coastal environment by phytoplankton, (2) a shift from unicellular to multicellular, and (3) a shift from largely asexual to mostly sexual life cycles that involve a drifting and/or motile planktonic dispersal phase.

By this definition, the simplest seaweed is an anchored unicellular organism, although at least some of these unicellular seaweeds can be surprisingly large and complex with a distinctly plant-like form. Water movement and the abundance of light, mineral nutrients, and carbon dioxide more than compensate for the loss of motility. Seaweeds are primarily members of the red, green, and brown algae, and while adaptations to the coastal environment evolved independently in each lineage, and probably more than once in the green algae, some generalizations can be made.

The first step in the transition from oceanic phytoplankton to coastal seaweed is accomplished when a formerly free-drifting unicellular organism invades the coastal environment by anchoring itself to a substrate. Motile cells can adhere to substrates by their flagella. If you have ever observed flagellated cells under a microscope, you may have chanced to observe a cell spinning because it was "stuck" to the cover slip by flagellar adhesion. The cell spins because the rotary motor of the stuck flagellum remains running while the flagellum is held in place. Motile unicells could simply anchor themselves to a substrate by flagellar adhesion followed by the development of a more substantial anchor. Dispersal in many seaweeds employs a motile planktonic stage called zoospores (ZOH-oh-spoars) that use flagellar adhesion to initiate anchoring to a substrate.

All seaweeds have a holdfast, an aptly named adaptation for anchoring the organism to its substrate. A holdfast can take several forms and can function in at least two different ways (Fig. 5–3): either anchoring the organism using rootlike rhizoids or cellular outgrowths, or secreting a substance that cements the seaweed to its substrate. Rhizoids and cement can be used in combination as well. Clearly as seaweeds grow into larger organisms or occupy more dynamic coastal habitats, holdfasts must become larger and must turn into stronger, rootlike multicellular structures themselves. Solid substrates offer the best anchorage sites, and no large seaweeds can grow on sandy or muddy sediments unless, like coral sands in a reef, they are protected and in areas with only gently moving water. Those small seaweeds that do anchor themselves in sediments

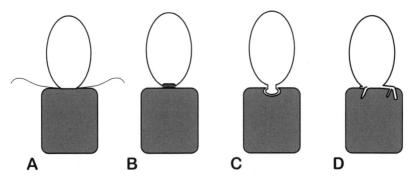

Fig. 5–3 Illustration of different types of holdfasts. A. Flagellated cell adhering to a substrate (gray). B. Cell secreting an adhesive holdfast. C. Dovetailing outgrowth locks cell into irregularity in substrate surface. D. Rhizoidal outgrowths of cell grasp surface of substrate. (Image source: The Author)

often grow as short clumps or tufts using rootlike rhizoids to penetrate the substrate. Some produce a network of horizontal axes bearing short aerial branch shoots or "leafy" blades that arise periodically from their upper surface and anchoring rhizoids that grow down into the substrate from their lower surface.

Shifting to a sessile lifestyle allows an organism to stay in a favorable patch. Being sessile provides new opportunities and also imposes new constraints and problems. Once anchored, a seaweed is free of the size constraints imposed by the oceanic environment. More raw materials can be obtained by an increased surface area, so now a larger size is favored by natural selection. Seaweeds spread thin, leaf-like sheets of cells (blades), or their largely filamentous bodies branch again and again, generating lots of surface area. The larger seaweeds have light-capturing arrays of lateral axes or leafy blades arranged along a main axis for optimal light harvesting. Rather than functioning like the stem or trunk of a land plant, the main axis of a seaweed, the stipe, functions as a tether between a massive holdfast and gas-filled floatation bladders at the base of blades. This places the stipe under tension rather than under compression. Rather than being massive, the stipe of a seaweed requires tensile strength like a kite string.

Yet here in an adaptive response to the coastal environment is found the first appearance of what can be called "plant form," an organism anchored by rootlike structures and bearing leaf-like structures upon a vertical or horizontal axis. Did you already guess that land plants had such ancestors? Well, that was completely logical but wrong. Land plants developed a similar form but separately from different algal ancestors and in response to somewhat different demands.

Many of the animals that adapted to the coastal region became sessile like seaweeds did. Such sessile animals, particularly those that feed by capturing drifting food particles or organisms—for example, anemones, corals, sea feathers, sea pens, sea fans, and hydra, can be described as "plant-like," a convergence in form caused by similar demands upon a sessile organism to produce considerable surface area for the capture of something diffuse, in this case small, drifting prey organisms.

GETTING BIGGER

A limited commodity needed by many generates competition. So, small patchy habitats generate crowding and competition among neighboring organisms for space, and for green organisms space equals light. Increased size can be an advantage in some types of competitions, like when 22 guys all want just one football and they disagree upon the direction it should move. For an anchored unicellular organism, two strategies exist for getting bigger to get more space and light: increase your cell size or become multicellular. Writing such statements makes it sound like the organism has and makes a choice, but that is not how evolution works. Natural selection must operate on inheritable variations that result in larger cells or a multicellular organism. If a particular organism has a genetic basis for somewhat larger cells and as a result leaves more offspring, then somewhat larger cells become more common in the population. As demonstrated in the last chapter, under selection pressure of predators, a unicellular organism became multicellular. With so many different types of organisms involved, it should come as no surprise that nature has tried both strategies with some success.

To attain a larger cell size, why not just balloon out, so to speak, grow outward and get bigger in all dimensions? As cells grow bigger, increasing both their surface area and volume, they quickly reach both functional and physical limitations, which can counter the selection pressure for larger size. The cell membrane is the interface between a living cell and its surrounding environment. All the materials the cell needs must pass across the cell membrane, so a functional ratio must be maintained between the surface area of cell membrane and the cell's volume of cytoplasm. When a cell with a roughly regular shape doubles its diameter, its cell membrane surface area would increase by a factor of four (2^2), but its volume would increase by a factor of eight (2^3). At some point the surface area of the cell membrane becomes a limiting factor, incapable of transporting enough of some basic resource to supply the volume of cytoplasm. Imagine a store with a single revolving door. Only so many customers can come and go in a given amount of time, so if the store gets too big not enough customers can get in and out. Since the cell membrane is a very thin, flexible boundary, its strength places a physical limitation on cell volume as well. The principle is simple: small water balloons are sturdier and more stable than large water balloons. Larger cells also help green organisms obtain more of the dilute, diffuse resources they need, so to some extent larger cells would be much favored.

One solution for having larger cells is to reduce the amount of metabolically expensive cytoplasm, which is largely protein. Having less cytoplasm helps maintain a functional ratio with the surface area of the cell. A cheap solution is to fill the volume of the cell with a large vacuole, a membrane-bound sack of water, leaving only a small volume of cytoplasm around the periphery. Many algal and most land plant cells have such a central vacuole. Over time, a central vacuole has taken on other functions—for example, housing water-soluble pigments and as a compartment to house metabolic waste materials. But a big water balloon inside another big water balloon is not going to impress any structural engineer.

Another part of the solution to having structurally sound larger cells is to reinforce the cell membrane with a cell wall. Cell coverings among unicellular organisms are pretty diverse, ranging from glass walls to assorted mineral plates to bands of proteinaceous materials to various carbohydrate polymers, of which the best known is cellulose. Cellulose is a polymer of glucose molecules but with the monomers (glucose sugars) arranged differently than in that other familiar glucose polymer, starch, the familiar food storage molecule and our primary source of sustenance. Long, thin molecules of cellulose are woven together, sometimes in an orderly lattice-like array, to form microfibrils, which are matted together to make a strong, thin, flexible cell wall to help stabilize the cell membrane and keep it from bursting. Together, a central vacuole and a cell wall allow a cell to be larger while functionally and structurally sound, and these cellular features, considered typical of land plant cells, were inherited from algal ancestors. Such cells have a greater surface area for absorbing dilute resources, but such an adaptation comes at a cost. Large, anchored, water-filled cells in a cellulose box have given up mobility. The evolutionary responses of algae competing for space and light in the coastal environment produced the first key characteristics of plants: photosynthetic organisms "rooted" in place consisting of larger cells with a central vacuole and a cellulosic cell wall.

Up to a point, seaweeds can become bigger and can develop more complex forms without being multicellular. A few notable green algal

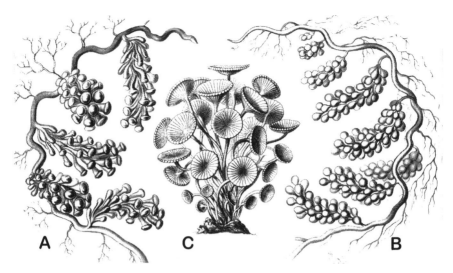

Fig. 5–4 Complex form in unicellular green seaweeds (A-B) *Caulerpa* and (C) *Acetabularia*. *Caulerpa* has horizontal axes (here artistically rendered upright) up to 3 mm in diameter that can be up to a meter long from which branched rootlike rhizoids and "leafy fronds" arise. These complex forms are basically tubular in construction, containing thousands of nuclei in a protoplast undivided by cell walls, so they are technically unicellular. This seaweed has become an exotic invasive weed in some areas when dumped from aquaria. This cluster of *Acetabularia* is composed of parasol-like seaweeds, each a single cell. A basal portion of the cell (not shown) produces rhizoidal outgrowths that function as a holdfast. Radiating branches are formed atop a filamentous stem-like portion. These cells stand 4–5 cm tall. (Image source: After Haeckel 1904, Wikimedia Creative Commons)

seaweeds are surprisingly large and complex looking although they remain unicellular organisms (Fig. 5–4). The strength of their cell walls is probably the primary limitation on their size, so most live in relatively quiet, shallow tropical waters. Although at least two orders of magnitude separate the largest unicellular seaweed from the largest multicellular seaweeds, nonetheless seaweeds like *Caulerpa* (Fig. 5–4 A-B) represent one of the largest, most complex unicellular organisms known. The other large complex unicellular organisms would be Ascomycete (as-koh-MY-seat) molds,[2] filamentous organisms lacking cross walls. Both of these types of organisms have many nuclei, and a slender, elongate, cylindrical cell maintains a constant surface area–to–volume ratio. Of collateral interest developmentally, *Caulerpa* has rootlike and leaf-like portions and yet without any cellular boundaries between. Clearly up to a point a larger size and complex form does not require a multicellular condition. This is quite at odds with traditional cell theory (Kaplan and Hagemann 1991), which assumes multicellularity precedes complex form. Presently the developmental basis for such organisms remains unknown. Nonetheless the more common route to larger size has been via multicellularity.

Multicellular organisms avoid the functional constraints of a very large cell size by being composed of numerous smaller cells, whose individual size is well within functional constraints. Rather than getting bigger by pushing the limits of cell size, multicellular organisms combine many small cells to generate larger size. Multicellular organisms range from those with two or four cells to those composed of millions and millions of cells. In simple multicellular organisms, all cells are similar in structure and function. The simplest multicellular condition is colonial, where a group of identical cells lives together more or less as one organism, and many phytoplankton organisms are colonial, so multicellularity was not invented by seaweeds. In many colonial organisms the total number of cells is determinate (fixed), and the colony or organism can reach only a certain size. For example, a 16-celled colony of *Pediastrum* (Fig. 4–1) reproduces asexually when one cell in a colony divides four successive times, producing 16 cells that form into a new daughter colony within the parental cell wall. When released they grow into an adult size, but no new cells are added in the process. When cells retain the ability to divide as the organism grows, the number of cells and the size of the organism is indeterminate. It is wrong to think of unicellular organisms as simple at the cellular level. Unicellular organisms are both cells in function and organisms "by virtue of molecular gradients and structural polarity" (Korn 1999); thus unicellular organisms occupy two different levels of complexity (cell and organism) simultaneously. As cells become specialized in function they tend to become simpler. A unicellular organism must by necessity be a Swiss army knife of cells, capable of performing all life functions, while in comparison a specialized single-function tool like a bottle cap remover is quite simple. The only example of cell specialization provided thus far is the heterocysts in filamentous cyanobacteria, a cell simplified for nitrogen fixation.

Even the simplest multicellular seaweed, a filament, has one specialized cell that functions as a holdfast. As multicellular organisms get larger, it leads to more specialized cells. Basal and central portions of seaweeds develop thicker cell walls or rhizoidal outgrowths to strengthen the organism, while thinner-walled, peripheral cells continue to function in photosynthesis and in growth. Basal holdfasts and lower portions are shaded by their own light-harvesting portions, so food molecules must be moved from photosynthetic portions to interior cells and basal portions of the organism. In central regions, longer cells or cells with long rhizoidal outgrowths have shapes that help move materials. Natural selection accomplishes such specializations by acting on variable cell shapes.

Multicellularity affects reproduction as well as form. When unicellular organisms participate in sexual reproduction, the individual ceases to exist because the whole organism functions as a gamete. Multicellular organisms with specialized reproductive cells can produce gametes and still survive to reproduce again. Also, a multicellular organism can produce more gametes and potentially more offspring because they can have many cells involved in reproduction. Lastly, because multicellular organisms can grow to sexual maturity and can survive reproduction, multicellularity is correlated with longer individual life spans. This is in stark contrast to asexual reproduction, which leads to short-lived individuals[3] but long-lived genotypes, clones.

Most unicellular organisms have a distinct polarity such that anterior and posterior portions of the cell can be readily identified, and in virtually all such organisms, cell division is parallel to this polar gradient. This way the two daughter cells each receive the polarity of the parental cell. If simple seaweeds had such a unicellular phytoplanktonic ancestor, then after finding a suitable coastal habitat and anchoring itself to the substrate by its anterior end, a series longitudinal divisions, parallel to the cell's polar gradient, would have resulted in a multicellular organism, a prostrate seaweed (Fig. 5–5 left). The plane of division is the same as in a unicellular organism, but the earlier onset of cell wall formation results in the multicellular condition. However, very few multicellular organisms develop this way. The cell divisions that take place during the early developmental stage of large organisms are usually perpendicular to an internal cellular polarity. A reorientation of the plane of division perpendicular to the polar gradient (a transverse division) produces a filamentous seaweed, one that has a holdfast at one end and a dividing apical cell at the other (Fig. 5–5 right). This suggests that one purpose of multicellularity is to establish a greater polarity from one end of the organism to the other than possible in a single cell (Korn 1999). In a very dynamic habitat, the prostrate seaweed is less likely to be damaged or dislodged by moving water, but in a competition for light, the filament can shade a prostrate seaweed (Fig. 5–5, gray triangle). Neither form is an absolute winner; it all depends on the environment. Many complex seaweeds and even land plants like moss start out their development with a filamentous growth. The prevalence of apical growth among land plants indicates this type of growth and development had tremendous potential.

To advance beyond simple filaments or colonies, to grow considerably larger, a degree of integration is also needed. Cellular communication and

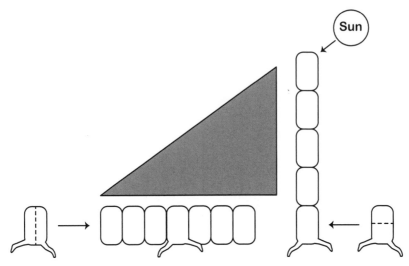

Fig. 5–5 Competition for light between prostrate (left) and erect (right) simple seaweeds. Development of each form arises from an initial cell dividing either in a longitudinal (left) or transverse (right) plane relative to the polarity of the cell. Although the prostrate seaweed occupies more space in competition for light, the erect, filamentous seaweed wins by casting a shadow (gray triangle) over the prostrate seaweed. (Image source: The Author)

interaction require molecular transmissions, signals, passing from one cell to another. When two cells are separated by a compound cell wall composed of the wall of each cell and an adhesive layer between them, a signal must diffuse across this barrier. Some loss of message into the surrounding aqueous habitat is inevitable. Direct cellular interconnections across cell walls are called plasmodesmata (plaz-moe-DEZ-mah-tah), and they are very fine cytoplasmic channels between cells that develop in some brown algae, some green algae, and all land plants. Cytoplasmic connections also form between cells in some species of *Volvox*, one of the largest and most complex colonial organisms. In such organisms the cytoplasm is one continuous unit, although still housed in individual cells.

How does a unicellular organism acquire the instructions to become multicellular? When you are studying genetics, genes for specific traits or characteristics—for example, tall or short pea plants, yellow or green pea seeds, wrinkled or smooth pea seeds—are the focus, but the great majority of a genome consists of developmental genes, which provide instructions controlling growth and development. Thinking of development as a set of instructions, something like a computer program, is useful conceptually, even if an inaccurate description of the mechanism. Unicellular organisms have a fairly simple developmental program that can be described as follows: when large enough, divide into daughter cells, separate, synthesize new cell wall material, and grow until you are big enough to divide again. Only a very minor developmental shift, an earlier onset of cell wall synthesis, is needed to produce multicellularity. After dividing, daughter cells usually separate before synthesizing new cell wall material, but if the two daughter cells begin secreting cell wall material while remaining in contact, the cell wall material effectively cements them together. Multicellularity results from a shift in the timing of events, not from new developmental instructions. Rather than requiring a whole new set of instructions, developmental shifts can occur from natural selection acting on preexisting variations in the onset of cell wall synthesis. Shifting the timing of developmental events can have some profound effects although the genetic changes are relatively minor. Such developmental shifts occur in nature for other reasons. An example was given in the previous chapter where the unicellular alga *Chlorella* became multicellular in the presence of a predator and thus became too big to consume. Remove the predator, and selection for smaller size shifts *Chlorella* back to unicellular.

An unbranched filament, a linear row of identical cells, can grow and add new cells in two different ways. Each of the cells in the filament may be capable of dividing when large enough, thus producing an intercalary (in-TURK-a-lair-ee) growth, or cell division may be restricted to the endmost cell, thus producing an apical (A-pick-el) growth, a product of having a greater cell-to-cell polarity so the organism "knows" which end is which. Under conditions where growth rate is important, intercalary growth can be an advantage because the filament is adding new cells to its length at many points in any given period of time. But new daughter cells have thinner walls, so new cells produce a weaker zone in the filament. This may not matter in a fairly quiet water environment, but the filament could fragment if subjected to the forces of moving water. In apical growth only the terminal cell divides, so new cells are always added at the far end of the filament and the cells get progressively older, thicker-walled, and stronger towards the base. Although slower than intercalary growth, apical growth results in a stronger filament. Again, environmental conditions will dictate which developmental program will be more successful. All land plants grow apically, a condition they inherited from their green algal ancestors.

Filaments become branched when divisions parallel to the filament's long axis are combined with transverse divisions. Branching combined with apical or intercalary growth produces very different forms (Fig. 5–6). Even when all the branching is apical, small variations in the ratio of transverse to longitudinal divisions result in very different forms (Fig. 5–7 left). The apical cell may also divide equally or unequally, which produces

Apical growth

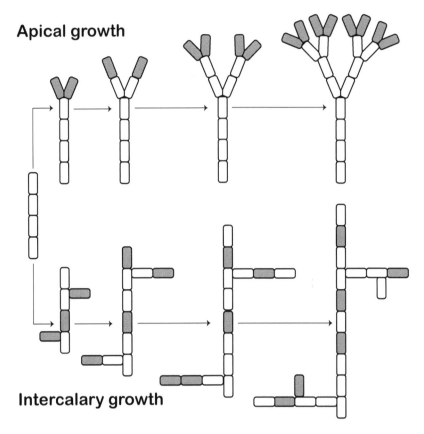

Intercalary growth

Fig. 5–6 Apical versus intercalary growth and branching of filamentous algae. Apical growth (top) limits all cell division to the apical-most cell (both transverse divisions, which adds new cells to the filament's length, and longitudinal divisions, which causes the filament to branch). When growth and branching is intercalary (bottom), both new cells and branching can occur anywhere in the filament. The forms produced are very different although both involve the same two types of divisions. Gray cells show cells about to divide. (Image source: The Author)

bigger main filament cells and smaller branch filaments (Fig. 5–7 right). This illustrates how small variations in a small set of developmental instructions can produce many diverse forms. Some rather large, complex-appearing seaweeds—for example, red algae (Fig. A51)—are fundamentally filamentous and develop very diverse forms through different combinations of a relatively few basic patterns.

When a third plane of division is introduced at right angles to the transverse and parallel divisions, a 3-dimensional tissue with length, width, and depth can be produced. Such a form is quite strong because of the cell walls intersecting in three dimensions. Such block-constructed

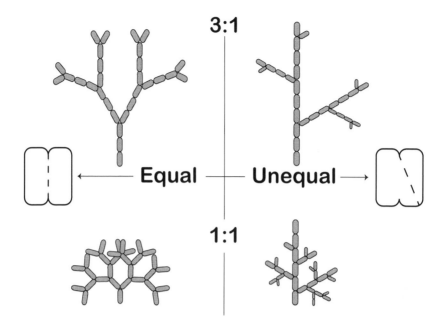

Fig. 5–7 Different forms generated by equal and unequal apical cell division and different ratios of transverse and longitudinal cell divisions. Changes in two variables, equal division (left center) versus unequal division (right center) of the apical cell, and different ratios (3:1 top, 1:1 bottom) of transverse divisions, which add new cells to the filament, to longitudinal divisions, which branch the filament, produce very different forms. Equal division of the apical cell produces a dichotomous branching pattern (left). Unequal branching produces monopodial forms (larger main axis and smaller branch axes) reminiscent of trees. More branching relative to growth in length produces forms that more densely cover an area, while less branching produces open forms that occupy more space. Each of the 3:1 filaments branches 3 times; each of the 1:1 filaments branches 4 times. By varying just two parameters, organisms can make different forms adaptive in different environments. (Image source: The Author)

tissue is called parenchyma (par-EN-keye-mah), and such 3-dimensional tissue becomes the mainstay for constructing larger seaweeds and all land plants.

The green algal seaweed *Ulva*, sea lettuce, begins development as a simple anchored filament, which then becomes 2-rowed by divisions parallel to the long axis of the filament. Another set of divisions parallel to the long axis of the filament, but at right angles to the former, produce a second layer of cells. Cells keep dividing longitudinally until a broad 2-layered sheet of cells has formed (Fig. 5–8). Some of the central cells produce long rhizoidal outgrowths that grow downward between the two cell layers providing some strengthening of the blade. Similar rhizoidal

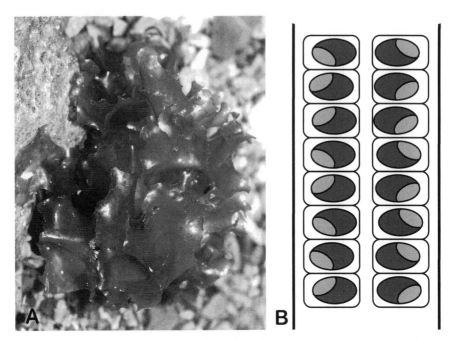

Fig. 5–8 General form and cellular organization of *Ulva*. A. This seaweed, *Ulva lactuca*, commonly called sea lettuce, consists of an irregular, broad, thin, bright-green blade. The specific epithet, *lactuca*, refers to the generic name of lettuce, *Lactuca*. The basal portion forms a small rhizoidal holdfast. B. The blade shown in diagrammatic cross-section consists of two layers of cells. One cup-shaped chloroplast is found in each cell. (Image source: A—Courtesy of H. Krisp, Wikimedia Creative Commons; B—The Author)

processes also produce the holdfast. However, because the cells lack plasmodesmata, *Ulva* is not constructed of true parenchyma. Even though *Ulva* is one of the most complex green algal seaweeds, it requires only a few developmental instructions beyond a filament.

Multicellularity is not a singular evolutionary event; multicellular organisms have evolved from unicellular ancestors many times in the distant past. A fossil of what appears to be a red algal seaweed dates to about 2.1 bya, and this represents the oldest large—that is, visible to the naked eye, multicellular organism known (Han and Runnegar 1992). Large is a relative term, but in comparison with microscopic algae, a seaweed 4–5 cm tall is huge. Small multicellular algae, both phytoplankton and seaweeds, undoubtedly predate macroscopic seaweeds, but the fossil record is mute about this. The period of time between the appearance of eukaryotic cells/organisms and the first large multicellular organism is a very long time, at least 500 million years. The entire history of terrestrial life fits into a similar timeframe. General biology books so often refer to the Cambrian explosion, the sudden appearance of many different types of large, conspicuous organisms, that people sometimes think this is literally true. As this seaweed fossil indicates, the history of large conspicuous organisms is actually almost four times older than the Cambrian, although the fossils are much less common and more difficult to interpret. Critics of science have often seized upon the misconception that all large, multicellular organisms appear suddenly in the Cambrian to argue that the fossil record fails to support the concept of evolution. The only failure is in their knowledge of the fossil record.

Once you have multicellular eukaryotic organisms, not all that much changes from that point to the present sizewise. On the basis of the volume of organisms calculated from fossils, living organisms as a whole increased greatly in size in two major episodes, one that took place during the early Proterozoic, more than 2 billion years ago, and one that took place during the late Proterozoic leading up to the Phanerozoic, the Cambrian "explosion" (Fig. 5–9). As a result of evolving the eukaryotic

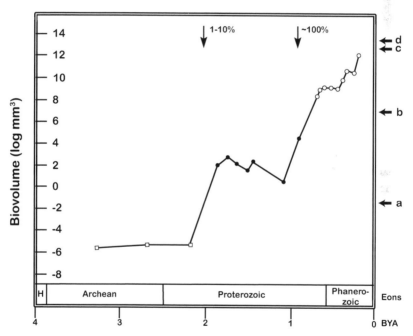

Fig. 5–9 Overall maximum sizes of fossils throughout Earth history for prokaryotes (squares), single-celled eukaryotes (black dots), and multicelluar organisms (circles). Arrows (top) show when the percent of oxygen in the atmosphere reached 1% to 10% and near 100% of present levels. Arrows (right) show the sizes of (a) the largest prokaryote, (b) the largest unicellular eukaryote, and the present day's largest animal, (c) blue whale, and largest plant, (d) giant sequoia. Size scale is logarithmic; each unit is an order of magnitude, a factor of 10, larger or smaller. Time scale in billions of years ago (BYA). (Image source: After Payne et al. 2009).

cell and multicellularity, organisms have increased in size by about 16 orders of magnitude (factors of 10) (Payne et al. 2009). These two episodes also roughly correspond to increases in the percentage of atmospheric oxygen to metabolically relevant levels (~1%) and to near-present levels. For plants, this size increase starts right in the coastal habitat with seaweeds, but it culminates with woody plants (trees), the largest organisms that have ever lived (chapter 8). But all of this still requires an invasion of land, so back to the shore.

STAYING WET

Limited space and competition would result in organisms exploring every available bit of real estate including the highest margin of the coastal environment, the intertidal zone. No one knows when this portion of the coast was first colonized, but modern stromatolite communities can occupy the intertidal zone. Perhaps the intertidal zone only became habitable, even for cyanobacteria, after a significant accumulation of oxygen in the atmosphere produced an ozone layer to filter out damaging ultraviolet radiation. Until then exposure was a double whammy of desiccation and sunburn. By the Silurian, when the first large terrestrial plants were appearing, coastal seaweed communities were well developed.

The intertidal zone presents special challenges to colonization. Sessile organisms living in the intertidal zone must stay hydrated when exposed during the twice-daily low tides. If you have ever clambered about on seaweed-covered rocks at low tide, you are well aware that seaweeds are very slippery. To coat their exteriors, seaweeds produce a number of colloidal substances, mostly carbohydrate polymers, macromolecules that bind water into a gel. Some of these substances are incorporated into the cell walls and others coat their surfaces. Agar, the gel used to prepare microbial growth media, is a polysaccharide[4] derived from red algal seaweeds (Appendix: Red Algae). These colloidal substances help maintain a hydrated coating on the seaweed's surface during low tides, and it is these gels that make seaweeds slippery. The gels get hydrated during high tides and then retain enough water to keep the seaweeds from dehydrating during low tides. Those seaweeds living in the upper reaches of the intertidal zone have the longest exposure. Gelatinous surface layers dissolve away on their outer surface and are replaced by new secretions from within the cells they coat. Many seaweeds living below the intertidal zone also have a thin gelatinous coating because the dissolution of such coatings has the effect of removing epiphytic organisms like bacteria, anchored protists, and diatoms that tend to colonize their surfaces. This may have been the original function of colloidal cell wall materials, and the rigors of tidal exposure would have selected for those seaweeds with enhanced production of gels.

If you think the intertidal zone is a tough place to stay wet, consider the problem for land plants. A gelatinous coating does not function well on land because colloids need frequent drenching to stay hydrated. In some wet tropical places, some parts of some lands plants produce a protective layer of colloidal gels—for example, the aerial roots of epiphytic plants like *Philodendron*. As they grow down from high in a tree crown to the ground below, these roots produce a slimy-gelatinous coating over the root apex. These roots can be the diameter of your little finger, and when hydrated by the frequent showers, their gel jackets bind a considerable amount of water about the root. Sometimes when walking in the rain forest you are abruptly reminded of this when something wet and slimy smacks into your cheek. The colloidal coating on aerial roots also protects the tender meristem from foraging animals, and it certainly functions as a means of capturing and holding water adjacent to the absorbing region of the root, which is just proximal to the apical growth region. Such coatings work only where frequent rainy inundations occur.

FROM THE SEA TO LAND

Ocean to coast to land presents a logical path, but the invasion of land by plants is not quite so simple. Seaweeds come in three colors, green, brown, and red, one for each of the three major lineages of algae, but land plants come in only one color, green. Neither the brown nor red algae have any terrestrial relatives even though some land plants possess purple-red pigments in addition to chlorophylls. All land plants are grass green and their chloroplasts contain chlorophylls *a* and *b*.[5] Land plants also have cell walls composed of cellulose, just like the green algae. Seaweeds are well adapted to the coastal environment, but seaweeds did not invade land directly up the shores, and most never adapted to habitats beyond the furthest reaches of sea water.

What prevented brown and red algae from invading land? Such questions are impossible to answer with certainty, but several factors probably played a role. Part of the answer involves the fact that organisms invading land are also invading air and fresh water. All three lineages have chloroplasts with different complements of pigments and different storage products. The accessory pigments of red and brown algae easily bleach out in reduced salinity and increased light, so no brown algae and only a few small red algae ever adapted even to fresh water. Chlorophyll is more resistant to bleaching; phycoerythrins, the accessory pigments of red algae, are the least resistant. Both red and brown algae incorporate gelatinous colloidal substances into their cell walls that must be regularly hydrated, and these are poorly suited to the terrestrial environment. The purely cellulose cell walls of green algae present no such problem, although further adaptations to resist desiccation are necessary. Red algae remained fundamentally filamentous and never developed plasmodesmata and true block-organized tissue, so they lack the structural basis for a more robust form. All together, such factors could limit the potential of red and brown algae to invade freshwater and/or land.

If land plants did not invade land directly from seaweed-covered coasts, then the ancestors of land plants must have taken some other route. A big clue to solving this problem resides in our knowledge of green algal diversity. Although botanists have been quite certain that the ancestry of land plants would be found among the green algae, they are a large and diverse group of organisms with a complex evolutionary history. Only one small group of freshwater algae, the charophyceans, shares a common ancestry with land plants. This group includes a diversity of organisms from motile unicells to filamentous algae to the largest of freshwater seaweeds, *Chara* and *Nitella* (Appendix: Green Algae) (Lewis and McCourt 2004). Since land plants have a common ancestry with freshwater green algae, this suggests the invasion of land came about via fresh water, the rivers and streams formed by freshwater runoff of the land. This idea will be pursued in the next chapter.

When considering past events one is tempted to think events like multicellularity were inevitable, but this is not necessarily so. Large multicellular organisms have appeared in only five lineages; most lineages have remained unicellular. Large multicellular organisms (macroscopic) became common only in the late Proterozoic, about 570 mya. Small multicellular organisms (plankton) may be nearly four times that age. The appearance of multicellular organisms remains a very significant event because this type of organization provides vast new possibilities, for greater size, for specialized cells, for redwoods and blue whales. It will take a few more chapters to get to redwoods; whales are left to my colleagues. Multicellular plants invade the land in the very next chapter.

SIX

The Great Invasion

*Wherein the challenges and colonization of the terrestrial environment are examined
so as to understand the adaptations of land plants, especially their life cycle.*

We are waiting for the long-promised invasion. So are the fishes.
—Winston Churchill, October 21, 1940

Visualize a seashore whose wave-washed rocks are covered with seaweeds.
Tidal pools and the deeper water beyond contain even more seaweeds:
brown ones, red ones, and green ones. Snails, mussels, barnacles, crabs,
shellfish, and other invertebrates abound. You might even observe some
fish in the tidal pools. If you were transported back in time to the Middle Silurian, some 450 mya, the fossil record indicates that a rocky shore
would look just about the same. A Silurian coastal community would not
seem strange or foreign at all because all of the above-mentioned groups
of familiar organisms were present, and many were sufficiently similar to
living species that, unless you were a taxonomic expert, you would hardly
notice any differences.

Only if you were very observant would you realize what was missing. Other than fish, there would be no vertebrate animals, no birds, no
reptiles, and certainly no mammals. But much more is missing from the
Silurian than just higher vertebrates. If you turned around and looked
inland, what a strange, hostile, and alien scene you would behold, a
bare and barren-looking landscape of rocky surfaces and coarse weath-
ered rock fragments, as far as you could see, and further, stretching across
entire continents. As hard as it is to imagine continents without their

familiar mantle of soil and plants, this is the setting for another significant evolutionary event in the history of life, the invasion of land, the terrestrial environment.

The barren appearance of the continents does not mean they were completely lifeless. Microbial mat communities would have existed in low wet areas of sediment accumulation for millennia before other terrestrial organisms appeared. The cyanobacteria forming the upper layer of these mats have gelatinous sheaths that retain water and slow the rate of drying when exposed. Moreover, these sheaths contain pigments that filter out damaging UV radiation (Graham and Wilcox 2000). Cyanobacteria acquired this adaptation long ago because intertidal stromatolites were exposed during low tides (Fig. 2–2). This ancient coastal marine adaptation works equally well in shallow freshwater habitats.

Even what appears to be bare rock can harbor microscopic life. Cyanobacteria and unicellular green algae can live in the tiny spaces between particles within sedimentary rocks, protected and buffered from the harsh terrestrial environment. Communities of algae forming a green layer beneath rock surfaces were discovered to be common in the red sandstone areas of the southwestern United States (Bell 1993). In this endolithic (within stone) habitat algae are protected from the "outside" extremes of temperature and the drying wind. Similar algal communities have been found within Antarctic rocks in places otherwise completely devoid of life. The invasion of land by green organisms begins with such small, tough colonizers that could deal with a barren landscape, and in the process these pioneers changed conditions, gradually ameliorating the harsh terrestrial environment, which provided more colonization opportunities. As the terrestrial habitat changed two things happened. One, organisms diversified and colonized new habitats, and two, each new group colonizing a new habitat produced new opportunities to exploit, and as a result a succession of new organisms appeared, because once organisms gain a toehold on land, opportunities abound.

If streams and rivers were the main avenues of invasion for green organisms, and if early terrestrial organisms are transitional between an aquatic and terrestrial environments, then a logical conclusion, in this case backed up by fossil evidence, is that low mat-like or crust-like wetland communities predated communities of more complex terrestrial plants (Tomescu and Rothwell 2006). Lower Silurian fossils document wetland communities that were composed of a diverse array of low-growing organisms forming a "well-developed, though discontinuous, groundcover" some 10–15 million years before the earliest appearance of more complex land plants. These communities formed crusts over the substrate comparable with modern soil crust communities, which are often dominated by lichens. Fossils also indicate the presence of cyanobacterial mats (Tomescu et al. 2006), demonstrating how important these ancient green organisms remain throughout Earth history.

Lichens are some of the toughest pioneer organisms known. Modern lichens, especially those with a low, crusty form, can colonize bare rock surfaces and such lichens are legendary for their ability to survive in harsh, exposed habitats like the extreme conditions common at high altitudes and high latitudes (Brodo et al. 2001). Lichens are plant-like symbiotic organisms whose "thallus"[1] looks like 3-dimensional tissue (parenchyma), but actually it is composed of densely woven fungal filaments forming enclosed spaces wherein algal cells reside, usually either cyanobacteria or a trebouxiophycean (trey-bucks-ee-oh-FEYE-see-an) green algae (Appendix: Green Algae). In the lichen-forming symbiosis both organisms retain their individual identities, but together they form a unique organism unlike either partner individually. Fungal associations with algae may have begun in the coastal environment, where they occur with brown and red seaweeds that live in the upper intertidal zone, a symbiosis that appears to protect the seaweeds from desiccation (Selosse and Le Tacon 1998). Presently fungi form symbiotic interactions with 90% of land plants, and these photoautotroph-fungal associations may have played a role in the invasion of the terrestrial environment (Selosse and Le Tacon 1998). Studies of liverworts, one of the most ancient lineages of land plants, show that they can grow better and faster when associated with fungi, and such fungal-plant associations have been dated to at least 400 mya (Humphreys et al. 2010). Phylogenetic data suggest that filamentous fungi arose no later than 400 mya and probably before 500 mya, so they are more than old enough to have played a role in the invasion of land.

Presently, lichens are the predominate organisms forming soil-crust communities in some harsh, exposed places where they endure thin, rocky soils, strong solar irradiation, aridity, and big temperature fluctuations. Soil crust communities may have helped build early soils, thus preparing the way for larger organisms to follow (Graham 1993) (Fig. 6–1). Soil-crust communities, although small and low by present standards and often barely noticed by hikers crunching them (and destroying them!) beneath their boots, are virtual forests, towering 1 to 2 cm or more above their substrate, tall enough to host a fauna of tiny animals like springtails and tardegrades.[2] Such "forests" are 10 to 20 times taller than microbial mat communities, which generally measure less than 1 to 2 mm in thickness.

Fragmentary microfossils, which are often difficult to interpret, suggest that simple land plants invaded land by the mid-Ordovician, 475 mya

Fig. 6–1 A soil crust community (dark-topped bumps) in western United States providing microhabitat for germination and growth of a few vascular plants. (Image source: Courtesy of Nihonjoe, Wikimedia Creative Commons)

(Wellman et al. 2003), some 50 my before land plant macrofossils appear. These tiny fossils consist of sheets of waxy cuticle bearing the imprint of cells, tubular cell fragments like the rhizoids of mosses or liverworts, and spores identical to those of land plants. Most land plants produce a waxy cuticle that covers their epidermal cells, a waterproofing very different from the colloidal gels of seaweeds. Waxy cuticles are very resistant to decomposition and they preserve the pattern of the cells upon which they were formed. In particular, the spores suggest these earliest land plants had affinities to liverworts. At one extreme, liverworts are hardly more complex than Charophycean seaweeds, so little structural evolution was required. From these little beginnings, the land turned green with plants. Unfortunately our knowledge of early terrestrial life remains fragmentary because the conditions involved and the types of organisms involved were not conducive to producing a rich fossil record, but even as fragmentary as it is, taken as a whole, these bits and pieces record the first evidence of plants upon the land.

Somehow aquatic green organisms gave rise to terrestrial green organisms—that is, land plants, organisms adapted to a harsh new environment. As land plants became larger and more diverse, they too changed the terrestrial environment because plants are obstructions, trapping sediments, capturing light, and interfering with the movement of wind and water. As plants grew larger and more numerous, they became bigger obstructions, capturing fine sediments, building and accumulating soil, and slowing the runoff of water. In the process plants created even more habitats for animals and for other land plants. Ultimately great forests and later vast grasslands covered the expanses, although in many places, in an eyeblink of geological time, both have now been much reduced or completely destroyed by human activities.

The fossil record provides one perspective of how this took place. Our phylogenetic hypotheses of relationships among organisms provide another perspective. The structure, physiology, and reproduction of living land plants and algae provide still another perspective, and no one perspective provides all the answers. A synthesis of data is necessary to construct the best overall explanation of how the invasion of land took place. And this account must also explain the unique land plant life cycle that produces an embryo, which is why they are called the embryo plants, embryophytes.

All three lineages of seaweeds, green, red, and brown, were well developed residents of the coastal region prior to the invasion of land. Such seaweeds possess a plant-like form, and some are certainly as big and as complex as simple land plants. And there they are on the doorstep to land, but these seaweeds invaded no further inland than the farthest distance salt water regularly reaches. The ancestors of land plants are not found among marine seaweeds, and botanists have long known that land plants have no relationship to red and brown seaweeds at all. However, land plants have long been thought to have had a common ancestry with green algae, a hypothesis now well confirmed, so clues to the origin of land plants and their life cycle must be sought within the green algae. But which of the many groups of green algae are most like land plants? Before considering this question, and its implications for the land plant life cycle, the logistics of this invasion must be examined more fully. What is the invasion route? What obstacles will the invaders have to surmount?

INVASION ROUTE

A pretty substantial boundary exists between an aquatic environment and the terrestrial environment even along coasts, one that neither seaweeds nor many other organisms have transgressed. Vertebrate animals accomplished marine to land only once but managed land back to water on several occasions (whales, penguins, manatees, seals, plesiosaurs, etc.). Only arthropods with their tough exoskeletons of impermeable chitin make it look easy to go back and forth between water and land. Embryo-producing plants invaded land only once (a single lineage[3]). Only once has a land plant (eel grass, a flowering plant) made the transition back to a marine aquatic environment. Land plants have invaded shallow freshwater environments a number of times. At least one major flowering plant lineage (monocots) may have arisen from freshwater ancestors that reinvaded land. On the whole it must be concluded that direct transgressions from marine to terrestrial environments, and vice versa, have been fairly rare. Seaweeds failed to march inland from their coastal habitats, but everywhere that rivers and streams reach sea level another type of coastal habitat exists, an aquatic one that provides an avenue inland. As part of the water cycle, precipitation deposits fresh water upon the land, where it accumulates in depressions of various sizes (puddles to great lakes). Much evaporates again, but the excess ultimately runs off and flows back to mix with the oceans. When water courses reach sea level two things happen. One, nutrient and sediment-laden fresh water washing from the continents mixes with seawater in estuaries and bays, which locally reduces the salt concentration by dilution, creating brackish water. Two, as their rate of flow slows, rivers and streams begin dropping their sediment load, which further reduces their rate of flow, resulting in more sediment deposition and soon forming spreading deltas.[4] Shallow delta waters and bays provide light-rich aquatic environments combined with nutrient-rich sediments. Deltas, estuaries, and bays are very productive aquatic habitats, but all together these regions represent only a tiny, tiny portion of the coastal habitat.

Like the rest of the coastal environment, deltas, estuaries, and bays are quite patchy and truly dynamic. These shallow-water coastal areas are subject to tidal fluctuations, alternately exposing and inundating any organisms growing in shallow places. And while the tides pulse in and out, on the whole, a current flows outward, which tends to wash any free-floating organisms, their swimming sex cells, or their offspring out to sea, unless they have excellent timing. Floating or swimming reproductive cells or offspring released at high tide will head out to sea with the retreating tide. Those released at low tide will head inland with the returning tide, but they have only a few hours to anchor themselves before the tide reverses. Anchoring is made more difficult because the current and tides constantly shift fine sediments. The combination of tide and currents contributes to a complex mixing of water producing a fluctuating, patchy continuum from fresh to salty marine.

As in other dynamic coastal habitats, to stay in place, to remain in a favorable patch, organisms must be sessile, anchored in place while still retaining a means of dispersing and colonizing new patches. Anchoring is an additional challenge in this habitat because fine, soft, shifting sediments do not provide a solid, stable substrate upon which to hold-fast. New depositions of sediments and the constantly shifting substrates threaten to bury and dislodge any low-lying organisms anchored in the silty sediments. In this environment, spreading rootlike rhizoids provide better anchoring than other kinds of holdfasts. Algae with both prostrate and aerial axes can spread and anchor themselves while raising photosynthetic portions above shifting sediments. The changeable nature of these habitats favors smaller, faster-growing algae rather than large seaweeds.

Spreading deltas and estuaries, with shallow channels cut through fine sediments, provide marine organisms a means to invade the land, sort of, because such mucky places are not exactly what you envision when we say land. As will be seen on an upcoming field trip (chapter 8), such habitats are thought to be places where marine green organisms began adapting to both fresh water and land. But before taking that field trip and looking for early land plants, consider what it means to adapt to land.

ELEMENTARY CHALLENGES FOR LIFE ON LAND

We've established the route of invasion; what obstacles need to be overcome? The ancients thought that some combination of four basic "elements" (earth, wind, fire, and water) in different proportions composed everything. In many respects the classic Greek "elements" represent the challenges presented by the terrestrial environment. "Earth" represents the rocky surfaces and, as a result of weathering and erosion, the ever finer pieces thereof that ultimately will form the inorganic component of soil. But the invasion of land is more correctly an invasion of air, so "wind" represents air and the constant loss of water. A jellyfish or seaweed lying on a beach is a pitiful sight, limp, helpless, drying, and dying (Fig. A1 C). Deprived of the buoyancy of water, seaweeds and other marine creatures lay limp and unable to extend their appendages. Without a constant bathing in water they dry out rapidly. But while air is unsupportive and drying, carbon dioxide is readily available because as a gas it diffuses up to 1,000 times faster in air than when dissolved in water.

"Fire" represents sunlight's visible light and ultraviolet radiation, both abundant without an aquatic filter. With no buffering mantle of soil or vegetation, the heating and cooling of rocks would occur rapidly, producing much greater temperature extremes than in aquatic environments. Daily and seasonal temperature fluctuations must have been extreme, akin to what we find in rocky deserts today. Such fluctuations would create storms and wind raging unabated across the rocky continental surfaces. Weathering under such conditions would have been rapid, and without vegetation and soil to sponge up water, the rain that fell upon the continents would wash towards the seas in sediment-laden torrents, rather like the streams formed by glaciers and heavy snowmelt or rainfall today. Such a sediment laden runoff would produce huge deltas and sediment-filled shallow bays, and sedimentary rocks derived from such formations are evidence that this was happening.

And "water" is water, or the lack thereof. Water composes 80% to 90% of cytoplasm even in terrestrial organisms. But terrestrial organisms are not continuously bathed in a watery medium, so organisms must remain hydrated against a tremendous gradient or become tolerant of drying. To make matters worse, terrestrial water derived from precipitation is fresh, not salty, which greatly changes how osmosis (oz-MOH-sis), the diffusion of water across cell membranes, affects organisms. Many marine organisms have mechanisms for ridding themselves of excess salt. But in fresh water, cells have a higher concentration of solutes, which would make water diffuse in, filling the cell until it bursts its cell membrane—so the cells of many freshwater unicellular organisms have mechanisms for expelling excess water.[5] But water becomes a severely limiting factor on land, and terrestrial organisms must consume or absorb enough water to maintain an adequately hydrated condition while offsetting the inevitable losses to the atmosphere. For aquatic organisms, water is also their dispersal medium; on land, plants must reproduce and disperse in the dry, gaseous atmosphere, an environment not at all conducive for swimming gametes or zoospores. But no matter how hostile the terrestrial environment, light and space abound, and mineral nutrients and substrates exist where sediments accumulate.

HALLMARKS OF A LAND PLANT

Land plants must have adaptations for surviving these elemental challenges of the terrestrial environment. So here is a seemingly simple question. How do you know when you have a land plant fossil? In a field guide to living plants, this question would sound a bit silly. Choice 1 in your dichotomous key[6]: plants growing on land or plants growing in water? What could be

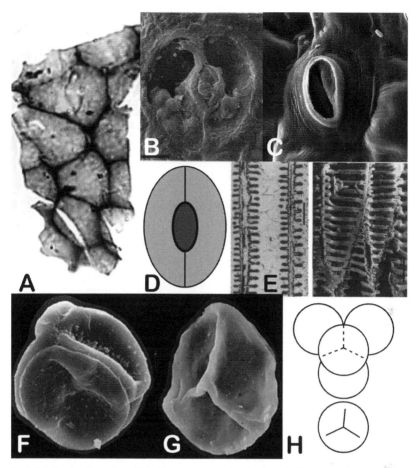

Fig. 6–2 Hallmarks of land plants. A. Fossil sheet of waxy cuticle showing a tile-like pattern of epidermal cells. B. Fossil cuticle with imprint of a stomate (pore) surrounded by a pair of guard cells. C. Scanning electron microscope image of stomate and guard cells under cuticle on a tomato leaf; orientation as in B. D. Diagram of a stomate (dark gray) and flanking pair of guard cells (light gray). E. Xylem cells (tracheids) showing thickenings on the walls (dark carbonized material) protruding into the cell center (gray rock matrix)(left side). Xylem cells (tracheids) with rock matrix dissolved away showing that the wall thickenings are helical (right side). F. Tetrad of fossil spores, lateral view. G. Fossil spore with trilete scar. H. Diagram of spore tetrad (top) and single spore with trilete scar produced by contact with the three adjacent spores. Ages: A, B, F, G—Silurian; E—Devonian; C—Present day. (Image sources: A, F, G—After Wellman and Gray 2000; B, E—After Edwards 1993; C—Courtesy of the U. Dartmouth Electron Microscope Facility)

more obvious? But when dealing with fossils, how could you tell if a plant was aquatic, growing in water, or was growing on land and then fell into watery sediments? And since most sediment accumulates in water, that is where most fossils form whether the organism grew there or not. The hallmarks of land plants should be distinctive features related directly to their adaptation to the terrestrial environment (Fig. 6–2), features that aquatic green algae, the ancestors of land plants, would not have.

Most land plants have a discrete epidermis, a layer of cells covering the organism that fit together like floor tiles without any intercellular spaces where water could escape. Aerial portions of most terrestrial plants have a waxy, waterproof coating (cutin) forming a layer, a cuticle, upon the epidermal cells' outer surface. When you polish an apple or cucumber, you are buffing a waxy cuticle.[7] Cutin is relatively resistant to decay and therefore cuticles are good candidates for fossilization, so fossil cuticles, complete with cellular imprints of an epidermis, are solid evidence of land plants.

The body of land plants is constructed of a multicellular, block-organized tissue called parenchyma, but in and of itself parenchyma is not a hallmark of land plants because many brown and some green algal seaweeds possess similar tissue. A block organization slows water loss because the surface area–to–volume ratio is greatly decreased. However, when an epidermis with an impervious waxy cuticle is placed upon the surface of such tissue, gaseous exchange must take place internally, upon the moist surfaces of internal cells, and this requires openings for gas diffusion. Some water will be lost through the openings for gas exchange, but to limit water loss and regulate gaseous exchange, access to the internal tissues of land plants is provided by stomates, pores surrounded by a pair of guard cells, specialized epidermal cells whose change in turgor (osmotic pressure) can open or close the pore. The presence of guard cells and stomates, also recorded in fossilized sheets of cuticle, is evidence of a parenchyma (block-organized tissue) underlying an epidermis and of an ability to regulate gaseous exchange and water loss, something aquatic organisms lack (Fig. 6–2 B-D).

Elongate, thick-walled, cylindrical cells provide both structural support and water conduction to aerial portions of the plant. Some brown algal seaweeds have elongate food-conducting cells, but like phloem (its vascular plant counterpart), these food-conducting cells are thin-walled and provide little if any support.[8] Water-conducting cells, a vascular tissue called xylem (ZEYE-lehm; from the Greek for wood), have thickened rings, helices, or layers of secondary wall material deposited upon and within the primary cell wall to strengthen and reinforce the cells against hydraulic collapse. Hydraulic collapse can occur because of the force needed to move water and the cohesion of water to the cell walls. You have probably caused the hydraulic collapse of a thin plastic straw when sucking up a thick milkshake. Xylem cells would collapse like the thin plastic wall of the straw if the thin outer xylem cell walls were not reinforced by a thick inner secondary wall. Obviously a reinforced cylinder is also an excellent form for structural support. The cellulose composing the thick secondary cell wall of xylem is impregnated with lignin, a polyphenolic[9] substance that is quite resistant to decay, so lignified xylem cells are very good candidates for fossilization.[10] Since neither algae nor, for that matter, any nonvascular land plant are known to synthesize lignin, fossilized tubular cells with decay-resistant inner thickenings would be good evidence of vascular land plants (Fig. 6–2 E).

Lastly the land plant life cycle has a diploid phase characterized by multicellular reproductive structures (sporangia) within which numerous cells undergo meiosis, each producing four spores packed into a tetrad (Fig. 6–2 F-H). Because of their small size and necessary exposure to the drying atmosphere during dispersal, the spore walls are thick and impregnated with the most decay-resistant biological material known, sporopollenin, another polyphenolic substance. As a result, spores fossilize extremely well. The spore wall bears three flat facets separated by three ridges radiating from a central point in the middle of the three facets, a so-called trilete scar, resulting from the packing of the four spores into a spherical tetrad, a clear indication that this cell is a haploid product of meiosis. Thick-walled, trilete-scarred spores are unknown among algae, and all of the reproductive structures of algae are unicellular. Fossilized spores with trilete scars, alone or in a multicellular sporangium, would be very definitive evidence of land plants. Unfortunately, while common enough, dispersed fossil spores do not tell us what plant they came from, although in some instances the spore wall sculpturing is distinctive enough to suggest a relationship.

If a fossil organism exhibited (1) a cuticle-covered epidermis, (2) stomates and guard cells, and (3) multicellular reproductive structures containing spore tetrads or thick-walled spores with trilete scars, it would without doubt be a land plant. However, it isn't quite so easy. First, intact fossil organisms are very rare, so finding even three of the four hallmarks together is not expected very often. Second, only vascular plants possess thick-walled, lignified xylem (xylem is the fourth hallmark). Land plants traditionally have been treated as two groups, the vascular plants (tracheophytes) and the bryophytes or nonvascular plants (mosses, liverworts, and hornworts). However, lack of a character, in this case xylem, is not a novel shared character (e.g., you lack xylem, too), unless it can be shown that the character in question was possessed by common ancestors and then subsequently lost. Since bryophytes have no novel shared characters uniting them, they can no longer be treated as a single lineage equal to tracheophytes, so at this point land plants consist of four lineages: liverworts, hornworts, mosses, and vascular plants.[11]

The only feature possessed by *all* land plants is trilete-scarred spores formed in a multicellular sporangium, and this and the other hallmarks are features of the sporophyte generation only. A cuticle-covered epidermis and stomates with guard cells are not possessed by most bryophytes, and if present then they are only in the small sporophyte generation. Bryophytes lack these hallmarks because they solve the problems of gaseous exchange and water conduction in different ways than vascular plants (chapter 7). This also means that most bryophytic plants lack these decay-resistant features that readily fossilize; however, they may have other decay-resistant parts that have produced enigmatic microfossils and evidence of their existence prior to vascular plants (Kodner and Graham 2001; Graham et al. 2004).

The order in which the hallmarks of terrestrial plants appear in the fossil record becomes significant when considering the phylogenetic relationships among all land plants, particularly the relationship between vascular plants, mosses, liverworts, and hornworts. Sheets of cutin and dispersed trilete spores were the first hallmarks of terrestrial plants to appear in the fossil record, followed by stomates. Cuticular sheets indicating a plant body consisting of parenchyma are common starting in the middle Ordovician, but they lack evidence of pores or stomates. Some have a cell pattern quite similar to that found on leafy organs of *Sphagnum*, an early diverging moss lineage (Buck and Goffinet 2000; Goffinet 2000). Dispersed spore tetrads and spores with trilete marks are found, which clearly indicate land plants, but paleobotanists remain uncertain about the group of land plants that produced them. Some of the spore tetrads and spore walls have sculpturing similar to spores produced by liverworts (Edwards et al. 1995). Tubular structures with and without helical thickenings bear a superficial similarity to xylem, but actually they more closely resemble the walls of epidermal cells from the sporangia of bryophytes (Graham and Gray 2001).

Other enigmatic fossils (tubular structures and distinctive arrays of epidermal cells) are similar to degradation-resistant portions of mosses and liverworts (Kodner and Graham 2001; Graham et al. 2004). While none of these fragments is convincing evidence individually, all together these fossils suggest the invasion of land began in the middle Ordovician with small, low-growing land plants of relatively simple organization, something organizationally akin to simple liverworts, hornworts, and basal mosses (Graham and Gray 2001). True xylem is the last of the four hallmarks of land plants to appear in the fossil record (Edwards and Wellman 2001). The earliest fossil to show unequivocal evidence of all four hallmarks of terrestrial plants is *Cooksonia* from the lower Devonian (Fig. 6–7).

Cooksonia appears in the middle Silurian either lacking full-fledged xylem, or as fossil impressions lacking anatomical detail. A recent report (Steemans et al. 2009) of much older trilete spores in the late Ordovician suggests vascular plants or their immediate predecessors appeared earlier than previously thought. The biggest problem with this analysis is the identification of the plants producing such spores; while without a doubt they were embryophytes (land plants), their designation as tracheophytes remains equivocal.

Such little bits and pieces are not the most impressive or exciting fossil record. The first unequivocal macrofossils of mosses and liverworts don't appear until the late Devonian and early Carboniferous, respectively, long after vascular plants appeared, and on the basis of the record of reasonably intact macrofossils, this has generated an alternative vascular-plants-came-first hypothesis. Perhaps vascular plants appeared first and gave rise to bryophytes via loss of xylem and a flip-flopping of the life cycle, a reversal of the more common hypothesis. Interestingly enough, phylogenetic studies of living land plants do not rule out this possibility, but they do not offer any strong support for a vascular-plants-came-first hypothesis either. The vascular-plants-came-first hypothesis generates a life cycle problem that will be explained presently, one easily avoided by the bryophytes-first-hypothesis. However, one point must be made clear: bryophytes are well-adapted and successful terrestrial organisms (chapter 7). Too often bryophytes have been regarded as an evolutionary way station between green algae and vascular plants, neither aquatic nor full-fledged terrestrial plants. Such a perception is simply wrong.

THE LAND PLANT LIFE CYCLE AND WHY IT IS IMPORTANT

Look at the two very typical ferns in Fig. 6–3. Chances are you are thinking that one fern (Fig. 6–3 A) looks typical enough with its fiddlehead and ferny fronds, but the other fern (Fig. 6–3 B), if indeed it is a fern, does not look like any fern you have ever seen. Be assured that this fern is both quite typical and relatively common. One reason such strange-looking ferns may have escaped your attention is that they are rather small. The ferny fern is at least a meter high, and the unferny fern is about one-hundredth that size at about 1 centimeter across at most, yet both indeed are free-living green organisms. This small size, together with a preference for dim, moist habitats, is probably why you have never before seen such a fern. Now for some more real surprises. One, this unferny fern possesses none of the hallmarks of a land plant, not one; it has the same level of

Fig. 6–3 Two ferns. A. This fern is a typical vascular land plant consisting of roots, a stem, and leaves with vascular tissue, a cuticle-covered epidermis with stomates and guard cells, and multicellular sporangia (size—1 meter tall). B. This fern consists of a thin, undifferentiated thallus with rhizoids. It lacks all the land plant features of a "typical fern," but is no less a fern and no less a land plant (size—6 millimeters across). In fact, these two ferns could even be the same species. To figure out how this is possible requires an understanding of the land plant life cycle. (Image source: A—The Author; B—Courtesy of Random Tree, use licensed under the Wikimedia Creative Commons)

organization as a liverwort and might even be mistaken for one. Two, these two ferns could be the same species! But how can that be? The answer is to be found in the land plant life cycle.

Here is a quick review of the basic biological rules of sexual life cycles (Fig. 3–9). (1) All sexual life cycles alternate between haploid and diploid. (2) Gametes are haploid. (3) Zygotes are diploid. (4) The haploid phase changes to diploid as the result of gamete fusion, fertilization. (5) The diploid phase changes to the haploid phase by meiosis. Once you have these five rules memorized or tattooed on the palm of your hand,[12] you have the keys to understand any sexual life cycle.[13]

Here is a reminder about what you need to know about land plants for this explanation to make sense. Land plants constitute a single lineage, the embryophytes, because they have a life cycle that produces an embryo. Embryos are formed because land plants are oogamous, meaning their gametes are differentiated into large, nonmotile eggs and much smaller, motile sperm, and because the egg does not disperse, even after fertilization, the resulting zygote begins development within the tissues of the maternal parent from which it receives nourishment. This lack of a dispersal event just prior to or after fertilization has some interesting consequences for sessile organisms like land plants.

The typical fern with which you are familiar makes spores (initially in tetrads) by the thousands in sporangia. Spores are tiny, almost dust like, so they can be dispersed widely by the wind. When a spore disperses and lands in a favorable location, it grows into a "fern," but a fern that looks just like the one on the right (Fig. 6–3 B). Such tiny ferns are as common, probably even more common, than the larger, more familiar ferns, but they don't live nearly as long. This tiny simple fern makes sperm and eggs, and it can either self-fertilize or be cross-fertilized if another such fern is growing nearby. The resulting zygote grows into an embryo that will in just a few weeks outgrow its mother and establish itself as an independent plant, which when mature will look like a typical spore-producing fern (Fig. 6–3 A).

All of these things have been known for a long time, but then about 100 years ago, botanists figured out that these two ferns also differed in chromosome number. All familiar ferns are diploid (2n); they have two sets of chromosomes. But the tiny fern is haploid (n), having a single set of chromosomes. And the life cycle rules spelled out above are based upon these facts. Fertilization is accomplished by the fusion of an egg and a sperm, so the resulting zygote, and the embryo, and the fern that develops from it must be diploid too. A spore grew directly into a haploid fern, so spores must be haploid also. To make haploid spores, the spore mother cells of a diploid fern must divide by meiosis. And if you know and understand all of that, ta da! you know and understand the land plant sexual life cycle.

The lack of a dispersal event just before, at, or after fertilization means that every typical diploid fern grows in exactly the same spot as its haploid mother, at least initially. Many ferns have rhizomes and can grow laterally to colonize new space. It works out rather well that haploid ferns are adapted for small, moist, shady habitats because fern sperm require free water to accomplish fertilization[14] and in such microhabitats, dew or drops of rain will be common enough for this to be accomplished. The tall diploid fern with its fronds arching a meter or more above the ground has all the land plant hallmarks, so it is better adapted to this drier habitat, which is more conducive for spore dispersal. So there you have it; the fern life cycle involves two different ferns because each is adapted to a different habitat for a different reproductive function.

Free-sporing land plants like ferns are limited by this life cycle. Diploid ferns are well adapted to the terrestrial environment, so they can grow in places where haploid ferns could never survive, but because of their life cycle, diploid ferns can grow, or at least start growing, only in the exact place where conditions were favorable for the growth and reproduction of its haploid mother fern. Conditions favorable for the growth and reproduction of a haploid fern need only have existed every now and again for just a few weeks because haploid ferns can reach sexual maturity pretty quickly, and then the well-adapted diploid fern can persist for decades. But this means the distribution of diploid ferns is constrained by the requirements of the sexual phase of their life cycle. And to some extent this is true for horsetails, clubmosses, and whisk ferns too. The innovation that finally allows the tougher diploid land plants to escape this maternal tyranny is the seed (chapter 9).

Textbooks call this life cycle an "alternation of generations," but this phrase obscures the fact that all sexual life cycles alternate between haploid and diploid phases. Alternation of generations also is rather misleading because technically both a diploid organism and haploid organisms are part of one generation, but this unfortunate phrase is firmly entrenched in the lexicon of biology. The land plant life cycle differs from all other sexual life cycles by having multicellular organisms representing both haploid and diploid phases, the latter of which starts out as an embryo. Quite a few seaweeds have haploid and diploid organisms composing their life cycle, but none produce an embryo. In the sexual life cycles of all other organisms, either the haploid (gametes) or diploid (zygote) phase is unicellular.

The biological relationship between the two ferns (Fig. 6–3) can now be illustrated by drawing an arrow labeled "meiosis and spores" from the diploid fern (left) to the haploid fern (right) and by drawing an arrow labeled "gametes and fertilization" from the haploid fern to the diploid fern. Both bryophytes and seed plants have an alternation-of-generations life cycle too, but each with some interesting differences that will be explored, sooner for the bryophytes and later for the seed plants.

These haploid and diploid generations, and their reproductive structures, have labels and using them allows botanists and botanical writers to avoid the continued use of awkward phrases. The problem is that these labels are jargon, and like all such terms, when unfamiliar they seem difficult and confusing. Unlike in textbooks, here the explanation of the land plant life cycle came first to provide some understanding prior to the introduction of terminology. Educationally this was an attempt to keep the horse in front of the cart. The reason for this is simple: the right way to teach science is the way science is done. These ferns and their reproductive functions were observed and understood before any scientific terminology for them was invented. Textbooks and many biology teachers work backwards by trotting out a cartload of terms and definitions before

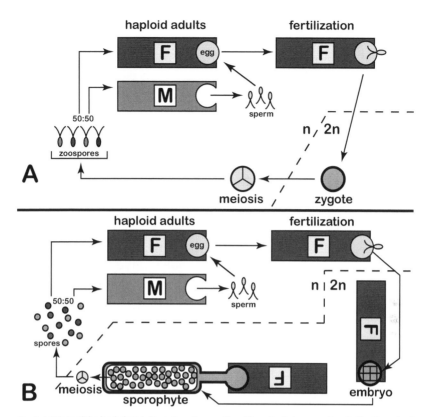

Fig. 6–4 Origin of the land plants' alternation-of-generations life cycle. A. In comparison to the life cycle of charophycean green algae, which share a most recent common ancestry with land plants (Fig. 6–5), the land plant life cycle (B) has a new multicellular diploid generation, a sporophyte (bottom), resulting from a delay in meiosis and development of the zygote into first an embryo. Here the sporophyte is shown dependent upon the gametophyte as in bryophytic plants. As a consequence each zygote, each fertilization event, can produce thousands of spores. Dashed line separates haploid (n) from diploid (2n) generations. F = female, M = male. (Image source: The Author)

any understanding is possible, and that leaves rote memorization as the only alternative. [15]

Logically a plant that produces gametes is called a gametophyte (Fig. 6–4 B, F and M, female and male), a gamete-producing plant, and its multicellular reproductive organs are called gametangia (gamete houses). Gametangia of land plants always have one or more layers of sterile jacket cells surrounding the gamete-producing cells. Gametangia in algae are always unicellular although a unicellular gametangium may produce more than one gamete. Since both gametophytes and gametes are haploid, gametes are produced by simple cell division, mitosis.[16] In land plants gametangia are of two types, archegonia (ark-eh-GOH-nee-ah), each of which produces an egg, and antheridia (an-ther-ID-ee-ah), which produce lots of sperm. Gametophytes can be of separate sexes or hermaphrodites having both types of gametangia.

A diploid fern produces nonflagellated, haploid reproductive cells by meiosis, and since these reproductive cells do not fuse but grow directly into a new haploid organism, they are not gametes. Such nonfusing reproductive cells are called spores, or if motile, zoospores (ZOH-oh-spores). Thus the diploid fern is called the sporophyte (Fig. 6–4), a spore-producing plant, and its reproductive organs are sporangia (spore houses), wherein many spore mother cells divide by meiosis to produce spores.

In all land plants the gametophyte and sporophyte generations are different in form, with perhaps one ancient exception. Bryophytes and vascular plants differ in whether it is the haploid or diploid organism which is the longer-lived, conspicuous, and "dominant" generation. In vascular plants, the conspicuous, long-lived organism is diploid (the sporophyte), so when you visualize a fig, fir, or fern, it's the sporophyte in your mind's eye. In addition to ferns, the rest of the free-sporing land plants (clubmosses, whisk ferns, and horsetails) also have a small, free-living haploid gametophyte generation, although with considerable variation in form and biology. In seed plants the gametophytes are so very reduced in size and completely dependent upon the sporophyte such that they remain hidden from casual observation, a matter that will be brought to your attention in a few chapters. The organism you recognize as a moss, liverwort, or hornwort is a haploid organism, the gametophyte. The sporophyte of bryophytes is small, simple, attached to and more or less wholly dependent upon the gametophyte for nutrition.

ORIGIN OF THE LAND PLANT LIFE CYCLE

Either land plants inherited their life cycle from green algal ancestors with a similar life cycle or the land plant life cycle represents an innovation derived from an ancestral life cycle of a different sort, two clear possibilities. The green algae provide lots of possibilities because all types of sexual life cycles are found among various members of this large, diverse group, so it becomes important to know which green algae have a common ancestry with land plants when formulating an explanation. But for the longest time the green algae that shared a common ancestry with land plants remained uncertain, and two different explanations of the land plant life cycle were developed. But they were not equally popular among botanists.

Charophyceans include some fairly large, fairly complex freshwater "seaweeds" like *Chara*, *Nitella*, and *Coleochaete*, and they have long been considered prime candidates for common ancestry with land plants (Bower 1908). Charophyceans are haploid organisms, and the only diploid part of their life cycle is the zygote (Fig. 6–4 A). This means the land plant sporophyte, which starts out as an embryo, would be an innovation. However, these charophyceans already have oogamy and retention of the egg, and in the case of *Coleochaete*, retention and nourishment of the zygote (Graham 1993). The haploid gametophyte with its lack of terrestrial adaptations and swimming sperm directly represents this algal ancestry.

While a charophycean common ancestry with land plants has long been favored, cytological, biochemical, and molecular studies in only the past 40 years have confirmed this to virtually everyone's satisfaction. Thus it was discovered that charophyceans have cell division unlike other green algae but just like land plants.[17] The charophyceans form a clade that includes land plants, a lineage now called the streptophytes (strepto- means division) (Fig. 6–5) (Becker and Marin 2009). Ultrastructural studies also found that the motile cells of the charophyceans (unicellular organisms, sperm, zoospores) and the motile cells of land plants (sperm) all have an asymmetrical insertion of the flagella (as opposed to the symmetrical flagellated cells of other green algae) and similar cytoskeletal microtubular "roots" associated with the basal bodies of their flagella. Members of the streptophyte clade also recover lost photosynthate[18] using glycolate oxidase, while other green algae use other enzymes for that purpose. *Chara* and *Coleochaete* have circular rosettes of cellulose synthesizing protein complexes on their cell membranes, another feature they share with all land plants (Graham and Wilcox 2000). The strong concordance among novel characters, each for a different function, has generated considerable confidence that charophyceans represent those organisms that share a most recent common ancestry with land plants (Mishler and Churchill 1985; Graham 1993; Karol et al. 2001).

Like land plants some charophyceans have apical growth and are composed of parenchyma. Charophyceans are all found in freshwater environments, some adapted to temporary or ephemeral shallow-water habitats. Some charophyceans form sporopollenin[19]—impregnated, thick-walled zygospores to survive both cold and dry conditions of near-terrestrial environments (Graham et al. 2012). The wind-dispersed spores

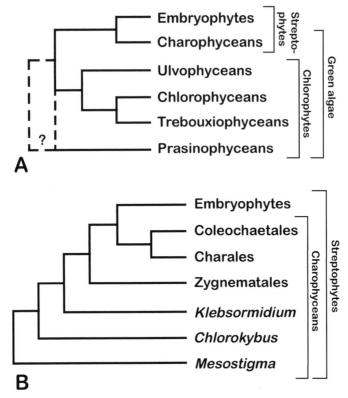

Fig. 6–5 Streptophyte clade (A) shows how land plants are nested within the green algae. Whether the Prasinophyceans share a common ancestry with the rest of the green algae and land plants remains uncertain. An expanded streptophyte clade (B) shows three basal genera (names in italics) and three orders, Zygnematales, Charales, and Coleochaetales. The latter two orders form a small clade that is sister group to the land plants. As shown, the charophycean algae do not form a clade. (Image source: After Graham and Wilcox 2000)

of land plants have a similar spore wall composition. Such charophycean zygospores are common in sediments of prairie sinks, shallow depressions that can be wet or dry for years depending upon variations in the climate. Desiccation resistance is a valuable adaptation for organisms living in ephemeral or variable aquatic habitats and this adaptation would function just as well in terrestrial habitats. The rigors of shallow fresh water and moist near-terrestrial environments may well have played a significant role in supplying the selection pressures to produce land plants (Graham 1985; Graham et al. 2012).

Cytological studies have identified some unusual basal members of the streptophyte clade, including a prasinophycean-like organism, some unicellular forms, and a small simple filamentous algae, *Klebsormidium*, that resembles ulotrichalean green algae. This means other members of the streptophyte clade may exist, but they remain unrecognized because they resemble other algae so closely. As with other microorganisms, similarity in form does not reflect phylogenetic relationships well at all.

Two changes are required to change the charophycean life cycle into the land plant life cycle (Fig. 6–4 B). First, the zygote, a fertilized egg, must be remain attached and intimately associated physically with the gametophyte that produced the egg. Second, meiosis must be "delayed" developmentally, allowing the zygote to divide by mitosis and thus form a new "generation," a multicellular diploid organism, the sporophyte. Eventually some diploid cells of the sporophyte must undergo meiosis to produce spores, which disperse and grow back into the haploid generation. The diploid generation is shown here attached to the haploid generation, which means as an embryo; in the case of bryophytes, throughout its life, the sporophyte remains nutritionally dependent on its haploid maternal parent. In charophyceans no developmental events intervene between fertilization and meiosis, although a considerable time lag can intervene if the organism is surviving inclement conditions as a zygospore.

This basic idea has been around for some time. Long before the common ancestry between charophyceans and embryophytes was well established, Bower (1908) proposed the interpolation hypothesis, although at the time, as was the custom, he used the term theory instead of hypothesis. According to Bower, a multicellular diploid phase, or sporophyte, was interpolated into the life cycle of a charophycean-like algae. Since the alternation-of-generations life cycle of all living land plants involves haploid and diploid organisms of very different forms, such a life cycle was referred to as an "antithetic alternation" because the sporophyte and gametophyte are the antithesis of each other. Indeed, Bower's interpolation hypothesis (Fig. 6–6 A-F) is often referred to as the antithetic "theory" in older literature.

What if the data showing a common ancestry of charophyceans with land plants were not so good? An alternate hypothesis existed and it was based on the idea that land plants got their alternation-of-generations life cycle directly from seaweeds that had a life cycle with alternating gametophytes and sporophytes (Fig. 6–6 A′). In red and brown algae the alternate generations can be either quite similar to each other or quite different in form, but in the green algal seaweeds like the ulvophyceans (*Ulva*, sea lettuce, Fig. 5–8) the haploid and diploid generations look identical. Unlike Bower's interpolation hypothesis, here the sporophyte is obtained intact from a common ancestor, which on the face of it seems to make good

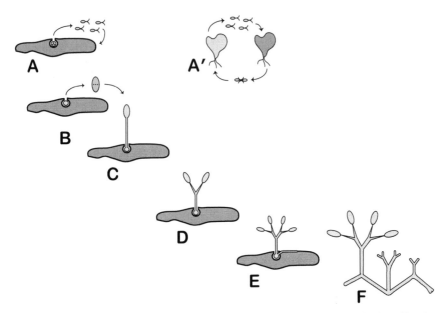

Fig. 6–6 Origin of the land plant life cycle. A-F. Interpolation hypothesis. A. Charophycean haplontic life cycle. Zygote divides by meiosis, producing zoospores that grow back into a haploid adult. B. Zygote retained; meiosis delayed and zygote divides by mitosis, producing an embryo. C. Bryophyte stage with unbranched sporophyte (light gray) dependent on gametophyte (darker gray). D. *Cooksonia* stage with partially (?) dependent, determinate, branched sporophyte. E. Early branching of apex produces horizontal axis and branched, determinate aerial shoot. F. Indeterminate horizontal axis continues branching, producing determinate aerial axes while perpetuating the horizontal axis. Sporophyte outgrows the gametophyte and becomes completely independent. A'—Ulvophycean seaweed alternation of similar generations. Haploid generation (dark gray) produces gametes that fuse to produce a diploid zygote. Diploid generation (light gray) produces zoospores by meiosis which grow into haploid generation. Transformational hypothesis requires that the life cycle of *Ulva* change in three different ways: similar motile gametes must become egg and sperm (A), the egg and zygote must be retained (B), and the form of the diploid generation must change (C). (Image source: The Author)

sense. However, the transformational (or homologous[20]) hypothesis, as it was called, must account for the shift from similar to different forms, a shift from isogamy (all gametes alike) to oogamy (larger nonmotile egg, smaller motile sperm), and retention of the egg and zygote together with the loss of this dispersal event.

A big problem for the transformational hypothesis was that you cannot account for bryophytes and vascular plants in the same way. When you start with two identical generations, and shift toward two unlike generations, you have to go two different "directions" for bryophytes (large gametophyte, small dependent sporophyte) and vascular plants (big sporophyte, small gametophyte). The transformational hypothesis also raises the possibility that bryophytes and vascular plants arose from different green algal ancestries. If true, then the embryophytes might not be a single lineage, so this becomes an important hypothesis to evaluate.

Two competing explanations, fair enough, but in the minds of most botanists, the older interpolation hypothesis (Fig. 6–6 A-F) posed fewer problems than the transformational hypothesis,[21] and because of this it has been favored ever since Bower proposed it. No one could suggest what selective force(s) could have acted to shift similar gametophytes and sporophytes in two different directions to produce both bryophytes and vascular plants. For decades no definitive evidence could tilt the decision about which was the correct hypothesis, even though one was certainly favored over the other. Once the data showed that land plants had a common ancestry with charophyceans, this fixed the starting point as a life cycle something akin to Fig. 6–6 A. The other reason that Bower's interpolation hypothesis was favored was that, even in the absence of evidence, most biologists thought that land plants of a bryophytic nature must have preceded vascular plants and that ulvophyceans were marine seaweeds while charophyceans were found in shallow fresh water. Now microfossils support this idea too.

The transformational hypothesis is not mentioned much anymore, a not uncommon happening in science. Once falsified, competing hypotheses often quickly fade from discussion, which is too bad in the sense that it fails to show how science makes progress through falsification. The transformational hypothesis had its supporters, but most scientists concede when the evidence overwhelmingly supports an alternative explanation. Nobody holds any news conferences or makes any concession speeches; the alternative explanation just fades away. And thus it very often appears that competing explanations for many biological phenomena do not exist. Biologists just don't waste time with the myriad of falsified explanations that have fallen to the wayside, unless, like here, they serve an illustrative purpose. Statements like "land plants share a common ancestry with charophyceans" can sound a bit dogmatic to outsiders with no knowledge of the history of such explanations. And of course this very phenomenon has been exploited with great success by deniers of evolution and critics of science to make biology sound dogmatic to the general public who are used to fair play and equal time. These are great ideals for dealing with people's different opinions, but science is not about fairness to all ideas at all.

Until rather recently, the fossil record had rather stubbornly refused to provide support to the interpolation hypothesis, although it did not falsify it either. The earliest land plant fossils were clearly vascular plant sporophytes; they had sporangia containing spores with the hallmark scars indicating they were the product of meiosis, so the sporangia-bearing plant was diploid and grew from a zygote. Nothing whatever was known of the gametophytes. A number of fossil discoveries conform to, and therefore support, the interpolation hypothesis. First, as already shown, fragmentary fossils taken as a whole are pretty convincing evidence that plants of a bryophytic nature preceded vascular plants by millions of years, thus suggesting that the scenario presented in Fig. 6–6 is valid.

The oldest fossil of the vascular plant lineage is *Cooksonia*, a sporophyte with Y-branched aerial axes (Fig. 6–7). Such fossils are known to be sporophytes because they have sporangia, and they are known to be sporangia because they contain spores with the hallmark of having been produced by meiosis (Fig. 6–2 F-H). This means the plant was diploid and grew from a zygote. *Cooksonia* sporophyte axes have been found arising from what might be a gametophyte thallus (Gerrianne et al. 2006), suggesting a heteromorphic life cycle in vascular plants is as old or older than those vascular plants with a supposed isomorphic life cycle. The "gametophyte" portion of the fossil is very indistinct, but the investigators significantly point out none of the axes continues beyond this common point, that is, no subterranean axes seem to exist. If interpreted correctly, this newly described fossil of *Cooksonia* from the late Silurian of China is a perfect intermediate between plants of a bryophytic organization and vascular plants.

Looking at this problem in a creative way has generated some calculations that corroborate the interpolation hypothesis. While similar, the variously preserved fossil remains of *Cooksonia* are not identical. In terms of axis diameter *Cooksonia* in the early Devonian was a more robust plant than the earlier-appearing late Silurian *Cooksonia*. Even though the earliest fossils of *Cooksonia* lack anatomical detail, a careful analysis of their diameter suggests that when the portions of the axis necessary for support, conduction, and prevention of drying are taken into account, the volume of tissue remaining for gas exchange and photosynthesis is inadequate for *Cooksonia* to have been a totally independent autotrophic organism (Boyce 2008). If true, the *Cooksonia* sporophyte must have been nutritionally dependent upon a gametophyte generation much like the sporophytes of mosses. Far from being a problem, this physiological analysis suggests exactly the type of intermediacy hypothesized (see Fig. 6–6 D). The earliest *Cooksonia* were small, dependent sporophytes whose primary function was to produce and disperse large numbers of spores. The basal portions would have been a foot firmly anchored in a

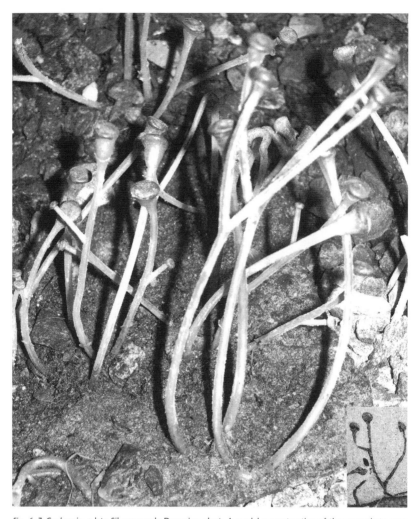

Fig. 6–7 *Cooksonia*, a late Silurean-early Devonian plant. A model reconstruction of the sporophyte axes. Inset lower right: Fossil of the aerial shoots of *Cooksonia* standing about half an inch tall. Naked axes show Y-branching and terminal sporangia but no roots or rhizomes. (Image sources: A—The Author. This model "grows" in the Darwin House of the Royal Botanical Gardens at Kew, UK; Inset—After Edwards and Wellman 2001)

maternal gametophyte, which strongly supports the idea of the earliest member of the vascular plant lineage (or polysporangiate linage) having an alternation of generations between very different-appearing gametophytes and sporophytes, exactly what was expected given a common ancestry with bryophytic plants (Kenrick 1994).

Biologists can be so picky! Simply saying the evidence suggests that the land plant life cycle arose because a zygote, rather than undergoing meiosis, developed into a multicellular diploid sporophyte, is not sufficient. A better explanation is needed because biologists want to know both how and why a sporophyte generation appeared. As always, biologists frame such questions in an evolutionary context and ask, "Of what reproductive benefit is a sporophyte?" Or to state it differently, "How did having a multicellular sporophyte generation contribute to the reproductive success of charophycean-like organisms on land?"

The benefit of the sporophyte is fairly obvious. Reproductive success in the very harsh and patchy terrestrial environment requires a different type of dispersal unit (spores) and lots of them. In charophyceans, zoospores produced by the meiotic division of the zygote are the primary dispersal stage of the life cycle (Fig. 6–6 A), the life stage that actively searches for suitable patches in which to anchor. But in the terrestrial environment, swimming dispersal units are not going to be very useful. In many charophyceans, the zygote also forms a desiccation-resistant spore that can withstand dry periods, delaying its meiosis until favorable conditions return. A spore combines such a desiccation-resistant wall inherited from its ancestors' zygospores with a small size that enhances its ability to disperse by wind rather than water. How this transition took place remains unknown.

The size of the zygote, which is determined by the size of the egg, also affects the viability or survivability of the zoospores. When zoospores are produced by meiotic division of the zygote, the cytoplasm is partitioned into four portions. To make twice as many zoospores requires that each of these divide in half, so each doubling of the number of zoospores halves their size. This is a classic size/number tradeoff. Doubling the number of zoospores reduces their size and therefore halves their energetic reserves. Since the viability or survivability of the zoospores is directly proportionate to their energetic reserves, the reproductive choice is between a few big, highly viable zoospores or 8, 16, 32, or even 64 zoospores, more chances to disperse but each with decreasing survival longevity.

The only way around this size/number tradeoff is to increase the energetic reserves possessed by the zygote. Perhaps by retaining the zygote longer before release, the maternal organism could transfer additional food reserves to the zygote. Some species of *Coleochaete*, a charophycean alga, do exactly that. But there must still be a size limitation. Now suppose that the zygote divides by mitosis, and both diploid cells get stocked up by mom before eventually dividing by meiosis. As a result of a pre-meiosis division of the zygote, together with maternal contributions, the potential number of cells undergoing meiosis doubles, leading to a doubling of dispersal units without sacrificing size. Once that mitotic division takes place, the embryonic sporophyte generation exists even if in its initial state all the resulting cells undergo meiosis to produce spores.

Meiosis immediately following (in a developmental sense) the zygote is like cutting up a pie. The more people you serve, the smaller the pieces. If the zygote gets some postfertilization maternal nutrition, then that is like baking a bigger pie. If the zygote undergoes a development into two or more cells that can be nourished, then that is like baking more pies. And that is what the sporophyte generation does for land plants, it bakes more pies, in a reproductive sense.

Natural selection would greatly favor any organism that made this developmental "mistake." And if the developmental events that previously led to the production of a zygospore were expressed after meiosis, then these new haploid cells would end up encased in a desiccation-resistant spore wall. This becomes a reproductive advantage only if the diploid generation gains nutrition from the maternal gametophyte organism as an embryo, and this occurs during the embryonic phases of all land plant sporophytes.

Liverworts and some mosses retain what we envisioned as the initial stage of this new life cycle. The sporophyte remains small and wholly dependent upon materials transferred from the maternal gametophyte in whose tissue the embryonic sporophyte is embedded. In hornworts and many mosses, the sporophyte gets large enough to nutritionally contribute materials to its own spore production, although it remains dependent upon the gametophyte for water and mineral nutrients. The sporophytes of mosses, liverworts, and hornworts are also unbranched so they produce only one sporangium per fertilization, although each of them may produce thousands of spores. In vascular plants, apical, indeterminate growth results in the sporophyte outgrowing the maternal gametophyte and establishing itself as a long-lived independent organism. *Cooksonia* shows an intermediate stage where the sporophytes are still determinate, still dependent upon a maternal organism for some nutrition, but branched apically so as to produce more terminal sporangia.

The adaptive value of the land plant life cycle has generated considerable conjecture. The survival value of the diploid condition can be traced all the way back to bacterial transformation. Because the sporophyte generation came to dominate land, its toughness and success might derive from having two copies of each gene. When diploid you can possess different forms of each gene (alleles), and this may be an advantage in a patchy, variable environment. However, biologists have questioned how much of the success of vascular land plants can be attributed to the genetics of the diploid condition. In contrast, haploid organisms have only a single copy of each gene.

The advantage of having an embryo is clearer. Rather than dispersing an egg or a zygote into the environment, the maternal gametophyte gives nourishment and protection to the diploid embryo. This allows the diploid generation to grow and produce more cells to undergo meiosis, thus producing more spores, a tremendous advantage in a very patchy terrestrial environment. And of course, if fertilization via swimming sperm inherited from aquatic ancestors was an iffy prospect for early terrestrial plants, which is a reasonable assumption, then producing the maximum number of dispersible offspring from each fertilization event, from each zygote produced, would be quite important for survival.

To enhance dispersal, the sporophyte needs to be as tall as possible, and so sporangia are placed terminally on a tall upright axis or on a short axis that is held aloft by a branch of the gametophyte. These are not dazzling heights here, but just a few centimeters can greatly enhance dispersal distances. Selection for more spore production and greater height for dispersal would select for a larger sporophyte from a bryophytic ancestor, and in chapter 8 the origin of a sporophyte-dominant life cycle of vascular plants will be considered further.

NO LITTLE BLUE PILL CAN CURE THIS SPERM PROBLEM

Land plants replaced swimming zoospores with wind-dispersed spores, but swimming sperm, inherited from aquatic ancestors, remain a problem. Terrestrial animals solve this reproductive problem by either returning to an aquatic environment for fertilization (amphibians) or by putting their sperm in a sea of bodily fluids transferred from male to female during copulation. These options work because animals are capable of moving to mates and to aquatic mating places. Neither option is available to land plants "rooted" in place. For land plants, the ultimate solution to the problem of swimming sperm is the seed habit, but seeds are still several chapters and about 100 million years away from appearing. So until then, another solution is required.

The successful dispersal of spores via wind and the successful transport of swimming sperm to an egg for fertilization operate best under very different circumstances, one dry and one wet, one as high as possible and one low to the ground. The best means of achieving such diametrically opposite goals is to have two very different growth forms, one tall and one low-growing, one adapted for wind dispersal and one adapted for fertilization with swimming sperm—in other words, a life cycle with very different sporophytes and gametophytes, each adapted to a different function (Bell 1994). Tall, well-spaced plants will best disperse spores in dry conditions, but low-growing, clustered plants will best accomplish fertilization in wet conditions (Keddy 1981). Land plant gametophytes are generally small, low-growing organisms adapted for wetter habitats and often found growing clustered together in a small space. Like bryophytes, they don't need vascular tissue because they have diverse specializations (discussed in chapter 7) for conducting water along their surfaces, which also provides a film of water needed by swimming sperm.

Lastly, the land plant life cycle also can be considered an adaptation to a terrestrial environment based on its genetic and selectional consequences (Bell 1994). When a haploid organism produces gametes, they are all genetically uniform and identical to their parent's genotype because they were produced by mitosis. All the gametes from a single parent will perform similarly, but those performances will vary from good to poor depending upon environmental variables. Very often winners in fertilization contests are determined by speed, but performances of individual genotypes will vary under different environmental conditions. Some horses run well on a muddy track and others do not, and fast-in-the-mud horses are not the fastest horses on a dry track. All sperm of the same genotype will be winners, or losers, depending upon current "track" conditions. Genetically diverse sperm would provide a safer betting strategy because it ensures some winners no matter what the conditions, but an individual haploid plant can produce only one genotype. Of course genetically diverse gametophytes exist, so genetically diverse sperm exist too. The horse racing analogy is most apt because biologists have discovered that sperm do compete with each other. Sperm really do race each other and the winners achieve fertilization, which directly results in passing those winning genes on to the next generation, the definition of Darwinian fitness.

When a sporophyte produces spores, they are genetically diverse because independent assortment during meiosis shuffles the genetic deck of cards. Under conditions of unusually strong selection, like some sort of extreme environmental condition, circumstances may result where spores of only one particular genotype will survive. The resulting gametophytes all produce very similar gametes with the effect that a great deal of genetic variation would be lost from the population if it passed through a narrow environmental filter. Needless to say, this is not desirable because maintaining genetic diversity is of tremendous value in a patchy and changeable terrestrial environment. Loss of genetic diversity reduces a population's ability to deal with future fluctuations and variations in their environment.

The land plant life cycle provides a means of avoiding this genetic canalization because loss of genetic diversity through competition among gametes in the haploid generation is offset by, and alternates with, independent assortment, which generates new gene combinations, in the diploid generation (Bell 1994). Bell envisioned a situation where a multicellular sporophyte would be favored (selected for) if genes existed that affected gamete competitiveness and vegetative growth in a diametrically opposite manner. This means a genotype that affords high gamete competitiveness (winners in fertilization contests) would result in limited vegetative growth (small, slow growers), and vice versa. These two different and opposite selectional forces cannot be optimized in a haploid organism, so if a multicellular diploid sporophyte were added into the life cycle it would allow gamete competition to continue in each generation without one particular haploid genotype becoming genetically fixed by strong selectional forces. To test this hypothesis, experimental studies were conducted on *Chlamydomonas*, a unicellular chlorophycean alga, and these experiments demonstrated that such genes exist in green algae and were subject to selection (Bell 1994). This genetic/selectionist argument complements other perspectives on the adaptiveness of the land plant life cycle.

The foregoing discussion was a bit esoteric and probably not completely understandable, so its inclusion in a book for nonbiologists is rather questionable, but it provides an interesting example of how science works. The hypothesis of genes with diametrically opposite effects on gamete competitiveness and vegetative growth was constructed logically by Bell on the basis of his knowledge of genetics and natural selection to account for what we have (an alternation of very different haploid and diploid generations) from what was proposed to be the ancestral condition (the haplontic life cycle of charophycean ancestors). This hypothesis arose from pure "what if" thinking based on assumptions of how evolution operates and what happened (charophycean green algae gave rise to land plants). Bell's hypothesis explains why the evolutionary mechanism of natural selection would have favored (increased the reproductive success of) any organism producing a multicellular diploid generation. Bell tested one genetic assumption of his hypothesis by demonstrating such genes exist in green algae, thus increasing our confidence in his, and therefore Bower's, interpolation hypothesis. Bell's hypothesis was sheer speculation, although consistent with sound biological knowledge, but at least some aspects of this hypothesis were testable. This also demonstrates why biologists are confident in the process of evolution to the point of certainty. Predictions based upon presumed processes of evolution are repeatedly found to be true.

This example also demonstrates another way science works. Bell used *Chlamydomonas* as his experimental organism even though this algal organism has only a distant common ancestry with charophyceans. So why was it chosen? *Chlamydomonas* was the closest relative to land plants on which the necessary experiment could be conducted. More is known about the genetics of this genus than just about any other green algal organism, and as a unicellular organism it lends itself to fairly easy culture and experimental manipulation. Since *Chlamydomonas* was demonstrated to have the hypothesized genes, it becomes reasonable to assume the common ancestors of charophyceans and land plants also had such genes. Biologists must often make such decisions about how to test their ideas, particularly when designing experiments. Now back to the discussion of the land plant life cycle in progress.

Land plants share another novel character not present in their algal ancestors: multicellular sex organs, called archegonia and antheridia. Gametangia grow from an initial cell that develops by a series of well-organized cell divisions into a sex organ that includes sterile jacketing cells enclosing the gamete-producing cell(s). In algae, all sex organs are unicellular, each gametangial cell giving rise to exactly one or two gametes. The traditional explanation of multicellular sex organs is that the sterile jackets protect developing sex cells until such time as they are released, but this does not seem a very compelling argument because the unicellular sex organs of seaweeds seem to function quite well at protecting algal sex cells. Adaptionist thinking always entails the danger of assuming everything is the result of natural selection, and this may not always be the case.

Perhaps the developmental genes that produce multicellular sporophytes and sporangia have a similar effect on gametangia. In other words, clearly genes exist for the production of unicellular reproductive organs in algae, and the type of reproductive cell produced (gamete or spore) may depend upon whether the organism is haploid or diploid. This is not an unreasonable assumption. For example, *Ectocarpus*, a brown algal seaweed, has an alternation of identical haploid and diploid phases, and its gametangia and sporangia look exactly the same (plurilocular gametangia/sporangia) (Appendix: Brown Algae), indicating that sporangia

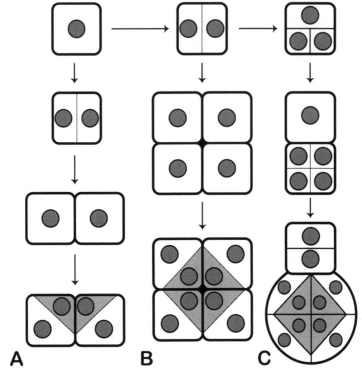

Fig. 6–8 Development of antheridial cells and antheridia. A. In some species of *Coleochaete* an initial cell divides transversely, and then each daughter cell divides unevenly to produce one spermatogenous cell (gray) and a sterile adjacent cell. B. If two initial cells were in adjacent to each other in a single layer of tissue, the four spermatogenous cells (gray) would be surrounded by sterile cells. C. A similar development in three-dimensional tissue, in a parenchyma, begins with a first division (center top) of the initial cell that establishes a polarity where the basal cell (on the top in this diagram) produces stalk cells and the apical cell produces the antheridium proper. Additional divisions take place in the four antheridial initials in a plane at right angles to this diagram such that the sperm-producing cells are surrounded by sterile jacket cells. (Image source: The Author)

and gametangia are homologous structures, so the same developmental program produces reproductive structures in both haploid and diploid phases. Even the flagellated reproductive cells are similar in form, but they behave as gametes, fusing with the opposite mating type (+ or -) when haploid and acting as zoospores when diploid. So perhaps multicellular gametangia are a developmental consequence of multicellular sporophyte/sporangium development and have no particular "advantage." Developing techniques in molecular developmental genetics may soon allow this hypothesis to be tested.

Jacketed gametangia also may have arisen as a consequence of the parenchyma, the block-constructed tissue of land plant gametophytes.

Those species of *Coleochaete* whose thallus consists of a single layer of cells have an interesting developmental pattern for producing antheridial cells (Graham 1993) (Appendix: Green Algae, charophyceans). A single initial cell divides transversely, and then each daughter cell divides unevenly, producing a sperm-generating cell and a vegetative cell. The latter would be developmentally homologous to jacket cells of a land plant antheridium (Fig. 6–8). If a similar developmental program were expressed in a multilayered thallus, sterile jacket cells would surround the sperm-generating cells. Again this hypothesis proposes that certain features result from ancestral genes controlling development being expressed in a new environment—in this case, a 3-dimensional tissue called parenchyma.

The land plant life cycle through its innovation of a sporophyte generation provides both reproductive and genetic advantages and, ultimately, new adaptations to the terrestrial environment (cuticle, stomates and guard cells, vascular tissue). This chapter ends with lichen-bryophyte crustal communities dominating many terrestrial areas while building soil and slowing water run-off. The rhizoids of the earliest vascular plants, or their immediate ancestors, are firmly embedded in the soft soils of Silurian deltas turning low-lying coastal areas green with a short forest of sporophytes towering to some 10 to 20 cm above the surface. While not terribly impressive by today's standards, these forests were sufficiently tall to disperse spores widely and easily compete with microbial mats, soil crust communities, and bryophytes for light. The plants composing these ancient communities will be introduced in chapter 8 and their relationships with other fossil and living land plants explored.

SEVEN

The Pioneer Spirit

Wherein liverworts, hornworts, and mosses are examined to demonstrate their
adaptations to terrestrial life and their relationships to each other and vascular plants.

Bryophytic
Graminaceous my lawn was, and not
Bryophytic—that's how it has got.
I should stop, cut my losses,
Embrace all these mosses,
'Cause I can't grow much grass in this plot.
—John Critten

To some extent, bryophytes, liverworts, hornworts, and mosses are the
Rodney Dangerfield of land plants: "I don't get no respect!" The botanists
who study these organisms call themselves bryologists, and some of them
claim they get no respect either. Why might this be the case? Well, they
are small and frequently just trod upon without notice (the bryophytes,
not the bryologists). So for average people, many of whom suffer from
plant blindness[1] anyways, bryophytes simply fall below the perceptual
horizon. Even avid naturalists and accomplished field biologists, the kind
of people who easily identify vascular plants, often completely ignore the
bryophytes or can get no more specific than "It's a moss." Small size is
only part of the problem; good field guides for bryophytes seldom exist.
Even when a field guide does exist,[2] the plethora of technical terms, the
jargon, can be rather off-putting even for botanists if they are unfamiliar
with bryophyte descriptive terminology. An illustrated glossary[3] can help,

but how many books do you want to carry around? Unfortunately, most botany books, most botany courses, and, indeed, most botanical staffing decisions place an emphasis on seed plants,[4] so perhaps you can understand why bryologists feel a bit trod upon, a bit overlooked, themselves. As a result it comes as no surprise that most people, and even most botanists, know relatively little about liverworts, hornworts, and mosses. Yet according to our prevailing hypothesis, the first land plants were bryophytic organisms. Their meager and fragmentary fossil record means we must turn to their living descendants to get clues about how such organisms invaded and adapted to the terrestrial environment.

In trying to accurately characterize bryophytes, some widely held misconceptions must be corrected. The hypothesized phylogenetic intermediacy of bryophytes between charophycean green algae and the vascular plants and the preference of many bryophytes for wetter habitats have generated the perspective that bryophytes are poorly or incompletely adapted to the terrestrial environment, representing and occupying an evolutionary waystation between water and land. But thinking of bryophytes as intermediate between green algae and vascular plants in size, stature, and structural complexity is an error not because you are comparing apples and oranges but because you are comparing apple pollen to apple trees, comparing gametophyte to sporophyte, comparing haploid to diploid generations. Remember how different haploid and diploid ferns were? When you make a fair comparison on a gametophyte to gametophyte basis, liverworts, hornworts, and mosses are much larger and often more complex than the gametophytes of any vascular plants (Fig. 6–3 B; Fig. 7–2 A). The familiar moss, liverwort, or hornwort is always the haploid gametophyte, and many textbooks say the gametophyte is "dominant" in their life cycle (Fig. 7–1), but calling them "the dominant generation" just does not do them justice. Liverworts, mosses, and hornworts represent the epitome, the zenith, the absolute pinnacle of haploid organisms—the largest, most complex haploid organisms in the entire history of life! And no bryologists paid me for that, although they should have.

With respect to the diploid generation, the sporophytes of mosses, liverworts, and hornworts are intermediate in size and complexity between the charophycean zygote, a single cell and the only diploid stage in their life cycle, and the sporophyte of vascular plants. Bryophytes have small, unbranched (uniaxial) sporophytes of determinate growth that remain largely dependent upon the larger gametophyte for nutrition and water. Determinate growth means that the sporophyte develops as an embryo until essentially all the cells have been produced, and then upon their enlargement and differentiation, growth is done and a mature size is reached;

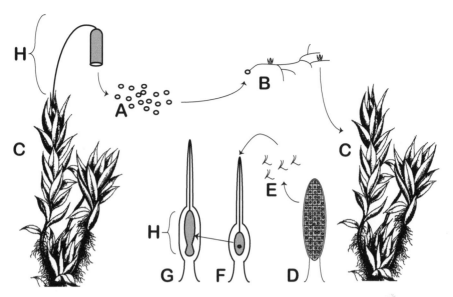

Fig. 7–1 Sexual life cycle of a typical moss. The gametophyte generation starts with a haploid spore (A) begins growth as a filamentous algal-like stage (B), which branches, spreads, and produces leafy aerial branches and anchoring rhizoids. At maturity these gametophyte axes (C) are independent and no longer interconnected. Gametangia are borne upon the apices of these axes, either antheridia (D) or archegonia (F). Swimming sperm (E) must make their way to an archegonium upon another axis. The diploid (2n) generation is represented by the zygote, embryo (H) , and mature sporophyte (H) consisting of a placental foot embedded in the gametophyte, a stalk, and a terminal sporangium. (Image sources: The Author; C—leafy gametophyte axis of *Bryum* after Grout 1903)

no further growth is possible because they do not retain a meristem. Uniaxial means the sporophyte consists of a single axis with a placental foot at the proximal end and a sporangium, often called the capsule, at the distal end. The largest and most complex sporophytes are found in hornworts and some mosses, and they approach the structural complexity of vascular plant sporophytes by exhibiting a discrete epidermis with cuticle, guard cells and stomates, a cortex of ground tissue, and in the case of some mosses, a central cylinder of conducting tissue. However, these mosses do not directly share a common ancestry with vascular plants.

Traditionally land plants, the embryophytes, were divided into two groups, vascular plants and nonvascular plants, but classifying mosses, liverworts, and hornworts as "nonvascular" plants is an error on several different levels even though they do lack vascular tissue. Animals and fungi lack xylem too, so my cat is just as nonvascular as a moss. To define a lineage by a character it never had (vascular tissue) is just poor practice.[5] A subtle conceptual error is generated by the implication that bryophytic land plants lacking vascular tissue are less well adapted to terrestrial conditions. Quite simply, this is not true; however, mosses, liverworts, and hornworts function quite differently than vascular plants.

Lastly, the concept of nonvascular plants as a taxonomic group—that is, a phylum Bryophyta based on their "shared" lack of vascular tissue—is a phylogenetic error because on the basis of all current hypotheses, mosses, liverworts, and hornworts fail to form a single lineage. Many people have the misconception that evolution means newer organisms replace older or earlier organisms,[6] so if early land plants were bryophytic in nature, and everything we know suggests they were, then why were they not displaced by vascular plants? Why were they not condemned to extinction? This is still another conceptual error. Actually the exact opposite has happened. Vascular plants are big environmental obstructions (Corner 1964), and in the process of affecting the movement and distribution of wind, water, and soil particles, they have generated many microhabitats to which bryophytes have adapted. Vascular plants and bryophytic plants are adapted to and occupy truly different habitats, so they coexist. So rather than dying out in the presence of vascular plants, bryophytic plants gained more opportunities from vascular plants, and some groups, some lineages, have become more diverse. With nearly 19,000 described species (mosses: more than 10,000; liverworts: more than 8,000; hornworts: 800), the combined diversity of bryophytes is topped only by flowering plants. But in spite of their prevalence in moist environments, bryophytes are nonetheless land plants. A very few are aquatic in fresh water, a couple grow on wave-soaked coastal rocks, but none are truly marine.

HOW DO BRYOPHYTES FUNCTION?

Since the dual functions of xylem are support and water conduction, it comes as no surprise that bryophytes are small in stature and inhabit moist environments. The tufts, mounds, and overlapping sheets of tissue formed by bryophytic plant bodies do not expose as much surface area to a drying atmosphere as do the aerial axes of vascular plants, which typically stand aloft with leaves like flags on poles. A low, substrate-hugging plant presents a minimal obstruction over which the atmosphere moves largely unimpeded. Rather than conduct water through their bodies, bryophytic organisms have scales, hairs, channels, rhizoids,[7] flat overlapping bodies, or leafy organs that form capillary spaces between them. These features serve to conduct water along and over the surface of the plant body. Their ability to capture and maintain a film of water externally provides a means of rapidly hydrating themselves, and this film of water also provides a means by which swimming sperm can reach an egg. In addition, many bryophytic plants have established symbioses with nitrogen-fixing cyanobacteria and with the same type of fungi that produce mycorrhizae in vascular plants. These associations enhance the ability of these plants to obtain water, nitrogen, and mineral nutrients even though they lack roots. Since the earliest vascular plant fossils exhibit evidence of mycorrhizae, this association may have evolved much earlier and may have been a significant factor in early colonization of the terrestrial environment (Selosse 2005).

Because of their adaptations and relatively small size, bryophytic plants are common and have their highest species diversity in warm and/ or wet/moist habitats. In some places like bogs, tundras, and cloud forests, and among epiphyte[8] communities in wet forests, mosses and liverworts may form a significant and even dominant portion of the plant community. Perhaps the best known, most important, and most extensive bryophyte-dominated communities are *Sphagnum* (peat moss) bogs (Fig. 7–2; Fig. 7–3) (Lappalainen 1996), wetlands characterized by extensive mats of *Sphagnum*, whose leafy organs are modified to hold large volumes of water by capillary action (Appendix: Fig. A48). Peat bogs have a low pH; they are acidic, so organic matter, the dead *Sphagnum*, accumulates and ultimately makes peat, a high-carbon soil that when dried can be burned as a fuel that lends its smokiness to Scotch whiskies. Peat bog communities tend to be at high latitudes or high altitudes, and if the climate warms, these bogs are subject to earlier spring thawing and longer,

Fig. 7–2 Sphagnum moss and bog community. A. Close-up of *Sphagnum*, peat moss, showing apical view of aerial shoots forming a mat. B. *Sphagnum* bog community with extensive mats of peat moss (foreground). (Image sources: A—Kristian Peters, Wikimedia Creative Commons; B–Pavlova Hut' Nature Preserve, Czech Republic, Courtesy of David Palouch, Wikimedia Creative Commons)

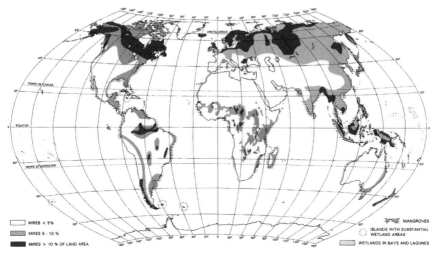

Fig. 7–3 Distribution of mires, wetland communities dominated by *Sphagnum*, peat moss. In high latitudes and high altitudes bogs and peat lands make up greater than 10% of the total land area (dark gray). Lighter gray areas have peat lands between 0.5% and 10%. (Image source: After image provided courtesy of the International Peat Society)

warmer summers, conditions that promote decomposition. The relatively rapid release of carbon from these long-term reservoirs may contribute to the accumulation of carbon dioxide in the atmosphere and oceans. This possibility is drawing more biological attention to these communities than usual.

Even the idea that bryophytic plants are delicate proves to be an error of overgeneralization. Many bryophytes are adapted to tough places: deserts, alpine zones, rocks, and tundra, areas too tough for most woody plants.[9] Their low stature and substrate-hugging growth forms allow bryophytic plants to grow upon thin rocky soils and hard substrates like rocks or bark that cannot be colonized by vascular plants, whose roots require a deeper soil substrate. And interestingly enough, the flowering plants that occupy alpine zones and tundra have adapted to these harsh habitats by adopting forms similar to mosses, small-leafed plants that form dense mounds or tussocks. Their common names reflect this resemblance—for example, moss champion and moss phlox.

Bryophytic plants are misunderstood because they function very differently from the vascular plants humans routinely interact with. Of the four hallmarks of land plants (Fig. 6–2)—stomates with guard cells, epidermis covered by a waxy cuticle, xylem, and trilete spores—the first three are hallmarks of homeohydric (hoh-ME-oh-high-drick) plants, adaptations for maintaining a hydrated state by preventing water loss to a drying atmosphere. So whereas most vascular plants minimize water loss, many bryophytic plants, like many microorganisms, can dry out without harm. This is a profound difference. Many bryophytes are poikilohydric (poy-KEY-low-high-drick; *poikilos* = various) and/or tolerant of drying (Proctor 2000; Graham and Gray 2001). Poikilohydrous organisms maintain an equilibrium with atmospheric humidity, gaining or losing water readily, sometimes hydrating and partially dehydrating on a daily basis, hydrating with a morning dew or a misty, foggy, or cloudy condensation, whenever the relative humidity is high, then drying out enough to stop metabolic activity each sunny afternoon, and rehydrating in the evening as the relative humidity goes back up. Many mosses and leafy liverworts demonstrate this ability. Poikilohydrous organisms are capable of surviving dehydration for hours or days but usually not longer, so they are found in habitats where a saturated atmosphere is a regular event. Such organisms may or may not be tolerant of drying as well. Organisms tolerant of prolonged drying can survive dehydration for longer periods of time in a metabolically dormant state. Both poikilohydrous organisms and those tolerant of drying are capable of rehydrating and recommencing active metabolism quickly, although poikilohydrous ones are fastest, sometimes taking only 15 minutes to revive.[10] Organisms adapted to ephemeral freshwater habitats or those that grow at or near the water surface are frequently tolerant of drying, and so the basis of these physiological adaptations might have been inherited from charophycean ancestors (Graham et al. 2012).

The ability to readily gain or lose water requires a small, thin plant body. If constructed of a thicker tissue, the plant body could not gain or lose water quickly enough. Large-volume bodies with thick, block-organized tissues are characteristic of homeohydric plants. An epidermis or a cuticle would interfere with water loss and uptake for poikilohydric plants, and their thinly constructed body tissue does not require xylem to stay hydrated or stomates for internal gaseous exchange.[11] Poikilohydric and homeohydric biologies produce plant bodies of rather divergent construction. In general, when a bryophytic plant possesses a thicker body (a block construction), it will also exhibit adaptations consistent with homeohydry and/or live in a continuously moist environment. Those with thin, delicate tissues are poikilohydric, which, rather counterintuitively, may live in less continuously moist environments.

Parenchyma and block construction have been mentioned but without much explanation. Parenchyma is the generalized primary cell type that forms the body or ground tissue of all nonwoody land plants. This includes the pith and cortex of stems, the cortex of roots, and the photosynthetic tissue of leaves. In addition to being in land plants, parenchyma is found in some charophycean green algae and some brown algal seaweeds. Parenchyma cells are live and thin walled and generally have a large central vacuole. Photosynthesis, water and food storage, and gas exchange are all functions of parenchyma, although not necessarily all at once. To be called parenchyma, the plant body must be a 2-dimensional sheet of cells or a 3-dimensional block-organized tissue. As a tissue, parenchyma is characterized by intercellular spaces and intercellular cytoplasmic connections called plasmodesmata, which result from nonfurrowing[12] types of cytoplasmic division (Cook and Graham 1999). Recall that all land plants share one such type of cell division. A multicellular organism can have plasmodesmata but not form parenchyma—for example, some filamentous algae. Some seaweeds have a block construction without plasmodesmata—for example, *Ulva* (Fig. 5–8), so although this seaweed's thallus consists of two cell layers, it is not parenchyma. Some species of *Coleochaete* are composed of a single layer of parenchyma (Appendix: Green Algae, charophycean). The simplest bryophytes have a body composed wholly of parenchyma, a structural complexity only slightly greater than charophycean algae. The general term for an undifferentiated plant body is thallus. Although lichens have a very different construction, basically densely woven filaments, those that are broad, flat, and green-appearing are sometimes called a thallus, as if a bryophyte, rather than more correctly called a mycelium (my-SEAL-ee-um), the organized mass of filaments making up a fungal body.

Some bryophytic plants may possess a discrete epidermis, but if so, a cuticle is usually not formed. Thicker-bodied liverworts have some dorsal to ventral differentiation of the parenchyma: more photosynthetic cells dorsally and more storage cells ventrally (Appendix: Liverworts). No sharp demarcation occurs, and in part this differentiation may be a result of decreasing light penetration, so as expected, thinner plant bodies tend to be more uniform throughout. A few liverworts have parenchyma organized into air chambers for floatation (Fig. 7–4) or photosynthesis where the chambers are filled with filamentous cells functioning like the leaf mesophyll of vascular plants (Fig. A45).

THE BRYOPHYTIC PLANT BODY

The predominate growth forms of bryophytic plants are a dorsoventrally flattened body (Fig. 7–4 A; Fig. 7–6 B) (hornworts and some liverworts) or a

Fig. 7–4 Plant body of the liverwort *Ricciocarpus*. A. Flat, Y-branching plant body (white arrows), a thallus with a central furrow visible as a line on the dorsal surface. B. Scanning electron micrograph of a section through the thallus showing that the upper portion of the plant body is organized into air chambers (ac) for flotation and light penetration. Lower central portions of the plant body have denser storage tissue (st). Overlapping scales and rhizoids on the ventral (lower) surface (vs) serve in uptake, movement, and absorption of water. Archegonia are situated deep within the thallus at the base of the dorsal furrow (df) and their neck canal opens into this space. (Image sources: A—Courtesy of and with permission of Alan Cressler; B—After Kronestedt 1981)

vertical or prostrate leafy axis (mosses and most liverworts) (Fig. 7–5 A-B). Dorsoventrally flattened bodies are referred to as or described as thalloid, but as noted above this generates some confusion because thallus is used to describe all largely undifferentiated plant bodies. Leafy bodies look more complex than a leafless, flat thallus, but generally the "leafy" plants are simpler in construction and internal organization. The early growth of many bryophytes following spore germination is filamentous, a developmental stage called a protonema (pro-tun-EE-mah). The protonema stage of mosses resembles green algae so much that they can be easily misidentified even under the microscope.[13] After initially dividing in a single transverse plane, cells in the protonema then divide in such a manner that it develops into a 2-dimensional thallus or an aerial axis. In mosses, the protonema stage can branch and proliferate extensively, spreading across its substrate. Many aerial branches are formed (Fig. 7–1 B-C) and grow into a cluster, clump, or tuft of vertical leafy axes. While not physically connected at maturity, these axes are a clone, one genetic individual, and therefore, a single sex. However, two or more protonema may interlace such that axes of different sexes are interspersed within a single tuft.

The vertical axes of some mosses display three distinct tissue regions: an epidermis, a cortical region of parenchyma, and a central cylinder of conducting tissue with phloem-like cells but no xylem. Since these mosses do not appear to be the direct ancestors of vascular plants, the logical conclusion is that conducting tissues evolved independently in the vascular plant lineage. Phloem-like conducting cells have had several independent origins; such conducting tissue also occurs in the stipes of large brown algal seaweeds.

The leafy appendages of bryophytes are called enations (ee-NAY-shuns) (Fig. 7–5). Although enations function to increase surface area like leaves, they lack internal organization, often consisting of just a single layer of cells. A few mosses have enations with more complex organizations to enhance either water uptake and retention (*Sphagnum*) or photosynthesis (*Polytrichum*) (Appendix: Fig. A46; Fig. A48). Like their algal ancestors, bryophytes possess rhizoids, either unicellular or multicellular filamentous outgrowths that serve to anchor the plant body. In the terrestrial environment, rhizoids have an additional function: water absorption and conduction. Rhizoids can be scattered across the ventral surface or produced in discrete zones, often along a central ventral region, particularly in some liverworts. Masses of rhizoids, sometimes enclosed within

Fig. 7–5 Leafy thallus of liverworts and mosses. A. Thallus of a leafy liverwort. The enations are arranged in three ranks or rows, one to each side, and a third reduced row, hidden from view, down the middle of the lower surface. A portion of each enation is folded around to the back side. The lower rank can function to help the thallus adhere to surfaces. B. Thallus of a prostrate moss whose enations are oriented in two lateral ranks and a third rank of smaller enations along the top of the axis. C. An upright aerial axis of moss with its enations arranged helically. (Image sources: A—Courtesy of Haynold, Wikimedia Creative Commons; B and C—Courtesy of and with permission of Niels Klazenga)

overlapping scaly outgrowths of the thallus, function like bristles on a brush wicking up and holding water by capillary action, and channeling it to and along the plant body (Fig. 7–4). Capillary spaces also function in the uptake and distribution of water along the plant surface.

THE SPOROPHYTE

The sporophytes of liverworts, hornworts, and mosses are the smallest and simplest diploid phase among land plants. Among the bryophytic plants all sporophytes remain attached to and partly to wholly dependent upon the gametophyte nutritionally. Sporophytes consist of a single unbranched axis with a basal foot, which has a placental function, embedded in the gametophyte tissue from which the sporophyte obtains nutrition and water. The sporophyte may be green and photosynthetic, thus capturing some of its own energy (mosses and hornworts). In some mosses the sporophyte is photosynthetically independent by the stage of spore production, but water is still obtained from the gametophyte. The distal end of the sporophyte axis is terminated by a single sporangium, which is called the "capsule" in mosses and liverworts. A stem-like stalk forms the axis between the foot and sporangium. The cells composing the stalk may or may not elongate as the sporophyte nears maturity, and in some species an axis between the foot and capsule may be largely lacking. Sporophytes display a discrete epidermis, and in mosses and hornworts a cuticle, stomates, and guard cells are present. Stomates differ from pores because the stomate is related developmentally to and circumscribed by a pair of guard cells, which can open and close to regulate the stomate. In some mosses the guard cells of their stomates open only once when the sporangium is mature to dehydrate and open the sporangium quickly. This may be the original function of stomates rather than regulating gas exchange and transpiration. Again, bryophytes function differently than vascular plants.

The hornwort sporophyte is a long, narrow cylinder that tapers to an apex when immature, thus a "horn." Spore-forming tissue occupies a cylindrical portion of this "stem-like" sporophyte (Fig. 7–6). External to the spore-forming tissues is a cortex of photosynthetic tissue and in the center is a small sterile core, something like a pith. Hornwort sporophytes also exhibit a limited degree of indeterminate growth from an intercalary meristem that produces new tissue between the foot and the base of the axis, which pushes up more immature tissue as the apex matures.

Fig. 7–6 Hornwort gametophyte with sporophytes. A. Growth habit showing cylindrical sporophytes (bracket) , the "horns" of the hornwort, growing up and out of a flat irregularly lobed gametophyte thallus. The sporophytes grow basally so the apical portion is the oldest, and the oldest apical portions (upper arrow) have reached maturity, dried out, and dispersed spores. Note the column of gametophyte tissue growing around the base of the sporophyte (lower arrow). B. Detail of young sporophytes (arrows) emerging from the gametophyte thallus. C. Cross-sections of nearly mature sporophytes showing stem-like organization (left—light microscope; right—scanning electron microscope). The epidermis has stomates and guard cells (e). A cylinder of spore-forming tissue (s) occupies the center; a green, photosynthetic cortex occupies the space between. (Image sources: A—Courtesy of and with permission of Bob Klips; B—Courtesy of Benutzer Oliver, Wikimedia Creative Commons; C—After Shaw and Renzaglia 2004, with permission and courtesy of the *American Journal of Botany*, Botanical Society of America)

Under some circumstances, hornwort sporophytes can survive independently for up to a year or more after the surrounding gametophyte thallus has died (Delwiche et al. 2004). This is as close to independent growth as a sporophyte gets in bryophytic plants.

As the bryophyte sporangium reaches maturity, hundreds of spore mother cells undergo meiosis producing four haploid spores per mother cell. When mature, the sporangium dehisces, opening by one of several mechanisms and dispersing the spores by wind. Since the height at which spores are released is a primary factor influencing dispersal distance, in many species the sporophyte stalk or a specialized outgrowth of the gametophyte thallus, a pseudopodium, functions to elevate the sporangium. Mosses elevate their sporangium by having long-stalked sporophytes. The sporophytes of liverworts have short stalks if they have any stalks at all, so the only way to elevate the sporangium is to elevate the archegonium; this is because after fertilization of the egg, the zygote within the archegonium is where the sporophyte will develop. The most dramatic of such structures are specialized gametangia-bearing aerial branches of the liverwort *Marchantia* and related genera (Fig. 7–7). Elevating both arche-

Fig. 7–7 The liverwort *Marchantia* showing specialized gametangia-bearing branches. A. Lobed aerial branches bearing archegonia and later, after fertilization, sporophytes. B. Flat-topped aerial branches bearing antheridia in sunken pits. Flat, prostrate, Y-branched gametophyte thallus can be observed 1.5 to 2 cm below. (Image sources: A—Courtesy of Galwaygirl, Wikimedia Creative Commons; B—Courtesy of and with permission of Roland Barth, Fontenelle Nature Search)

gonia and antheridia does not seem likely to enhance fertilization, but the advantage of enhanced spore dispersal must more than compensate for any reduction in the rate of fertilization that comes from releasing swimming sperm from antheridia raised aloft on similar branches. You may wonder then why evolutionary processes produced such an ungainly situation because having antheridia on a raised branch does not seem to make much sense, but things like this happen because the developmental genes being acted on by natural selection are not that specific. Under proper conditions a developmental pathway produces an aerial, gametangia-bearing branch rather than a prostrate branch. In other words, both types of sex organs develop on elevated branches because the genes involved are not sex specific. It's like having genes for five digits, which develop into fingers on your front limbs and toes on your hind limbs. While the stalks look stem-like, they are more like a folded thallus with a ventral groove containing rhizoids and thus providing a vertical watery channel for sperm dispersal. Quite frequently in biology, evolution has produced different solutions to the same problem, differences that arise from constraints based upon what can and cannot be altered by natural selection.

A few centimeters may not seem like much gain in elevation, but research has demonstrated that height and dispersal distance are nonlinear, meaning that small increases in height can greatly enhance dispersal distance (Niklas 1992). Because of spore dispersal mosses can colonize new habitats at a considerable rate (Miller and McDaniel 2004). Specialized cells called elaters may differentiate from an initial division of the spore mother cells. Elaters have unevenly thickened secondary cell walls and are dead at maturity, but the cell wall changes its shape with changes in humidity, and the writhings of these dead cells function to break up clumps of spores and assist in spore discharge.

WHAT'S IN A NAME?

Foregoing a taxonomy in phylogenetic flux, the familiar and unambiguous common names—liverwort, hornwort, and moss—remain very useful because the taxonomic names used in botanical literature can be quite daunting for all but the specialist. Here is the problem. Several names exist for each group, and none are unambiguously correct. Not only have the authors of different classifications used different taxonomic ranks, each with its own designated suffix, but the taxonomic names have been formed in both regular (following taxonomic rules) and irregular ways. In older literature, the division Bryophyta (suffix -ophyta = phylum in zoology) consisted of three classes, the Hepaticae, Anthoceridae, and Musci. These taxonomic names are irregular in construction[14] because they use nonstandard suffixes and two of the three are not based upon a root of a generic name. Hepato- is a prefix meaning "liver,"[15] and liverworts are sometimes called the hepatics. Class Hepaticae, if formed by convention, would use the suffix -opsida, forming Hepatopsida, sometimes spelled Hepaticopsida. If liverworts were treated as a division, the taxon would be Hepatophyta. If a strict adherence to typification is followed, higher taxa are named after either a characteristic genus or the oldest genus. Thus *Marchantia* is used to form a root for naming liverworts (class—Marchantiopsida, division—Marchantiophyta). Within the class would also be order Marchantiales (-ales) and family Marchantiaceae (-aceae) along with other orders and families named for other genera.

The genus *Anthoceros* forms the root of taxonomic names for hornworts: family Anthoceraceae, order Anthocerales, class Anthoceropsida, phylum Anthocerophyta. *Musci* is simply the Latin word for moss. Forming a taxonomic name based on the genus *Bryum* results in Bryaceae, Bryales, Bryopsida, Bryophyta—but Bryopsida and Bryophyta, thus formed, can be misinterpreted as referring to all bryophytes, so you must be careful. Just as algae remains a useful general term, the term bryophyte remains convenient to refer to all three groups of nonvascular land plants because phrases like "plants of byrophytic organization" and "bryophytic plants" get a bit awkward and tiresome.

The class/phylum/kingdom higher category confusion arises from the way in which the taxonomic hierarchy was established. In a sort of taxonomic grassroots movement, a poor pun, the taxonomic classification of flowering plants was constructed upward from binomial species names (genus + specific epithet = a species name). Genera that shared more general characters were grouped into families, many of which have familiar names (grasses, orchids, mustards, lilies, mallows, beans, nightshades, roses, asters, etc.). Families were in turn grouped into orders, and likewise, orders were grouped into the traditional two classes, monocots or dicots, depending upon whether the embryo had one or two cotyledons. Angiosperm taxonomy was organized upward from the species, rather than downward from the kingdom, and as a result flowering plants were usually treated as the phylum Magnoliophyta[16] of the Plant Kingdom. But according to current phylogenetic hypotheses, the flowering plant lineage is nested within the seed plant lineage (spermatophytes), which in turn is nested within a woody plant lineage (ligniophytes), which in turn is nested within the vascular plant lineage (tracheophytes), which is in turn is nested within the embryophytes, streptophytes, and chlorophytes (green algae), all of which are in the plant kingdom. This means at least six nested levels occur between Phylum and Kingdom, and no taxonomic categories exist in between unless you construct subkingdoms, infrakingdoms, superphyla, and so on (see Kenrick and Crane 1997 for just one such an exercise). Even if embryophyte is taken as the phylum, then working downward, the rest of traditional taxonomy gets crunched into too few categories, making a different kind of mess. In many current phylogenetic hypotheses, mosses, liverworts, and hornworts form three lineages equivalent to vascular plants. This naming problem cannot be resolved in any satisfactory manner presently, and so liverworts, hornworts, and mosses they are.

FALSIFYING A TAXONOMIC HYPOTHESIS: BRYOPHYTA IS NO MORE

Along with their lack of vascular tissue, bryophytes share a gametophyte-dominant life cycle, and this long seemed an adequate justification for grouping them as division Bryophyta. Given the many morphological and anatomical differences among mosses, liverworts, and hornworts, in both their gametophytes and sporophytes, their classification as three distinct groups has seldom been in question (Crandall-Stotler 1980). The debate has been, and continues to be, how these groups relate to each other and to the vascular plants.[17]

Many recent phylogenetic studies (over twenty at this writing, from 1985 to 2010), using different data sets, have generated a total of six different hypotheses for relationships among the four lineages of land plants (Duff and Nickrent 1999; Goffinet 2000; Qiu et al. 2006). Data sets are derived from ribosomal DNA sequences, chloroplast DNA sequences, sequences of several nuclear genes, cell structure, developmental studies, and morphology, alone or in various combinations. Since the sporophyte is a novel character shared by all land plants, no charophycean has sporophyte features that can be used for outgroup comparison, so the inability to polarize sporophyte characteristics using an outgroup complicates the problem of using morphological data. Although the exact phylogenetic relationships among embryophytes are very much in question, one issue is by now pretty clear: Bryophyta as a taxonomic group no longer exists. Only one of the six phylogenetic hypotheses for embryophytes results in a bryophyte lineage (Fig. 7–8 A), and this one is considered unlikely to be correct. Other hypotheses variously arrange the three bryophyte lineages as a basal grade to vascular plants. Either liverworts or hornworts are placed as the basal or earliest divergent land plant lineage (Fig. 7–8 B-F). The various hypotheses differ as to which of the three lineages of bryophytes is sister group to the vascular plants. One hypothesis (Fig. 7–8 F) even places the hornworts as a basal lineage and the tracheophytes as a sister group to a moss-liverwort clade.

These hypotheses are not just a matter of paying your money and taking your choice, nor are these hypotheses just the opinions of different authors. Each hypothesis is based upon one or more data sets, and the phylogenetic patterns recovered from the analysis of these data differ. Each hypothesis requires different assumptions about the origin and sequence of different characters, and science will proceed as biologists attempt to falsify each of these hypotheses, until a consensus is reached, and until diverse data sets produce enough congruence that our confidence is higher for one hypothesis than for the others. A study of the mechanisms of auxin regulation in the sporophytes of mosses, liverworts, and hornworts (Poli et al. 2003) supports hypotheses B and E. Auxin is a major plant growth hormone, and it appears that all three bryophyte lineages have different mechanisms of regulation. However, mosses have the same mechanism as vascular plants; it is more likely they obtained the mechanism from a common ancestor than that they each independently

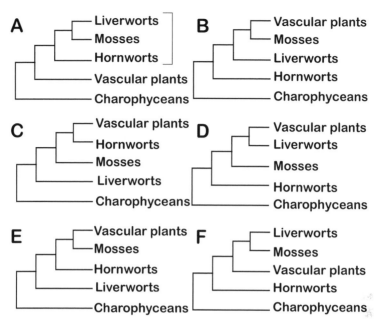

Fig. 7–8 Six phylogenetic hypotheses of relationships among the four embryophyte lineages using the charophyceans as an outgroup. A. Bracket to right indicates a bryophyte clade. The data sets used were derived from: A—ribosomal RNA, ultrastructure; B—morphology, ribosomal RNA; C—chloroplast DNA, nuclear genes, and a supergene matrix; D—sporophyte morphology, nuclear gene, ribosomal RNA; E—morphology, ribosomal RNA; F- ribosomal RNA, nuclear gene, morphology, chloroplast DNA, ontogeny. (Image sources: After Duff and Nickrent 1999; Goffinet 2000; Shaw and Renzaglia 2004; Qiu et al. 2006; Ligione et al. 2012)

hit upon the same mechanism by chance, so hypotheses where mosses and vascular plants share a most recent common ancestry are consistent with inheritance of the same auxin regulator. And so it goes as research continues.

Nonetheless some general conclusions can be drawn. Vascular plants always form a single lineage that shares a common ancestry with one or more lineages of bryophytes. In all but one of these hypotheses, mosses, liverworts, and hornworts do not form a single lineage, so as a taxonomic group Bryophyta has been falsified. Several independent characters support liverworts as the earliest divergent lineage of land plants (Graham and Gray 2001), which agrees with the microfossil record. At least some data support hornworts as the oldest extant land plant lineage (citations in Renzaglia and Vaughn 2000), and this agrees with the many similarities between the hornwort gametophyte and the charophyceans. Mosses are the lineage placed as sister group to tracheophytes most often, although the latest and most extensive molecular study places hornworts as sister group to the vascular plants. Even though some mosses have an organizational level similar to vascular plants, the vascular plant lineage does not end up nested within the mosses. This means that while mosses and vascular plants may share a common ancestry, they independently developed a three-tissue[18] system anatomy.

What is at stake is how we interpret the evolution of characters. A couple of demonstrations will suffice. Consider the chloroplast situation. Hornworts have a single chloroplast per cell that contains a pyrenoid, just like *Coleochaete*, a charophycean algae. So hypotheses that place hornworts as the basal lineage (Fig. 7–8 B, D, F) would suggest hornworts retained an ancestral character (single chloroplast per cell with pyrenoid[19]), and that all the remaining land plants shared a derived character of having many ovoid chloroplasts per cell lacking pyrenoids. Another possibility is that the many-chloroplast condition was inherited from a *Chara*-like ancestor, and either hornworts reverted to a single chloroplast or they had a different ancestor. If this were true, then land plants are not a single lineage, although the streptophytes (charophyceans and all land plants) would still remain one lineage. Some mosses and liverworts have a single chloroplast in apical cells, and in derivative cells the chloroplasts repeatedly divide to produce the many-chloroplast condition (Graham 1993), which suggests that many chloroplasts per cell is derived from a single chloroplast in each of these lineages. This is perfectly consistent with hornworts as the earliest diverging lineage. While reasonable, the actual scenario cannot be determined presently, and of course, other characters display other patterns.

Now consider sporophytes, one of the land plant hallmarks. The sporophytes of liverworts and some basal lineages of mosses (e.g., *Andreaea*) lack stomates. The sporophytes of hornworts, most mosses, and vascular plants have stomates. If the hypotheses shown in Fig. 7–8 D or F are correct, stomates with guard cells were acquired in the ancestors of all land plants but then lost in liverworts and some mosses. This is possible because those sporophytes that lack stomates and guard cells develop enveloped within gametophyte tissues so their ability to conduct photosynthesis is limited or lacking. Under such conditions, loss of stomates would not be surprising. Now that the gene for stomate development is known, perhaps liverworts and *Andreaea* can be probed to see whether they have this gene or not, and to help determine whether perhaps they "lost" stomates but left a genetic "fossil" in their genome. This example is complicated by *Sphagnum*; it has stomates on its sporophyte capsule but these stomates open only once to hasten dehydration and aid in spore dispersal. These stomates do not open upon interior intercellular spaces to exchange gases for photosynthesis. Perhaps the original function of stomates was in rapid dehydration, a condition retained in *Sphagnum* and then co-opted for photosynthesis later. In those phylogenies where liverworts are the earliest divergent lineage of land plants, the hornworts, mosses, and tracheophytes form a clade whose sporophytes possess stomates, which are now regarded as homologous and acquired in a common ancestor, forming a lineage that could be called the "stomatophytes" (Ligione et al. 2012). An indeterminate sporophyte growing from an apical meristem was the key innovation of "polysporangiophytes," and it was in this lineage that the large, long-lived sporophyte developed xylem and became homeohydric.

This discussion was long, too long, but its purpose was to demonstrate that the study of diversity is an active field of research and significant to our understanding of biology. You would be impressed to hear how thoughtfully, cordially, and politely these fine points of plant phylogeny are discussed among botanists, and how much they admire the hypotheses of others. This correspondent is not one of the combatants, but he likes watching the verbal jousting. Although deniers of evolution like to pretend otherwise, nothing in such a scientific controversy casts any doubt upon the theory of evolution.

None of the phylogenetic hypotheses presented above suggests vascular plants had mosses, liverworts, or hornworts as a direct ancestor, although the vascular plants might share a common ancestor with either one (Fig. 7–8 B, C, D, E), two (Fig. 7–8 F), or all three bryophyte lineages (Fig. 7–8 A). With the exception of Fig. 7–8 A, these hypotheses suggest that the immediate ancestor of vascular plants was an organism with a bryophytic level of organization and with bryophyte biology. The scenario presented in the next chapter suggests vascular tissue arose only once in the vascular plant lineage, and although the underlying reason for suggesting this is quite logical, it is not the only possibility. Fortunately the vascular plants have left a significant fossil record of their history. To see what these fossils tell us requires a field trip back to the Devonian. Grab your boots and let's go!

EIGHT

Back to the Devonian

*Wherein a field trip to the Devonian introduces early vascular plants and examines
how, from such small beginnings, xylem and new ways of branching helped
plants produce leaves and roots and grow into trees, Earth's first forests.*

I think that I shall never see
a poem as lovely as a pseudosporochnalean cladoxylopsid . . . tree.
—Robert Titus, in *Kaatskill Life*, Winter 2007–8

If anyone ever invents a time machine, paleobotanists will be among the
people first in line to buy tickets. First, they would borrow a plant press
from a colleague, get a quick lesson on its use (paleobotanists already
understand the concept of compression), and then select the Devonian
as their destination. Oh, a few would opt for the early Cretaceous or the
coal swamps of the Carboniferous, which are pretty this time of year, but
every seat on the Devonian field trip would be filled because this was one
eventful geological period in history of green organisms.

The Devonian is a fairly short Phanerozoic period, beginning some
416 mya, lasting a little less than 60 million years, so constituting only
slightly more than 1% of Earth history (Fig. 1–5). At the beginning of the
Devonian, there existed one or maybe two lineages of vascular plants,
and by the end of the Devonian every major lineage of vascular plants
had appeared except for the flowering plants, which would not appear
for another 250 million years or so. The earliest vascular plants were just
little naked stems, but during the Devonian naked stems became leafy
and stems found new ways to branch, some of which became roots and
some of which eventually became a new type of leaf. Stems got taller and

stiffer, and plants got bigger because of xylem; some became woody, which takes lots of xylem year after year. So during the Devonian the first trees and the first forests appeared, and last but not least, the first seeds appeared. Wow! The Devonian was one big green explosion of diversity in terms of the history of green organisms. This is really when the land turned green, and who wouldn't want to see that! Who could pass up a chance to see some of that firsthand?

One specific destination for a field trip would be the region that would eventually be Aberdeenshire, Scotland, the location of the Rhynie chert, at a time of about 410 mya, the early Devonian. The area is flat and low with lots of shallow pools and small water courses, the result of flooding and deposition of sediments from a braided river running through the region. Nearby, upland just a bit, volcanic hot springs are spilling silica-rich water down onto this marshy area. The mineral water interacting with the sediments produced a very fine-grained chert, and in the process resulted in the best-preserved early land plant fossils known. Field trippers wouldn't have too much difficulty finding plants because analysis of the fossil-bearing rocks indicate vegetation covered over 50% of the land area.

So here we go! The plants being found probably would not impress you, mostly looking like little green stems standing no more than 40 centimeters tall. Ah, groups gather to examine these green stems. The discussions involve whether these axes are gametophytes (Remy and Remy 1980a, 1980b; Remy et al. 1993; Taylor et al. 2005) or sporophytes, and indeed, just exactly what type of plant is it anyways? But what excitement among the field trip participants as more specimens are spotted! Is that a sporangium? Are those rhizomes or roots? How is that axis branching? What's this thing? Are there any leafy organs? Does it have vascular tissue (Edwards 1986)? The Swiss army knives, hand lenses, and plastic bags—the constant companions of field botanists of all sorts—come out. Lots of field work will follow as specimens are collected for careful study much later—that is, back in the future. The bag lunches provided by the tour organizers would be forgotten.[1] Nothing distracts botanists from new plants.

In a manner of speaking, paleobotanists regularly take such field trips back in time via a geological time machine consisting of all the many specimens of earlier life that were preserved willy-nilly and haphazardly by nature in the form of fossils. These fossils provide us with a distant window through which we must attempt to view and understand the biology of extinct organisms and the environment in which they lived. Fossils are our time machines. A great deal more can be observed from such specimens than readily meets the eye, especially when they come from the Rhynie chert. One of the early Devonian plants growing here actually gets named *Rhynia*, along with such genera as *Horneophyton* and *Asteroxylon* (Appendix: Fig. A11). The thing that was most exciting about these fossils is how well preserved their tissues were. Older stems are found in the fossil record, but they just consist of impressions, almost like the organic shadow of an organism (see Fig. 6–7 insert), so whether they had vascular tissue or not remains a question.

But not so with these Rhynie chert plants. Here's a cross section of the stem of *Rhynia*, but rather than being cut by a pocket knife, this one was cut with a diamond saw and then polished down to a wafer of rock and then examined using a microscope (Fig. 8–1). The center of the stem is occupied by a small core of vascular tissue; this central core of dark xylem cells is surrounded by a larger column of phloem. A not-large plant with minimal surface area growing in a very wet habitat has limited need of water conducting/supporting tissue. Stomates in the epidermis indicate the need for gas exchange and there are intercellular spaces among the cortex cells, so the stems were photosynthetic organs. And so it goes. The detail of these studies and others should impress you greatly. Too bad that time travel field trips are not possible. The Devonian would have been a great time to visit marshy areas; mosquitoes had not yet appeared.

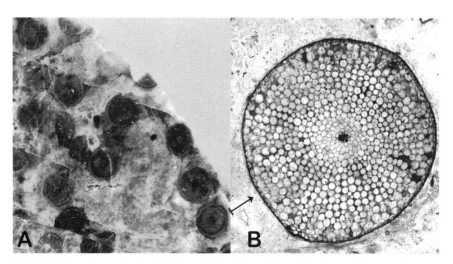

Fig. 8–1 Early Devonian vascular plant fossil from the Rhynie chert of Scotland. A. Stems of *Rhynia* in rock matrix (stems about 1.5 mm diameter). B. Stem cross-section as seen in a thin, polished slice of the rock magnified about 7 times larger than in A (arrow). (Image sources: A—Courtesy of Peter Coxhead, Wikimedia Creative Commons; B—Courtesy of Plantsurfer, Wikimedia Creative Commons)

As a result of Devonian geological studies, one thing is now clear: the earliest vascular plants (and their immediate ancestors) grew in wet, low-lying, swampy, deltaic areas. Estuarine and deltaic environments leading in to low-land braided rivers provided a means, both in terms of access and adaptations, for marine organisms to invade freshwater and terrestrial environments. Fossils of many early vascular land plants of the Late Silurian and Early Devonian are found in geological settings associated with braided streambeds, river-deposited sediments, and deltas, which are complex and varied patchy environments that include brackish marshes, sandy and muddy tidal flats, levees, freshwater lakes, and marshes (Hotton et al. 2001; Kotyk et al. 2002) (Fig. 8–2). Such low-lying marshy areas are good candidates for habitats where green organisms began adapting to fresher water, rapidly changing environments, and fine, soft sediments. If land plants did not arise directly in estuarine or deltaic environments, then they most likely arose in shallow fresh water, even ephemeral, aquatic environments (Graham 1993; Graham et al. 2012), but still their ancestors must have left the seas via rivers and streams.

Fig. 8–2 A complex coastal environment in the early Devonian. A. The geological setting of fossils of early vascular plants, primarily zosterophylls, and their immediate ancestors suggest they grew in such moist, semiaquatic habitats. The zosterophyll illustrated is *Sawdonia*, characterized by small leafy appendages (B—fossil imprint of stem) and "fiddlehead" shoot apices. (Image sources: Drawing courtesy of D. H. Griffing, from Hotton, Hueber, Griffing, and Bridge 2001, used with permission of Columbia University Press; B—Courtesy of Verisimilus, Wikimedia Creative Commons)

The "explosion" of vascular plant diversity that occurred in the Devonian is sometimes called an adaptive radiation, which means a lot of evolutionary experiments were taking place over a relatively short period of time, although a period of time still measured in millions of years. An adaptive radiation can happen whenever organisms invade a whole new habitat or uninhabited area, or whenever a lineage of organisms develops a significant new adaptation allowing it access to a previously unused or underused resource. Thus, in the early Devonian vascular plants began exploiting and exploring the terrestrial environment, and because they lacked competition, just about any variant would be successful, at least for a while. So this opened the door to biological "exploration" resulting in lots of new plant forms for new opportunities (Graham et al. 2000). Some experiments produced whole new lineages that dominated communities for millions of years before ultimately declining in importance and diversity as environmental conditions and/or competition from still other later-appearing plants with new innovations.

The trick to reading the fossil record is to decipher such a story using fragmentary evidence. Imagine that all the world history books had gotten ripped into chunks of pages, most of which are destroyed, and then the remaining chunks dispersed, hidden, and variously preserved. Now from this fragmentary evidence you must try to build an understanding of human history. Lots of little bits, single pages, would be lost and destroyed, but sometimes you might find a set of pages connected, or if very fortunate, two chunks from a single book identified because they have overlapping pages in common. Of course care must be taken because not all the books were identical to begin with, but if you collected enough book fragments, and pieced them together diligently, you would begin recovering this history. You would be most likely to understand the common events, the broad trends and major sequence of events of world history, as these would be the easiest to reconstruct. A world war would be hard to miss, although small details like a single battle might be missed. Small details and events are far more likely to be lost (or if found, not understood) because they are out of context.

So it is with the fossil record. Big, broad general events are far more certain. Large, longer-lived, and more common organisms will be better known than smaller, more ephemeral, less common organisms. Things like stems will be more common than reproductive organs, and only occasionally are two different parts found connected. It would be hard to know how many species were involved, but the number of major groups would be much more certain. In spite of its fragmentary nature, the fossil

Fig. 8–3 Zosterophyllum, *an early Devonian plant. This model reconstruction shows a rhizomatous plant growing as a semiaquatic. It exhibits H-branching (arrow, see Fig. 8–13) where a lateral branch produces both an upward aerial shoot and downward rhizome. The aerial axes are naked except for at their apices they laterally bear overlapping, flattened sporangia (inset: illustration of sporangia). See Fig. 8–4 for size scale. (Image source: Photograph of model from National Botanic Garden of Belgium, Courtesy of Daderot, Wikimedia Creative Commons)*

record of vascular plants is a very impressive set of data that cannot be dismissed lightly.

Here come some of the field trippers now. Let us examine one of their early Devonian specimens a bit more closely. Some of these little green sticks standing just 10 to 20 centimeters above the partially submerged sediments look a bit grassy at a distance (Fig. 8–3), but once you look closely you observe some critical differences. Rather than flat blades, the green shoots are clearly slender, cylindrical stems. These axes are naked and leafless. Some stems clearly grow along or just beneath the surface of sediments, and even in standing water. The horizontal shoots (called rhizomes) Y-branch[2] evenly at the apex; sometimes one of the two turns upward and becomes an aerial shoot. The aerial shoots seem to branch more irregularly. Some of the lateral branches come out almost on the horizontal and then produce both an upward and downward growing axis, forming an H-shape (Fig. 8–3; Fig. 8–4; Fig. 8–13). Some of the aerial shoots have a terminal cluster of somewhat flattened, rounded append-ages, most probably sporangia elevated to enhance spore dispersal. A slit across the top of each sporangium indicates where it will split open to discharge spores. Such an apical clustering of sporangia is one of the first "cones." All the axes seem similar, although the horizontal ones have fine rootlike hairy outgrowths called rhizoids. These specimens look like a species of *Zosterophyllum* (Fig. 8–4). To better indicate the size scale, there's my hand span next to it. This little plant may not look like much, but this organism is one of the first plants in the lineage of clubmosses, a group that will dominate the Carboniferous before going into decline. This truly spectacular fossil was recently discovered in China.

The paleobotanists will be at this for days, so let me make a critical point. The early Devonian is when vascular land plants began to diversify. From these humble beginnings more than one evolutionary path ultimately will lead to those truly remarkable organisms, those behemoths of biomass and beauty[3] called trees. The significance of such changes and their impact upon the environment cannot be underestimated. Even small vascular plants like *Zosterophyllum* tower over the average moss or liverwort, and especially in big stands, such plants are much bigger obstacles than bryophytes, organisms that adapted by minimizing their environmental profiles. Therefore the impact of vascular plants upon the environment is much greater. By the Late Devonian, woody plants, the first

Fig. 8–4 *Zosterophyllum*, an early Devonian plant. A. Interpretive illustration and B. actual fossil. Stems are leafless, naked, except for flattened, kidney-shaped sporangia (sp) borne laterally on the upper portions of the aerial stems (compare with Fig. 8–3). Lowest branches show H-branching rhizomes (h) that both spread the plant and produce new aerial shoots (see Fig. 8–13). This unusually intact fossil shows a dense central mass of rhizoids anchoring the plant (rh). The illustration is based upon studying the fossil on many levels as the stone matrix is removed. To better judge its size, a large hand span (thumb to tip of little finger) is shown at about the same scale (right). (Image source: After Hao et al. 2010)

trees, will form the first forests, growing several meters above the soil, and will push their roots a meter or more into the soil, where they will accelerate weathering and accumulate sediments (Elick et al. 1998; Algeo et al. 2001). Such big obstacles slow water runoff, trap more sediment-building soil, impede the wind, hold more moisture, and gradually ameliorate the harsh terrestrial environment. In making terrestrial habitats more suitable for themselves, vascular plants created more opportunities for their own diversification and more habitats for other organisms (Algeo et al. 2001). In the process trees totally transformed the world.

As vascular plants got larger and woodier, and forests became more widespread, covering more of the land, they sequestered more carbon for longer periods of time. Carbon dioxide is a greenhouse gas, and its removal from the atmosphere can produce a long-term cooling effect on the climate. The sequestering of carbon in terrestrial plant biomass had begun in the Ordovician when nonvascular plants invaded the land, but the innovation of xylem with its massive cell walls meant vascular plants were much bigger, longer-lasting reservoirs of carbon (Fig. 8–5). So from the middle Devonian through the Carboniferous (450–300 mya), removal of carbon dioxide from the atmosphere increased because vascular plants were so much larger.

The evolution of roots is difficult to pinpoint with any accuracy because they left relatively few good fossils. Without anatomical detail roots are hard to distinguish from masses of rhizomes—that is, stems with a rootlike demeanor—especially since roots appear to be derived from stems. Both could function in anchoring, and both could function in absorption especially if a symbiotic relationship had been formed with fungi. In the early Devonian only *Zosterophyllum* showed some evidence of rootlike structures, rhizoids. As the clubmosses got bigger and as other groups appeared, like the seed ferns and their immediate predecessors, particularly those that became woody trees, more extensive rooting systems penetrating deeper into the soil were required. By the end of the Devonian roots were down a meter or more and beginning to hold a substantial mantle of soil in place (Elick et al. 1998). As the Earth's plant mantle spread from low wetlands to better-drained uplands, the presence of plant roots greatly increased the weathering of calcium and magnesium silicates, and their runoff into oceans generated a geochemistry that produced dolomite, a limestone composed of calcium magnesium carbonate (Berner 2001). This had a major impact on the atmospheric concentration of carbon dioxide because limestone is a low-turnover reservoir on the order of millions of years. So not only did the rise of land plants result in more carbon dioxide being fixed into organic molecules by photosynthesis, but also their biology changed soil weathering, both with the same result, removing more carbon dioxide from the atmosphere.

The solubility of carbon dioxide in water is in direct proportion to its concentration in the atmosphere, so the CO_2 concentrations in the oceans and in the atmosphere are in a dynamic equilibrium. The removal of CO_2 from the oceans as carbonate would allow more atmospheric CO_2 to dissolve, replacing the CO_2 thus removed, further lowering the CO_2 concentration in the atmosphere. The proliferation of land plants caused a reduction in the atmospheric CO_2 concentration, leading to an extended period of continental glaciation, an ice age, during the Carboniferous (Fig. 8–5). Curiously enough this means that presently, in an age when global warming is taking place, the Earth's average temperature and the atmospheric CO_2 concentration are near all-time minima. This does not mean that it is a good idea to increase the CO_2 concentration because humans and all our domesticated plants and animals are not adapted for the changes that would occur.

For a second time[4] green organisms had produced a worldwide change in the environment, and from each event a cascade of changes followed. Once you begin studying biological diversity from a historical perspective, you realize that an evolutionary mechanism is a biological necessity

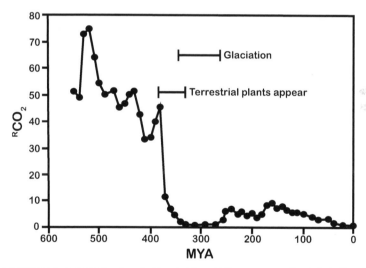

Fig. 8–5 Changes in atmospheric carbon dioxide coinciding with the appearance of land plants. The relative concentration of atmospheric CO_2 (Y axis—referenced against present day levels = 0) begins dropping in the late Cambrian, which coincides with the development of large seaweeds followed by the appearance of land plants. The rise of woody plants took place in the late Devonian (350–360 mya), leading to the only time when CO_2 levels and temperatures were as low as they are today. The sudden drop in atmospheric CO_2 levels coincides with a prolonged ice age. (Image source: After Berner, Berner, and Moulton 2003)

because organisms themselves change their environments and then organisms must change to adapt. When someone tells us, "Evolution is not a fact," biologists should respond, "Evolution is a necessity." Life is an ongoing interaction between organisms and their environment, each causing changes and influencing the other in turn. A history of biological change is not possible without an evolutionary process, and this is why so many critics of evolution work so hard to ignore, dismiss, or misrepresent everything that is known about the history of life.

TREES OF LIFE

The evolutionary history of green organisms can be thought of as building vegetation, an idea directly from Corner's (1964) influential little book, *The Life of Plants*. But the history of green organisms actually started so far back, with organisms so small and so simple that vegetation seems an inappropriate term. Nonetheless a path does lead from those cyanobacteria that became intracellular symbionts to the grandest of all vegetation, trees. Just as small unicellular plankton built seaweeds by way of adapting to coastal environments (chapter 5), some charophycean algae had a seemingly insignificant change in development (a delay in meiosis) that resulted in the development of a multicellular diploid phase, the sporophyte (chapter 6). What began as a single transverse cell division of a zygote ended up producing trees.

The key innovation that allows for trees was the vascular tissue xylem. Xylem allows plants to conduct water more efficiently and to grow larger because xylem doubles as support tissue. Robert Hooke (1665), a famous early microscopist, had this to say about xylem: "so prodigiously curious are the contrivances, pipes, or sluces by which the *Succus nutritius*, or Juyce of a Vegetable is convey'd from place to place." But before trying to figure out what got this wooden ball rolling, consider the end product, trees, the biggest of land plants, the biggest biological obstructions that have ever evolved. This form is so successful among vascular plants that virtually every lineage has given rise to trees, from the first tree to appear in the late Devonian to the present day with around one-quarter of all vascular plants species (ca. 400,000) living as trees (Raven and Crane 2007).

There are trees and then there are TREES, big to the point of being majestically awesome. The largest trees are gigantic columns of wood over 10 meters in diameter towering over 100 meters above the ground. Really big trees are impressive, like the sequoias and Australian ghost gums (eucalypts), and as something of a tree connoisseur I have seen both. It took hours of travel to see baobabs in Africa and bottle trees in Queensland. And then, for similar reasons, but for quite the opposite in adaptations, I hiked and climbed to the top of the Pyrenees to see a dwarf buckthorn and a pygmy willow. But hands down the most impressive trees are figs, the so-called banyan trees. In northern Queensland, near the town of Yungaburra and the Lake Eacham Hotel, the Curtain Fig spreads a solid wall of roots and trunks across the forest for 15 or so meters, hoisting aloft a massive emergent crown that spreads above the forest canopy (Fig. 8–6). Some banyan trees, also figs, are so remarkably big a single tree creates its own forest community. Fig trees with upwards of 800 trunks are known to cover and occupy over 20,000 square meters of land, something on the order of seven football fields.[5] Figs accomplish this because they produce roots that grow down from horizontally oriented limbs. Once these aerial roots contact the ground, they become woody, ultimately forming new trunks.

Tropical forests are a tough place for trees to grow to maturity. Tree seeds fall onto the perpetually dim forest floor beneath a tall, dense, evergreen canopy, a very inauspicious place to commence growth. Spindly tree seedlings manage to gamely persist under the overbearing, ever-present canopy for some time, in some cases for years, but their growth is anything but impressive. A tree sapling will succeed in reaching the forest canopy only if a fortuitous tree-fall produces a gap in the canopy above it, and then only if it outgrows all its competitors and wins the race to bask in the sun. Strangler figs take another route to the canopy. Rather than waiting in the dense shade for good fortune, strangler fig seedlings begin growth as epiphytes on the trunk or on the branches of a canopy tree, their seeds having been conveniently deposited there by bird or bat dispersers.[6] The fig's roots grow down and around the host tree's trunk innocently enough, but after reaching the ground they begin thickening and fusing with each other into a massive woody braid that eventually can strangle and kill the host tree. By this time the fig's braided trunk and canopy are big enough to take over its host's place in the canopy. No wonder these figs are called *matapalo* in Central America, the "tree killer."

Trees are the longest lived of all individual organisms, some reaching 7,000 to 10,000 years old.[7] Clonal trees can be even bigger and older than any single-trunk tree, although no one part of a clone is so old. Clonal trees grow from a single individual by spreading roots that send up new trunks, or if a fig, spreading branches that send down woody roots. Clones may look like a whole grove of trees, but actually many may be interconnected below ground, and they are all one genetic individual.

Fig. 8–6 Curtain Fig (Queensland, Australia) showing its wall of woody roots and lower portion of the crown above. A diagonal trunk continues downward to the left well out of the picture. This is presumed to be the original tree that fell to its right up against another tree. The fig roots grew down and around the prop tree and eventually the fig took over completely. The emergent crown is completely out of this picture, as is about half the wall of roots. A couple of colleagues who are more than six feet tall provide some size scale. (Image source: The Author)

King's *Lomatia* in Tasmania consists of hundreds of woody trunks spread across several square kilometers. This clone has been estimated to be at least 40,000 years old. King's *Lomatia* is an endangered species; in fact, it is as endangered as a species can get. Only two "groves" exist, both are clones, and both are one and the same genetic individual. Since it cannot pollinate itself, it cannot set seed, so there is no hope for this species to recovery, but such a clone may persist for a very long time. The Pando clone of aspen in Utah is estimated to be at least 80,000 years old, and some think perhaps even ten times older, which makes this clone as old as our whole species. Nonetheless, sooner or later such individuals encounter an environmental disaster that dooms them. And if that isn't bad enough, because a clone is a single genetic individual, such clones cannot pollinate themselves and set seed, and because mutations accumulate in direct proportion to the number of cell divisions, very long-lived organisms like these clones accumulate mutations that begin to reduce the viability of their pollen[8] (Ally et al. 2010). Very old clones have little to no chance of producing more offspring via pollen or seed, so they will just persist until they meet the inevitable.

Trees have evolved independently in numerous vascular plant lineages, and as a result we can ask, why? Two good reasons come to mind. First, an arborescent growth form, consisting of a substantial main axis (a trunk) supporting a crown of branches and/or leaves, competes very effectively for space, light, and water. Plant life operates by intercepting diffuse resources, and the bigger you are, the more resources you intercept. Second, trees are also the longest-lived organisms; they have long reproductive lives measured in decades to centuries and, in a few cases, millennia. Their immense size and long life allows them to produce and widely disperse a mind-boggling number of offspring. But remember, for a population to remain stable, exactly one offspring on average must reach reproductive maturity. The overwhelming majority of tree offspring die. Nature is tough, and those are the odds, but trees can afford to play the long odds. No wonder this plant form has appeared in so many lineages.

In the early Devonian, some 410 mya, our field trippers found no terrestrial vegetation more than knee high, although in comparison with the bryophyte and lichen crustal communities that preceded them, these early vascular land plants were a towering forest. But before the Devonian ended, real forests existed. *Archaeopteris* (Fig. 9–3), a common ancestor of seed plants, has reigned as the oldest known tree since its discovery some 50 years ago. But recently an older tree was discovered to have formed Earth's first forests (Stein et al. 2007). The earliest tree on Earth was a pseudosporochnalean cladoxylopsid ("sue-doe-spor-och-NAY-lee-an kla-doxy-LOP-sid") "fern," a member of a group, the trimerophytes, that is thought to be ancestral to the ferns (Fig. 8–7). The fossil stumps of these forests, named *Eospermatopteris*, had long been known, but what grew on these stems was unknown. The "foliage" branches of cladoxylopsid ferns also

Fig. 8–7 Eospermatopteris. A reconstruction of an arborescent middle Devonian pseudosporochnalean cladoxylopsid "fern," the oldest tree that composed the oldest forests on Earth. A. Fossil axis. B. Reconstruction. Bar equals 1 m. (Image source: After Stein et al. 2007)

had been known for a long time, but only recently were the foliage branches found physically connected to fossil tree trunks, demonstrating conclusively that they were parts of the same plant. The "foliage" of this tree, named *Wattieza*, consists of a helical whorl of much dissected branches on trunks that could reach 8 meters in height, a tree and arborescent by anyone's definition, but not actually woody. It was constructed of tough ground tissues like modern tree ferns. On the basis of the scars on the trunks, it appears the "foliage" branches were regularly, perhaps seasonally, shed. Forests of cladoxylopsid ferns appeared 385 mya, some 15 million years earlier than *Archaeopteris*, a true woody plant. A bit later in this chapter, a discussion of the origin of leaves will show how modifications of "foliage" branches like these gave rise to the leaves of ferns and seed plants.

Trees are such a long way both developmentally and phylogenetically from the minute vegetation of the early Devonian towering a few centimeters above sprawling gametophytes that it is hard to imagine how such small, dependent sporophytes gave rise to trees. And as descendants of arboreal ancestors we have trees to thank for many aspects of our human condition, but more on this later.

We have already considered how a zygote became the uniaxial sporophyte of bryophytic plants, both determinate and dependent; what further transformations were needed to produce still bigger independent sporophytes and trees? A uniaxial sporophyte with a foot at its basal end and a sporangium at its apex produces only one sporangium from each fertilization event, so the number of spores produced, while in the thousands, remains limited by the number of fertilizations. Fertilization itself may have been a limiting factor for early land plants and their swimming sperm, so getting as many dispersible offspring as possible from each successful fertilization event would be very advantageous.

The polarity of these uniaxial sporophytes is established by the initial transverse division of the zygote, which results in a bottom cell that gives rise to the foot, and an upper cell that gives rise to the sporangium and any axis in between. Two solutions are possible: make a bigger sporangium or make more sporangia. Since the spore-producing cells inside a sporangium are nourished by the jacket cells surrounding them, there appears to be an upper limit on the functional size of sporangia. If the sporangium got any larger in diameter, spores in the central region might not receive enough nutrition from cells around the margin. Small increases in the diameter increase the volume of the sporangium by a cubic factor, but the jacket layer increases its area only by a square function, so increases in sporangium size may quickly exceed a functional ratio between the jacket layer and the volume of spores within. This leaves land plants with only one option: produce more sporangia from each zygote, and the only way to accomplish this is to branch apically.

Early embryonic development in many moss sporophytes involves division of an apical cell, but this soon ceases and the remaining growth takes place by cell division throughout the embryo. As a result these sporophytes display a determinate development.[10] If the young embryonic apex could divide parallel to the foot-sporangium axis, it could Y-branch, and each of the two apices could produce a sporangium, thus doubling spore production. With each such Y-branching of the apices, the number of potential sporangia doubles. Apical branching requires apical growth from either an apical cell or an apical meristem (group of cells). This occasionally happens during the embryonic development of some moss sporophytes, resulting in two sporangia (Bower 1935), so apical branching of such sporophytes is by no means a ridiculous suggestion. If such "mistakes" have a genetic basis and the plants possessing them produce more offspring as a result, then apical growth and Y-branching sporophytes spread through the population. Apical growth occurs in charophycean algae, mosses, liverworts, and hornworts, in the gametophyte generation, so the genes are present. How such genes become expressed in the diploid generation is not yet understood.

All vascular plant embryos are initially bipolar, like those of bryophytes, but then both basal and apical cells divide in a plane perpendicular to the initial bipolar axis, thus establishing an embryo with four quadrants. The two basal quadrants become the foot and a lateral axis, and the two distal quadrants become an aerial shoot and a leafy organ. But this is an embryo of an indeterminate leafy plant. The sporophyte of *Cooksonia*, as well as any hypothetical intermediates, would be leafless and determinate; its foot was still firmly embedded in the gametophyte, so there would be no lateral axis and no leafy organ. An embryo of this simpler sporophyte might have begun when a single division of the upper cell parallel to the embryo polarity produced two cells, each of which then acted like the original apical cell, both going on to produce an axis terminating with a sporangium. Any subsequent longitudinal divisions of the apical cell would produce more dichotomies, more apices, until finally each apical cell or meristem is consumed in the production of a sporangium. This illustrates how plants grow larger by adding modules, duplicating and repeating, module by module until you have a large plant.

The earliest-appearing (Late Silurian), apically branched sporophytes of *Cooksonia* lacked true xylem and were scarcely taller than many modern moss sporophytes, which strongly argues that sporophyte branching preceded xylem. Their supporting/conducting tissue has been interpreted to be of the same level of organization as that found in the axes of some moss sporophytes. This makes sense because the functional demands of a taller, branched sporophyte with two or more sporangia would select for conducting/supporting tissue, thus supporting the hypothetical sequence of events presented earlier (Fig. 6–6) from a presumed bryophyte-like ancestor to an apically branched sporophyte, to a taller sporophyte to enhance dispersal, and finally to conducting/supporting tissue, xylem. The innovations that started the long road toward trees in terms of building vegetation were sporophyte branching and xylem. To make a long story short, the history of vascular plants involves two reciprocal trends: bigger, independent, and more elaborate sporophytes, and smaller, simpler, and more dependent gametophytes, a trend that culminates in the seed.

The primary dichotomy among land plants was long considered to be between vascular and nonvascular plants, but the difference between land plants with uniaxial, unisporangiate sporophytes (mosses, liverworts, and hornworts) and those with branched sporophytes producing two or more apical sporangia (all the rest) is more basic. This conclusion has resulted in the hypothesis of a polysporangiophyte clade, a lineage that includes all plants with branched sporophyte axes and within which all vascular plants, tracheophytes, are nested (Kenrick and Crane 1997). Of course, those embryophytes with uniaxial sporophytes, the bryophytic plants, still do not form a single lineage.

Aerial axes of sporophytes are determinate, whether they bear apical or lateral (clubmosses and whisk ferns) sporangia. They grow only so tall because of physical restraints; if they continued to grow they would outgrow the structural limitations of their components. The axes would collapse or buckle, and spores would not get dispersed very far at all. Indeterminate growth of the sporophyte becomes possible when the initial branching of the young sporophyte establishes an asymmetrical orientation of the two resulting apices: one becomes an aerial axis of determinate growth and the other becomes a horizontal axis of indeterminate growth. The horizontal axis, a rhizome, continues to grow, branching dichotomously, such that one branch perpetuates itself and the other produces an aerial shoot (Fig. 6–6 F). Living clubmosses grow in just such a manner. Ferns also grow in a similar manner, but instead of an aerial shoot, their rhizome produces leaves. Why this is similar will be explained below when the origin of leaves is considered.

Support and conduction are key concepts for plants adapting to the terrestrial environment. The differences between growing in an aquatic and a terrestrial environment are profound (Niklas 1997). The key ideas are pretty straightforward. Suspended in water, you float; suspended in air, you fall. Suspended in water, you remain hydrated; suspended in air, you dry out. A conducting and supporting tissue like xylem functions to deal with both problems. Of course, xylem is not the only means of support found in land plants. Cylindrical shapes, internal cellular pressure called turgor, and thick-walled ground tissue (collenchyma or sclerenchyma) can all provide considerable structural support. Everyone has probably observed the wilting (loss of turgor) of nonwoody vascular plants, perhaps after letting the house plants go a bit too long without watering. Well-hydrated, membrane-bound cells exert an outward force upon their flexible cell walls called turgor pressure, and the loss of turgor causes wilting. When all the cells in a block-constructed tissue like parenchyma push outwards against each other, turgor pressure provides hydrostatic support (Niklas 1997), and we say the tissue has turgor. This principle explains what happens when limp carrot or celery sticks are refreshed in cold water, becoming crispy and firm (turgid).

Vascular plants with slender, nonwoody stems obtain most of their structural support from turgor pressure of parenchyma rather than from their small central core or cylinder of xylem, so the amount of xylem does not limit the sporophyte size, allowing for as much as 10-fold increase in height without needing more xylem for support. Not only does a branching sporophyte produce more spores from each fertilization event, but also its upright stature and greater height extremely increases the distance spores will disperse. A larger sporophyte axis exposes more surface area for the absorption of those dilute necessities, carbon dioxide and sunlight, but an upright stature also exposes far more surface area to the dry atmosphere, subjecting the plant to more water loss, which is why even the sporophytes of mosses and hornworts are photosynthetic and have a waxy cuticle, stomates, and guard cells to allow gas exchange while limiting water loss.

Even though a considerable increase in size can be accomplished structurally without xylem, hydration is another matter. Water diffuses from cell to cell according to an inverse square law. The time it takes for water to diffuse across two adjacent cells is four times greater than the time it takes for water to diffuse across a single cell. As the series of cells increases in number, cell-to-cell diffusion quickly becomes an inadequate means for replacing water lost to the atmosphere. This means xylem is needed for conduction before xylem is needed for support because without specialized water-conducting cells, even a small increase in size quickly reaches the limits of cell-to-cell transport of water.

Elongate cells whose long axis is parallel to the direction of water movement are more efficient at diffusing water than a series of shorter cells because cell membranes and cell walls are the primary impediments to diffusion. Natural selection would favor any organism whose water diffusion is more efficient because of somewhat elongated cells, and the evolution of conducting tissue begins. Large seaweeds have no need for water-conducting cells, but their need to move carbohydrates poses a similar problem. Brown algal seaweeds like the kelps have elongate cells forming a phloem-like food-conducting tissue in their stipe to help move carbohydrates synthesized in the blade cells down to the holdfast cells in the dim depths below. Basal and subterranean portions of early terrestrial plant sporophytes would have the same need for a food-conducting tissue.

Providing physical support and conducting massive amounts of water up to the crown to replace the kiloliters lost each day represent a significant engineering problem in building a tree. Xylem performs the dual functions of support and conduction because the same form, a thick-walled, semirigid tube, is both strong and efficient at moving liquids. When many long, skinny, thick-walled tubes are bundled together, the resulting tissue is both strong and flexible. Unlike the stipes of seaweeds, which are under tension, pulled upward by gas-filled floats and tugged on by tides and currents, the stems of land plants bear a load under compression. As such they are subject to buckling both under the weight of their own crown and from the lateral forces impinging upon these aerial parts by wind (Niklas 1994). The axes of land plants must be strong enough to bear the weight of the branches above, and flexible enough to prevent snapping or buckling from the lateral force of wind hitting the crown. And of course this is why humans cut down so many trees; they are constructed of an immensely useful tissue, wood, an accumulation of annual growths of xylem.[11]

The basic cell composing xylem is the tracheid (TRAY-key-id), an elongate cell with long tapering ends (Fig. 8–8 A-B). After enlarging to its full size, the cell protoplast secretes a thick secondary cell wall within the thin, stretchable primary wall, and then the cell dies leaving its cell wall as the functional conducting/supporting unit. The overlapping walls of two adjacent tracheids contain aligned pits, pit pairs, through which water may pass. For water to move through a vertical series of tracheids, a force must be applied, a push, a pull, or both. The conduction of water through xylem operates by a combination of one push and two pulls. An osmotic (turgor) pressure generated by parenchyma cells in the root and/or rhizomes push water into the xylem cells and upward, but that can only move water vertically a few meters. Capillary action caused by the physical attraction between water molecules and the xylem cell wall helps pull water upward, but this is the lesser pull. The transpirational loss of water from the aerial portions of the plant exerts a strong pull upwards on the column of water.[12] Of the three, transpirational pull is by far the greatest force, and without it plant height would be greatly limited.

Transpirational water tension exerts a tremendous inward pull upon the cell walls in the process. Consider what can happen when you try to suck a thick milkshake through a thin-walled plastic straw. The partial vacuum you generate to pull the liquid upward also pulls inward on the

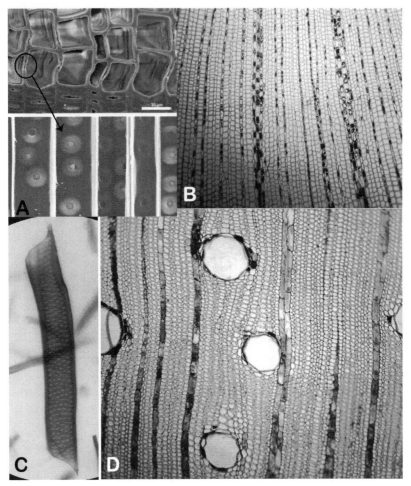

Fig. 8–8 Conducting elements of wood: tracheids and vessels. A. Tracheids in cross-section (top) and in lon-gitudinal section showing their radial walls (bottom). Water moving through a tracheid must pass through a pair of circular bordered pits where the long, narrow cells overlap. Circular bordered pit pair in cross-section (circle). White bar is 30 micrometers long, 3% of a millimeter. B. Wood of *Drimys* whose vertical conducting elements are composed entirely of tracheids. Note the straight files and how similar in size and shape the tracheids are in cross-section. This angiosperm with "gymnosperm" wood belongs to one of the ANA grade basal lineages. C. A single vessel element from maple wood shows large diameter and big round open pores at each end. Stacked up in a vertical sequence they form a thin-walled, large diameter vessel for efficiently conducting water. D. Wood of maple showing large diameter vessels among smaller diameter, thicker-walled tracheids in straight files, each providing different functions, conduction and support, respectively. (Image sources: A—Courtesy of Dr. Michael Rosenthal, Technische Universität Dresden, Wikimedia Creative Commons; B-D—The Author)

straw's wall, and this force can collapse the straw's plastic wall, rendering it useless in conduction. Although water is not pulled upward by a vacuum, similar forces are exerted upon the cell walls of xylem by the pull of water molecules upon each other, and the thick reinforced walls of xylem cells function to prevent their hydraulic collapse.

Under some summer conditions plants frequently will wilt in the midday heat because water is being lost from the leaves (transpiration) faster than the water can be replaced, resulting in a loss of turgor in the leaves. Loss of water reduces the volume of the central vacuole in each cell such that the cell protoplasts no longer push outward against the cell walls, so the whole block of tissue goes limp. If this water deficit is not corrected later, it can lead to drying, cell death, and ultimately to leaf and then plant death. However, midday wilting has a beneficial aspect because it lowers the position of leaves, reducing their absorption of solar energy, cooling the leaves, which in turn reduces transpiration and water loss. Midday wilting is not something to worry about as long as the plant recovers in the late afternoon. If it does not recover, start watering.

Plants also have need of water-conducting xylem during growth, and this poses a bit of a problem. Plants grow by cell division, first in the embryo, then in meristems for all subsequent growth. Tissues derived from meristems are termed primary. Cells enlarge, producing a noticeable growth of the plant, and finally the cells differentiate and mature according to their cell type. During its development a tracheid elongates, stretching its primary cellulose wall, so the first-maturing tracheids that function during plant elongation have inner walls composed of reinforcing rings or helices so that the intervening primary cell wall can stretch (Fig. 8–9). Other tracheids remain immature during elongation and without reinforced inner walls. Since they reach full size and maturation after stem elongation has ceased, such tracheids can deposit a thicker, solid secondary cell wall within the original or primary cell wall. This inner secondary wall can have up to three layers of cellulose fibers oriented in different spiral planes, which then become impregnated with and strengthened by lignin. At maturity the tracheid dies, and its pitted cell wall remains as the functional cell. The earliest-maturing xylem cells may get stretched to the point of ripping, but by the time they are destroyed, later maturing xylem has taken over conduction. In herbaceous plants this primary xylem may be the only xylem the plant produces.

Herbaceous plants at maturity possess vascular tissue derived from embryonic cells or apical meristems only.[13] A few large herbaceous annuals, like sunflowers, can produce a considerable amount of wood in

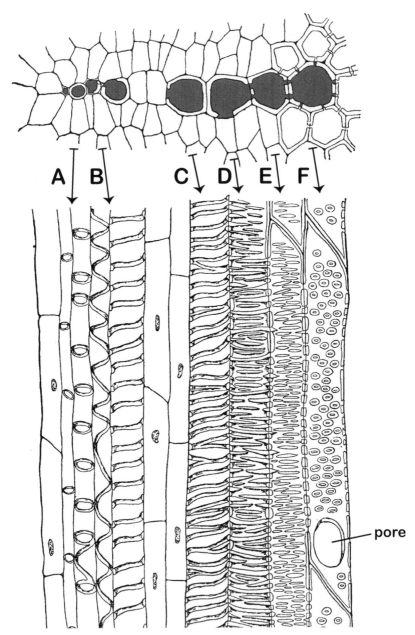

pore

Fig. 8–9 Xylem vessel elements formed during herbaceous growth (top cross-section, bottom longitudinal section). A. Vessel elements with thickened inner wall in the form of rings; thin wall in-between. B-C. Thickened inner wall in the form of helices; thin wall in-between. D-F. Thickened inner wall solid with pits. A-F. Vessel elements increase in diameter from first to last to mature. Rings and helices of early vessel elements show stretching because stem was still elongating. F. Vessel element shows large end wall pore (bottom right; cf. Fig. 8–8 C). (Image source: After Eames and MacDaniels 1925).

their aerial stems in a single season of growth, but they are not perennials and technically not woody. The perennial parts of herbaceous plants are at or below the soil surface. In seasonal climates the aerial portions of such plants often die back to the ground, leaving the perennial portions to grow next year, so even though the plants grow for many years, aerial stems function for only a single season. The aerial portions of such plants often have phloem fibers that contribute considerably to the support function. Bundles of phloem fibers in such plants as flax and cannabis have considerable value as fibers (linen and hemp, respectively).

Woody plants (trees and shrubs) accumulate an annual or seasonal layer of xylem (and phloem) formed by a tubular, lateral meristem, the vascular cambium. Such tissues are termed secondary, and the annual accumulations of secondary xylem are called wood. In addition to tracheids, all secondary xylem has parenchyma cells forming rays, strands of living cells along stem radii for lateral movement, for starch storage, and for wound reactions. Wood that has living parenchyma cells and can conduct water is called sapwood. In some woody plants, only the current year's secondary xylem is sapwood, which is usually light in color. Other species can have several years of sapwood. Sapwood becomes heartwood when it stops conducting and the parenchyma cells die. Heartwood is darker in color because of the deposition of materials and is always in the center or heart of the stem because the vascular cambium adds new growth rings of wood to the outside. Heartwood continues to function in support.

In most gymnosperms, the vertical conducting and supporting tissue consists wholly of tracheids. When one cell type must perform two functions, support and conduction, it results in a functional compromise. Tracheids best at conducting have the widest diameter and thinner walls, but such cells are not the strongest. The strongest tracheids have narrower diameters and thicker cell walls, but they are not best at conducting because depositing a thicker cell wall necessarily reduces the cell volume. Woody plants produce larger-diameter, thinner-walled tracheids early in each growth season (springwood) for maximum conduction, and then smaller-diameter, thicker-walled tracheids later in the growth season (summerwood) (Fig. 8–10). The abrupt transition between last year's summerwood tracheids and the following year's springwood tracheids demarcate growth rings, which consists of the springwood and summerwood of one growing season.[14]

In the xylem of flowering plants, xylem cells have become specialized for both support (fibers) and conduction (vessel elements). Fiber cells are

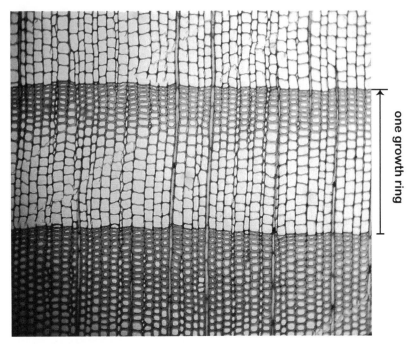

one growth ring

Fig. 8–10 One whole (arrow) and two partial growth rings in a conifer (western hemlock) showing larger diameter, thinner-walled springwood tracheids (in straight files) at the beginning of the year's growth (in direction of the arrow), changing fairly abruptly to late season, smaller diameter, thicker-walled tracheids (summerwood). At the end of the growing season the vascular cambium ceases its activity and begins the next year by producing large diameter springwood tracheids again, making a very abrupt demarcation of the growth ring. The arrow shows the direction of cambial growth and points toward the outside of the stem.

longer, narrower, and thicker-walled than tracheids, and so they function better in support. Vessel elements are shorter, broader, and thinner-walled than tracheids, and they have open pores at their ends rather than pits, and so they are better at conducting (Fig. 8–8 C-D). Vessel elements line up vertically to make very efficient conducting vessels that range in length from a meter or so to the height of a whole tree. As a result, angiosperm wood is anatomically more complex than gymnosperm wood (compare Fig. 8–8 D with Fig. 8–10), but a few basal angiosperms retain a gymnosperm-type wood composed solely of tracheids (Fig. 8–8 B) (Carlquist 1996; Carlquist and Schneider 2002).

At one extreme, vessel elements are similar to tracheids in size, shape, and form, except for having pit-sized pores on its ends; and at the other extreme, vessel elements are shaped like broad little barrels with one large open pore at either end (Fig. 8–8 C). In the latter case, such vessel elements are large enough in diameter that they can easily be seen with the naked eye—for example, in oak or elm. These large-diameter vessels appear only in the springwood and for this reason the wood is called ring-porous (because you can see a ring of "pores" in the springwood of each growth ring). While efficient, such large vessels are subject to air embolisms that render them nonfunctional, especially during hotter summer months. Both fibers and vessel elements are derived evolutionarily from tracheids, so it will come as no surprise that at the other extreme (the former case), many flowering plants have longer, narrower vessel elements that are shaped very much like tracheids, and some members of the basal lineages have wood that lacks vessel elements—for example, *Drimys* (Fig. 8–8 B). Evolutionary trends in wood, like the artfully arranged vessel elements in Fig. 8–11, tracheid-like to specialized, bottom to top, are well supported by both the fossil record and phylogeny. Vessel elements have evolved independently in the xylem of ferns and gnetophytes as well.

Phloem is the vascular tissue that conducts photosynthetic materials from the leaves to other parts of the plant. Analogs of phloem occur in the

Fig. 8–11 General trend in specialization of angiosperm vessel elements. From bottom to top this illustrates vessel elements with an increased diameter, shorter length, more transverse end walls, and a change from many barred pores to one big simple pore. Vessel elements most like tracheids (bottom half) are found in trees of the ANA grade and the magnolialean clade. Short, larger diameter vessel elements with big open end pores (upper half) represent more specialized woods. This artful illustration was drawn by B. G. L. Swamy in honor of his long collaboration with Prof. I. W. Bailey. (Image source: End piece illustration by B. G. L. Swamy, from Bailey 1954)

long axes of some moss sporophytes and in the stipes of brown algal sea-weeds because efficient distribution of photosynthetic materials becomes a necessity as plant size increases and photosynthesis becomes a function of leafy organs. Phloem is a living tissue usually positioned outside of or around the xylem, and like xylem, primary phloem derives from apical meristems or embryonic cells and secondary phloem is produced by the vascular cambium. To produce both xylem and phloem, the vascular cambium of woody plants is a two-sided meristem, producing cell derivatives for secondary xylem from its inner side and cell derivatives for secondary phloem from its outer side. Secondary phloem does not continue to accumulate year after year throughout the life of the tree like wood, although it can accumulate and function for more than one year. After two or three years, older, nonfunctional phloem becomes incorporated into the corky tissues of the bark produced by yet another cambium, the cork cambium, and these dead tissues are eventually sloughed off with the older bark tissues. If you look closely you can see annual layers of cork tissue in the bark of many temperate zone trees. Although the protective waxy cells of the bark are generally called cork, only the spongy, waxy bark of the cork oak, *Quercus suber*, produces commercial bottle corks.[15]

The conducting tissue of phloem consists of thin-walled sieve cells, the functional counterpart of tracheids, or sieve tubes, the functional counterpart of vessels. Although phloem sieve cells and sieve tube members are living cells, they lack nuclei and are always developmentally and functionally associated with small, adjacent, nucleated companion cells. As its different structure suggests, phloem conducts in a different way from xylem. In a leaf, parenchyma cells adjacent to phloem load sugar into the sieve cells or sieve tubes, and the increase in sugar concentration causes water to diffuse in rapidly from nearby xylem. Incoming water increases the volume within the phloem cells, and with nowhere else to go, the sugar-laden water is flushed toward more central and basal parts of the plant where parenchyma cells actively unload the sugar to maintain the osmotic differential between the source and sink, keeping the flow going. Photosynthetic regions are sources of the carbohydrates producing the phloem flow, whereas storage tissues, developing fruits, or other growing regions are resource sinks toward which the photosynthetic products are directed by sugar unloading.

All three parts of the "typical" plant body, stems, roots, and leaves, are constructed of three discrete tissue systems, dermal, ground, and vascular (Fig. 8–12), but each part differs in tissue organization. Parenchyma and a block construction of the ground tissue appeared first in charophyceans and bryophytes.[16] The dermal tissue system appeared next, covering the

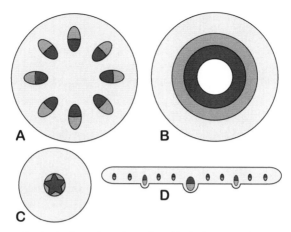

Fig. 8–12 Diagrams showing the three primary tissues, dermal (outline), ground (off-white background), and vascular (phloem light gray, xylem dark gray) as they are distributed in stems (A-B), roots (C), and leaves (D). A-B. Stems have a ring of vascular bundles or a solid cylinder of vascular tissue. The central ground tissue is called the pith; the peripheral ground tissue is called the cortex. C. Some stems and all roots have a solid cylindrical core of vascular tissue where the xylem is central and alternate with the phloem. The cortex is broad and there is no pith. Earliest vascular plant stems had a solid core of xylem like roots. D. Leaves have vascular bundles embedded in ground tissue; larger vascular bundles may produce protuberances, veins, on the lower leaf surface. Xylem is on the upper side of vascular bundles, the morphological inner side.

plant body with a single layer of cells, the epidermis, which also includes specialized cells like guard cells, diverse types of hairs, and secretory cells. The epidermis appears in some bryophytes, simple at first, and then with a waxy cuticle. Stomates with guard cells appear after cuticle, having become necessary for gas exchange after a waxy, waterproof cuticle covers the epidermis. The last tissue system to appear is the vascular tissue (xylem and phloem).

Elongate conducting cells and phloem-like cells appear earlier, with true xylem appearing last, an innovation that defines the tracheophytes, the vascular plants. As explained in the previous chapter, gametophytic organisms like mosses, liverworts, and hornworts grow and function in such a way that a cuticle-covered epidermis and vascular tissues are not needed. It's clear that bryophytic organisms have the genes for cutin production because a waxy cuticle is produced around the pores of photosynthetic chambers of some liverworts and on top of photosynthetic lamellae of moss enations. In both cases this topping of cuticle functions to prevent water from entering and filling up capillary spaces needed for gas exchange. This shows how evolution works. Organisms adapt by using former solutions in new ways. Cuticle functions in moss and liverwort gametophytes to keep water out, rather than keep water in, but cuticle adopted its familiar waterproofing function as their sporophytes got larger.

Vascular tissue forms a central cylinder in stem axes; it can be either solid, which is the ancestral condition (Fig. 8–12 C), or hollow (Fig. 8–12 A-B). If the vascular tissue forms a hollow cylinder or a ring of vascular bundles, then the ground tissue that fills the central region is called pith. Roots have an organization like the primitive stems they are hypothesized to have arisen from, a solid core of xylem (Fig. 8–12 C; compare with Fig. 8–1). Ground tissue also fills the peripheral regions of stems and roots outside the vascular cylinder, a region called the cortex. A layer of epidermis covers the ground tissue on the outside of the axis or root. Basically ground tissue fills all the volume between the vascular tissue and the epidermis, thus composing the bulk of an axis in nonwoody stems. Leaves are basically a sandwich of ground tissue between an upper and lower epidermis with a network of vascular bundles embedded in the ground tissue (Fig. 8–12 D).

The primary cell type of ground tissue is parenchyma, the basic block-organized tissue, which functions in photosynthesis as well as food and water storage. Ground tissue can also differentiate into strands of elongate collenchyma (coh-LINK-eh-mah), flexible cells with unevenly thickened primary walls, and sclerenchyma (sklair-INK-eh-mah), dead cells with thick secondary walls. Collenchyma functions in support and is located just beneath the epidermis, particularly where the stem is angled or ridged[17] and along leaf veins on the underside of leaves. Sclerenchyma can form long, narrow fibers, or smaller, more isodiametric stone cells. Sclerenchyma can form very hard tissue functioning in protection and support—for example, seed coats and the fruit walls of nuts and stone fruit pits.

The stems of early vascular plants had a small, solid core of vascular tissue surrounded by a broad cortex of ground tissue (Fig. 8–1). Only clubmosses and whisk ferns retain this primitive stem anatomy among living plants. The same tissue organization is still found in roots, which comes as no surprise because roots are hypothesized to have arisen from stems that performed all the functions of both stems and roots. Roots serve to anchor plants, like holdfasts, but they also function in absorption and conduction of water and mineral nutrients to aerial portions of the plant, in food storage, and sometimes in asexual reproduction. Indeterminate horizontal axes (rhizomes) and axes with a rootlike function appear to have preceded and led to true roots. Two Devonian plants in the clubmoss lineage show H and K branching of their axes. After the Y-branching of a rhizome, one axis grows up to form an aerial shoot while the other continues growing horizontally. Sometimes the aerial shoot then H-branches, sort of a wide Y-branch followed by a branching at nearly 180 degrees to produce two aerial shoots and a downward growing axis (Fig. 8–13). This certainly suggests roots were derived from stems by a difference in orientation and function (Gensel et al. 2001). Exactly when downward-growing stem axes become true roots is difficult to determine because the anatomical details were seldom preserved, and a continuum of forms, beginning with true stems and ending with true roots, is expected. Without preservation of anatomical details paleobotanists are unable to determine whether something is truly a root or is a stem and their descriptions refer to the rootlike or rooting structures as a result. No sharp demarcation exists even when going from shoot to root in the same plant. Shoot anatomy gradually transitions to root anatomy.[18] In the clubmoss *Selaginella*, specialized rootlike axes grow out of the stem and down, often Y-branching one to several times, and when one of the apices reaches a

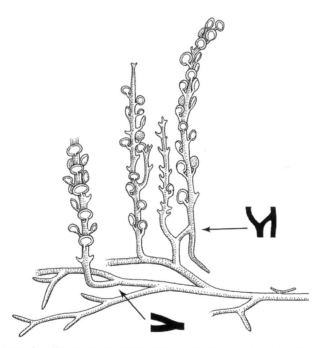

Fig. 8–13 Origin of rootlike axes via H-branching. This reconstruction of *Zosterophyllum* (Fig. 8–3) shows Y-branching (bottom) where one shoot grows upward and the other remains horizontal, and H-branching (upper right) where following a Y-branch, the aerial shoot branches again at nearly 180 degrees, a combination of two branchings that produces two aerial shoots and a rootlike shoot growing down. Shoots that grow downward rather than upward or horizontally are thought to have given rise to roots. (Image source: After Gerrienne 1988)

substrate it produces roots. So far nothing in the fossil record provides any clues about the origin of root caps and the internal origin of branch roots, both distinctive features (Gensel et al. 2001).

Roots arose independently at least twice, once in the clubmoss lineage and once in the lineage leading to the rest of vascular plants. At least one small group of vascular plants, the whisk ferns, remains rootless, but since they lack a fossil record, it cannot be determined whether this is a primitive or derived (loss of roots) condition. The whisk ferns are no longer considered primitive relicts, leftovers from the Devonian, and their small leafy organs are more likely reduced and vestigial than arrested in an early stage of leaf evolution. Their closest relatives appear to be the adder's tongue ferns, a basal lineage, but they share few similarities beyond a subterranean gametophyte. However, this phylogenetic position argues that their roots were lost.

During the Devonian the aerial shoots of vascular plants underwent substantial modification resulting in the appearance of leafy appendages, which is a considerable change from naked stems. Harvesting diffuse solar energy requires surface area, so optimal light harvesting favors an increased surface area. An axis is basically a cylinder. Increasing axis length generates more surface area, but axes must remain fairly slender since light can penetrate only the outer portions of the cortex. Larger-diameter stems increase the surface area but also increase the volume of internal tissues that must be maintained. Doubling the stem diameter will increase surface area 4-fold, but the volume of this larger stem increases 8-fold, so this quickly becomes a losing proposition unless you develop hollow stems like those of horsetails. Slender axes maintain a higher proportion of photosynthetic cortex to nonphotosynthetic inner tissues, but long, fairly slender cylinders have a structural limitation; they are weak, so such axes can grow only so tall before they easily buckle at the slightest lateral force.

Branching is another way to increase surface area. An equal branching of the axis at its apex (Y-branching[19]) produces two axes of a similar diameter diverging at similar angles (Fig. 8–14). Among living land plants, only the whisk fern *Psilotum* has Y-branched aerial shoots as its primary photosynthetic organ (Appendix: Fig. A53), but these plants stand only about 30 cm tall.[20] A spreading crown of ever more slender branches quickly begins to droop under its own weight (Fig. 8–14 F). The aerial shoots of some clubmosses and the whisk fern *Tmesipteris* have Y-branching, but these axes bear leaves. And they also droop under their own weight (Fig. A7, A53, A54). Other plants where the green stem is leafless (or nearly so) and functions as the primary photosynthetic organ are mostly desert-adapted

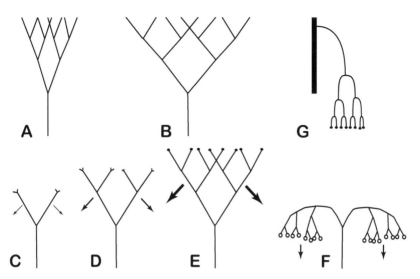

Fig. 8–14 Y-branching axes. A. Narrower angle of Y-branching produces a narrower, more erect crown of axes. B. Wider angle Y-branching produces a broader crown of axes. C.-F. As an aerial shoot system grows larger via apical Y-branching, the increasing mass of the crown of axes increases the force (arrows) that could bend or buckle the basal axes (F). G. If a plant grows as a pendant branch system, the increasing mass of the aerial shoot is supported by the tensile strength of the basal axes. Plants that grow this way are largely epiphytes "borrowing" larger size from a woody plant. (Image source: The Author)

succulents, flowering plants like cacti and euphorbs. But these plants have a vertical main axis or trunk, which is often little branched, and this condition evolved from more typical leafy shrubs to conserve water loss. Some smallish tropical cacti have slender, much branched stems forming a photosynthetic crown that droops downward rather than standing erect. Most of these grow as epiphytes primarily upon trees.

A leafy appendage to an aerial axis, a broad, flat, laterally oriented sheet of tissue, produces a very large surface area for absorbing solar energy. But a thin, flat leaf has a high surface area–to–volume ratio, so it is also efficient at losing water.[21] Broad, flat leafy appendages are so obviously superior to naked cylindrical axes that it comes as no surprise to find that leafy organs evolved among land plants at least six times independently: in liverworts, mosses, clubmosses, horsetails, ferns, and seed plants (or their immediate ancestors) (Kenrick and Crane 1997).

As an aerial shoot system branches and gets larger other aspects of construction come into play. In the whisk fern *Psilotum* each successive apical branching is rotated 90 degrees from the plane of previous branching to produce a spreading, 3-dimensional array of aerial branches. Size constraints are imposed upon such growth forms by increasing moments of force[22] bearing down upon the branch junctions (Fig. 8–14 C-F). The amount of force applied to a branch junction is a function of the mass of axes borne by the branch and the angle of the branch junction. A narrower branching angle results in a more vertical force borne by the axis under compression and a lesser lateral force that would tend to buckle the axis (Fig. 8–14 A-B).[23] However, this produces a narrower cluster of more-vertically oriented branches, resulting in more shading and interference. A wider branching angle results in a broader aerial shoot that will absorb more light, but this shape shifts more weight outward, making the lower axes more prone to buckling. These two different constraints, light absorption, best accomplished by laterally oriented axes, and resistance to buckling, best accomplished by vertically oriented axes, result in natural selection producing a functional compromise, an optimal shape of the aerial shoot system based upon the constraints of its construction materials. Both living and fossil plants with apical Y-branching demonstrate an intermediacy of form. Branching angles are generally acute, more vertical than horizontal, and the aerial axes grow only to a certain height and then stop, a determinate development that limits the maximum load upon the branch joints.

An increased axis height also selects for more conducting/supporting tissues. In general horizontal axes (rhizomes) are indeterminate in development and can keep growing and branching indefinitely. They need no support because they grow resting upon a substrate. Determinate growth is the result of very strong selection limiting the size of aerial shoot systems to keep them well within the physical parameters of the axes. Woody plants adopted a different solution, allowing them to grow bigger each year; still trees can reach a size where the weight of their branches exceeds their strength. As trees get older and bigger, the more-horizontal limbs begin breaking under their own weight or when they are subjected to an increased load caused by snow, wet leaves, or epiphytes.[24]

Aerial shoots of the whisk fern *Psilotum* reach only 2 to 3 decimeters in height (~12 inches), at which point its apical meristems simply stop growing. The central shoot apex divides itself in half six to seven times, and the apical meristem of each branch axis continues to get smaller until it no longer exists. This tough shoot system can persist for a considerable time, but it is no longer capable of growing and will not get any taller. If damaged or browsed, it can be replaced only by the growth of another new aerial shoot. This limited size keeps the slender axes from exceeding their structural capabilities. The small central core of xylem provides very little in the way of structural support, but a cylinder of thick-walled ground tissue in the cortex provides considerable support. Such axes also tend to grow in a cluster so that they provide some mutual support for each other.

A species of *Psilotum* with flattened stems presents more light-absorbing area per unit length of stem without increasing the volume of ground tissue, but these flat axes are lax and drooping. This plant and the closely related *Tmesipteris* are both epiphytes growing upon the stems of tree ferns. By growing aloft a meter or two, their somewhat lax, 15-to-20-cm–long shoots can hang down and still be exposed to light. Some of the clubmosses represent the largest Y-branching plants alive, and they use the same strategy; they grow as pendent epiphytes (Fig. 8–14 G). Pendent shoot systems can be considerably larger than an aerial shoot system without having to invest in more structural support tissues because it takes considerably more force to pull an axis apart lengthwise (tensile strength) than to buckle it laterally. Even a limp noodle requires some force to pull it apart from the ends while it collapses under its own weight. But this raises the basic question of how to get bigger because the epiphytic solution can only work if larger plants already exist.

The fossil record shows that during the Devonian land plants found another adaptive solution to the combined problems of producing more spores, dispersing them more widely, and absorbing more light. By shifting from equal apical branching toward unequal apical branching at different angles, plants could produce one larger and more erect axis that bears weight under compression and one smaller and more laterally oriented axis that is not so hard to hold out laterally. This produces a larger vertical main axis with smaller lateral branches, an orientational shift in branching called overtopping (Fig. 8–15 A-D). The larger, more vertical branch, the "main" axis, continues to grow upwardly, while the smaller lateral branches retain Y-branching, but they are determinate (limited in growth and size), rather like the aerial shoots of smaller plants. This also shifts their functions such that the main axis functions more in support and the lateral branches function more in light capture and reproduction. A larger, vertical main axis can bear a much greater load because it is bigger in diameter and under compression by a vertical force. Even very lightweight cylinders can bear a tremendous force under compression and yet collapse under a slight lateral pressure.[25] Lateral forces from wind increase as plants get taller, so the main axis must become more massive to counteract lateral forces upon it. Retaining flexibility helps disperse such forces and prevent breakage. Taller plants compete successfully for light and disperse their spores further. Continuing this trend produces a vertical main axis, a trunk, and lateral branches that function both as

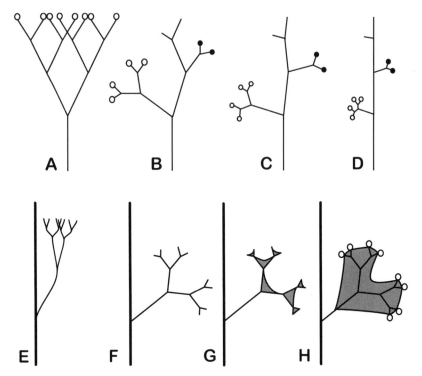

Fig. 8–15 A.-D. Hypothetical transformations of Y-branched aerial shoot system to produce a shoot system with a central axis and lateral branches via increasingly unequal branching and overtopping. E.-H. Hypothetical transformations of determinate lateral branch system into the type of leaf called a megaphyll. E. Three-dimensional branch. F. More lateral orientation and planation of branches. G. Tissue webbing between branches. H. Leaf with marginal sporangia.

photosynthetic and as reproductive sporangia-bearing organs. Rather simple selective forces appear to have reshaped a small, Y-branched axis system into a monopodial (a single axis) growth form that became the predominate form among woody plants, an evolutionary trend that occurred separately in the clubmosses, the horsetails, and the ancestors of the seed plants. Arborescent ferns and monocots (palms) also evolved, but neither are woody plants.

Leaves appear to have arisen as modifications of lateral branches in the common ancestors of horsetails, ferns, and seed plants (Fig. 8–15, E-H). Initially the lateral branches were still a 3-dimensional array of axes (Fig. 8–15 E). Because of the limited size, lateral branch systems can be oriented more horizontally to capture more light and still remain well within structural constraints. Some of these lateral branches might be shed seasonally like the leaves of deciduous trees (e.g., *Wattieza*, Fig. 8–7).

If the 3-dimensional array of axes composing the lateral branches became flattened into a single 2-dimensional plane either by constricting the plane of branching or subsequent reorientations, the branch would present more surface area for light absorption (Fig. 8–15 F). Any webbing of tissue between the axes generates still more surface area (Fig. 8–15 G-H). Such a process has been hypothesized to have produced a type of leaf termed a megaphyll, literally a big leaf, although not all are actually of a large size. According to the megaphyll hypothesis, a leaf is a determinate lateral branch that became flattened into one plane and webbed.

An interesting consequence of the megaphyll hypothesis is that if sporangia were terminal on lateral branch axes (Fig. 8–15 B-D), sporangia would become marginal on a leaf (Fig. 8–15 H), or as some fossils demonstrate, there would be fertile branches bearing terminal sporangia that have an unwebbed "ancestral" form (Fig. 9–3). Thus lateral branch systems and this type of leaf are homologous structures, and this becomes important to remember when interpreting some reproductive structures. Both overtopping and megaphyll development involve branch systems called telomes, so these hypotheses have sometimes been called the telome theory, and references to telome theory are common in the morphological literature of the last century. The Devonian fossil record has a number of vascular plants that can be interpreted as intermediates supportive of telome theory, plants showing overtopping and determinate lateral branches, others with planated lateral branches, others with webbing among the telomes; and many have sporangia terminal on lateral branches and sterile megaphylls—for example, *Pseudosporochnus*, a fossil fern. Unfortunately all these fossils occur in about the same geological time frame, and so are arranged into a sequence only by the dictates of this hypothesis and not temporally. Nonetheless, the homology between megaphyllous leaves and lateral branches is not in doubt. The fossil plants purported to show megaphyll evolution all appear to have a common ancestry among trimerophytes, lineages basal to a megaphyll-bearing clade that includes ferns, horsetails, whisk ferns, and seed plants.

In the clubmoss lineage leaves are referred to as microphylls because they generally are smaller and simper than megaphylls, and they are thought to have had a separate and different origin. Microphylls are flat and mostly linear with a single strand of vascular tissue extending from the leaf base to the tip of the leaf. Microphylls are narrow because the width of the blade is limited by the distance water can efficiently diffuse cell-to-cell from the central vascular strand to the margin of the leaf, but some fossil leaves were 25 to 40 cm long. Among living clubmosses, microphylls are mostly no more than 1 cm long, and some are reduced

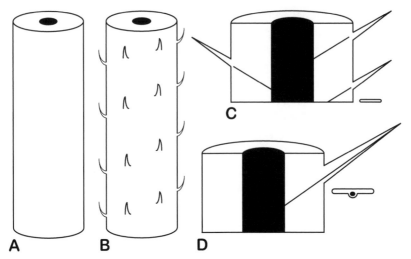

Fig. 8–16 Hypothetical origin of microphylls via the vascularization of enations. A. Naked axes with solid core of vascular tissue. B. Enations increase surface area. C. Enations get bigger with strands of vascular tissue extending across the cortex to their base. D. Microphyll with a single, central vascular strand running the length of the leaf. Fossils representing all four "stages" occur in Devonian members of rhyniophytes and clubmosses.

to scaly vestigial structures. The aquatic clubmoss *Isoetes*, quillwort, has round, quill-like microphylls up to 25 cm long. In fact, microphylls and megaphylls broadly overlap in size, particularly when leaf size is reduced as an adaptation to drier conditions, but the terms are now used to refer to their different origins.

The long-standing and popular hypothesis for microphyll origin proposes that microphylls originated by elaboration from enations, the same kind of nonvascularized leafy organ found on the gametophytes of liverworts and mosses. The idea is that you begin with a naked sporophyte axis such as those of the earliest vascular plants. Leafy outgrowths of the axes (i.e., enations) increase surface area, but they remain limited in size because they lack vascular tissue. Gradually they became vascularized and transformed into a new leafy organ (Fig. 8–16). Devonian fossils exist with naked axes, axes with enations, axes with enations vascularized to their base (*Asteroxylon*), and then clubmosses themselves with microphylls; but again, just because fossils can be arranged in a particular order does not mean the resulting sequence actually represents an exact evolutionary history. Clubmosses have a common ancestry with naked-stemmed rhyniophytes and *Asteroxylon* is a stem group on the clubmoss lineage, so both the fossil record and phylogeny provide some support for this hypothesis. It is clear that vascular plants did not directly inherit leafy organs from byrophytic ancestors because it is the haploid, gametophyte generation in liverworts and mosses that possess leafy enations; their sporophytes have leafless, naked axes.

Another hypothesis suggests microphylls are actually reduced, simplified megaphylls, and in horsetails this appears to be true, but this does not appear to be true for the microphylls of whisk ferns. As a result, at one time clubmosses and horsetails were grouped together as the microphyllous plants, but both molecular phylogenies and the fossil record have falsified this grouping. No one has figured out how the microphyllous leaves of whisk ferns arose or how the whisk fern leaves fit into the megaphyll-bearing horsetail and fern clade.

Other recent fossil finds have produced a very different hypothesis for the origin of microphylls in the clubmoss lineage. Lateral appendages are produced by the apical meristem, and the first apically produced lateral appendages to appear are sporangia—for example, *Zosterophyllum* (Fig. 8–3; Fig. 8–4; Fig. 8–13), and lateral sporangia are a characteristic of the clubmoss lineage. Shifting sporangia to a lateral position increases spore production on a single axis without any branching. Some early Devonian fossil plants in the clubmoss lineage have sporangia clustered in ranks (rows) at the apices of axes, and the sporangia are associated with sterile, leafy structures of a somewhat larger size but of a shape very similar to the sporangia themselves (Fig. 8–17). Otherwise the axes appear naked. These two observations together suggest that microphylls may have arisen by sterilization of sporangia (Kenrick and Crane 1997; Shou-Gang and Gensel 2001), perhaps produced to protect and nourish fertile sporangia by placing a photosynthetic organ close by. This hypothesis also explains the developmental association between sporangia and leafy organs seen in clubmosses. In addition, sporangia have a vascular strand extending into their stalk. The fossil record supports the sterile sporangium hypothesis as strongly as the vascularized enation hypothesis, but either way microphylls are of a very different origin than megaphylls.

If microphylls and megaphylls have had separate and different origins, then leafy organs among vascular plants are not all homologous structures. Kaplan (2001) argued that it does not matter because in terms of morphology, leaf-stem continuity, arrangement on the axis, and their origin at the shoot apex, the distinctions between enations, microphylls, and megaphylls are arbitrary. Furthermore, he argued that the common distinction between nodes, points of leaf attachment on a stem, and internode, that portion of a stem between nodes, presents an inaccurate view of leaf-stem organization, one that is based on seed plants where

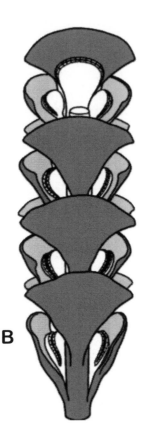

Fig. 8–17 Hypothetical origin of microphylls via sterilization of sporangia. A. Reconstruction of a fertile axis of a zosterophyll named *Adoketophyton*, an early Devonian stem group of the clubmoss lineage showing an apical clustering of sporangia where each sporangium is above and paired with a similarly shaped, sterile leafy organ. B. Enlargement of a portion of the fertile axis showing similar shapes and sizes, although the sterile leafy appendages are larger (shaded gray) than the sporangia (white). (Image source: After Shou-Gang and Gensel 2001).

this distinction is very clear. But in many vascular plants, the stem is not a separate entity from its appendages. Leafy appendages begin growing on the sides of either a dome-shaped apical meristem or just below a single, domed, apical cell. Cell lineages demonstrate that the basal portions of leaf primordia contribute to the outer portions of the axis, producing a leaf-shoot continuum that becomes obscured in seed plants, the group upon which the nodal concept was based (Fig. 8–18). When all cells arise from a single apical cell, cell lineages can be identified very clearly, and leafy appendages can be developmentally traced to a single derivative cell. This means the outer peripheral portions of stem axes actually are composed of leaf bases extending downward, and rather than having a shoot bearing a leafy appendage, there is a leaf-shoot continuum. Nonetheless, enation, microphyll, and megaphyll remain useful terms helping distinguish the very real phylogenetic differences among leafy organs of different origins.

A developmental similarity among gametophyte enations and vascular plant leaves can be traced back to the development of the branching filamentous axis of *Chara*, a charophycean green algae (Fig. 8–18 A) that is part of the lineage having a common ancestry with land plants. In *Chara* and some land plants, a single apical cell divides to produce a derivative cell that then divides perpendicular to the long axis of the stem. In *Chara* and mosses, this division produces an lower internodal cell (i) and an upper nodal cell that will give rise to branch filaments in *Chara* or leafy organs in land plants. In horsetails, the derivative cell divides the same way, but both cells then contribute to the leaf and axis. The leaf or branch

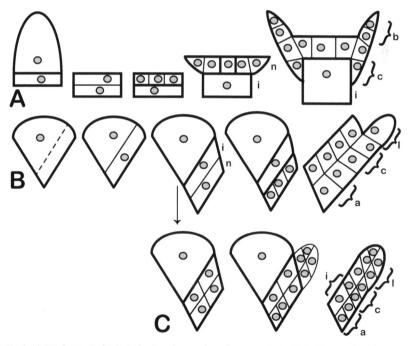

Fig. 8–18 Development of a central axis and appendages from an apical cell. A. *Chara*, a charophycean algae. The apical cell divides perpendicular to the filament axis, producing a basal derivative cell, which divides again to produce an internode cell (i) and a nodal cell (n) that will give rise to appendages, a whorl of branch filaments (b). Downward-growing cells (c) produce cortical cells that surround the internodal cell. B. *Sphagnum*, a moss. An apical cell divides parallel to one of its lower faces (dashed line), and the derivative cell divides in the same plane to produce two cells, an upper one that will give rise to internode cells (i) and a lower one that will give rise to a leafy organ (enation) (l), as well as both inner (a) and cortical (c) regions of the axis. C. *Equisetum*, a horsetail. The development pattern is similar to that of *Sphagnum* except that both derivative cells contribute to the internode (i), leaf (l), and axis (a, c).

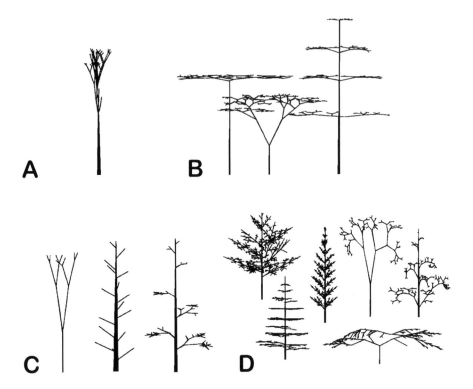

Fig. 8–19 Optimal tree forms for different functions. A. Upright form to optimize spore production and spore dispersal. B. Broad, flat-crowned forms to optimize light harvesting. C. Forms that optimize structural stability. D. When optimized for all three functions the forms appear very similar to tree forms observed in fossil and living trees. (Image source: After Niklas 1997, used with permission of University of Chicago Press)

filament initially divides to produce cells that contribute to the central axis (a), the outer axis or cortex (c), and the leaf or branch (l). As the ability to find and compare genetic components of development improves, it will be interesting to determine whether similar genes are functioning as these developmental patterns suggest.

Computer modeling provides a different approach to examining the morphological problems of the conflicting functional demands placed upon an aerial shoot system. Shoot systems optimized for spore production (and dispersal), structural stability, or harvesting of light have very different shapes (Fig. 8–19 A-C), but in the real world where shoot systems have multiple functions, natural selection is expected to produce forms that are optimal for two or more of these functions, so-called adaptive peaks. When programmed to solve for forms that were optimal for light harvesting, spore production, and structural stability, the models produced appear very much like the form of plants found in the fossil record and alive today (Fig. 8–19 D) (Niklas 1997). This result is not surprising because variation in form is genetically based and selective forces are acting upon them to generate adaptive forms.

VASCULAR PLANT RELATIONSHIPS

Having now constructed some vegetation starting with stems and sporangia, making leaves two or more ways, and making roots, we can turn our attention to the broad phylogenetic relationships among vascular plants (Fig. 8–20). All early Devonian fossil plants originally were placed in a group called the rhyniophytes. As more details have come to light, the rhyniophytes now appear to be a diverse assemblage of plants best treated as stem groups to the tracheophytes, the vascular plants. The sequence of fossil organisms presented here generally supports the origin-of-vascular-plants hypothesis presented above. Rhyniophytes and tracheophytes all have branching sporophytes and so all are polysporangiophytes (Fig. 8–20), but not all polysporangiophytes were vascular plants. The aerial stems of many early rhyniophytes showed a small central core of vascular tissue and a peripheral band of thick-walled cells called a sterome, a structural adaptation that functions in support in axes with little xylem. Whisk ferns are the only living plants possessing a sterome.

Conducting tissue, although not xylem, is found in both the gametophyte and sporophyte axes of one group of mosses (Appendix: Mosses, Polytrichopsida). While you might think polysporangiophytes shared a common ancestry with this group of mosses, this does not seem to be the case. If conducting tissue was acquired by the common ancestors of mosses and vascular plants (Fig. 8–20 character #3), it also means that conducting tissue was lost in most mosses, a most unlikely event, so this character is misplaced on this diagram. If the moss clade were expanded to show major lineages, the character of conducting tissue would appear on only one branch, which would suggest an independent origin. If vascular plants and mosses with conducting tissue shared a common ancestor, then vascular plants would be nested within the moss clade. Molecular studies do not support such an ancestral relationship, so conducting tissue arose separately in mosses and the ancestors, whatever they may be, of vascular plants. Conducting tissue (character #3) should move up one node beside branched sporophyte (character #4) since both originate in

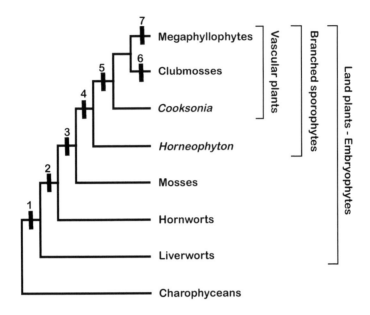

Fig. 8–20 Broad phylogenetic relationships among land plants (embryophytes). The five main clades of land plants are liverworts, hornworts, mosses, clubmosses, and megaphyllophytes (plants with megaphylls: ferns, horsetails, and seed plants). Liverworts, hornworts, and mosses have uniaxial sporophytes. Branched sporophytes produced two or more sporangia. *Cooksonia* and *Horneophyton* are late Silurian and early Devonian rhyniophytes and they form stem groups to the rest of the land plants with branched sporophytes. Some (*Horneophyton*) lack true xylem; others (*Cooksonia*) had true xylem. Charophycean algae forms a sister group to all land plants. Appearance of novel characters: 1. embryo, uniaxial sporophyte; 2. stomates and guard cells on sporophytes; 3. conducting tissue; 4. branched sporophyte; 5. xylem; 6. dorsoventrally flattened, lateral sporangia, microphylls; 7. megaphylls. Compare with Fig. 7–8 E. (Image source: After Pryer et al. 2004)

the ancestor of this lineage. In this diagram land plants (embryophytes) consist of five clades: liverworts, hornworts, mosses, megaphyllophytes (plants with megaphylls for leaves and their immediate common ancestors = pteridophytes + seed plants), and clubmosses.

Clubmosses are the earliest divergent, the most ancient lineage, of living vascular plants. The living genera of clubmosses have a somewhat relictual status exemplified by *Selaginella*, the oldest living genus on Earth. In addition to microphylls, clubmosses have dorsoventrally flattened sporangia that are borne laterally rather than terminally. All the remaining vascular plants form a lineage that possess megaphyllous leaves, which includes the ferns, horsetails, whisk ferns (megaphylls?), and seed plants. Both the clubmoss and megaphyll plant lineages have stem groups that are not shown in Fig. 8–20; these are organisms that have some but not all of the characters that define these lineages (e.g., *Zosterophyllum* and *Asteroxylon* are clubmoss stem groups). Stem group organisms can be thought of as transitional, but while stem groups can be placed in such positions as hypotheses to make sense out of the characters, the fossil record in this case neither confirms nor refutes the hypothesis of intermediacy because all appear more or less simultaneously. Not all members of the megaphyll lineage[26] have megaphylls, but those lacking true leaves are clearly stem group organisms that have leaf-like lateral branches (e.g., Fig. 8–7, *Wattieza*). The megaphyll clade consists of two main groups, the pteridophytes (ferns, horsetails, and whisk ferns), now minus the clubmosses, and the spermatophytes, the seed plants, which share a common ancestry among the trimerophytes.

Pterido- (terr-EE-dough) means fern, and the pteridophyte clade includes the ferns and those plants referred to as the fern allies or the so-called lower vascular plants; the clade comprises all free-sporing vascular plants (as opposed to seed plants). The pteridophyte clade is a phylogenetic hypothesis supported by a variety of morphological, biochemical, paleobotanical data (Kenrick and Crane 1997; Pryer et al. 1995, 2001, 2004). Traditionally, pteridophytes included the clubmosses, but their lineage is now firmly excluded, occupying a lineage basal to all other living vascular plants. Rather than redefine pteridophyte as ferns and fern allies, minus clubmosses, this clade is being called the moniliophytes (moh-KNEE-lee-oh-fights) (Kenrick and Crane 1997; Pryer et al. 2004), which follows from a new practice of naming clades after a novel shared character. The clade should really be called the monilioformophytes because *moniliformis* is Latin for necklace-like, a reference to the bead-like appearance of "mesarch protoxylem confined to lobes of the xylem strands." Almost nobody except plant anatomists and paleobotanists knows what that means. Moniliophytes, or the protoxylem-appearing-like-a-necklace plants, isn't exactly a name anyone except a botanical wonk can hang his hat on, and in the present context of botany for nonbotanists, it makes more sense to retain and redefine the familiar pteridophytes than to insist on using a name that simply adds to a learner's memory burden. Perhaps the use of an old familiar name means the author is a bit of an old fogy, an admission that will cost him a pint of beer if certain colleagues ever read this, but such are the educational decisions that often must be made. Recent publications in the botanical literature, and even the most recent introductory botany textbooks, demonstrate that moniliophyte is being used increasingly by professionals, but it remains very impractical

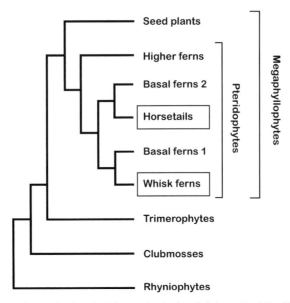

Fig. 8–21 The vascular plant (tracheophyte) lineage showing broad phylogenetic relationships. The extinct Devonian rhyniophytes form the basal lineage. Clubmosses are the earliest diverging lineage with living members. Ferns, whisk ferns, two lineages of basal ferns, and the horsetails form the pteridophyte clade. The pteridophyte clade is sister group to the seed plants, and together they form a clade of megaphyll-bearing plants, megaphyllophytes. The trimerophytes form a stem group to the megaphyllophytes. See Appendix: Ferns for more detailed phylogenetic information about the pteridophytes. (Image source: After Pryer et al. 2004; Smith et al. 2006)

for amateurs and in general education. If you like betting on sure things, you can bet that some of my botanical colleagues will give this book poor reviews because so much of their beloved jargon was purposely avoided.

Within the pteridophyte clade (Fig. 8–21), the relationships of the horsetails and whisk ferns to the ferns was a surprise. Many prior classifications placed the horsetails and whisk ferns in separate lineages. In one phylogenetic study (Rothwell 1999), whisk ferns and horsetails formed basal lineages to a clade consisting of two lineages, seed plants and ferns, so no pteridophyte clade was formed. Most recent molecular studies have nested both the horsetails and whisk ferns within basal lineages of ferns (Pryer et al. 1995, 2004), producing a pteridophyte clade (see Appendix: Ferns). To call all these organisms "ferns" really abuses the concept, but either "fern" is equivalent to the whole pteridophyte clade or the name fern is more narrowly applied to just those ferns with more specialized sporangia, thereby leaving out the basal ferns, so perhaps the label fern, or maybe, eufern, will be reserved for the higher ferns, while other labels are applied to the basal fern lineages that include the horsetails and whisk ferns. The whisk ferns have no fossil record to assist although the structure of their sperm and their subterranean gametophyte is similar to the adder's tongue ferns with whom they share a common ancestry (Renzaglia et al. 2001). In the case of horsetails, fossil evidence suggests the horsetail lineage is at least as old as the ferns or older and forms a distinct lineage arising separately from a trimerophyte ancestry. While many diagrams show the cladoxylopsid ferns to be ancestral to the pteridophytes, the cladoxylopsids so formulated include the Hyeniales ("high-in-ee-AYE-lees"), organisms generally considered ancestral to horsetails. The reasons for lumping Hyeniales with the other cladoxylopsid ferns are not at all clear other than to make things look simpler than they really are. To reconcile the fossil record with the molecular phylogenetic studies on living members does not seem possible at present.

Horsetails display tremendous fossil diversity and were a major component of the Carboniferous flora, but presently there is only one living genus, *Equisetum* (Appendix: Horsetails). The whisk ferns (Appendix: Whisk Ferns) are a very a small group with only two genera that were once thought to be direct living descendants of rhyniophytes, but their resemblance to rhyniophytes is now considered superficial. Ferns are a large, diverse group or groups (Appendix: Ferns); some fern lineages have long fossil histories, including cinnamon fern, which is presently the oldest known living species on Earth.[27] Both the cinnamon ferns and the Marattioid ferns have fossil records extending back into the Carboniferous. Other fern groups have increased in diversity along with the flowering plants.

TRIMEROPHYTES

This only leaves one group of organisms in need of some introduction, the trimerophytes, a fossil group first proposed by Banks (1968). Trimerophytes (Fig. 8–22) retain some rhyniophyte features while providing intermediates to the rest of megaphyll-bearing plants. In other words, trimerophytes are interpreted as transitional between the earliest vascular plants and the pteridophytes and seed plants. Thus when examining the phylogenetic relationships among ferns and horsetails, we see that trimerophytes possess features presumed to be ancestral to both lineages, and their position in the geological column places them in the right time frame, appearing in the early Devonian and continuing into the late Devonian.

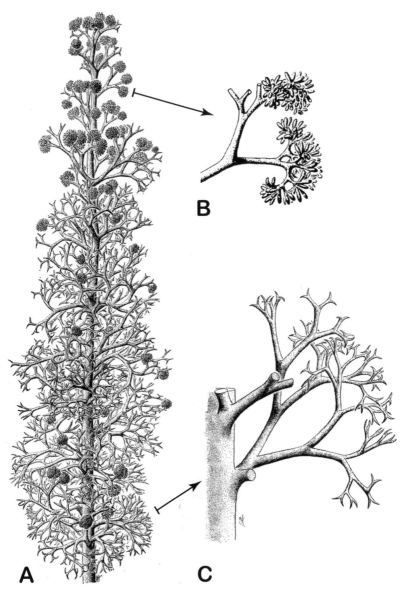

Fig. 8–22 Trimerophytes. A. *Pertica*, a genus with a strong central axis and Y-branched laterals. This plant is estimated to have grown 1 meter tall. B. Detail of fertile branch bearing terminal sporangia. C. Detail of sterile lateral branch showing Y-branching in three dimensions. Note that front branch of the 3-parted lateral branch is not shown. (Image source: Kasper and Andrews 1972, used with permission of the Botanical Society of America)

Trimerophytes were rhizomatous plants with either naked stems or stems bearing small enations. The aerial axes exhibited a main axis with distinctly lateral branches. Spirally borne lateral shoots Y- or tri-branched (in threes) close to the main stem, a pattern that gave the group its name (trimero-). Sporangia were borne terminally on lateral shoots in clusters that bend downward or inward. Two well-known genera, *Trimerophyton* and *Psilophyton*, represent this group, but it is here illustrated by a nicely reconstructed genus *Pertica*. Lateral shoots that are much reduced in size, branching just a few times, are exactly the first intermediate stage hypothesized in the evolution of megaphylls (Fig. 8–22 B-C).

This chapter broadly presented the botanical innovation that took place in the Devonian. In the span of this relatively short geological period, one small group of land plants with little, naked, branched sporophytes developed vascular tissue and diversified into all of the other major lineages of land plants. Before the Devonian ended, the first trees, forests, and seeds had appeared. So much change in so little time can be attributed to an open door to so many new unexploited habitats as well as to a positive feedback loop where the increasing size and diversity of plants altered the environment and provided many opportunities for new species. This is a well-understood evolutionary phenomenon: an adaptive radiation. The usual example is Darwin's finches, a group of finch species that occupy some rather un-finch-like ecological roles in the Galapagos Islands. This resulted from a situation where a rather typical finch dispersed to a group of volcanic islands lacking the usual complement of terrestrial birds. This allowed the founder stock to diversify into lifestyles not typical of finches. The Devonian fossil record now reveals the results of an unfettered diversification of land plants onto the largely empty continents.

The conquest of land resulted in cladoxylopsid trees, clubmoss trees, and horsetail trees. Although the ferns failed to become woody, even the ferns produced arborescent forms, tree ferns. However, as you will see in the next chapter, in a manner of speaking, a lineage of ferny plants was woody and became trees. This lineage, the ligniophytes or woody plants, did become a dominant component of terrestrial communities because this lineage includes the seed plants, the spermatophytes, our familiar (and not so familiar) gymnosperms and angiosperms.

NINE

Seeds to Success

*Wherein the nature of seeds, their impact on the land plant life cycle, their history,
and the diversity of seed plants is investigated.*

Though I do not believe that a plant will spring up where no seed has been,
I have great faith in a seed. Convince me that you have a seed there, and
I am prepared to expect wonders.
—H. D. Thoreau

It's March[1] in the northern temperate zone and gardeners are purchasing lots of packets of seeds to plant in their gardens. Planting seeds and watching them grow is a very satisfying activity because something almost magical happens when these seemingly inanimate, lifeless motes start growing. Seeds are planted with great expectations of what will happen, expectations encouraged by the promissory pictures on the seed packets or in catalogs. Some of the plants will be grown for vegetables, so we can eat their roots (carrot), stems (kohlrabi), or leaves (lettuce). Some of the plants will be raised for their fruits: tomato, cucumber, zucchini, beans, and sweet corn.[2] Some of the plants, from *Alyssum* to *Zinnia*,[3] will be raised for their attractive flowers. And, lastly, the most amazing part is that eventually seeds beget seeds, so it can be done all over again. Thus both tangible and intangible rewards are reaped from planting seeds.

While seemingly simple, seeds are quite complex and few people understand their biological nature, an ignorance that shall be weeded out. In fact seeds should be nominated as the most familiar yet least-understood biological entity.[4] As is often the case, our biological ignorance has not in any significant way hindered humans from using seeds for our particular purposes. Indeed, human existence totally depends upon our interactions with seed plants. Perhaps this is why our distant ancestors regarded the agriculture upon which their very lives depended, a combination of seeds and sex, with a religious awe (Heiser 1981). Seed plants now dominate most terrestrial communities, so the evolution of seeds must be considered one of the singularly most important events in the history of green organisms, or even of life itself, since so much of present-day biological diversity depends upon seed plants.

The seed plants had their origin back in the late Devonian when the first seeds appear in the fossil record. The seed is a novel shared character that defines the seed plants or spermatophytes as the third major lineage of vascular plants in addition to the clubmosses and pteridophytes. Traditionally seed plants have been treated as two groups: gymnosperms, meaning naked seeds, and flowering plants or angiosperms, meaning housed or covered seeds. The significance of "naked" versus "housed" will be explored a bit later, but for now it is only important to indicate that cycads, conifers, and *Ginkgo*, plus a number of less-known plants, both living and fossil, are gymnosperms. Many fossil gymnosperms are lumped into a category called seed ferns or pteridosperms, woody seed plants with ferny foliage, but this grouping is a bit of a catch-all that includes considerable diversity. Some pteridosperms are more like cycads, some more like conifers, and some, perhaps, more akin to flowering plants. Flowering plants, and this includes grasses, trees, and other plants whose small and insignificant flowers may have escaped your notice, are the last major group of plants to appear, and while a common ancestry among gymnosperms is a given, the closest gymnosperm relatives of flowering plants remains quite uncertain.

FREE-SPORING PLANTS

To understand seeds, seed plants, and their success, you have to fully understand the limitations the land plant life cycle places upon free-sporing plants. Free-sporing vascular plants disperse their spores and each spore develops into a free-living gametophyte, which has a bryophyte-like structural organization and reproduction unless the gametophyte has been so reduced that it develops endosporically, completely within the spore—for example, *Selaginella*. If heterosporous like *Selaginella*, free-sporing plants disperse both microspores and megaspores at the same time. These free-living gametophytes are the weak link in the adaptation of these plants to terrestrial conditions. The spores of free-sporing plants disperse largely by wind, although a couple of exceptions exist. The distance a spore disperses involves many factors: wind velocity and direction, weather conditions, topography and obstructions, spore size, and height of release. Wind is cheap, readily available, and capable of dispersing spores widely across considerable distances, but spores end up being deposited across the Earth's very patchy terrestrial landscape largely at random, although local physical features may enhance or reduce spore deposition in a particular location. The best strategy for dispersal success in such a broadcast sweepstakes is to invest in vast numbers of individually cheap spores to improve the odds of one landing in a favorable patch of terrestrial habitat. Most spores are doomed; they will fail to land on a good piece of real estate or fail to even land on land.

Spores land everywhere, but what limits the distribution of clubmosses, ferns, horsetails, and whisk ferns are the requirements of their free-living gametophytes for a moist habitat, both because their "bryophyty" gametophytes lack vascular tissue and because their swimming sperm require free water for fertilization.[5] Some clubmosses, horsetails, and ferns have pretty tough sporophytes, well adapted to moderately dry conditions, but no matter how tough they are, their distribution is constrained by the biology of their gametophytes because *no dispersal event* follows fertilization. Remember, an egg is an immotile gamete, so the new sporophyte begins growing exactly where a gametophyte was successful in mating. When you look at a fern can you visualize the other fern, the haploid gametophytic fern that by necessity preceded it in that location? (Recall Fig. 6–3.) A fern growing out of a crack in a rock is there because a gametophytic fern successfully got an egg fertilized and raised a sporophyte to the point of self-sufficiency. At times you wonder how it can be possible unless you recognize two things: one, these gametophytes are tiny and a favorable habitat may be equally small, and two, the sporophytes of free-sporing plants are all potentially long-lived perennials. To produce a winner from the long-shot dispersal lottery, you have to keep playing the spore dispersal game. In marginal habitats conditions may only rarely be conducive for the reproductive success of the gametophyte generation, but that may be enough to establish a new sporophyte that can persist for decades, even centuries, producing large numbers of spores, season after season.

Every now and then nature provides us with some particularly good examples of the consequences of the free-sporing life cycle. The Great Basin is a high, dry plateau covered largely by sagebrush desert, an area which receives less than twenty-five centimeters of rain a year. Here you might chance to visit a lovely area in Idaho with the intriguing name of

"Hell's Half Acre," a rather desolate landscape formed by an immense lava flow several meters thick covering the high valley floor. Hell's Half Acre is easily visited because an interstate highway cuts right through it and thoughtful planners placed a rest stop conveniently in the middle although few people venture beyond the asphalt Hades of its parking lot. Although not visually appealing, such places hold a strange fascination for biologists who are always curious to see what organisms can live in different places (besides, the driver needed to stretch his legs). The surface of the lava flow presents an extremely arid landscape whose rough, dark, rocky, and fractured surface is hot and baked dry in the summer sun, then windswept and frozen in the winter. The thin soil atop the lava field supports only sparse, scattered vegetation, a community dominated by a few very tough dry- and cold-adapted desert plants, seed producers all. The largest plants are small, sparsely scattered juniper trees, whose roots can penetrate deeply into cracks leading to moisture below the lava field. But an intrepid visitor[6] need not walk very far on the lava field's surface to observe some non-desert plants thriving.

The lava field contains some deep, narrow fissures, cracks that formed when the molten lava cooled and contracted. At the base of these narrow cracks, some 10 to 20 feet below the level surface of the lava field, can be found rather luxuriant growths of ferns. These basaltic crevasses create moister and shadier microhabitats protected from the summer sun and winter winds. No fern can grow on the lava flow's surface for miles and miles. People unfamiliar with fern biology might wonder how ferns came to grow in such a spot.

Wind-dispersed fern spores rain down upon Hell's Half Acre and all other areas at a rate of so many spores per square meter per year. The fern spores raining down on Hell's Half Acre come from ferns growing a considerable distance away, perhaps from the forested mountains just visible on the western horizon. All the fern spores landing upon the surface of the lava field are doomed, losers in the dispersal lottery, but a spore falling into a moist, shady crevasse stands a chance, every now and then, of finding sufficient moisture to grow and develop into a mature gametophyte, fertilize itself, and establish a new fern sporophyte. Who knows how long ago or how often this happens, because fern sporophytes can persist for decades. Either through clonal growth or because of the numerous spores now being deposited within the crevasse, populations of ferns now grow in the middle of a desert.

Just south of the Illinois State University campus is a less exotic example. An inhospitable, xeric, and nearly barren rocky ridge slices across landscape. No, this isn't the terminal moraine of the Wisconsin glacier, which is located three miles further south, but the rocky ballast of the Illinois Central Railroad tracks. A colony of *Equisetum arvense*, a common horsetail, grows in one section, thrusting its aerial shoots up from among the rocks while its rhizomes remain hidden deep within the ballast. You can walk the tracks for many miles without finding another horsetail growing in the ballast, although satisfying your curiosity will probably get you arrested by Pinkertons for trespassing on railroad property. They have no respect for horsetails, botany, or botanists.

Horsetail spores must disperse onto the railroad ballast regularly. But just once, well over a century ago,[7] a spore encountered moist enough conditions for a long enough time that a gametophyte reached sexual maturity, and probably fertilized itself.[8] The resulting horsetail sporophyte became established, and with its tough adaptations for terrestrial life it has found the railroad ballast a fine habitat with little competition from bigger and taller plants.

In both of these examples, the environmental needs of the gametophyte constrained the distribution of the sporophyte even though the latter could grow and persist in even harsher places. The bottom line is simple: the gametophyte generation is less adapted to the terrestrial environment than the sporophyte, but in the free-sporing life cycle of pteridophytes, no dispersal takes place between the gametophyte and sporophyte phases, so the adaptability of the gametophyte constrains the sporophyte's habitat choice.

Ferns and other free-sporing vascular plants can adapt to drier environments where there are shorter and less frequent moist episodes, by having smaller, faster-developing gametophytes. Of course evolution does not work this way, and what that actually means is that those free-sporing vascular plants whose gametophytes develop a bit faster will succeed in more marginal places thus extending their range into drier places. Such a selectional force is what leads to endosporic gametophytes, which are very small and relatively quick to develop. Although considered free-living, these endosporic gametophytes are dependent, wholly or nearly so, upon the sporophyte that produced the food-packed spore. Thus endosporic gametophytes completely reverse the bryophyte situation where the sporophyte is wholly or largely dependent on the gametophyte. In fact, with endosporic gametophytes, one sporophyte generation nourishes the next sporophyte generation in its embryonic stage, albeit indirectly, by providing nourishment for the completely dependent female gametophyte. Endosporic gametophytes are nothing new; they evolved long ago in the Devonian along with heterospory in the clubmosses and in the ancestors to seed plants.

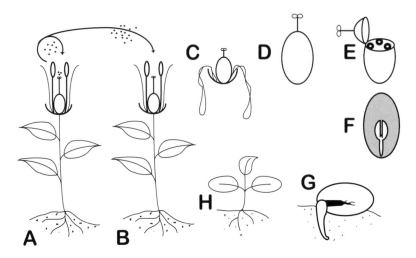

Fig. 9–1 How a seed plant's life cycle appears. A-B. Mature flower (or cone) self- or cross-pollinates. C-E. Postpollination development of ovary and ovules produces a fruit containing seeds. F. Each seed contains a plant embryo. G. After germination, this embryo grows back into another plant, which when mature will be like those whose flowers made the seed (A and B). Compare this seed plant life cycle to the general land plant life cycle (see Fig. 6–4). No alternation of generations is evident, leaving the impression that seed plants have a life cycle like animals (see Fig. 3–9 for animal life cycle).

UNDERSTANDING SEEDS

The evolution of seeds from a free-sporing life cycle and the advantages of the former over the latter are not immediately obvious because many misconceptions about seeds arise from what we perceive and do not perceive. Plant a radish seed and you get a radish plant, which then produces more radish seeds.[9] Fantastick![10] If you have ever been curious enough to dissect a radish seed, you would discover it contained an embryonic radish plant complete with a root-shoot axis with an apical meristem at each end, and a pair of embryonic leaves, cotyledons, which appear above ground after germination to function photosynthetically. When looked at superficially, the life cycle of seed plants seems very simple (Fig. 9–1) and very much like that of animals. When a rooster and a hen, both diploid adults, mate, the result is a chicken embryo within a well-provisioned egg. Pollination between two diploid plants results in an "ovule" maturing into a seed containing an embryo of a diploid plant. Biology textbooks commonly define a seed as a "mature ovule," which certainly leaves all misconceptions intact!

The botanical use of the term "ovule" for an immature seed is an interesting example of an entrenched terminological error. Ovule literally means egg, and seed plant "ovules" mostly do look like little eggs, but they are not. And a pollen grain acts as if it was a gamete, but this is not the case either. The correct answers to these questions have been known since 1868 when Wilhelm Hofmeister surmised that seed plants have the same life cycle as free-sporing plants (Kaplan and Cooke 1996), although at the time no one understood that this was an alternation between haploid and diploid phases. This was quite an amazing insight since chromosomes and chromosome numbers were unknown and in fact the function of the nucleus was unknown too. Nowhere in the radish-seed-to-radish-seed cycle is the haploid gametophyte phase of the land plant life cycle evident. But present they are! Working backwards from what is known leads logically to correct interpretations. A radish embryo, found inside a radish seed, arises from a zygote, so a radish plant must be a diploid, a sporophyte, and sporophytes produce spores in sporangia. One type of sporangium, a microsporangium, produces microspores, which become pollen grains, male gametophytes really. Pollen grains disperse widely, ideally into the immediate vicinity of a female gametophyte, sometimes with some active assistance of the female gametophyte herself. Of course, most of these little males are doomed to lose the dispersal lottery. This remains the same whether the plant is free-sporing or produces seeds. Betting on any particular male is a long shot.

Another type of sporangium, a megasporangium, produces a megaspore. The structure called an ovule is a megasporangium enclosed within a jacket of tissue. Even though inaccurate and wildly misleading, the term "ovule" will continue being used because who wants to keep saying "jacketed indehiscent (non-opening) megasporangium"? As a consequence of delaying the dispersal of the megaspore, the female gametophyte develops protected within its megasporangium and nourished by a big sporophyte plant. After this well-nourished endosporic female gametophyte has matured, produced an egg, had it fertilized by a nearby male gametophyte's sperm, and produced an embryo, the entire package disperses as a seed.

Ah ha! Seed plants have a dispersal phase, one that follows the formation of an embryonic sporophyte! This means junior does not actually "leave home," but along with his mother and her house, complete with a loaded pantry, he does disperse and, if proper conditions prevail, grows out of and outgrows his mother and her house. Many ecological studies have demonstrated that seed mass is correlated positively with seedling viability. With seeds, bigger is better. And not only that, but a seed provides a very handy dormancy stage for avoiding seasons that are too cold or too dry. These advantages more than anything else are the key to the success of seed plants. What a contrast to free-sporing plants where the sporophyte is constrained to living its independent life in a location well adapted to its nonvascular, bryophyty mother. The seed plant life cycle provides two major advantages: a female gametophyte that is sheltered from the environment and nourished by a big, well-adapted sporophyte, which helps insure that lots of female gametophytes will mate successfully, and a seed dispersal phase that follows production of an embryo.

The seed habit accompanies and takes advantage of the increasing size and dominance of the sporophyte, coupled with the reduced size of the gametophyte and its increasing dependence upon the sporophyte. Once the sporophyte plant begins nourishing (and protecting) both the female gametophyte and the next generation of sporophyte, some interesting consequences occur. Since size is no longer a constraint for a nondispersing megaspore, you might logically conclude that seed plants will produce really big robust megaspores. But they do not. Seed plant megaspores are actually pretty small; you need a microscope to see one. In some cases the megaspore is actually smaller than the microspore.[11] While that sounds a bit dopey, it is another example of why the seed plant life cycle makes no sense unless understood within a phylogenetic context. The nourishment provided by the sporophyte means a big, fat megaspore is no longer needed. For the exact same reason, the eggs of placental animals are microscopic and chicken eggs are big. Females of placental animals continually provide nourishment to their developing embryo for weeks or months, whereas the chick embryo has to have enough lunch packed just after fertilization to last it until hatching.

Seeds are not the whole story. Botanists often refer to the "seed habit," meaning the entire suite of changes associated with seeds. Pollination, as opposed to fertilization, involves the dispersal of one or more male gametophytes (pollen grains) into the vicinity of an ovule, commonly over distances far too great for sperm to swim. Thus producing a lot of tiny, readily dispersed males is part of the solution to the problem of having swimming sperm on land. Although pollen is a plant, these males are not sessile but free to disperse, seek, and compete for mates just like animals. And they do!

A female gametophyte located within her jacketed sporangium is a hard lady to find and get close to. Somehow males must get their sperm to her there within the ovule. Two solutions to this problem exist, one used by gymnosperms and one used by angiosperms. The jacket around the megasporangium is not completely closed; a tiny apical opening remains (the micropyle = tiny hole). In gymnosperms, entire pollen grains pass through this opening into the ovule. In angiosperms, the pollen is deposited at some distance from the ovules and an outgrowth of the male gametophyte, a pollen tube, grows to the ovule and in through the opening to the female. In both cases fertilization takes place within the ovule and while the ovule is still attached to the sporophyte. In gymnosperms, either the intervening sporangial wall tissue undergoes dissolution, providing both a liquid for swimming sperm and access to the female gametophyte, or the male gametophyte grows a pollen tube through the sporangial wall to the female gametophyte. In angiosperms, the pollen tube actually penetrates the female gametophyte, delivering sperm directly. Either way, delivery of sperm and fertilization no longer depends upon free water. Penetration of the female gametophyte by a pollen tube is a seed plant equivalent of animal copulation.[12]

All seed plants are heterosporous, and their male and female gametophytes are tiny, endosporic organisms. The critical difference between free-sporing plants and seed plants relates to when in the life cycle the megaspore is dispersed and, most importantly, its condition at the time of dispersal. Seed plants disperse their microspores when mature as pollen, each containing a partially to fully developed male gametophyte. But the megaspore of seed plants is not dispersed at this time; in fact, the megaspore resides within a jacketed megasporangium that does not even open to allow for spore dispersal, ever. Seed plants disperse their megaspores, each still housed within its sporangium, weeks to months after pollen is dispersed. And by the time of megaspore dispersal, an endosporic female gametophyte has developed and produced one egg (or more), which after being fertilized has grown into an embryo. At this stage the whole kit and caboodle, a jacketed sporangium containing remnants of a megaspore, a fully developed but endosporic female gametophyte that is carrying an embryonic offspring, is dispersed, and this entire package is a seed. Ta da! Far from being a simple structure, a seed encompasses three generations: a sporophyte's sporangium and spore, a female gametophyte, and a new embryonic sporophyte.

THE EVOLUTION OF SEEDS

Our hypotheses about the origin of seeds and the seed habit must come from both living and fossil organisms. So the early stages in the evolution of seeds involved a change from homosporous to heterosporous, the development of endosporic gametophytes, a reduction in the number of megaspores per sporangium with a concomitant increase in the size of the megaspore. The ultimate expression of this trend would be an uneven meiosis resulting in only one viable megaspore (Fig. 9–2 A-F). But what about the jacket layer to make an ovule? Our hypothesis about jacketing the megasporangium involved the "condensing" of sterile branches to form the jacket (Fig. 9–2 J) until the megasporangium is accessible only via an apical pore.

Fortunately seeds fossilize fairly well because of their tough, resistant seed coat derived from the jacket layer, and the fossil history of seeds dates back to the late Devonian. Unfortunately seeds disperse, so often the problem is to discover or determine what plant should be associated with each fossil seed. The late Devonian fossil *Archaeosperma*, meaning "ancient seed," was not completely jacketed, so technically *Archaeosperma* was not an ovule and therefore by definition not a seed,[13] a pre-seed stage of seed evolution. The jacket left the apex of the sporangium exposed and surrounded it with a whorl of finger-like lobes (Fig. 9–2 J4). Internally, *Archaeosperma* has one functional and three small abortive megaspores, the result of an unequal meiosis (Fig. 9–2 E). The whole fossil consisted of a twice Y-branched axis bearing four terminal "ovules" each adjacent to a leafy organ, the whole branch unit forming a cupule, a cup-like structure around the ovules (Pettitt and Beck 1968) (Fig. 9–2 J5).

Since the initial description of *Archaeosperma*, slightly older fossils of a similar nature have been found (Stewart and Rothwell 1993), and they differ in the degree of fusion of the jacket's lobes around the megasporangium, from none to a partial fusion, and from 4 to 5 lobes to 8 to 10 lobes. Like *Archaeosperma* each "ovule" is erect and paired with a leafy organ, and four such units construct a cupule. Fossils such as these are consistent with our hypothetical interpretation of the ovule's jacket as originating from the fusion of a whorl of sterile stem/leafy structures. Unfortunately a geological sequence of such fossils does not exist; from our perspective they all appeared about at once, so while they can be arranged in a sequence showing fusion, this sequence does not exist, but the diversity of partially to wholly jacketed megasporangia does exist.

The jacket's apical lobes and the leafy cupules surrounding the ovules function to capture pollen grains by causing eddies in moving air that enhance pollen deposition upon the apex of the ovule. This is not a guess, a maybe, or a perhaps. Scale models were constructed and tested in a wind tunnel (Niklas 1981, 1992). In living gymnosperms, the nearby foliage needles and the whole seed cone structure aids in capturing pollen in a cupule-like manner.[14] Prior to Niklas' experiments the chief function of a jacket layer was thought to be protection of the ovule from insects (Rothwell and Scheckler 1988).

Even with the structure enhancing deposition, a small apical pore is a pretty challenging target, but pollen doesn't have to fall into this little tiny opening to get inside the ovule. Ovules of living gymnosperms exude

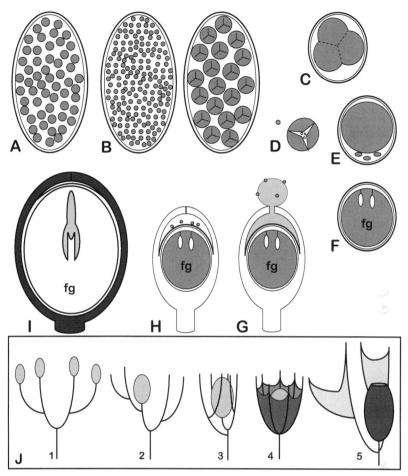

Fig. 9–2 Diagram showing trends in seed evolution. A. Homosporous and free-sporing; spores produce hermaphroditic gametophytes. B. Heterosporous and free-sporing. Microspores produce male gametophytes and megaspores produce female gametophytes. Size differences represent trade-offs in numbers, resources, and ability to disperse. C. Reduction of megaspore mother cells to one produces four megaspores. D. Dispersed microspore and megaspore with endosporic development of gametophytes like *Selaginella* (Appendix Fig. A9). E. Reduction to one megaspore by uneven meiosis. F. Retention of megaspore within megasporangium and endosporic development of the female gametophyte (fg) with eggs. G. An ovule consisting of a jacket layer around a megasporangium with a single megaspore and endosporic development of the female gametophyte. A sticky pollen drop exudes out through apical opening in the jacket layer to trap pollen grains. H. Pollen drop reabsorption pulls pollen into a pollen chamber between the jacket and megasporangium. Male gametophytes penetrate megasporangium by growth of a pollen tube to reach the mature female gametophyte, fertilizing at least one egg. I. Mature seed. The jacket layer matures into a seed coat while the zygote grows into an embryo within the tissue of the female gametophyte. J. Diagram showing hypothetical stages in development of a jacket layer. 1. Sporangia borne terminally on fertile branch. 2. When combined with later stages of heterospory (C-D), fertile shoots bear smaller number of megasporangia. 3. One or more leafy branches encircle the megasporangium to enhance number of microspores/pollen grains landing on the sticky surface of the megasporangium. 4. Leafy branches fuse to form a jacket layer around the megasporangium, leaving only the sticky apex exposed for pollen deposition. 5. Jacket layer leaves only an apical pore open for pollen entry; modified leaves surround ovule to enhance pollen deposition. (Image source: A-I—The Author; J—After Stewart and Rothwell 1993)

a bubble of sticky fluid from the apical pore, a pollen drop, which captures pollen grains upon its surface (Fig. 9–2 G-H). Pollen drops are easily seen on yews and *Ginkgo* in the spring because their ovules are exposed. When reabsorbed by the ovule, the captured pollen grains are pulled into the ovule. This is definitely not the action of a passive female! Something like a pollen drop does not readily fossilize, but suppose the distal end of *Archaeosperma*'s sporangium produced a sticky coating to better capture pollen. As apical fusion of the jacket continued, natural selection might increase the volume of the exudate until the sticky fluid emerged as a pollen drop from among the nearly fused lobes. But no question about it, ancient seed plants did have pollen drops. A fossilized pollen drop has been found as part of a Lower Carboniferous seed fern ovule as well as male gametophytes growing toward the female from within the pollen chamber (Rothwell 1979).

Pulled into the ovule by the pollen drop, pollen is deposited in the aptly named pollen chamber, a space that exists between the jacket and apical portion of the sporangial wall, which can be thought of as a sort of bachelor pad. The male gametophytes develop and stay here while waiting for the female gametophyte to mature. Possessing few stored provisions, the male gametophyte produces an outgrowth that invades the sporophyte tissues to obtain nourishment.[15] In cycads, the male gametophytes do exactly this and thus are nourished until their huge flagellated sperm are released to swim to the female gametophyte after dissolution of the sporangium wall. In conifers the outgrowth serves double duty by eventually growing through the sporangium wall to deliver the sperm to the female gametophyte within. Thus it also functions as a pollen tube. In flowering plants, because the pollen is deposited at a site spatially separated from the ovule, the pollen tube grows much further, much faster. This also suggests that the pollen tube of angiosperms evolved from a nutritional outgrowth via an intermediate stage that functioned to both nourish and deliver sperm as in conifers, but this is not suggesting conifers share a common ancestry with flowering plants.

But what kind of plant were the dispersed seeds and pre-seeds attached to? The immediate ancestors of seed plants were a complete mystery for some time. Paleobotanists knew that by the late Devonian a considerable diversity of ferny foliage had appeared in the fossil record, but detached ferny leaves could have come from several different groups of plants in addition to ferns. *Callixylon* was a late Devonian woody stem whose wood consisted of tracheids with circular bordered pits, a very gymnospermous anatomy, but what kind of tree was this? Undisputed gymnosperms appear in the Carboniferous, but *Callixylon* was much younger. Both ques-

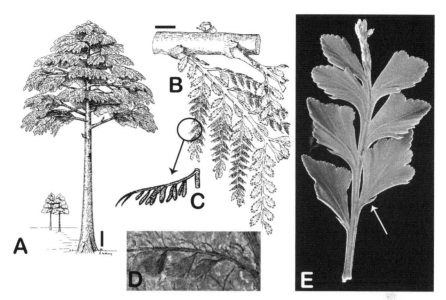

Fig. 9–3 Reconstruction of the progymnosperm *Archaeopteris*. A. *Archaeopteris* as a conifer-like tree. Based upon a stem diameter of 1.5 m these trees could grow in excess of 10 m in height. Bar = 2 meters. B. Foliage branch bearing both sterile leaves and fertile branches (circle). C. Fertile branches are Y-branched apically and bear one or two rows of sporangia. Although similar in appearance some sporangia contain microspores and some contain megaspores. D. Fossil showing several leaves along an axis. E. Foliage of a living conifer *Phyllocladus* where modified branches function as leaves. The true leaves are reduced to tiny bracts (arrow). (Image sources: A-B, D—Beck 1962 with permission of the Botanical Society of America; C—After Phillips, Andrews, and Gensel 1972; E—Courtesy of and with permission of Bob Hill)

tions were answered more or less simultaneously when the fossil fern foliage named *Archaeopteris* ("ancient fern")[16] was found physically connected to a *Callixylon* stem, absolute proof that they were one and the same plant (Beck 1960). Reevaluation of these and similar fossils, together with subsequent discoveries, determined that *Archaeopteris*, the priority name based on the oldest named part of the reconstructed plant, was a free-sporing, woody tree with fern-like foliage. *Archaeopteris* and an assemblage of similar fossils became recognized as the progymnosperms, a group possessing the predicted characteristics of seed plant ancestors. And very importantly, they appear in the fossil record immediately before seed plants and at the same time as pre-seed, pre-ovule "seed" fossils.[17] Progymnosperms provided the much sought ancestors of seed plants, and a "missing link" was found.

Preserved trunks of *Archaeopteris* reaching 1.5 meters in diameter suggest this tree could reach 10 meters in height, making it one of the largest trees of its era (Fig. 9–3). The wood of *Callixylon* is very similar to modern gymnosperms. The foliar organs were first interpreted as large, twice-branched, fern-like fronds bearing leaflets along both sides of the stem (pinnate), but later they were reinterpreted as a planated branch system bearing smaller, simple, spirally arranged leaves (Carluccio et al. 1966) (Fig. 9–3 B-D). Fertile branches were positioned among sterile leaves, and they Y-branched 2 to 3 times bearing one or two rows of sporangia (Phillips et al. 1972) (Fig. 9–3 C). The terminal Y-branch was sterile. You may not think *Archaeopteris* looks much like a coniferous type of tree based on this reconstruction, but actually the similarity to a conifer like *Phyllocladus*, where the leafy organs are flattened branches (phylloclades), is obvious. This does not mean *Phyllocladus* is a survivor of so long ago; these leafy organs had a different evolutionary history. *Archaeopteris* lived at that point in the history of plants when megaphylls were appearing, and many plants of the late Devonian had appendages to their vertical axes that were difficult to classify either as a branch system bearing simple leaves or as a compound leaf bearing leaflets.

Just as predicted, detailed examination of progymnosperm sporangia demonstrated that both homosporous and heterosporous species existed. Megasporangia contained 8 to 16 megaspores, the result of two, three, or four mother cells dividing by meiosis. Progymnosperms were not seed plants, so they were free-sporing and their gametophyte generation and sexual reproduction might have been similar to that of *Selaginella*. Progymnosperm fossils were widespread, forming ancient forests,[18] and on the basis of their paleolatitude, they had a circumpolar southern temperate distribution in the late Devonian and early Carboniferous (Beck and Wight 1988). Progymnosperms are now interpreted as consisting of three groups each treated as a separate order, but all were woody plants (trees or shrubs). Because the first character to appear in the seed plant lineage was woody arborescence, the progymnosperms together with the seed plants form an inclusive lineage or clade named the ligniophytes, the woody plants (*lignio-* = wood) (Rothwell and Serbert 1994; Kenrick and Crane 1997). The basic organization of progymnosperms suggests they had a common ancestry with trimerophytes.

LIVING SEED PLANTS

The living gymnosperms consist of the cycads, the conifers, ginkgo, and the gnetophytes (NEAT-oh-fights). While the conifers remain ecologically and economically important, and coniferous forests still cover vast areas of the Earth's surface, the other gymnosperms are relics of bygone days. Cycads, the oldest living lineage of gymnosperms, were much more common during the Triassic and Jurassic when they composed up to 20% of the world's terrestrial flora (Jones 1993), real components of Jurassic Park.[19] Ginkgoes were once widespread forests and now only a single species survives, and it was down to just a few sacred groves protected by monasteries before it resurged horticulturally as one of the most common urban trees in the world. Most of you have never seen a gnetophyte, or if you have, you didn't realize what it was. More specific information on each of these groups can be found in the appendix.

Cycads are more familiar in tropical areas where some members are commonly planted as landscape ornamentals. With their short, stout, seldom-branched stems and helical whorl of pinnately compound leaves, cycads look like palms. Common names like sago palm, referring to species of *Cycas*, only contribute to the confusion, but in most important respects cycads are more like ferns, or actually, more like pteridosperms. Cycad leaves and leaflets even develop by uncoiling from a "fiddlehead" just like fern fronds (Fig. A19). Today cycads are regarded as rather minor but interesting components of many tropical and subtropical communities. Cycad plants bear either pollen cones or seed cones, and both types of cones are helical whorls of sporophylls, modified leaves (Fig. A20). Cycads also differ from all other living gymnosperms in that many seem to employ insect pollen dispersers (Terry et al. 2005).

Of all the plants groups considered so far, conifers (see Appendix) are the first to have significant economic value, largely in the form of wood (for lumber and paper pulp) and byproducts of their resin-secreting structures. Humans also use conifers for ornamentation and landscaping. Their greenery decorates our homes during winter holidays,[20] an important symbolism. Still it can be quite dismaying to discover how little people know about these important and familiar plants. Conifers are all woody plants (trees and shrubs), and some remarkable individuals are among the largest and oldest individual organisms on Earth. They can grow more than 100 meters tall and have trunks several meters in diameter. Trees several centuries old are common; some are known to be several thousand years old (bristle cone pines), but old-growth forests and really big trees have disappeared because of the human demand for wood. Conifers are widespread geographically, and they form the dominant vegetation in many areas, particularly the boreal forests of the medium high latitudes and medium high altitudes.

The familiar conifers of the Northern temperate zone are members of the pine family (Pinaceae), the pines, spruces, hemlocks, firs, and douglas-fir, common trees with widespread distributions. Conifers include five other families with living genera and a seventh family of fossil conifers.

Hoop-pines and monkey-puzzle trees (Araucariaceae) and podocarps (Podocarpaceae) are as familiar in the southern hemisphere as pines are here in northern hemisphere (Fig. A13). In subtropical areas of North America some of these southern conifers are used as landscape ornamentals, and some are even used as house plants. Sequoias, bald cypress, and redwoods (Taxodiaceae) are found in eastern Asia and North America, and they are notable trees for their size. Junipers, arborvitaes, and cypresses (Cupressaceae) are widespread and include many ornamental trees and shrubs characterized by small scale-like leaves. The plum-yew family only has one Asian genus.

While sharing many similarities with conifers, both ginkgoes (*Ginkgo*[21]) and yews (*Taxus*) display some distinct differences too. As a result each is placed in its own family in orders separate from the other conifers. Yews are one of the most common ornamental evergreens, lovely if they are not poodle pruned (a pet peeve). *Ginkgo* trees grow like a conifer except their branches angle upward rather than being more or less horizontal or drooping, but *Ginkgo*'s broad, fan-shaped leaves (perfect form for a megaphyll) seem atypical to residents of the northern temperate zone because most are unfamiliar with broad-leafed tropical conifers.

As the name conifer, cone-bearer, suggests, most of these trees bear cones. The seed cones are the most familiar as they are larger, harder, and more persistent than pollen cones. In the spring small, delicate, short-lived pollen cones disperse pollen by wind, sometimes in copious amounts. The ovules and seeds of conifers are borne "terminally" upon modified stems aggregated into cones, or solitarily in ginkgo and yews. The nature of these structures and how they have been interpreted is explained in the appendices.

The gnetophytes (NEAT-oh-fights) were another component of Jurassic Park, having appeared in the fossil record in the Triassic and having their greatest diversity during the Jurassic and Cretaceous before declining in the Tertiary. Presently gnetophytes are represented by only three genera: *Gnetum*, *Ephedra*, and *Welwitschia*, none of which are commonplace or familiar. *Ephedra*, a shrubby plant, is encountered by people familiar with sagebrush communities of western North America where it is called "Mormon tea." All three have embryos with two cotyledons, sterile bracts that form a perianth of sorts around their ovules, and somewhat anther-like pollen-producing structures. All three have vessels in their wood and *Gnetum* has broad, net-veined leaves. All of these are "angiospermy" features, and thus gnetophytes have long been considered as possibly having a common ancestry with flowering plants.

Pteridosperms or seed ferns (see Appendix) are the first true seed plants. As their name suggests, pteridosperms are seed plants that have fern-like foliage, and generally they are similar to the progymnosperms vegetatively. Pteridosperms include several fairly well-circumscribed families and many less-known fossils of more uncertain affinities, overall a tremendous diversity of plants extending in time from the early Carboniferous to the Jurassic (Stewart and Rothwell 1993). For purposes of this narrative, the better-known and ancient organisms from the Carboniferous generally will suffice. Many pteridosperm reconstructions made the plants look similar to tree ferns—for example, *Medullosa*, but other reconstructions suggest they may have been low, scrambling, somewhat sprawling vine-like plants (Fig. 9–4 C). The leaves were large and compound fronds, and the ultimate leafy units, the pinnae, as leaflets are called in ferns, were distinctly fern-like in their appearance and venation (Fig. 9–4 A). As shown in this reconstruction, the leaves were arranged on the stem helically. The leaf stalks were branched several times, including a basal Y-branching. Like progymnosperms, pteridosperms were woody plants, and their stems and leaf stalks had a very distinctive anatomy that makes them fairly easy to recognize in anatomically detailed fossils.

Unlike in the progymnosperms, sterile and fertile leaves of pteridosperms had the same form. The diverse ovules of pteridosperms are well known but seldom which foliage they were attached to (Fig. 9–4 B). Pollen-producing structures ranged from small clusters of sporangia fused together, not unlike those of the Marattialean ferns (Appendix: Ferns), to some massive pollen-producing organs containing many microsporangia variously fused together.

From the first appearance of pteridosperms in the late Devonian, through the Carboniferous, and into the Permian, six groups, usually treated as orders, are presently distinguished, including the glossopterids. Three other groups of seed ferns appeared, and disappeared, during the Triassic, Jurassic, and Cretaceous, respectively. Some pteridosperms have similarities to living seed plants, and they include the cycadeoids (similar to cycads), Czekanowskiales (similar to ginkgoes), Cordaitales and Voltziales (similar to conifers), and Pentoxylales and Caytoniales (similar to angiosperms).

Glossopterids appear in the early Permian and continue through the Triassic before becoming extinct in the lower Jurassic. Glossopterids were a very prominent and common component of forests that covered the

Fig. 9–4 Reconstructions and fossil parts of pteridosperms. A. Fossil seed fern foliage called *Neuropteris* that was probably part of a larger, compound frond. B. Fossil seed. C. Reconstruction of two seed ferns, one vine-like (darker image) called *Pseudomariopteris* climbing on an arborescent *Medullosa*-type seed fern that could reach 3–3.5 m tall. (Image sources: A—Courtesy of Gunnar Ries, Wikimedia Creative Commons; B—Courtesy of Verisimilus, Wikimedia Creative Commons; C—From Krings et al. 2001, with permission of the Botanical Society of America)

southern temperate zone (30°–60° S latitude) of the ancient supercontinent Gondwana (Fig. 9–5). Cycads appear at a similar point in time, the early Permian, or possibly in the very late Carboniferous, and they reach a maximum diversity in the Jurassic before declining in the Cretaceous, which is when angiosperms appear and begin diversifying. Cycadeoids (also called the Bennettitales) appear in the Triassic and continue into the late Cretaceous. Caytoniales (pteridosperms) also appear in the early Triassic and disappear in the early Cretaceous (Fig. 9–6).

Cordaites and its relatives were pteridosperms that appeared in the Upper Carboniferous and continued nearly through the Permian. Some cordaites were trees that may have reached 30 meters in height; they grew in dense stands forming forests. Others have been reconstructed as small trees with stilt roots, perhaps with a mangrove habit, suggesting they grew on the margins of swamps and along shores of estuaries and seas. *Cordaites* had simple, leathery, strap-like leaves ranging from 2 to 100 cm in length. Their leaves show anatomical modifications well adapted for

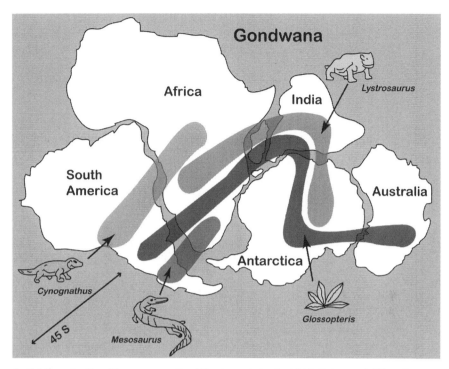

Fig. 9–5 Reconstruction of former supercontinent Gondwana during the Middle Cretaceous (~100 mya) consisting of South America, Africa, Madagascar, India, Australia, and Antarctica. Distribution of Triassic terrestrial and freshwater fossil animals (*Cynognathus*, land reptile; *Lystrosaurus*, land reptile; *Mesosaurus*, freshwater reptile) is evidence of these continental connections. Permian *Glossopteris* leaf fossils (pteridosperms) were distributed across all of Gondwana's continental subunits. Diagonal arrow (lower left) shows the approximate paleolatitude of 45 degrees south, a temperate climate. (Image source: After United States Geological Survey, Snider-Pellegrini Wegener map)

dry and/or seasonally cold climates, or for dealing with salt water. Modern needle-bearing conifers also have tough, hard leaves for dealing with the terrifically dry conditions of the winter season. Helical clusters of these strap-like leaves were borne at the ends of branches, suggesting that *Cordaites* may have looked a bit like a strange, modern conifer, the umbrella pine (*Sciadopitys*). Cordaitalean pteridosperms have very complex, compound, cone-like reproductive structures that are difficult to interpret because few well-preserved fossils have been found.

Voltziales appear in the late Carboniferous and reached their greatest diversity and prominence through the Permian and Triassic before waning during the Jurassic, a pattern very similar to the glossopterids. Voltziales are usually considered intermediates bridging the gap between Cordaitales and conifers, but there are lots of differences between these groups. Branches attributed to Voltziales are covered with a helix of short needle-like leaves so they looked something like our modern monkey-puzzle trees or Norfolk Island pines of the genus *Araucaria*. Cones were borne at the ends of branches. Seed cones were compound structures (like Cordaitales), while pollen cones were simple helixes of (presumably) microsporophylls, each bearing a pair of microsporangia on the lower surface like many modern conifers. Also like modern conifers, ovule-bearing shoots were associated with bracts aggregated into seed cones.

SEED PLANT RELATIONSHIPS

The preceding two sections were a bit like a playbill introducing the characters. To figure out and explain the evolutionary relationships among seed plants requires a step backward to their common ancestors: the progymnosperms, which is the basal lineage of the ligniophytes (the woody plants), a clade that includes all the seed plants. This lineage appears in the late Devonian with three groups of progymnosperms; the first true seed plants, pteridosperms, appear in the lower Carboniferous. Pteridosperms remain the predominate seed plants throughout the Carboniferous and into the Permian (Fig. 9–6). Ginkgoes and cycads are the two oldest living groups of seed plants. Both groups became diverse during the Jurassic and Cretaceous before waning in the Tertiary. Most of the other conifer families, including the yews, appear in the Jurassic. The pine family and angiosperms both first appear in the early Cretaceous, both diversifying throughout the Cretaceous and into the Tertiary, but on completely different orders of magnitude.

Phylogenetic relationships among all these seed plants are by no means certain. No one argues that pteridosperms are the basal lineage; the fossil record makes it clear pteridosperms are by far the oldest seed plants, but pteridosperms were diverse, composed of several very distinct groups, and quite likely their relationship to the rest of seed plants is more complex than depicted by a single lineage. Although living seed plants consist of gymnosperms and angiosperms, no current phylogenetic hypothesis results in a single gymnosperm clade. In fact, it is pretty certain that "gymnosperm" describes a grade, a level of reproductive development, not a single lineage or taxonomic group.

The striking resemblance between progymnosperms like *Archaeopteris* and the more conifer-like gymnosperms, and a similar resemblance between more ferny pteridosperms and the fern-like cycadeoids and cycads, suggests a polyphyletic (multiple lineage) hypothesis of seed plant origins. The more primitive fern-like members of the progymnosperms

| Dev | Carb | Perm | Tri | Jur | Cret | Tert | Today |

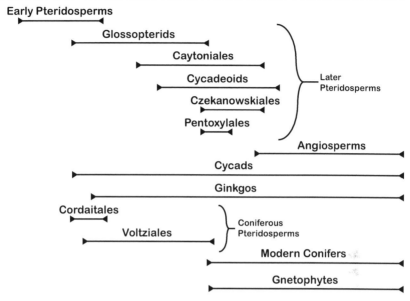

Fig. 9–6 Fossil history of seed plant groups. Seeds and the first seed plants, pteridosperms, appear in the late Devonian and later groups of pteridosperms persist until well into the Cretaceous. Cycads are the oldest living group of seed plants followed by the ginkgoes. Gnetophytes and modern conifers both appear in the Jurassic. Angiosperms do not appear until the Cretaceous. Dev = Devonian, Carb = Carboniferous, Perm = Permian, Tri = Triassic, Jur = Jurassic, Cret = Cretaceous, Tert = Tertiary (see Fig. 1–5 for ages). (Image source: After Stewart and Rothwell 1993)

were considered more likely ancestors to pteridosperms (Rothwell 1981), while *Archaeopteris* itself and similar progymnosperms would be ancestral to Cordaitales, Voltziales, conifers, and ginkgoes. If this hypothesis is true, then ligniophytes would remain a clade, but seed plants would be polyphyletic consisting of two lineages, each with a separate ancestry among pre-seed progymnosperms, an *Archaeopteris*-Cordaitales-conifer-*Ginkgo*-gnetophyte lineage with circular bordered pits on their tracheids,[22] and a ferny progymnosperm-pteridosperm-glossopterid-cycadeoid-cycad lineage with scalariform bordered pits on their tracheids, a lineage that would include the angiosperms. Wood with vessels appears in both lineages, but vessels have evolved in other vascular plant groups too. Even this is oversimplified because one gnetophyte (*Gnetum*) has scalariform pitting. In general, this hypothesis is rejected today in favor of seed plants being a single lineage.

Although all current studies of seed plant phylogeny result in a single seed plant clade with the progymnosperms as their sister group, their most

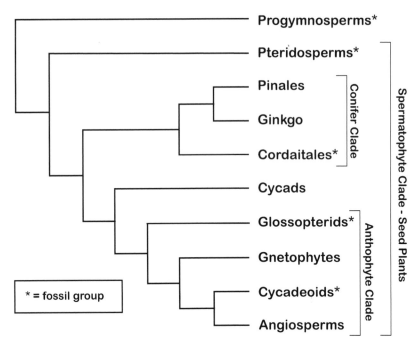

Fig. 9–7 A seed plant phylogenetic hypothesis showing a single spermatophyte clade with progymnosperms as a sister group. Pteridosperms (seed ferns) form the basal lineage. Cordaitales, Pinales, and *Ginkgo* form a conifer clade. Cycads are basal to an anthophyte clade consisting of glossopterids, cycadeoids, gnetophytes, and angiosperms/flowering plants. The conifer clade bears sporangia terminally on modified branches rather than marginally on leaves like progymnosperms, cycads, and the anthophytes. (Image source: After Doyle 1996)

recent common ancestor (Fig. 9–7; Fig. 9–8), various studies differ in other respects. The two phylogenetic hypotheses presented here were published only four years apart, and these were selected from among several other hypotheses that differ only in minor details. These two hypotheses can be used to illustrate the basic points of agreement and the differences that exist, including a polyphyletic seed plant hypothesis. One of the reasons for such different hypotheses is the lack of information about many fossil gymnosperm groups, particularly those without any apparent living relatives. Molecular phylogenies cannot be fully informative when so many key fossil groups are by necessity omitted. Thus phylogenetic hypotheses based on molecular data must be modified to square with the fossil record and then interpreted with care.[23]

In neither hypothesis is a single lineage of gymnosperms proposed. In the first hypothesis (Fig. 9–7), cycads and gnetophytes, plus the fossil glossopterids and cycadeoids, form a clade with the flowering plants. The glossopterids, gnetophytes, and cycadeoids together with the flower-

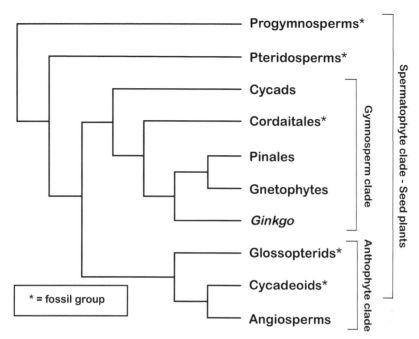

Progymnosperms*

Pteridosperms*

Cycads

Cordaitales*

Pinales

Gnetophytes

Ginkgo

Glossopterids*

Cycadeoids*

Angiosperms

* = fossil group

Gymnosperm clade

Anthophyte clade

Spermatophyte clade - Seed plants

Fig. 9–8 A seed plant phylogeny hypothesis showing a single spermatophyte/seed plant clade with progymnosperms as the sister group and pteridosperms as the basal lineage of the spermatophyte clade (same as in Fig. 9–7). In this hypothesis gnetophytes are placed within the conifer clade, which has the cycads as their sister group, thus forming a gymnosperm clade. The anthophyte clade in this hypothesis consists of angiosperms/flowering plants plus two fossil groups and is the sister group to the gymnosperm clade. Conifers no longer form a clade, but Cordaitales, Pinales, gnetophytes, and *Ginkgo* all have ovules borne terminally on modified branches. (Image source: After Bowe, Coat, and dePamphilis 2000)

ing plants form an anthophyte clade, plants with flower-like reproductive structures (antho- = flower). The Cordaitales, Pinales, and ginkgoes form a "conifer" clade. Pteridosperms are sister group to both the conifer and cycad-anthophyte clades, but as explained above, placing pteridosperms as a single basal lineage is quite an oversimplification even when pteridosperms like the glossopterids and cycadeoids are treated separately.

The alternative phylogenetic hypothesis (Fig. 9–8) shifts the cycads and gnetophytes from the anthophyte clade (angiosperms plus gymnosperms with angiospermy features) to a gymnosperm clade, which now includes all the living gymnosperms and their fossil relatives, but pteridosperms, glossopterids, and cycadeoids remain part of the other lineage. In this hypothesis the anthophyte clade includes flowering plants and their putative fossil ancestors, but no living gymnosperms (Chaw et al. 2000). Recent molecular data suggest the gnetophytes are nested within the conifer-ginkgo clade, but no question about it, gnetophytes are pretty strange-looking, much modified gymnosperms, and do possess a "peri-anth," sterile bracts around their reproductive structures.

An anthophyte clade that includes gnetophytes has been pretty well falsified by both anatomical and molecular data, but the rest is by no means resolved. Another alternative hypothesis shifts the cycadeoids (Bennettitales) to the conifer clade as sister group to the gnetophytes. They both have stalked ovules with long, tubular apices on their ovules but otherwise are very different. Presently both hypotheses agree that the immediate common ancestors of flowering plants must be sought among extinct groups of gymnosperms. If angiosperms turn out to have a common ancestry among Triassic-Jurassic pteridosperms, then perhaps the glossopterids and cycadeoids are misplaced in the anthophyte clade, and their flower-like reproductive features represent a parallelism, a convergence of form from similar function. Further functional differences between gymnospermous reproductive structures and flowers will be discussed in the next chapter, differences that are significant in explaining the rapid rise of flowering plants to ecological dominance.

ANGIOSPERMY GYMNOSPERMS

A general description of those pteridosperms with possible angiosperm affinities, introduced above, will help you understand these hypotheses. The glossopterids are so named because they have very distinctive spatulate leaves (*glosso-* = tongue, pteris = fern) with a well-defined midrib vein and reticulate secondary veins borne in helical whorls at the ends of branches (Fig. 9–9). Over 50 species of glossopterids have been described. Their fossil leaves are fairly common and have been found in South America, Africa, Madagascar, India, Australia, and Antarctica, the land masses that constituted the supercontinent Gondwana (Fig. 9–5). When the earlier groups of pteridosperms died out at the end of the Carboniferous, glossopterids and Voltzialean conifers became diverse and important during the Permian and Triassic. On the basis of the distribution of their fossils, glossopterids were adapted to the cool temperate high latitudes of Gondwana. Voltzialean conifers were adapted to warmer, drier climates. Both groups became extinct at the end of the Triassic as modern groups appeared (Fig. 9–6).

During the Permian and Triassic, glossopterids must have been one of the most important and widespread groups of forest trees. Glossopterids had unique pollen-bearing and ovule-bearing structures, and all were associated with vegetative leaves, an angiosperm-like feature. Ovules were

Fig. 9–9 *Glossopteris*, a fossil pteridosperm from Gondwana. A. Tongue-shaped fossil leaves (glosso- = tongue) from Antarctica showing distinctive midrib and venation. Bar = 5 cm. B. Living broad-leafed conifer (*Podocarpus*), providing some idea of how the leaves of *Glossopteris* would have looked when arranged on a branch. (Image sources: A—P. Rejcek, National Science Foundation, USA; B—Courtesy of Abu Shawka, Wikimedia Creative Commons)

associated with and borne upon bracts, modified leaves. In some cases the bract's enrolled margins partially enclosed the ovules, producing a cupule and suggesting some sophisticated pollination mechanisms. This is important because gymnosperms whose ovules are borne upon modified leaves approach the organization of a flower where the modified ovule-bearing leaf is called a carpel (Taylor and Kirchner 1996).

At least three other orders of fossil gymnosperms are known to have reproductive structures with flower-like characteristics and a fossil record leading up to the Cretaceous, thus making them potential ancestors of flowering plants: the cycadeoids, the Caytoniales, which are pteridosperms of dinosaur age, and the Czekanowskiales (check-an-ow-ski-AYE-lees). Cycadeoids, as their name suggests, look like cycads in their stems, leaves, and basic growth forms, but they differ from cycads in having complex bisporangiate reproductive structures (Fig. 9–10) that are somewhat reminiscent of many-parted *Magnolia*-like flowers. Based on just vegetative features, a close relationship between cycads and cycadeoids might be justified, but the complex reproductive structures of cycadeoids indicate any relationship is very distant (Stewart and Rothwell 1997).

Cycadeoids appear during the Triassic and had their greatest diversity in the Jurassic before declining in the first half of the Cretaceous as the flowering plants became dominant, which makes them of a proper age to have spawned flowering plant ancestors. The cones consisted of sterile, subtending bracts, a perianth, and whorls of microsporophylls with embedded microsporangia, which are the features very reminiscent of a flower. In other specimens the microsporophylls quite obviously have a

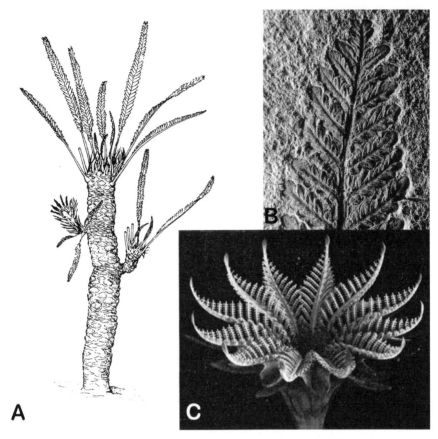

Fig. 9–10 Reconstructions of a Jurassic cycadeoid pteridosperm. A. The cycadeoid *Williamsonia* reconstructed with a cycad-like growth form. B. Fossil impression of a cycadeoid leaf/frond. C. Model of a reconstructed cycadeoid cone. Unlike cycads (Appendix Fig. A24), cycadeoid cones were bisporangiate. The whorl of bipinnately compound pollen-bearing structures has a flower-like appearance. Ovules were borne on the apical portion of the central axis within. (Image sources: A—After Andrews (1961), from Rothwell, Crepet, and Stockey 2009, with permission of the Botanical Society of America; B—Courtesy of Ghedoghedo, Wikimedia Creative Commons; C—Courtesy of the USA National Park Service)

pinnate organization and might even be considered fertile fronds. Stalked ovules were borne laterally on the central axis of the cone packed among sterile scales, which is not flower-like. While it is fairly easy to imagine such a cone opening into a flower-like display, a careful study of developing specimens led to the conclusion that the microsporophylls could not unfold in the way they are depicted in some reconstructions (Delevoryas 1963, 1968).

A permanently closed cone does not afford much opportunity for either wind or animal pollination. This conclusion is most perplexing because most land plants have mechanisms to promote outcrossing, and nonopening bisporangiate cones would seem to require self-pollination. The very nature of fossilization introduces a bias in reproductive structures because it dictates that most specimens are going to be either developing buds or the postpollination structures of the developing-seed stage, because both of these stages last much longer than the relatively brief pollination stage. Furthermore, both pre- and postpollination stages were probably much tougher and more easily fossilized. Guessing how a flower is pollinated from examining only flower buds or developing fruits is nearly impossible, and figuring out the shape and functional arrangement of floral parts from the bud stage is quite difficult to do too, because many flowers unfold, rearrange, and enlarge considerably. So our understanding of cycadeoid cones probably remains very incomplete.

The sterile scales tightly packed around the ovules/seeds provide protection (Stewart and Rothwell 1997), and they may have primarily functioned after pollination, after the microsporophylls and bracts were shed. Several observations suggest the sterile scales and ovules are homologous structures (Stewart and Rothwell 1997), but it is uncertain whether the stalked ovules and sterile scales represent foliar organs or modified shoots. Both ovules and scales are lateral appendages on a modified shoot, so interpreting them as leafy organs is certainly a reasonable conclusion. The detailed study of some fossils was made possible because of excellent fossilization of cellular detail. This revealed one additional salient feature: in early stages of development ovules had a linear tetrad of megaspores, a feature like flowering plants, and very unlike the tetrahedral megaspore arrangement that has been found in ancient seeds like *Archaeosperma* (Crepet and Delevoryas 1972).

The fossil record of Jurassic gymnosperms tells of considerable diversity. A number of less-known organisms can best be termed "enigmatic and intriguing." The Caytoniales are a group of pteridosperms with leaves or leaflets that have the reticulate venation of ferns, but no one knows exactly their phylogenetic affinities (Harris 1964). Ovules were borne on a modified pinnate leaf (megasporophyll) whose lateral appendages (leaflets?) were recurved cupules enclosing several ovules. Pollen was produced in a "fused" structure consisting of four microsporangia, a structure that resembles a flowering plant anther.[24] Pollen grains had the same type of sac-like wings as conifer pollen that function to orient pollen grains in a pollen drop, and this pollen has been found within the micropyle of the ovules. If Caytoniales used a pollen drop mechanism, the organization of the cupule suggests the pollen drops of the enclosed ovules may have coalesced into a single common pollen drop for the entire cupule. Such a cupule functions like an angiosperm pistil where a stigmatic surface is spatially removed from the micropyles of the enclosed ovule or ovules. However, Caytonialean cupules are folded or rolled top to bottom rather than side to side as in the angiosperm carpel, and the Caytonialean megasporophyll is a compound rather than a simple single leafy organ. This suggests the similarity is more likely a parallelism rather than an indication of a direct relationship between Caytoniales and angiosperms.

Lastly, the Czekanowskiales is a very little-known group of fossil gymnosperms, which originally was thought to be related to ginkgoes because of their slender dichotomously lobed leaves borne in a helical whorl on a short shoot (Harris 1951). Numerous detached fossils suggest *Czekanowskia* was a deciduous tree, as is *Ginkgo*, where the entire short leafy shoot was shed as a unit, as in bald cypress (*Taxodium*) or dawn redwood (*Metasequoia*). The vegetative parts have been associated via anatomical similarities (as opposed to a physical connection) with ovule-bearing axes with spirally arranged scales at the base, similar to those at the base of the vegetative shoots. Each long slender axis bears pairs of cupules that face each other rather like clam shells. Each cupule encloses recurved ovules, and if indeed the ovules are upon a leafy structure, then it becomes a megasporophyll. If one half were sterile and one half fertile they might be similar to some glossopterid ovulate structures. A pair of cupules completely encloses the ovules and pollen grains have been found among epidermal hairs that form a "stigmatic" suture between the two cupules (Harris 1976). Such an arrangement has considerable similarity to basal flowering plant pistils that are composed of a single "enrolled" megasporophyll where the two lateral margins form an unsealed suture that functions as a stigmatic surface, except again the Czekanowskian cupules are oriented top to bottom, not side to side, and constructed of a pair of leafy organs rather than a single foliar organ as in angiosperms.

What has this quick romp through seed plants told us? This narrative just cannot do them justice. Fossil spermatophytes present an intriguing array of features, a fascinating diversity that we only partly understand, but from the Carboniferous to the Cretaceous, seed plant diversity was tremendous. Molecular studies may not come to the rescue because too many critical groups are extinct. Several groups have leaves and reproductive structures that have features suggesting angiosperms, but none so similar as to suggest an immediate common ancestry. As a whole, pteridosperms show experimentation with leafy reproductive organs and pollination, suggesting how the gymnosperm mode of reproduction may have shifted to what we now consider the angiosperm mode. Perhaps someone will discover a critical fossil that will provide an answer, but once you grasp the differences between angiosperms and gymnosperms such a fossil seems very unlikely because the differences are so slight and not of the type readily fossilized. All the latest data place the gnetophytes as more closely related to conifers than to angiosperms, which reverses the position prevalent for many years. Lastly, the seed habit clearly spawned its own adaptive radiation, although presently we have living representatives of only three or four lineages, unless some intrepid botanical adventurer in the remote recesses of temperate South America or Australia, parts of ancient Gondwana, happens upon a glossopterid.[25]

TEN

A Cretaceous Takeover

Wherein the quite singular ecological dominance and species diversity of the flowering plants are examined in light of their novel features and their gymnospermous ancestry.

Many eyes go through the meadow, but few see the flowers in it.
—Ralph Waldo Emerson, *Journals*, 1834

We like flowers. The very mention of flowers brings to mind many familiar images of loveliness, a delight to eye and nose, but each of these images actually functions to attract some nonhuman pollinator. Most of the flowers that humans find aesthetically pleasing are adapted for pollination by bees, butterflies, moths, or birds. The floral displays of these flowers are adapted for the sensory organs of the pollen disperser, so why humans should receive pleasure from images and odors designed to attract these particular pollinators is an interesting question. Perhaps it is because the primary attractants are visual, and since humans are very visual organisms, we like the variation, the patterns, the colorful interruptions to the background of green foliage. Even more curiously, we seldom see exactly what the pollinator sees because of differences in visual acuity. For example, bees see ultraviolet wavelengths and in ultraviolet light, bee flowers often have very distinct dark "target" patterns or bright spots that are invisible to us. Whatever the reason, humans so love the way these flowers look that we decorate our abodes and bodies, celebrate special events, and even declare our affections with flowers. Flowers help express our greatest joy and sadness. Our connection to flowers is so ingrained in our psyche that the failure to appreciate the aesthetics of flowers is a sign of clinical depression (Pollan 2001). Human appreciation of flowers is nothing new. Pollen from bundles of flowers found in Neanderthal graves shows that honoring people in death has not changed very much over the past 150,000 years.

Perhaps human fascination with flowers had an even earlier origin. When did our primate ancestors develop the intellectual ability to understand that fruits and seeds appear after flowers? Thus big floral displays are indicators of feasts to come, a future resource to be recognized and remembered. Indeed, the demands of efficiently foraging for islands of fruit (individual trees) in a very patchy sea of tropical forest trees, a mosaic that changes season to season and year to year, may have started our primate ancestors toward our current intellectual status. And very recently humans have returned the favor by increasing the reproductive success and ecological prominence of those flowering plants that feed, cloth, and please us. This was accomplished through agriculture, a human lifestyle adopted only about 10,000 years ago. Although agriculture has kept our ever-increasing human population fed, while at the same time allowing its increase, many of our modern problems can be traced to possessing the brains and biology of Pleistocene hunters and gatherers, a lifestyle very few humans now lead (Shephard 1998). Our technology and ability to manipulate nature has decoupled our cultural evolution from our biological evolution.

The agricultural interaction between humans and flowering plants is only the most recent plant-animal interaction that flowering plants have engaged in. Mutually beneficial interactions with animals are a hallmark of flowering plants, and these interactions help us explain how flowering plants burst upon the scene and succeeded so well. Plants bearing flowers and fruits appeared in the early Cretaceous and quickly displaced ferns and cycads from the ecological roles they had played for the previous 150 million years. Somehow, some way, communities of flowering plants displaced communities dominated by gymnosperms and ferns, and no geological catastrophe was involved.[1] Their Cretaceous takeover is even more confounding because no major lineage-defining innovation arose in flowering plants, nothing as momentous as eukaryotic cells, multicellular bodies, a new life cycle, wood, or seeds. Rather a suite of small differences, none seemingly of singular importance, separates angiosperms from gymnosperms, and this includes flowers and fruits. Yet somehow these changes resulted in an ecological revolution.

This raises another point of tremendous interest. Flowering plants are

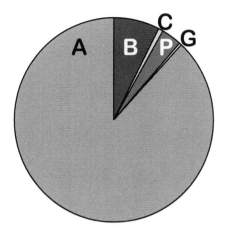

Fig. 10–1 Proportionate distribution of land plant species. A—angiosperms 85.5%, B—bryophytes (mosses, liverworts, and hornworts combined) 8.8%, C—clubmosses 0.5%, P—pteridophytes 3.9%, G—gymnosperms 0.3%. (Image source: After Crepet and Niklas 2009)

the most recent group of green organisms to appear, and they have become the most diverse group of green organisms ever. With over 220,000 recorded species,[2] there are over seven flowering plant species for every species of nonflowering land plant. That is correct! Over 85.5% of all land plant species are flowering plants (Fig. 10–1). Not all groups of angiosperms are equally diverse. If graphed separately, the number of species in the bean family and in the orchid family would represent 5.5% and 6.7% of land plant species, respectively. That is correct! One out of every 15 land plant species is an orchid! So where are they all and why do we think orchids so exotic if they have so many species? Having lots of species does not necessarily translate into commonness, although orchids are more common than you probably realize, especially in the tropics. In the temperate zone most orchids go unnoticed because they live in largely undisturbed natural areas,[3] which are increasingly rare and remote from human areas of habitation, and most are rather small and nondescript—small plants with even smaller flowers, not at all like the big showy orchids seen at flower shows. Other families consist of a single genus or even a single species, some perhaps relicts of a bygone era, bits and pieces of old and previously more diverse lineages.[4]

How did flowering plants come to be so diverse? No small amount of time has been spent pondering the answer to that question, and several answers have been proposed, but they will have to wait until it is clearer what makes a flowering plant different from a gymnosperm. And while distinguishing conifers from familiar flowering plants is no problem, species exist that would pose a problem, and it only gets worse with fossils.

Fig. 10–2 Aril-covered seeds of a European yew (*Taxus baccata*). Arils (a) are a fleshy outgrowth that surrounds the seed partially or in total. The mature aril is red and fleshy and surrounds the seed (s—right) except at its apex, making a berry-like structure to attract avian seed dispersers. The developing aril (s—left) remains green and surrounds only the base of the ovule at this stage, looking a bit like a tiny acorn. (Image source: Courtesy of Didier Descouens, Wikimedia Creative Commons)

WHAT MAKES FLOWERING PLANTS UNIQUE?

How about flowers and fruits? Although distinguishing flowers and fruits from cones and other gymnosperm reproductive structures seems pretty easy, among the great array of seed plants, particularly those known only as fossils, there existed and there exist reproductive structures difficult to classify. That pink-red fleshy tissue surrounding a yew seed (Fig. 10–2) is an aril, not a fruit, although it functions in the same manner as a fruit by attracting and rewarding an avian seed disperser. In some angiosperms, the fruit is hard and/or dry, protective, but the seed is surrounded by a fleshy aril that functions just like the aril of the yew—for example, the red mace of nutmegs.[5] Even on a technical level flowers are not all that much different from the reproductive structures of gymnosperms. In fact, a flower consisting of whorls or helices of modified leaves on a short axis

Fig. 10–3 Many-parted flowers. A. *Magnolia virginiana*—sweet bay magnolia. Flower showing ten or so perianth parts surrounding numerous stamens and pistils; all floral parts are helically arranged. B-D. Flowers of the ANA grade from the namesake genera. B. *Amborella*. C. *Nymphaea* (water lily). D. *Austrobaileya*. B. These pistillate flowers of *Amborella* are small, about 1 cm wide, whitish-pinkish-green in color, consisting of some 5–8 smallish perianth parts, 1–4 sterile stamens, and 4–6 pistils. C. The water lily has a large flower some 10–12 cm across, with numerous helically arranged perianth parts that transition into stamens. D. *Austrobaileya* has flowers several centimeters wide, with relatively few helical perianth parts with a greenish color. These transition into leaf-like stamens with dark purple splotches of pigment (see Fig. 10–5). (Image sources: A—The Author; B—From Sangtae Kim, cover photo of *American Journal of Botany*, December 2004, used with permission of the Botanical Society of America; C—Courtesy of Fujnky, Wikimedia Creative Commons; D—Courtesy of and with permission of D. W. Stevenson, www.plantsystematics.org)

cannot be defined in such a way as to exclude all gymnosperm reproductive structures. All cycad cones and all conifer pollen cones are composed of modified leaves (sporophylls) arranged in helices on a short axis. Flowering plants bear their sporangia on modified leaves, sporophylls, too, and they are arranged helically in many basal angiosperms, so in this sense flowers share a basic similarity with the pollen cones of all seed plants and cycad seed cones. Flowers differ from these cones because they generally, but not always, have one or more whorls of modified sterile leaves, a perianth, literally meaning "around the flower" (Fig. 10–3), but gnetophytes also have a perianth of bracts (modified sterile leaves).

A flower's perianth most commonly consists of two whorls of modified leaves referred to as sepals and petals, which usually occur in a discrete and regular number. In many flowers either one or both members of the perianth are "fused" into a tubular whorl often bearing a characteristic number of lobes. In some flowers, particularly systematically basal ones, the perianth is not differentiated into sepals and petals, but rather it is helically arranged and composed of a large number of individual parts, and as a result, the parts can be numerous and of variable number, which is a rather cone-like organization, even if they look somewhat different (Fig. 10–3 A, C). Sometimes the perianth parts intergrade from what might be called sepals to what might be called petals, and even further, the innermost perianth parts may intergrade with stamens (Fig. 10–3 C-D). Other basal flowers may have few parts, and little to no perianth (Fig. 10–3 B, Fig. 10–6 A).

Flowering plants bear their ovules marginally on modified leaves, like cycads. In angiosperms the megasporophylls are called carpels, and one or more typically form a cylindrical pistil, each carpel appearing enrolled, such that the marginal ovules are enclosed in a tube,[6] thus the name angiosperm meaning housed or covered seeds (Fig. 10–4). Simple pistils are composed of a single carpel and a flower may have one to many such pistils (Fig. 10–4 A-C). Compound pistils are composed of two to many carpels (Fig. 10–4 D-E); a flower only has one compound pistil. On a functional level, not only does housing ovules within a modified leaf offer them protection, but also it means endosporic males, pollen grains, no longer have direct access to ovules and the female gametophyte within. Pollination occurs when pollen is deposited upon a portion of the carpel, its apex or margin (Fig. 10–4 B, E-G), sometimes at a considerable distance from the ovules. The implications for fertilization will be discussed in a bit.

Flowering plants are also fruiting plants, and following pollination and fertilization, each pistil further develops into a fruit, if there is a single pistil, or, if there are two or more individual pistils per flower, fruitlets, which can be quite separate or fused into an aggregate fruit like a raspberry.[7] Fruits can be dry or fleshy at maturity, popping open at maturity to release the seeds, or remaining closed, containing one to many seeds. In fact one definition of a fruit is a flower at the stage of seed dispersal. Perianth parts can persist following pollination, although often the showy perianth and stamens are shed, and in some cases these and other floral parts develop along with or become incorporated into what functions as a fruit—for example, the pineapple and strawberry.

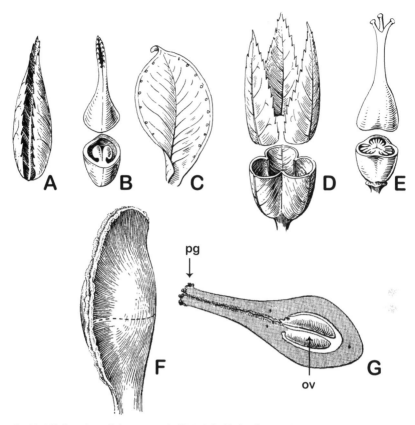

Fig. 10–4 Pistils and carpels (megasporophylls). A. Rolled leaf to illustrate the general view of a carpel as a modified leaf. B. Simple pistil consisting of one carpel with a row of ovules on each margin. The stigma is illustrated as occupying the apex. C. Simple pistil composed of 1 carpel at a fruiting stage, for example a follicle like a milkweed pod, split open along its marginal seam with seeds removed to show its leafy form. D. Three enrolled leaves connected by the basal margins. E. Compound pistil composed of three carpels whose margins meet centrally, each bearing multiple rows of ovules on each margin. The apex of three carpels extends into a style and a three-parted stigma for pollen capture. F.-G. Leaf-like simple pistil of *Drimys*, a member of the ANA basal lineages. F. Pistil showing the extended but unsealed margin forming a stigmatic margin. G. Median section (dashed line in F) showing two marginal ovules (ov) and pollen grains (pg) on the stigmatic margin. Pollen tubes grow between the carpel margins. (Image sources: A-E—After Porter 1967; F-G—After Foster and Gifford 1959)

HOW DO FLOWERS WORK?

Flowers primarily differ from gymnosperm cones in function. With the exception of cycads, gymnosperms employ wind for pollen and seed dispersal, although evidence is growing that a number of extinct gymnosperms had biotic pollen dispersers (Labandeira et al. 2007). Wind is free, but wind dispersal is not without its costs; it requires that the plant invest substantially in producing huge numbers of spores (pollen grains) because wind dispersal is pretty random. To be reasonably certain of landing some males in the vicinity of female gametophytes enclosed within their ovules, a plant must disperse a lot of pollen. Quite a number of flowering plants, particularly most grasses and many temperate deciduous trees, are wind pollinated and the causes of "hay fever" allergies, but they did not inherit this biology directly from gymnosperm ancestors. Many people think wind-pollinated flowering plants actually lack flowers because such flowers are either cone-like or lack attractive displays and are seldom noticed. Many flowering plants interact with animals to accomplish pollen and seed dispersal. Animals do not function as dispersal agents to be nice or to help flowering plants; they are enticed into providing this service because they receive a reward for their services that is often but not limited to food.[8]

Flowers signal their presence to pollen and seed dispersers by producing attractive displays. While the most familiar floral displays are visual, some flowers function largely by generating olfactory displays ranging from quite pleasant to rather disgusting by our sensory standards, but flies and beetles would beg to differ about what smells great. Attractants and rewards come at a cost, and evolutionary theory allows us to predict that the reproductive benefits of biotic dispersers must exceed the costs of using them. Many experimental studies have demonstrated that animal dispersers respond to differences in displays and rewards, and that plants can use their displays and rewards to move pollen in ways advantageous for their reproductive success.

Animals have behaviors shaped by natural selection to forage optimally for resources and to obtain the most reward for the least effort. To maximize their reward, animals must respond to and move efficiently from display to display, which results in a relatively precise, nonrandom movement of pollen, a reproductive benefit that more than offsets the cost of the displays and rewards. Plants use their displays to compete for pollinators, where the winners disperse and/or obtain more pollen, and as a result natural selection can affect the size of displays. Many field studies have demonstrated that pollinators are very sensitive to displays and rewards, and that displays are subject to natural selection. No wonder some flowers are so large, no wonder some plants produce such big flowering displays; plants are not trying to please us, these displays are competitions with those of other plants, other members of their own species. When pollen and ovules are borne on different plants, pollen-producing plants (often called "males" for convenience) and seed-producing plants ("females") can compete with and affect the displays of each other.[9]

The postpollination development of a flower produces a fruit (or fruit-lets), and to some extent what is true for pollen dispersal is also true for seed dispersal. Fruits are often protective as well as functioning in various means of dispersal. This means flowering plants enhance their spore dispersal through interactions with animals at two different developmental stages of the flower: when microspores are shed as pollen and when megaspores, and what they have become, are dispersed as seeds. Again there is an energy tradeoff. Resources used to produce protective, attractive, and rewarding fruits come at the cost of making fewer seeds, so better protection and more effective dispersal must more than offset this cost. Some plants invest in making big, well-protected seeds—for example, a peach or walnut; others make fantastic numbers of individually small, cheap seeds variously packaged in fruits—for example, a kiwi or milkweed.

When fleshy fruits mature, they develop a color, taste, texture, and smell signaling their ripeness. Such displays promise fruit eaters a safe, nutritious food. Prior to the stage of seed dispersal, fruits are protective by being hard, green, distasteful, toxic, and/or of limited food value. Color vision, which is limited to primates and birds, allows fruit foragers to quickly notice brightly colored fruits, in reds, yellows, oranges, bright blues, whites, or purples,[10] which contrast nicely against the constant, uniform background of green foliage. A pleasing taste and odor, and a substantial reward, assures many fruits will be consumed before the consumers move on, having discarded or ingested many seeds. Field research has shown that the mortality rate of seeds (think offspring) varies inversely with dispersal distance. The closer to their maternal plant seeds land, the less their chance of surviving. Of course, a lot of seeds can be found close to seed-producing plants, so they are easier to find and represent a bigger resource for seed predators. The mortality rate falls off as seeds disperse further, so at least some dispersal away from the maternal plant is highly beneficial in terms of survival.

A dry fruit often protects the seeds within, which is not unlike the function of gymnosperm seed cone scales, but dry fruits can also function in diverse dispersal mechanisms—for example, shakers, hitchhikers, catapults, floats, gliders, and whirligigs. Some dry fruits, like nuts, function in dispersal by attracting seed predators and counting on the difficulty of getting the seed within and the inefficiency of seed predators who "let some seeds get away." Some dry fruits open to release the seeds, which may have their own adaptations for enhancing dispersal like a winged or tufted seed coat or an aril, a fleshy outgrowth that partially to wholly surrounds a seed. Arils are often colorful, white or red, contrasting with both fruit and seed coat, both attractive and rewarding to bird and mammal dispersers. One of my favorite families of plants, the nutmegs,[11] has red, fleshy arils, which in the nutmeg of commerce constitute the spice called mace. Nutmeg arils form both attractive displays and a reward for "fruit"-eating birds once the tough, plum-like fruit splits open. Yews have similar arils (Fig. 10–2), and cycads have fleshy seed coats with a similar function. Whether the fleshy seed coat of ginkgo had such a function is unknown, but it has a singularly unpleasant odor to us, but then so does the aril of durian, and its seeds are dispersed by primate (orangutan) and fruit bat. Seed-dispersing animals tend to be fairly large, bird- or bat-sized or bigger, and capable of transporting a reasonable number of seeds some distance. Animal dispersers are very good at removing at least some seeds from the vicinity of the plant that bore them. Many rain forest trees that have seedlings that require dispersal into a forest gap to germinate and grow have primate, bird, or bat seed dispersers, organisms that forage widely and disperse lots of seeds over a considerable distance.

Until flowering plants appeared, the primary interaction between plants and animals had been very one-sided. Plants captured solar energy and synthesized biomass, and animals consumed as much of it as they could. But flowering plants ushered in a new era of mutually beneficial interactions, where the plant gets effective dispersal in return for giving animals their fill. When a symbiosis is characterized by a mutually beneficial interaction it is called a mutualism. As any gardener knows, many animals still engage in the traditional one-sided interaction, and plants have adaptations to evade or limit herbivory; so the war continues.

But there is still another "advantage" to animal pollinators and seed dispersers. Wind pollination operates most efficiently in communities where the wind-pollinated plants are dominant, common, and closely spaced—that is, in communities with relatively low species diversity and high species density, lots of individuals per unit area, like in a coniferous forest or grassland. Animal pollination allows plants to be smaller, more isolated, and less common because a foraging animal entrained upon seeking a particular display can move considerable distances between different individual plants in a community. Thus animal dispersal allows flowering plant communities to be more diverse, to include more species at a lower frequency, than does wind pollination. Another way of saying this is that animal pollen vectors allow flowering plants to reproduce successfully when they are farther apart, in smaller numbers, and mixed with more species. And this is true even when the dominant plants in the community are wind dispersed. While relatively few wind-pollinated grass species dominate our prairie communities, the majority of species are animal pollinated.[12] Those few gymnosperms that employ biotic dispersers are without exception found occupying a habitat where wind dispersal would be inefficient. For example, small cycads of the forest understory employ insect pollination, or to put it the other way round, insect pollination allows some cycads to occupy a forest understory habitat.

HOW DID FLOWERING PLANTS TAKE OVER?

Could mutualistic interactions between flowering plants and animal pollen and seed dispersers have played a role in the flowering plant takeover? This idea has been considered. While the fossil record of the Cretaceous tells us such an event took place (Hickey and Doyle 1977; Crane 1987), it did not record how it happened, so consider this scenario.

In a forest of gymnosperms, a broad expanse of territory is dominated by a relatively few species of trees. To avoid pollen interference, each conifer species releases copious amounts of pollen at different times to be carried along by the wind. The volume of pollen can be so great that lakes surrounded by coniferous forests accumulate a floating layer of pollen a couple of inches thick. Such a forest is a relatively simple community. The evergreen canopy makes the understory dimly lit year-round, so little more than a ground cover of shade-adapted ferns grows underneath the trees. Where could flowering plants obtain a toehold in such a community? One idea is that the first flowering plants were probably small shrubby plants, and perhaps somewhat weedy, especially in comparison with forest trees. Such plants would have been adapted to margins, small spaces at the edges of forest communities particularly along streams or coasts. Such plants would need to grow fairly quickly to deal with the more changeable marginal habitats. A recent analysis of leaf traits from three early Cretaceous fossil sites in North America suggests that early flowering plants lived along streams, were somewhat weedy, and had short-lived leaves, traits that support the idea of fast-growing plants and their subsequent ecological success (Royer et al. 2010).

Although long lived, trees are not forever and individual trees in the forest die (or encounter an environmental mishap) and fall, producing a gap in the forest canopy. Animal-dispersed seeds would allow flowering plants adapted to margins to invade tree-fall gaps, a bit of well-lit space. For reasons that will be discussed below, flowering plants possess some structural differences from gymnosperms that can result in more rapid growth to reproductive maturity. Thus a flowering plant of a somewhat weedy nature invading a tree-fall gap can outgrow gymnosperm seedlings or fern gametophytes, successfully winning the competition for light and space. If the gap invader were a wind-pollinated plant, it would have very little chance of getting pollinated because of its isolated location down in a forest gap. But animal pollen vectors capable of foraging across greater distances allow even somewhat isolated plants to be successfully pollinated. In this manner a tree-fall gap in a gymnosperm forest could be converted into an island of shrubby flowering plants rather than affording a reproductive opportunity for one of the dominant tree species whose seedlings could take one, two, or more decades to reach the canopy level.

Trees are long-lived organisms, nearly immortal, because they possess perpetually juvenile tissues that allow them to add new growth and new layers to themselves. Aging and senescence in trees is mostly the result of growing too big and too heavy for their construction material, wood, and then they begin to literally fall apart. The life expectancy of trees can be on the order of decades to centuries, and many or most die not of "old age" but because they encounter an environmental hazard: high winds, heavy snow, lightening, floods, storms, or fires. The longer a sessile organism lives the more likely it is to encounter some life-threatening event. But from our short-term, human perspective, forests have a look of permanence, unless we seek timber, but tree turnover in forests is much more rapid than you would guess by just looking at them. Fortunately biologists have actual data, and on average a forest will turn over every 100 to 300 years, faster for tropical forests, slower for temperate forests.[13]

Animal seed and pollen dispersal allows small, widely spaced flowering plants to invade forest communities dominated by wind-pollinated gymnosperms operationally. As more and more gaps are converted from gymnosperm to flowering plants, the gymnosperm forest first becomes an open network and then becomes fragmented until the efficiency of wind pollination is reduced. As a result, gymnosperms produce fewer seeds, making the flowering plant invasion of fragmented forest even easier. At some point a critical population structure is reached where the fragmentation of the forest and the reduced gymnosperm numbers render wind pollination very inefficient. Without replacement by juveniles, gymnosperms are replaced wholesale by flowering plants, and a whole new angiosperm forest appears. Freed from marginal habitats, selection favors larger, longer-lived trees, and lots of new tree species take over. This did not happen everywhere because forests dominated by conifers continue to exist, particularly in high latitudes and higher altitudes.

Such a scenario suggests the flowering plant revolution was caused by an ecological takeover rather than a climate change or a catastrophe. And this may not be the first such ecological takeover. Seaweeds may have outcompeted stromatolite-forming bacterial communities, leading to their drastic decline in prominence. Early seed plants apparently grew in marginal habitats (newly emerged land or newly deposited sediments), effectively outgrowing, outcompeting the dominant horsetails and club-mosses. Gymnosperms themselves may have invaded and transformed the forest communities of the Carboniferous because of the ecological advantage of seeds over free-sporing plants, a takeover assisted at the end of the Carboniferous by a climatic coup de grace when a hotter, drier climate favored the seed habit.

The loss of gymnosperm-dominated communities allowed some flowering plants to revert to wind pollination. Presently, most grasses and many temperate, deciduous forest trees are wind pollinated. The dominant species in both grasslands and deciduous forests exhibit relatively low diversity and high density, forming communities where wind pollination is effective. As a result the wind-pollinated flowers of grasses and deciduous trees feature no displays or rewards,[14] but as many hay fever sufferers know, such plants produce lots of wind-dispersed pollen.[15] Deciduous trees flower early in the growth season before their expanding leaves begin interfering with wind dispersal. Wind-pollinated trees become very uncommon in the tropics because species diversity increases, species density declines, and the evergreen condition of most woody plants, as well as the rainy climate, interferes with wind dispersal, all combining to make wind dispersal very inefficient.

POLLEN COMPETITION

Functionally flowers permit and encourage males to compete for females, which is something that does not occur on the same order of magnitude in gymnosperms. The housing of ovules within the carpel has the consequence of separating the site of pollination some distance from the ovule, whereas in gymnosperms pollination occurs when pollen actually enters the ovule through the micropyle. True, if an ovule captures two to several pollen grains, only one fellow will successfully sire the resulting offspring. But a flower's stigma can capture or receive a huge number of pollen grains, so rather than a shoving match in a singles bar, a contest more like the Boston Marathon occurs, and the winner sires an offspring. Having successfully dispersed to a welcoming stigma is only half the battle because the stigma is sometimes a considerable distance (several centimeters) from the ovules. Considering the tiny size of the male, this would seem a huge disadvantage, and perhaps it is from the male's perspective. However ,since stigmas are quite capable of capturing or receiving many more pollen grains than needed to pollinate all the ovules within, all these males have to compete in a race to the beckoning (via molecular cues) female(s).[16] Such competitions have been experimentally demonstrated (e.g., a study on buffalo gourd by Winsor et al. 2000). This contest assures fertilization by the most vigorous males and the resulting offspring demonstrate greater vigor.

The competition results from differences in pollen tube growth rates through the stigma and style, and along the inner surface of the carpel(s), to reach the ovules and grow in through the small opening in the jacketing and accomplish fertilization. Thus the housing of the ovules within a pistil not only protects the ovules but also functions in the evaluation of male genotypes. How appropriate that we human males compete for females by bringing them flowers! The interaction between stigma and pollen also inhibits or prevents self-fertilization and reduces or eliminates inbreeding by providing an environment unfavorable to closely related males. Both of these functions demonstrate a clear reproductive advantage over pollination in gymnosperms.

DOUBLE FERTILIZATION

In flowering plants sperm are reduced to cells consisting of little more than nuclei, and each male makes exactly two sperm. One fertilizes the egg and the other fuses with one or two other nuclei that are identical sisters to the egg, forming a unique tissue called endosperm, literally "within the seed." Endosperm develops into a nutritive tissue for the growing embryo, and at the time of dispersal, a seed will either have a small embryo surrounded by a large mass of endosperm (e.g., a coconut[17]) or a large embryo that has absorbed the endosperm (e.g., a peanut[18]). You may never have heard of endosperm before, but it is of singular importance to humans because endosperm provides 50% to 70% of <u>all</u> human calories (!) mostly in the form of cereal grains, the one-seeded fruits of the grass family (more on this in chapter 11). Endosperm is the biological twin of the embryo, so as a nutritive tissue it is most unusual and unique. In all other seed plants and in free-sporing plants too, the embryo is nourished by maternal tissues. The production of endosperm is a novel feature that defines the flowering plant lineage.

But of what advantage is endosperm? Maternal nourishment certainly has worked well enough for all the rest of land plants. Timing is the key to understanding the advantage of endosperm. The female gametophyte of gymnosperms must grow to sexual maturity before fertilization can take place, so in essence provisioning the seed begins with the growth of the megaspore into the female gametophyte. But in angiosperms the female gametophyte is quite small and occupies the same volume as the megaspore. Provisioning of the seed does not begin in any significant way until after the fertilization event. Some experimental evidence suggests selective abortion of ovules is a means of choosing favorable genotypes among the various offspring.[19] And of course if the fertilization is not successful, then no more energy is wasted upon that particular ovule. This means angiosperms "invest" in a sure thing only when it comes to producing seeds. This is a huge advantage.

THE BIOLOGY OF FLOWERS

My job allows me to study and to educate people about the workings of flowers. Initial interest often turns to thinly veiled disappointment when people discover that many flowers are not nice and pretty, but part of the explanation of why flowering plants are so successful involves the diversity of flowers and their equally diverse functions. For example, my favorite flowers belong to nutmegs and custard apples (Fig. 1–1), both magnolialean families that are largely composed of tropical trees whose flowers use olfactory displays, strong odors, to attract beetles, thrips, and flies as pollinators (Sharma and Armstrong 2013). Without visual attractants, such flowers can and do function at night, which can keep biologists, or their students, out late in the rain forest. While not the type of flowers that bring visual delight, such flowers are interesting because many engage in complex interactions with their pollinators, often not only providing a food reward for the adult pollinators but also functioning as a brood substrate providing food for their larvae as well. Providing insects with a place to reproduce and the resources to rear their larvae is a huge reward because in the life cycle of many insects, the larval feeding stage is when they gain the necessary mass to transition from egg to adult size. Some simpler types of beetle pollination, where only food is involved, like those of a wild nutmeg (Armstrong and Irvine 1989; Sharma and Armstrong 2013), are similar to the insect pollination found in cycads. Could such biotic interactions have occurred in the gymnosperm ancestors of angiosperms? Unfortunately most aspects of pollination do not fossilize, so we probably will never know, but some fossils suggest such a possibility.

After you have studied flowers a bit, you can begin to recognize what are called floral syndromes, characteristics of floral size and organization, attractants and rewards that are associated with a particular class of pollinators. When people enter our teaching greenhouse, they often attempt to sniff the gaudy red flowers of the hibiscus growing by the door. Humans tend to associate visual attractiveness with an expectation of a pleasing odor, but hibiscus flowers clearly display the hallmarks of a floral syndrome adapted for hummingbird pollination. They are large and red, they produce lots of nectar accessible through slots at the base of the perianth, the pollen dispersing and receiving parts stick well out of the flower, and the floral parts are fairly robust. But since nectar-feeding birds have no sense of smell, bird-pollinated flowers, while visually gaudy, have no scent.

Space does not permit more than a few brief examples of pollination syndromes to illustrate how flowering plants have surpassed gymnosperms. Eelgrass, an aquatic flowering plant, has water-dispersed pollen, a most unusual adaptation involving long, thread-like pollen that floats like "noodles upon the sea" (Cox 1985). This shape enhances the pollen grains' chances of making contact with floating stigmas attached to flowers submerged below. Some flowers mimic carrion, presenting the coloration, hairiness, and odor of a rotting carcass to attract fly and beetle pollinators. In such cases the pollinators are seeking food, both for themselves and their offspring, but they are cruelly deceived by these plants because in most cases the advertised reward of rotting flesh is not provided. A deception that seems less cruel involves flowers that mimic female insects. Male insects pick up or deliver pollen when they attempt to mate with the flower, so for pollination to occur, they must get fooled by these sex decoys more than once. The male insects make this task fairly easy because natural selection shapes reproductive behavior such that organisms attempt to pass along as many copies of their genes as possible. Some pseudocopulation flowers even produce a fragrance that mimics an insect sex pheromone. This deception costs the pollinating insect time and energy, but the cost cannot be too great, too detrimental to the pollinator's reproductive success, or the plant risks losing its pollinator. In the tropics, some flowers are pollinated by nectar-feeding bats. Bat-pollinated flowers usually open at dusk and have either a reflective white color, or a drab color and a sound-reflecting dish, both to assist the bats in finding a nectar reward. The reward for a bat has to be substantial, but bats can carry lots of pollen, so bat flowers tend to produce lots of seed. White or pale blue flowers that resemble long, narrow trumpets often open in the late afternoon. Such flowers are pollinated by hawkmoths, which are the only insects with a proboscis long enough and thin enough to reach nectar at the base of a narrow corolla tube. And some flowers are more like generalists, attracting an array of floral visitors of varying degrees of effectiveness as they vary in fidelity, frequency, and ability to vector pollen.

Animal pollination has another "advantage" at the community level that has already been discussed; it allows many more species to share an area than does wind pollination. Wind-pollinated species can partition themselves only temporally to avoid the congestion from other species' pollen. One-dimensional partitioning restricts the number of similar species that can occupy a given area. Biotic pollinators provide another dimension. The pollen-dispersing environment can still be partitioned in time, then partitioned by different pollinator species, and then partitioned even by different ways of using the same pollinator. For example, nectar-foraging bumblebees approach flowers right side up, while pollen-foraging bumblebees grab the flower and then flip upside down. Two species can coexist using the same pollinator by having flowers that differ in rewards and in the location where pollen is placed on the same pollinator's body. Some flowers are even asymmetrical, which forces insects to approach from a single direction, so that pollen is deposited on and delivered from either the left or right side of the pollinator, whereas most flowers would be placing pollen on either the dorsal or ventral side. A switch in floral handedness has never resulted in a new plant species, but it could. And of course such interactions have resulted in the evolution of lots of flowers possessing a dorsoventral orientation to precisely interact with an animal pollinator.

A tropical botanist at the Missouri Botanical Garden, Allyn Gentry[20] (1974) figured out that members of the bignon family, a group of large-flowered, tropical trees and vines,[21] had a certain number of species in each region, one species for each of the major groups of pollinators present. So there would be one bat-pollinated species, one big-bee pollinated species, one small-bee pollinated species, a bird-pollinated species, one hummingbird pollinated species, and so on, each with the appropriate floral form. Such patterns suggest that biotic interactions may determine how many plant species can coexist. Thus it comes as no surprise to learn that major groups of pollinators, bees and wasps (Hymenopterans), butterflies and moths (Lepidopterans), and nectar-feeding and fruit-eating birds, show an increase in diversity that parallels that of the flowering plants themselves. And of course, nowhere is this diversity of interacting mutualists more evident than in tropical rain forests, which is where this story started some ten chapters ago.

THE FIRST FLOWER

With all this floral diversity, everyone wants to know what the first flower looked like, but no one knows. Was the ancestral flower large or small, with lots of floral parts or relatively few? Were the most ancient flowers colorful and attractive or drab and insignificant, and what pollinated them? Were the oldest flowers unisexual like so many cones or was the ancestral flower perfect[22] (bisporangiate)? No one knows and hypotheses about angiosperm origin differ on this point (Endress and Doyle 2009). Two sources of information are available to us, the fossil record and hypotheses of flowering plant phylogeny. Phylogenetic studies identify which groups among living angiosperms should retain the most ancestral features and which are more derived. Everyone agrees that the flower-like inflorescences of daisies, asters, and sunflowers[23] are highly derived. But our concept of a primitive flower has changed as our phylogenetic hypotheses have changed. Since gymnosperms are all woody plants, the first angiosperms and any of their living relatives were expected to be woody as well.

A hundred years ago botanists thought the simple, inconspicuous[24] flowers of wind-pollinated trees were the archetype of the primitive flower. Such flowers generally were unisexual, meaning these flowers had stamens or pistils but not both, which was thought to be similar to all living gymnosperms. Pollen-producing flowers were crowded into helically arranged, pendent inflorescences called catkins or aments, which look a lot like pollen cones. Flowers bearing pistils, destined to become fruit, were solitary or in small groups usually surrounded by bracts, something similar to ovule-bearing structures found in *Ginkgo*, yews, and podocarps (Appendix: Conifers and Ginkgoes). This phylogenetic concept was tested by extensive comparative anatomical studies of wood that began in 1918 (Bailey and Tupper 1918) and continued over the next 30 years. These studies demonstrated that the wood of ament-bearing trees, a group then called the Amentiferae, which included willows, birches, and beeches, was highly derived and this at least in part falsifies the hypothesis that they were primitive among flowering plants. This also means that wind pollination in these forest trees was not directly inherited from wind-pollinated gymnosperm ancestors.

These same comparative studies discovered that those flowering plants with wood most like that of gymnosperms were found among the magnolias and their relatives, the so-called woody Ranales.[25] The hypothesis that members of the woody Ranales retained the most-ancestral characters was a viable concept for decades. The flowers of this group were relatively large, many-parted and helically arranged, cone-like features, and the stamens were leaf-like (Fig. 10–5), also suggestive of gymnosperm sporophylls. These flowers were reminiscent of gymnosperms' cones, albeit bisporangiate ones, and the correlation of such flowers with primitive wood characters convinced decades of botanists that such flowers were the primitive angiosperm condition (e.g., Bailey and Swamy 1948, 1949, 1951; Eames 1961).

Comparative wood anatomy is an exceptional example of scientific inference, correlation, and falsification of hypotheses, but it remains very little known outside a handful of us elders.[26] Rather than using some arbitrary criteria or presupposed ideas about phylogeny to establish primitive character states, the temporal sequence in which wood characters appeared in the fossil record was used to determine the ancestral condition. The wood of living vascular plants was then compared with fossil woods to determine which retained ancestral character states and which were derived. Sometime later this same approach would be applied to the venation patterns and leaf form of angiosperms, a sequential change toward more orderly patterns, which was also recorded temporally in the fossil record (Hickey and Doyle 1977).

Some features of the floral organs found among the woody Ranales were instrumental in furthering the concept that stamens and pistils were correctly interpreted as modified leaves (sporophylls). Some of these flowers have very leafy sporophylls, stamens with broad, flat blades with no differentiation into sterile filament and fertile anther (Fig. 10–5), and like some gymnosperm microsporophylls, some have two "sori," each with two sporangia. The primitive pistil was composed of a single leaf-like carpel not differentiated into an ovary, style, and stigma. Such a carpel was called conduplicate because it had the appearance of being a folded or enrolled leaf where the two margins and two rows of marginal ovules became adjacent (Fig. 10–4 F-G). The marginal suture of such pistils was wholly or partly unsealed, although closely appressed, and had a stigmatic function. This concept was developed by Bailey and Swamy (1948, 1949, 1951).

As an academic great-grandson of Bailey,[27] I was steeped in the Ranalean tradition, but the Ranalean family I chose to study was different. Nutmegs

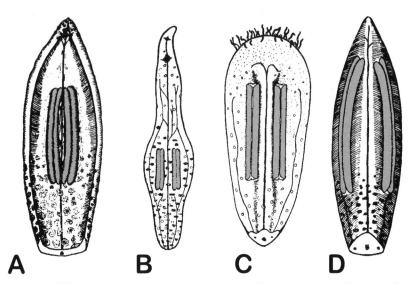

Fig. 10–5 Leaf-like stamens from four genera (A. *Austrobaileya*, B. *Himantandra*, C. *Degeneria*, D. *Magnolia*) in four different families of the former woody Ranales, each stamen showing no differentiation into anther and filament, four elongate microsporangia in two pairs (shaded gray), and three vascular bundles, one central and two marginal. *Austrobaileya* now belongs to Austrobaileyales of the ANA basal lineages of flowering plants (see Fig. 10–3 D); the rest are in the Magnolialean clade (Fig. 11–4). All would be 5–10 mm tall. (Image source: After Canright 1952, used with permission of the Botanical Society of America)

have small, simple, unisexual, three-parted flowers on tropical forest trees that grow like conifers with horizontal whorls of branches. Furthermore, nutmegs do not have a conduplicate carpel, so I never wholeheartedly accepted the idea that *Magnolia*-like flowers were primitive. In particular, the beetle pollination of nutmegs is a simple, less complex, less specific type, a pollination syndrome that could be interpreted as primitive (Armstrong and Irvine 1989), one that could function just fine in a gymnosperm, whereas the pollination syndromes of the large, many-parted flowers like *Magnolia* were comparatively complex (e.g., Thien 1974), which suggests specialization. The pollination and simplicity, with no evidence of a prior more complex condition, suggested the nutmegs might be primitively simple. Recent molecular phylogenetic studies place the nutmegs as the basal lineage of the Magnoliales, which to some extent concurs with this assessment.

Then along came another hypothesis. The Ranalean concept of a primitive angiosperm flower was challenged by the hypothesis that the basal angiosperms were small and herbaceous, maybe even weedy little plants, with small, simple, even unisexual, flowers. These so-called paleo-

Fig. 10–6 Few-parted flowers and oldest flower. A-B. *Chloranthus* family. A. Two flowers of *Sarcandra chloranthoides* showing their simple organization consisting of a single pistil (p) with a terminal stigma, a single fleshy stamen (s), and a subtending bract (b). The pistil is greenish at maturity, but the stamen is yellow-orange and the margins of the pollen sacs are reddish. B. *Chloranthus spicatus* showing spikes (inflorescences) of small, simple flowers very similar to those above borne in pairs. C. Fossil of *Archaefructus*, the oldest known flower. These "flowers" consist of helically arranged pistils/fruits on the terminal portions of the shoots. It is hard to determine whether each pistil is a flower as in the *Chloranthus* family, or whether the entire axis is a flower. (Image sources : A—Courtesy and with permission of P. K. Endress; B—Courtesy of and with permission of L. M. Kelly, www.plantsystematics.com; C—Courtesy of and with permission of D. Dilcher)

herbs were exemplified by members of the *Chloranthus* family and the Piper family, which has only a couple of relatively familiar members, black pepper (*Piper nigrum*), common enough on your table but the plant that produces those little fruits is unfamiliar, and the house plant *Peperomia*. The oldest known fossil angiosperm pollen from the Lower Cretaceous, *Clavitopollenites*, is essentially identical to the pollen of *Chloranthus*, a plant with small, extremely simple flowers consisting of just one pistil and one stamen and a subtending bract (Hughes 1976; Hickey and Doyle 1977) (Fig. 10–6 A-B). These flowers have no perianth of any sort, and the bract does not function as perianth either, although the stamen itself is colored. Fossil flowers attributed to the *Chloranthus* family have been found in the early fossil record of angiosperms (Friis et al. 2000, 2006; Doyle et al. 2003), and this is the only living family so represented. This became very significant when some molecular studies produced phylogenetic hypotheses where *Chloranthus* and other paleoherb lineages were basal to the Magnoliales, an order that is generally equivalent to the woody Ranales. The paleoherb hypothesis also fits well with the idea presented above that angiosperms originated in marginal, perhaps changeable habitats where their somewhat faster growth and animal spore vectoring would provide an advantage over slower-growing, wind-dispersed ferns and woody gymnosperms. The simple flowers of paleoherbs also suggested that the gnetophytes, whose cones were constructed of units that are similar to these simple flowers, were likely to be the closest living relatives of flowering plants. Just a decade ago, two of my colleagues were feeling pretty confident that the paleoherb hypothesis and its interpretation of ancestral flowers had "brought this historic puzzle, Darwin's 'abominable mystery,'[28] to the verge of a solution" (Hickey and Taylor 1996). Paleobotanists are an optimistic group in general, a required trait in their field.

The current hypothesis about the earliest angiosperms is based upon a molecular phylogeny that has identified several lineages of flowering plants that are basal to all other angiosperms, lineages of the so-called ANA grade. The first A stands for *Amborella* (Fig. 10–3 B), a small shrubby tree whose wood lacks vessels—that is, it has wood like most gymnosperms. Although long placed in its own family, *Amborella* was a member of the woody Ranales (Eames 1961). N stands for Nymphaeales, the water lilies (Fig. 10–3 C) and their relatives, which have long been thought to be at the transition between terrestrial and aquatic, between woody plants and herbs, and between dicots and monocots. In addition to their many-parted flowers, water lilies, which in all other respects are good "dicots," have embryos with only one cotyledon, and of course, they are aquatic, growing rooted in shallow water—where they need little support, so their stems produce very little wood. Other members of this aquatic lineage have small, simple flowers. The second A stands for Austrobaileyales, which consists of *Austrobaileya* (Fig. 10–3 D), named in honor of I. W. Bailey, and several other small families formerly considered members of the woody Ranales, of which perhaps *Illicium*, star anise, is the only familiar member. Some of the old woody Ranales are basal again, but most remain in the Magnoliales, a group that shares a common ancestry with the ANA grade just like the rest of angiosperms. Have we made progress? Why, yes, some of these hypotheses have been falsified with some certainty.

What does this new phylogenetic hypothesis tell us about ancient flowers? Both morphological (Doyle and Endress 2000) and now several molecular studies have found *Amborella* to be the basal-most lineage of flowering plants. The single species of *Amborella* is found growing only on the island of New Caledonia, which is a small fragment of Gondwana located some 1,600 kilometers due east of Queensland Australia, so it is both geographically and taxonomically quite isolated. The flowers are unisexual with 5 to 7 spirally arranged perianth parts surrounding either numerous stamens or a whorl of 5 to 6 pistils (Fig. 10–3 B). The pistils, each composed of a single carpel, have a sessile, terminal stigma and contain only a single ovule. Rather than being conduplicate, such pistils are sac-like because the carpel develops up and around what appears to be a terminal ovule.[29] If your head was an ovule, then pulling a hood up over your head and drawing the string to produce a stigmatic surface gives you a visual idea of how such a carpel develops. The ovule is not in a marginal position as predicted by the megaphyll/conduplicate carpel hypothesis. This is significant because if a sac-like pistil surrounds a terminal ovule, then it has a similarity to an encircling bract and the basal angiosperm flower begins to sound like one unit of a gnetophyte cone, or perhaps, a glossopterid ovulate structure.

Other flowers of the ANA grade plants (Fig. 10–3 C-D) provide some additional information about the nature of the primitive angiosperm flower. On the basis of a comprehensive floral survey (Endress 2001; Endress and Doyle 2009), the primitive flower was likely to have been relatively small, with a moderate to small number of floral organs spirally arranged or possibly whorled like *Amborella* and other members of the ANA grade. The pistils were without styles and sac-like with only one to a few ovules. The flowers were bisexual but easily becoming unisexual through abortive development of either the stamens or pistils, often leaving vestigial organs as evidence of their bisexual ancestry. Pollination was probably by small insects, especially flies, perhaps pollen-foraging thrips, moths, and beetles. Such flowers could have evolved from the reproductive structures of fossil organisms like the glossopterids or cycadeoids via an abbreviated or truncated development. Flowers of the ANA lineages are all insect pollinated.

What does the fossil record contribute to our understanding of ancient flowers? Already one of the problems has been hinted at. Flowers cannot be defined in such a way as to divide all known gymnosperm and angiosperm reproductive structures into two mutually discrete categories. This makes it difficult to know when an enigmatic fossil should be considered a flower. The fossil record has other inherent biases because the chance of fossilization is directly proportionate to how common an organism, organ, or developmental stage was, and some features fossilize more readily than others; in the case of flowers, both buds and fruits

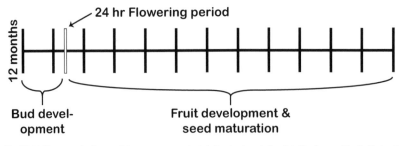

Fig. 10–7 Life span of a flower of *Anaxagorea crassipetala* (custard apple family). The flower (Fig. 1–1) develops during a bud stage that lasts 4–6 weeks. Flowering lasts 24 hours and includes a time period for both pollen accepting (23.5 hours) and pollen dispersing (0.5 hours). Petals and stamens are shed immediately. The postpollination fruit development stage lasts 9–10 months until the seeds are mature and ready for dispersal. (Image source: The Author. Based on data in Armstrong and Marsh 1997)

occur for longer periods of time and are harder and denser, making for better fossils. As a result, the fossil record does not tell us much about flowers at the stage of pollination. Delicate perianth parts rarely fossilize and they persist only briefly, often less than 1% of a flower's life, further reducing any chances of preservation at this stage (Fig. 10–7).

Magnolia-type flowers are found in the fairly early fossil record of angiosperms. *Archaeanthus* ("ancient flower") fossils from the mid-Cretaceous (Fig. 10–8) could be placed in a woody Ranalean/magnolialian family without much difficulty (Dilcher and Crane 1984), and the leaves are very reminiscent of a living member of the *Magnolia* family, the tulip tree. These reconstructions are based upon separate fossil parts, all part of the same fossil assemblage, and all having the same type of secretory cell, suggesting they all came from the same plant. At the fruiting stage, the axis shows scars below the pistils, evidence of stamens and perianth parts (Fig. 10–8 B). Since the floral axis usually elongates during fruit development, the flower is reconstructed with a shorter axis and the pistils packed together (Fig. 10–8 A).

The oldest known fossil flower, *Archaefructus* (Sun et al. 2002), contributes to our concept of a floral prototype but does not settle the issue. *Archaefructus'* flower is larger with multiple parts and its pistils are each composed of a single carpel looking rather like small pea pods (Fig. 10–6 C). On the basis of its morphological characters and age, these authors placed *Archaefructus* as the basal-most lineage of angiosperms, so ANA would become A'ANA (A' = *Archaefructus*). Some fossils of *Archaefructus* are remarkably intact and because of its fine detail, the reconstruction of the entire plant requires almost no educated guesswork, a very unusual situation in paleobotany. The foliage suggests an herbaceous, aquatic

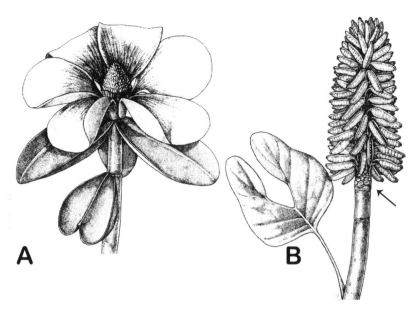

Fig. 10–8 Reconstruction of *Archaeanthus*, flower and fruit. A. This mid-Cretaceous flower had large perianth parts surrounding many stamens and pistils on a conical axis. B. Fruiting axis with numerous follicular fruits. Scars left from fallen perianth parts and stamens appear on the axis below the pistils (arrow). Fossil leaf, *Liriophyllum*, and perianth parts were found as dissociated parts in the same rock matrix, all having very similar secretory cells, suggesting that they were from the same plant. *Archaeanthus* and the leaves, as their name suggests, bear considerable similarity to members of the *Magnolia* family. (Image source: After Dilcher and Crane 1984)

plant and, together with the flower, makes *Archaefructus* most similar to the waterlilies (N). This fossil rather shifts the concept of a primitive flower back toward larger bisporangiate flowers with conduplicate pistils.

In *Archaefructus* the axis bearing floral parts is much more elongate than most living flowers, but an elongate fruiting axis is not unknown among living flowers—for example, the magnolia vine, *Schisandra*, part of the ANA grade, has many individual pistils on a short, dome-shaped receptacle, rather like a buttercup, but the receptacle elongates so much that the fleshy fruitlets end up looking like little grapes along a stem. The longer axis of pistils seen on *Archaefructus* appear to be a fruiting stage flower as suggested by the size of the pistils and the size of the ovules within them. Younger floral axes based on their position on the plant are shorter, placing the floral organs closer to each other. *Archaefructus* also has helically arranged pairs of stamens borne on the same axis below the helically arranged pistils, and below them there are no obvious perianth parts or even any scars left from them.

Perhaps rather than being a single flower with many parts, *Archaefructus* is an axis bearing lots of small, simple, unisexual flowers, something more like *Chloranthus* (Fig. 10–6 B). While these are clearly angiosperm flower parts, these axes are truly terminal, the apical-most ends of the shoot system. Flowers are generally lateral branches and located above either vegetative leaves or bracts. This helical array of flower parts occupies a position where an inflorescence, a group of flowers, is usually found. To further complicate things, each stamen has four microsporangia, typical enough, but the stamens are paired and each pair of stamens is connected to the main axis by a short stalk. Stamens found in the woody Ranales and ANA grade lack filaments (Fig. 10–5) or have short, stout basal filaments, so a filament is an odd character to appear in an ancient flower.[30] No living angiosperm has pairs of anthers on a single filament. Rather than being a filament, could the axis bearing a pair of *Archaefructus'* anthers be a lateral axis bearing a pair of terminal stamens? If so, then each pair of stamens would be a simple flower as would each pistil. Presently there is no way to tell which is the case. If each of these stamen pairs and each pistil were subtended by a bract, then this would be an inflorescence of very simple, unisexual flowers, much like those of *Chloranthus*, a paleoherb, but no bracts are in evidence. And of course, perhaps a terminal clustering of sporophylls may be how larger, many-parted flowers got started.

A recent study places the small strange *Hydatella* family of grass-like aquatic plants native to Australia and New Zealand as a sister group to the water lilies (Saarela et al. 2007). So not only is the *Hydatella* family not related to grasses, although they were long classified as monocots, but also their basal phylogenetic position in the ANA grade means their last common ancestry with the rest of flowering plants occurred prior to the divergence of monocots. Their flowers are not flowers at all but assemblages of pistils and stamens in small but variable numbers, surrounded by perianth-like bracts, except they are assembled backwards from flowers with pistils surrounding stamens in the middle (Rudall et al. 2009). The bracts might suggest a similarity to gnetophytes, but the nonflower assemblage they surround is better interpreted as an inflorescence, albeit a reduced one. This new member of the ANA grade argues for interpreting *Archaefructus'* flower as an inflorescence of simple, bractless flowers, and presently *Archaefructus* is placed in the Nymphaeales as sister group to the Hydatella family.

Aments or catkins, dangling, cone-like inflorescences of simple flowers, are also quite ancient, suggesting wind pollination was derived early and larger magnolia-type flowers are nearly as old. Curiously, very few fossils of angiosperm wood have been found from the early Cretaceous, which if absence of evidence counts for anything may support the hypothesis that early angiosperms were either herbaceous or small shrubby plants like most members of the ANA lineages. Neither the fossil record nor phylogenetic hypotheses have provided a definitive answer about the nature of the ancestral flower, but the research will continue.

WHAT MADE ANGIOSPERMS SO SUCCESSFUL?

What is it about flowering plants that have made them so successful both in terms of ecology and species diversity? The fossil record of the Cretaceous indicates the flowering plant lineage diversified early and rose to ecological dominance quickly. Using a molecular clock to date major phylogenetic divergences suggests 44.5% of living species are found in lineages that diversified and appeared from 130 to 102 mya in the lower Cretaceous (Magallón and Castillo 2009). This suggests that angiosperm evolution was well underway by the time the oldest unambiguous flowering plant fossil appears at 132 mya (Crane et al. 2004). Many more lineages appear to have originated between 102 and 77 mya in the Upper Cretaceous. The surprise is not the speediness of the diversification; after all, this is a 55-million-year period of time, and only 65 million years have transpired since the end of the Cretaceous to the present day. The surprise is that so many major lineages appear to have arisen so long ago. Conifers and ferns also diversified during the same period but on a much smaller scale.

Our human tendency is to select some key feature of flowering plants and declare it responsible for all their success. As previously discussed, the advantage of animal dispersal afforded by the flower and its carpel-enclosed ovules, at the stages of both pollen and seed dispersal, plays a key role in explaining the flowering plant phenomenon. But flowering plants have other features that can be regarded as an "improvement" over gymnosperms, and these include a range of leaf forms and functions (in addition to flower parts), double fertilization leading to endosperm, structurally complex wood, diverse life histories, and varied growth forms like epiphytes, vines, herbs, and parasites.

In some manner the answer must be "yes to all" because all these features play a role and in no single respect, except for endosperm, are these flowering plant features unique to this group. Similar features can be found in various other seed plants and even ferns, but only in flowering plants do they all come together in concert. This suite of characters all together is the key to flowering plant dominance, so let us have a look at the rest of these angiosperm features.

Leaves

In general flowering plants have broad leaves with a complex network of veins in a wide variety of shapes and forms. Some of the largest leaves can have blades nearly two meters in diameter. Like their diverse growth forms, flowering plants display many examples of highly modified leaves specialized for many other functions in addition to photosynthesis and reproduction (Fig. 10–9; Fig. 10–10). Some leaves are so modified that people are surprised to learn they are leaves. Flowering plants adapted to dry habitats can have tough, hard, leathery, scaly, spiny, needle-like leaves, like gymnosperms. Evergreen flowering plants tend to have broad, leathery leaves, which in the tropics taper to a long apical point. In wetter habitats an elongated apical tip[31] and veins in valleys help the leaves readily shed water (Fig. 10–9 A). Some flowering plants have leaves that can be attractive, aiding in pollination (e.g., *Poinsettia*) (Fig. 10–10 A) or seed dispersal; or protective when modified as spines (Fig. 10–10 B); or used to climb as tendrils (Fig. 10–10 D). Some highly modified leaves trap insects (Venus flytrap, pitcher plants) and use the digested or decaying bodies of insects as a source of nitrogen (Fig. 10–9 B). Some flowering plant leaves have become thick and succulent, modified for food and water storage (Fig. 10–10 C)—for example, the layers of onion bulbs are such modified leaves. Leaves function as floats for some aquatic plants (water hyacinth). Some leaves wrap around each other tightly to make a vase-like shape for capturing water (Fig. 10–9 C). And some leaves produce plantlets, outgrowths that can separate and grow, a means of cloning and asexual reproduction (Fig. 10–9 D); indeed, some leaves, if detached from the plant, readily grow new plantlets at their base—for example, the leaves of stonecrops (Fig. 10–10 C)—a trait humans have taken considerable advantage of.

Some leaves engage in symbiotic interactions with animals to protect themselves from herbivory or obtain nutrients. Some highly specialized leaves provide domiciles and food rewards for ants and mites that protect the plant from herbivores (e.g., ant acacias). In one case, an epiphytic vine of seasonally dry forests (*Dischidia*, milkweed family) has small succulent leaves for photosynthesis and water storage as well as pocket-like leaves that form ant domiciles. When a leaf gets filled with the carcasses of the

Fig. 10–9 Modified leaves with specialized functions. A. Yam—Leaf with vein furrows and elongated apical drip tip to efficiently shed water. B. Cobra lily—Tubular leaf traps insects that eventually die, drown, and decompose, releasing nitrogen that the plant absorbs. "Carnivorous" plants generally live in low-nutrient environments and trap invertebrates for nitrogen. C. Bromeliad—Whorl of closely spaced leaves with a thick, waxy cuticle forms a tank that traps water. D. *Kalanchoe*—Leaves form detachable plantlets on marginal teeth for asexual reproduction. (Image source: The Author)

Fig. 10–10 Modified leaves with specialized functions. A. *Anthurium*—A large, heart-shaped, bright red bract attracts pollinators to the whitish spike of small, non-showy flowers. B. Peach palm—Leaf forms hard spines along the bottom of the midvein for protection. C. Stonecrop (*Sedum*)—A dense rosette of thick, succulent leaves functions in water storage in addition to photosynthesis. D. Cat's claw vine—Trifoliate leaf has a terminal leaflet modified into a grappling hook tendril for climbing and supporting the vine. (Image source: The Author)

ants' prey, the ants move on to younger leaves. Then the leaf reorients from horizontal to vertical such that they capture any rain and become convenient compost-filled pots into which the plant's roots grow.

The oldest angiosperm leaf fossils were simple in shape with less-organized venation. Similar leaves are presently found in ANA grade and among members of the magnolialean clade. Among living gymnosperms, only the tropical viney/shrubby gnetophyte *Gnetum* (Fig. A33) has broad, reticulate-veined leaves that are similar to flowering plant leaves. Among those fossil gymnosperms that are considered possible angiosperm ancestors, glossopterids (Fig. 9–9) have broad, reticulate-veined leaves, although the veins end at the leaf margin, which is typical of ferns and cycads. Some tropical conifers (Fig. A13) and *Welwitschia* (Fig. A34) have broad, strap-like leaves with parallel veins. In flowering plants the veins loop at the margins and interconnect.

Growth Forms and Life Histories

Flowering plants have adopted more different growth forms than any other group of land plants. In fact, in terms of diversity of form, flowering plants are rivaled only by the green algae. Many flowering plants are woody (trees and shrubs), like most gymnosperms. However, it is generally unappreciated that during the heyday of gymnosperms, the Triassic, Jurassic, and Cretaceous, now-extinct lineages exhibited diverse growth forms including herbs, vines, thick-stemmed trees, stilt-rooted mangrove-like trees, stem succulents, palm-like trees, and trees with branching crowns of multiple axes (like flowering plant forest trees), as well as conifer-like trees with a single central trunk and whorled or helical branches (Labandeira et al. 2007). Clearly a diversity of growth forms cannot be the whole answer to flowering plants' success.

Nonetheless, flowering plants today exhibit a remarkable diversity of forms that goes well beyond that known for living gymnosperms and even extinct gymnosperms. Flowering plants grow not only as forest trees but also as shrubs of many types, herbaceous perennials, annuals, vines, both floating and rooted aquatics, epiphytes, several types of succulents, and even parasites. These diverse growth forms represent adaptations to many diverse environments. Ghost gums of Australia (*Eucalyptus grandis*) are so big they rival the massively huge gymnosperms like redwoods for title of largest organism ever. The smallest flowering plant is water meal (*Wolffia*), a plant reduced to a little green floating oval barely 1 mm in diameter.

This diversity of form and life histories allows plants to adapt to diverse habitats. But plants have still another trick, which is their ability to alter their form in response to different environments, a phenomenon called developmental plasticity. For some reason, flowering plants seem better at this than any other land plant. All organisms have some developmental plasticity. If you were born in Lima, Peru, the high altitude would affect your development such that you would end up with a significantly larger heart. If you move to Lima as an adult, your body will adapt physiologically, but your heart will not enlarge. Developmental plasticity produces a variety of forms or phenotypes from a single genotype. A genotype that would grow as an upright tree in an inland location may grow as a short, gnarled shrub in a coastal or high altitude habitat where form is influenced by exposure to storms and strong prevailing winds. Developmental plasticity allows plants to alter their form in response to the local environmental conditions, providing them with a bit of wiggle room in a patchy environment. Humans often take advantage of this ability for our use and pleasure by pruning tree genotypes into hedges, espaliered fruit trees, and bonsai trees, miniature versions of themselves.

Of course, natural selection does act on plant form as well. A species living along an environmental gradient will often form a series of distinct forms, called ecotypes, each genetically adapted to slightly different conditions. Such variation can lead to speciation if the gene flow between populations is disrupted by pollinator behavior or a physical barrier. Scientists demonstrated that ecotypes were genetically based and not the result of developmental plasticity by growing them in common gardens. Plastic differences will not be manifest, but genetic differences will remain. The final form of a plant results from an interaction of both the genotype and the environment, but unless you do the experiment it is hard to factor their relative importance.

Different growth forms and life histories allow plants to adapt to different conditions. The prevalence of evergreen flowering trees shifts from quite common in the low latitudes to rather uncommon as you move to high latitudes where deciduous trees are predominate. Most of the spring wildflowers that inhabit deciduous[32] forests of the temperate zone are herbaceous perennials. These plants grow new aerial shoots each spring as soon as the weather warms enough to allow metabolic activity. Many woodland spring flowers are called ephemerals because such plants flower and fruit quickly in the spring before the leafy forest canopy fills in overhead, cutting off most of the light.[33] They store surplus food in perennial parts at or just below ground level, stems, roots, or leaves (bulbs), and at the end of the growing season, or at the end of their reproductive season, their aerial portions die back to ground level, disappearing until next year. Stored food allows them to grow quickly in the spring. A few surprise us by separating their vegetative growth from their reproductive growth so their flowers appear long after the spring's photosynthetic foliage has died back. Herbaceous perennials occupy positions in forest communities similar to both the past and present habitat of ferns. But ferns tend to be adapted to low light, while the rapid growth and reproduction of these ephemeral flowering plants take maximum advantage of spring's window of opportunity. Like ferns, flowering plants in evergreen forests are adapted to low light conditions. In some communities like wet or dry grasslands, and high latitude or high altitude alpine tundra, herbaceous perennial plants and extremely dwarf shrubs dominate.

Some flowering plants are annuals (plants that complete a seed-to-seed life cycle in a single season, sometimes just a few weeks). Annuals tend to be weeds, plants adapted to disturbed habitats by being able to quickly establish themselves, grow, reproduce, and disperse lots and lots of seeds before dying. All of their available energy is focused on reproductive output. Some are winter annuals, plants that sprout in the fall and grow a low rosette of leaves that overwinters. In the very early spring, these plants flower and quickly produce fruits and seeds every time the temperature gets above freezing in their ground-hugging microhabitat. The seeds then sit dormant until the fall. The herbaceous weed is essentially a new life form among flowering plants, although a few mosses and a fern or two, and some trees, can be considered somewhat weedy too. When it comes to reproduction, weeds are reproductive speed demons compared with all other vascular plants. The slowest of flowering plants, trees like magnolias or nutmegs, take 6 to 12 months from pollination to seed, and then the seedling must germinate and grow for another 5 to 10 years before becoming reproductive. This is about the same speed as many gymnosperms and explains why early angiosperms are envisioned not as trees but as smaller, faster-growing shrubby plants. Flowering plant forest trees appear later, gradually increasing in dominance throughout the Cretaceous (Crane 1987).

Disturbed habitats occur as "islands" or patches of different sizes scattered here and there within the fabric of communities, and these disturbances are important for maintaining species diversity. The "live fast, die young" life history of weeds allows them to find and exploit such patches quickly and briefly. Ultimately, such patches undergo a succession of organisms, gradually reverting back to the characteristics of the surrounding community, a shift in which weeds play an initial colonizing role. Weedy species have become extraordinarily successful in the last 10,000 years because the primary effect of human activities upon plant communities is to create disturbances; in fact, ecologically, typical agriculture is one big disturbance, so it should come as no surprise to learn that a number of weedy species have evolved in response to human agriculture. And then in a bit of turn-around-is-fair-play, humans have domesticated some weeds for our purposes, like lettuce or spinach. Weeds are good bets for human food because they usually do not expend their limited resources on toxic chemical defenses.

Very few land plants have reverted to an aquatic life style. A couple of mosses, a leafy liverwort, the clubmoss *Isoetes*, and the water ferns are aquatic, but no gymnosperms at all. Among flowering plants many have become aquatic, growing submerged or rooting in shallow fresh water and producing emergent vegetation or floating leaves, and some have become free-floating (water-lettuce, water hyacinth, duckweed). Only one land plant, eel grass, a flowering plant, has managed to readapt to the marine environment.

Vines are rare among ferns and very rare among living gymnosperms (only *Gnetum*) but common among flowering plants. A lowland, rain forest community in Queensland, Australia, is called a vine forest because it has over 300 species of vines (lianas) whose stems reach more than 1 cm in diameter. The unparalleled pleasure of working in this vine forest cannot be understood without experiencing the viney rattan palms that make walking through the understory a dangerous and maddening experience. They climb by means of grappling tree bark with long, thin, whiplike inflorescences bearing three rows of recurved spines rather than flowers, which is sort of like stringing piano wire festooned with fish hooks across the dimly lit forest understory. Locally rattans are called "lawyer canes" because once they get their hooks into you, they never let go.

Numerous flowering plants live as epiphytes, a lifestyle also fairly common among lichens, leafy liverworts, and mosses, clubmosses, and ferns, but one virtually devoid of gymnosperms. Epiphytes account for over 30,000 species of flowering plants. Among the best-known families for epiphytes are the orchids, aroids, cacti, and bromeliads (pineapple), the latter with over 2,000 species including the rootless, draping "Spanish moss," which is neither from Spain nor a moss.

More than a dozen lineages of flowering plants have produced parasitic plants, a habit adopted by no pteridosperms and only one gymnosperm. Some parasitic plants lose their photosynthetic ability; others remain at least partially green such that you would not guess they are parasites.[34] Obligate root parasites live underground and are never seen unless they flower (Indian pipes, beech drops), whereas some leafy parasites like mistletoe are readily seen growing on branches where they grow haustorial connections directly into their host's xylem. The legendary status of mistletoe may arise because they remain evergreen while their deciduous host trees drop their leaves. Curiously, the world's largest flower at nearly a meter in diameter belongs to a root parasite native to Indonesia's tropical forests, *Rafflesia* (rah-FLEA-zee-ah). A few flowering plants are saprophytes, obtaining nutrients from decaying material via a symbiosis with fungi.

In desert areas several groups of flowering plants have adapted with a number of different water-storing, water-conserving forms called succulents. Succulence means juiciness, and succulent plants have thick, fleshy leaves, stems, and/or roots that provide a large volume for water storage and a reduced surface area to limit water loss. Leaf succulents tend to be low-growing shrubs or herbs. In some cases the leaves are nearly spherical a form that maximizes the volume–to–surface area ratio. Most look rather glossy because of a thick waxy cuticle, and some succulent leaves even have dark-looking patches that turn out to be transparent windows that let light penetrate the thick leaf to reach photosynthetic tissue deep within. In the case of stem succulents, a thick, fleshy stem functions to store water, and often leaves are reduced in size, vestigial (barely there), or even lacking, which leaves the outer portions of the stem to function in photosynthesis. Root succulents store water and food in very large fleshy roots at or below ground, often producing nonsucculent aerial shoots that die back during the dry season.

Plants with unusually thick stems for their size are sometimes referred to as pachycauls, literally thick stemmed, just as elephants are called pachyderms, thick skinned. The best-known stem succulents are members of the cactus family, but several other families, primarily the euphorbs or spurges and milkweeds, have produced very similar forms. This is an example of convergent evolution where unrelated organisms look quite similar because they were subjected to similar selection pressures and have adapted to similar environments. Most people fail to distinguish the differences among them and call them all "cacti," or they treat succulents as if they were a taxonomic group, as if they are all related. Most people encounter succulents as pet house plants because their slow growth, interesting forms, and adaptations allow them to prosper in the xeric environment of our homes. Because pet plants are seldom seen in their native habitats, and they have been moved around a good deal, many people are not aware that cacti are strictly neotropical, confined to

North and South America. In Africa, succulent euphorbs and milkweeds occupy deserts and thorn scrubs, but while members of these families are found in this hemisphere, none are stem succulents.

Members of the cactus family demonstrate how plants adapt to diverse environments, but those adaptations are not without limitations. Most familiar cacti are spiny, stem succulents that totally lack regular leaves. Actually their clusters of spines are modified leafy shoots. Prickly pears have small, fleshy, vestigial leaves that fall off shortly after being produced. Usually you will see them only near the top of a young branch, which is flattened into a leaf-like "pad." Other members of the family are spiny shrubs or sprawling vines that bear regular, broad, albeit rather thick and leathery, leaves. Such leaves are tough but still often shed during dry seasons. These leafy "cacti" are the primitive members of the family. But still another lineage of cacti live in tropical forests, a substantially different environment from the rest of the family. These cacti often have broad, fairly thin, leaf-like but leafless stems, or slender pendent stems, and they seldom bear spines. Many grow as epiphytes and are known as orchid cacti because they grow like orchids, and some even have large showy flowers like the familiar Christmas cactus. So why do tropical cacti not have thick, leathery leaves? Tropical cacti have desert-adapted cacti as their common ancestors, and so lack the developmental ability to produce regular leaves. When genes controlling leaf production somehow got turned off, or were modified to produce spines, or were just lost, offspring dispersing into wetter habitats or who found themselves growing in a different climate could not simply turn the genes back on again because they needed broad leaves or because it suited them. However, natural selection could operate on variability in stem developmental genes to produce leaf-like stems. The evolutionary history of organisms often places constraints on the potential to adapt and means of adaptation.

Wood

Flowering plant wood is more complex than gymnosperm wood, which is composed strictly of tracheids as the primary support and conducting element. Therefore, the tracheid represents a form adapted to perform two different functions, which is not optimal for either. In flowering plants vessel elements and fibers both have evolved from tracheids (chapter 8), the former specialized for conducting and the latter for support. Some temperate zone flowering trees produce very-large-diameter vessel elements in the spring to accommodate the demands for conducting lots of water and stored materials to grow a new crown. But such large-diameter conducting cells are prone to air embolisms in hotter weather, so after the spring growth, summerwood with smaller diameter vessels is produced. Their smaller diameter makes them less susceptible to embolisms, but they are sufficiently large to meet the demands of a fully formed crown of leaves. Gymnosperm wood is called softwood because it generally is less dense and more uniform than flowering plant wood, but different angiosperms make both much softer and lighter wood (balsa) and very much harder and denser wood (lignum vitae). With a specific gravity of about 1.6, lignum vitae,[35] a South American tropical tree, sinks in water (SG = 1). So just like with leaves, the more complex wood of flowering plants produces many more types of wood adapted to various different habitats and growth forms. Although it takes a specialist, years of study, and a big reference collection, a great many woody plants can be identified on the basis of their wood.

Many general references will say that the presence of conducting vessels in their wood is an angiosperm character, but this is a bit oversimplified. A few angiosperms, including some in basal lineages (the Austrobaileyales) and some magnolialeans, have wood composed solely of tracheids, a condition presumably retained from gymnosperm ancestors. Among gymnosperms, gnetophytes have vessels in their wood; *Gnetum* even has vessels of the angiosperm type but presumably of separate origin. Vessel elements are widespread in the pteridophyte clade (Schneider and Carlquist 1999, 2000; Carlquist and Schneider 1997, 1999), but they are quite certainly of separate origins, making it quite certain that vessel elements in xylem are another iterative character, one that has evolved multiple times in multiple lineages.

Double-Jacketed Ovules

By this point you may be wondering whether flowering plants possess any novel shared character that define the lineage or whether this is another case where a traditional taxonomic group ceases to exist. Gymnosperms have a single jacket layer around their ovule. The flowering plant ovule has two jacketing layers, and those flowering plants that have only a single jacket layer are phylogenetically derived from those that have two. The inner jacket, presumably inherited from a common ancestry with gymnosperms, creates the tiny hole through which pollen tubes grow. Together the jacket layers develop into the seed coat after fertilization. While some flowering plants develop a two-layered seed coat, an outer fleshy layer and an inner stony layer (e.g., pomegranate, passion fruit, kiwano), gymnosperms manage the same thing with but a single jacket layer. While a second jacket provides the flowering plant clade with a novel derived character, the adaptive value and the origin of the second jacket layer is uncertain.

Double Fertilization and Endosperm

Most general references will tell you that double fertilization is a novel shared character defining angiosperms as a clade, but this is technically incorrect. While both of the sperm produced by each male gametophyte are involved in a fusion event, the development of endosperm from the second fertilization is the novel shared character of angiosperms. In flowering plants one sperm fuses with the egg cell producing a zygote and the other sperm fuses with one or two other nuclei, both mitotic sisters (genetically identical) to the egg, producing a specialized 2n (diploid) or 3n (triploid) tissue called endosperm. In all other land plants the embryo develops embedded in the tissues of the maternal gametophyte, but in angiosperms the embryo develops within and receives nutrition from endosperm. Since endosperm is a genetic twin to the developing embryo, its relatedness to the embryo is 100% compared with a 50% maternal relatedness.

Does endosperm enhance embryonic development by reducing any risk of genetic incompatibilities between mother and offspring? Perhaps, but production of endosperm has one fairly clear reproductive advantage. Angiosperms do not start packing food reserves into a seed until after a successful fertilization, whereas in gymnosperms, a fairly large female gametophyte with food reserves had to be mature before fertilization could take place (Fig. 10–11). The typical female gametophyte of a flowering plant is not much larger than the megaspore, while the female gametophyte of a cycad has a volume about 600 times larger, and yet it may not get fertilized. Because of endosperm angiosperms waste very little energy on ovules lacking an embryo. The cost of big female gametophytes is more easily borne by big, long-lived, woody plants, like most gymnosperms, but the cost of loading food into ovules with no embryo would be proportionately much greater for smaller shrubby and herbaceous plants. Some evidence suggests that angiosperms can even select among their many offspring, provisioning some seeds and starving others.

While checking on the veracity of some classic observations, some ANA grade angiosperms were found to have female gametophytes consist-

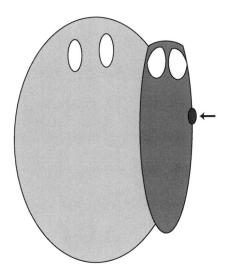

Fig. 10–11 Size comparison of mature female gametophytes of gymnosperms and angiosperms: cycad, pine, and lily (l-r). The female gametophyte of a cycad is largest (light gray), showing two archegonia. Partially superimposed on that is the female gametophyte of a pine (medium gray) with two archegonia. The female gametophyte of lily, a flowering plant, is tiny in comparison (dark gray, arrow). (Image source: The Author)

ing of just four nuclei (Friedman and Williams 2003, 2004; Williams and Friedman 2002, 2004) (Fig. 10–12), which results in a diploid endosperm. A developmental doubling of this tiny reduced female gametophyte results in a second endosperm nucleus, both of which fuse with one sperm nucleus, producing the more typical triploid endosperm (n + n + n = 3n). In some cases, a second twinning even leads to a female gametophyte with 16 nuclei but still only triploid endosperm forms. The female gametophyte usually illustrated in textbooks is the result of three successive mitotic divisions resulting in eight nuclei. Three nuclei become cellular and a fourth nucleus occupies the remaining megaspore cytoplasm. One of the three small cells functions as an egg, and the fusion of a second sperm with the fourth nucleus produces diploid endosperm, an exact genetic twin of the zygote. No archegonia are present and any haploid cell can function as the egg. One indication that people were uncertain of the nature of the female gametophyte is its common name, the "embryo sac," a term most often seen on microscope slide labels. Today, in retrospect, this seems a bit of a disrespectful term for a female gametophyte, sort of like a botanical equivalent of calling a plant embryo's mother an old bag.

Many gymnosperms also produce two or more sperm, and double fertilization has been found in gnetophytes and, to date, two conifers (a fir and an arborvitae [Friedman and Floyd 2001]), so in and of itself, double fertilization is not unique to angiosperms. Double fertilization in *Ephedra*, the basal lineage of the gnetophytes, is very much like that of

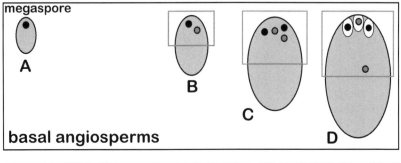

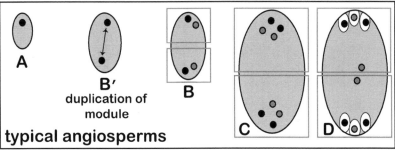

Fig. 10–12 Female gametophyte development of some basal angiosperms (top) and most other angiosperms (bottom). A. Both sequences start with a megaspore. B-D (top). The female gametophyte develops by two successive mitotic divisions to produce four nuclei, and three of the four form cells of which one will function as an egg; the remaining nucleus will fuse with the second sperm nucleus, resulting in double fertilization and diploid endosperm. B' (bottom). An additional step consisting of one additional mitotic division "twins" the megaspore nucleus, and then development of each module (gray boxes) proceeds normally (B-D), except the mature female gametophyte (D) now consists of seven cells and eight nuclei. The twinning results in two central nuclei that will fuse with the second sperm nucleus, producing triploid endosperm. (Image source: After Friedmann and Williams 2004)

the conifers, another fact that suggests the phylogenetic placement of gnetophytes near the conifers is correct. In gymnosperms double fertilization produces two zygotes, but neither a second embryo nor a diploid endosperm develop. Presumably double fertilization began as a fail-safe, a redundancy mechanism, to assure a viable zygote, but where only one of the twins safely develops. If, however, the abortive development of a second embryo became a nutritional asset for the remaining embryo, then natural selection could account for the nonembryonic development of a diploid endosperm.

As with other seed plants, the much reduced, endosporic gametophytes of angiosperms are minimal vestiges of multicellular organisms, a culmination of a trend observed throughout the history of embryophytes (Fig. 10–13) (Niklas 1997). The male gametophyte of angiosperms is highly uniform. The microspore nucleus undergoes two successive mi-

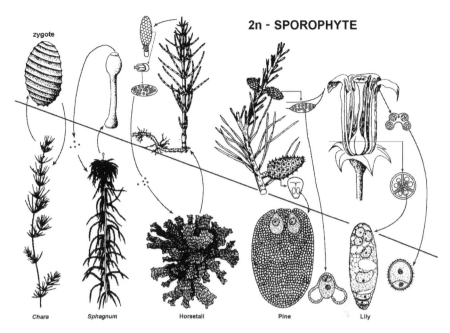

2n - SPOROPHYTE

zygote

Chara   Sphagnum   Horsetail   Pine   Lily

n - GAMETOPHYTE

Fig. 10–13 General trend in the relative sizes of the sporophyte and gametophyte generations in land plants and their immediate common ancestors. As a general phylogenetic trend, the diploid sporophyte phase (top) has become larger and more dominant, while the haploid gametophyte phase (bottom) has gotten smaller and more dependent. (l-r) *Chara*—charophycean green algae; *Sphagnum* moss—bryophyte; horsetail—pteridophyte; pine–gymnosperm; lily—flowering plant. The diploid phase of *Chara* consists only of a zygote. Downward arrows crossing the line represent meiosis. The sporophyte of mosses is small, simple, and dependent on the gametophyte. In the horsetail, pine, and lily the sporophyte is the dominant and familiar phase, whereas the gametophyte of *Chara* and *Sphagnum* is the common and familiar phase. The gametophyte of horsetail is inconspicuous and free-living with a bryophyte level of complexity. Both male and female gametophytes of seed plants (pine, lily) are very small in size and endosporic and therefore seldom seen other than in a botany laboratory classroom (female gametophyte to the left, male gametophyte [pollen] to the right). (Image source: After Niklas 1997)

totic divisions, the first resulting in a vegetative nucleus and a generative nucleus. The former is debatably a vegetative nucleus because it functions to direct the growth of the pollen tube as the tube nucleus. The generative nucleus divides, producing two gametes that are basically just nuclei that have only a small volume of cytoplasm and no flagella. Without prior knowledge of the history of microgametophytes, without knowledge of the history of the embryophyte life cycle, angiosperm pollen would be almost impossible to interpret as a haploid male organism.

Endosperm and the ovule's double-jacket layers are the only novel shared characters that distinguish angiosperms from gymnosperms. In terms of wood, leaves, gametophytes, and reproductive structures, angiosperms demonstrate a continuum with gymnosperms, which is rather expected. However, in terms of wood, leaves, reproductive structures, life histories, and life forms, angiosperms display considerably greater diversity and many adaptations. But unanswered as yet is this: What about angiosperms allowed them to diversify so rapidly into so many different forms? Why is their development so much more plastic than that of gymnosperms? Perhaps this question will be answered if and when the gap between molecular developmental genetics and morphology is closed, and the effort to do this is underway. This problem may be all the more difficult to solve because no living gymnosperm has a direct common ancestry with angiosperms.

In the previous chapter several groups of fossil gymnosperms were introduced as candidates for being common ancestors with angiosperms. Features shared in common among basal ANA lineages suggest a starting point from which to derive other flowering plant features, but while a clearer picture of angiosperm phylogeny is emerging, a gymnosperm sister group remains in doubt. Reconstructing seed plant phylogeny is rather like having a double-sided jigsaw puzzle and no pictures on the box to assist the sorting and reconstruction. Even when a piece fits, you are uncertain which side is up, angiosperm or gymnosperm. Paleobotany is a perfect career for real puzzle fans who like such frustration.

Many features of flowering plants have contributed to their ecological and evolutionary success and the success of many animal groups, too. But this story is far from over. The initial angiosperm takeover and diversification took place in the Cretaceous over 65 mya. Dinosaurs were still the dominant animals; mammals were small insect-eaters. The Earth was quite green with ferns, conifers, and flowering plants, but very little of it would look familiar. And then a bit of rock, a piece no bigger than a kilometer or so in diameter, landed on the Yucatan peninsula, setting off a sequence of events that led to the most recent (not counting human activities) environmental disaster. The extinction of the dinosaurs at the end of the Cretaceous marks a major turning point in the history of life, and while the impact on the floristic composition of the world was far less drastic, many changes would have to be forthcoming before arriving at a familiar-looking green world. In particular, even after the end of the Cretaceous, flowering plants remained largely woody; most herbaceous plants had not yet appeared. These floristic changes are quite significant as the interactions and influences between flowering plants and animals continued. Most recently, a particular interaction developed between certain flowering plants and a unique primate, and the rest is history.

ELEVEN

All Flesh Is Grass

Wherein the development of modern vegetation and recent interactions, like agriculture, and their impact on both plants and humans are examined.

Acorns were good until bread was found.
—Francis Bacon

Who would have thought that a kid fascinated by comic books and science fiction would become a scientist? Thinking about people in fantastic places and different times, even in other worlds and galaxies, spurred the imagination. But you grow up, and after a while you lose interest whenever whatever was depicted was clearly impossible, and now plants and animals in the "wrong place" ruin lots of movies. A favorite science fiction comic book was *Turok*, which was about a couple of Native Americans who stumble upon a "lost world" of dinosaurs and "cave men" leading to many harrowing adventures (Fig. 11–1). The idea of humans and dinosaurs co-existing is a popular theme in science fiction; it remains popular among some creationists who claim to have found human footprints alongside dinosaur footprints,[1] thus rendering geological time scales false, but how having dinosaurs around helps their case puzzles me. But humans and dinosaurs existing together never happened by a long shot, and the problem is one of understanding the time scale and how our own species' history fits within it. Nothing in our human existence assists us in really understanding the time scales involved, and even though millions and billions of years have been tossed around throughout this

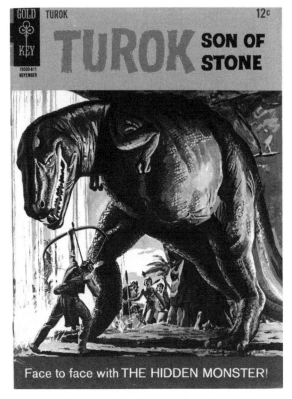

Fig. 11–1 Dinosaurs and humans lived together in the same geological period only in science fiction stories. (Image source: cover *Turok, Son of Stone* #54, *Dell/Gold Key Comics* November 1966)

book, somehow a million years and just how long it is to living organisms just does not register. So it is no surprise that we cannot imagine easily just how different and unfamiliar, with a few exceptions, the world still was as this last chapter opens upon the Cenozoic, even after flowering plants have established their dominance among land plants.

The Cenozoic began 65 million years ago as the Cretaceous ended and that eon brings us up to the present day (Fig. 11–2). This knowledge generates two diametrically opposing thoughts: 65 million years is a really, really long time, but when considering the whole geological record of life, the Cenozoic only represents the last little bit, just 1.5%, of Earth history. And of course, both perspectives are quite correct. This means both that a lot of ground still has to be covered, literally and figuratively, and that the history of green organisms is nearly done, which works out nicely because this is the last chapter.

Ceno- is a prefix meaning common, and of course, *-zoic* means animals, the eon of common animals, but from the green perspective, the Cenozoic should be called the Anthofloric, the eon of flowering plants. Speaking now as a member of *Homo sapiens*, the Anthofloric is particularly significant to us because not only is our species wholly a product of this eon, but also over the very recent past, our activities have greatly influenced the green world, particularly the flowering plants. At no time since a cyanobacterium took up residence inside an accommodating host cell has any single species ever had a bigger impact on the world's biosphere than we humans. Some species, some plants, have fared quite well; others, not so well.

Like several other geological boundaries, a considerable discontinuity marks the K-T (Cretaceous-Tertiary) boundary.[2] In this case, the K-T boundary marks the most recent mass extinction event,[3] the one that marks the disappearance of dinosaurs; this event might be linked to a meteorite impact on the northern margin of the Yucatan peninsula marked by the Chicxulub crater.[4] Diverse evidences suggest this impact, or massive volcanism, caused a 10° C (18° F) drop in average world temperature, a sudden, prolonged winter produced by the large amount of dust, debris, and smoke added to the atmosphere following the impact, a solar umbrella that reflected a lot of sunlight back into space. Compare this to a more recent event, the explosive eruption of Krakatau, a volcano in Indonesia, on August 27, 1883, which noticeably cooled the Earth's climate an estimated 0.3° C, but as big as this volcanic eruption was, the Chicxulub crater, at 180 km (120 miles) wide and 1,600 m (1 mile) deep, indicates how much bigger this event was. Following the Chicxulub impact, a lot of forest communities were destroyed, those near the impact by fire and those further away by the prolonged cold. Spores in post-impact sediments indicate that ferns flourished where forests had previously been. But plant groups apparently escaped major extinctions, and early Cenozoic forests recovered with their late Cretaceous plant diversity largely intact. Nonetheless, some changes took place—for example, Araucarian conifers, such as monkey-puzzle trees, disappeared from the Northern Hemisphere, while deciduous conifers related to bald cypress became abundant and widespread (Graham 1993).

For mammals, the K-T boundary can be viewed as a pretty fortuitous event. Flowering plants first appeared in the early Cretaceous when the dinosaurs were still at the peak of their dominance and diversity. Mammals were still small, insignificant insectivores, something like shrews, active nocturnally in forest canopies and understories. The largest mammal alive when dinosaurs still roamed the Earth had a mass of about 10 kg

(22 pounds), smaller than my daughter's Maine coon cat. Almost all of the animals and plants that we consider common arose during the Cenozoic, and in the absence of dinosaurs, mammals underwent a series of radiations producing, bit by bit, replacements for all of the vacated biological roles: various herbivores, various carnivores, various flying mammals, various aquatic mammals, and so on. Around 34 mya, the largest terrestrial mammals that have ever lived appeared and they were relatives of sloths and rhinoceroses, giant herbivores that weighed 15,400 kg (17 tons)! And still no humans were around.

At the beginning of the Cenozoic, the world would have still looked very strange. Only a very few species like the cinnamon fern have existed long enough to have been around back then. A few genera are of great antiquity, like dawn redwoods (*Metasquoia*) and horsetails (*Equisetum*), and they would have been recognizable, although different species existed then. A number of familiar genera of deciduous trees appeared in the Eocene: maple, hackberry, poplar, sumac, and sweetgum, and at this time deciduous forest still occupied many areas where grassland communities exist today. Even still the natural world would not have begun to look familiar until about 20 million years ago in the Oligocene (Potts and Behrensmeyer 1992; Graham 1993). All of the plant families recognized in the fossil record of pollen at 20 million years ago are still alive today. All of the plant genera recognized in the fossil record at 10 million years ago are still alive today, but still those would not have been our familiar species.

The fossil record indicates that most modern orders of mammals appear during a relatively brief period from 56 to 34 mya in during the Eocene (Fig. 11–2). Of course those orders comprising basal lineages and possessing many ancestral characteristics (Monotremes—echidnas and platypuses, Marsupials—possums and kangaroos, and Insectivores—moles, shrews, and hedgehogs) have much earlier appearances, around 100 mya or more. Our own particular order dates back to about 60 mya, the age of the oldest confirmed primate fossils. Early primates evolved from ancient nocturnal insectivores and resembled lemurs or tarsiers. Based upon their characteristic features (grasping hands, rotatable shoulders, binocular vision[5]) such animals were largely arboreal, living in the widespread tropical or subtropical forests. Primates in general have relatively large brains in comparison with their body size, but the first hominids, those primates in the lineage that includes humans, did not appear until the Pliocene some 5 to 1.8 mya.

Still other geological events have helped shape the recent history of organisms. As the continents moved into their current positions, floras

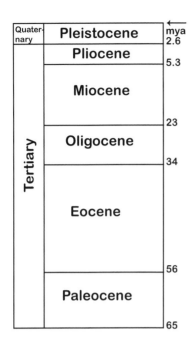

Quater-nary	Pleistocene	← mya 2.6
	Pliocene	5.3
Tertiary	Miocene	23
	Oligocene	34
	Eocene	56
	Paleocene	65

Fig. 11–2 Geological timescale of the Cenozoic. This eon consists of just two periods: the Tertiary followed by the Quaternary, which is ongoing. The Tertiary is subdivided into 5 epochs: Paleocene, Eocene, Oligocene, Miocene, and Pliocene. The Quaternary consists of the Pleistocene (the ice ages) and the Holocene, which began just 10,000 years ago. Even on this scale the Holocene is covered by the thickness of the line at the top (arrow). Time scale is in millions of years ago (mya). (Image source: After Geological Society of America's 2009 time scale, http://www.geosociety.org/science/timescale/timescl.pdf)

and faunas of Laurasia and Gondwana, the great northern and southern supercontinents, began exchanges along island archipelagos like those between Southeast Asia and Australia, and across land bridges like the Isthmus of Panama, and where the horn of Africa made contact with Asia, and where India plowed into Asia. Of course, not all organisms are good dispersers, so the distributions of some organisms still mark the old boundaries.

The breakup of the supercontinents also greatly changed worldwide climate and weather patterns. Presently those of us living here in Midwestern North America are quite aware that just 18,000 years ago the last ice sheet of the Pleistocene[6] glaciations covered two-thirds of Wisconsin and the northern third of Illinois. A colleague drilling a new well about 20 miles north of my location near the terminal moraine hit some wood about 30 feet below ground level, and it could be identified as a spruce even though the fine details that would identify the species had been lost to partial decomposition. The nearest spruces now grow about in the middle of Wisconsin 150–200 miles to the north. Ten miles south of my office an island of forest exists in a sea of prairie grasslands, or at least what was prairie before being converted to farmland, and this woodland is home to paw-paws (*Asimina triloba*), the northernmost member of the largely tropical custard apple family, and snow trilliums. No paw-paws or snow trilliums are found in forests north of the terminal moraine in this region.

Clearly a lot has changed, especially so in the last 10,000 years after humans invented a new way of life, agriculture. The communities of tall grass prairie that covered this area following the glacial retreat built up deep topsoil, which led to very productive agriculture once someone[7] invented a means of plowing it. This conversion, which began only 150 years ago, greatly affected the Midwestern landscape of North America. This emphasizes that some of the most dramatic and rapid changes in Earth history have just happened, or rather, are still happening. Without question the Cenozoic represents a significant period of evolutionary change, and that includes the first appearance of some very prominent and familiar groups of plants.

Early Cenozoic plant diversity, as read by the diversity of fossil pollen, recovered to late Cretaceous levels within 1 to 2 million years of the K-T boundary. After the sudden and brief descent into winter at the end of the Cretaceous, the global climate gradually underwent a long-term warming. The end of the Paleocene and early Eocene were characterized by a period of extreme global warming. The release of methane, a powerful greenhouse gas, from deep sea reservoirs is the primary suspect for causing this episode of global warming. At the beginning of the Eocene (56 mya) the climate was so warm that the equator to pole thermal gradient was only about one-half of the gradient we have today. This means that warm temperate vegetation grew as far north as the Arctic Circle, forests that included many familiar tree genera, beech, elm, chestnut, magnolia, and birch. Imagine that, magnolias in Alaska and Canada! Tropical forests extended as far as 50 degrees north and south from the equator, the latitude of Confusion Bay in northern Newfoundland. This indicates that the most drastic differences occurred at the very high latitudes where on average temperatures were some 30° C warmer then, while in the present day tropical latitudes, on average temperatures were only 5° to 10° C warmer then (Graham 1993). Such estimates of climatic changes are based upon many diverse forms of data, and they include such things as fossil woods that show, or do not show, distinct seasonality, and fossil leaves whose size (larger), margins (smooth), and shapes (drip tips, Fig. 10–9 A) are typical of those found today in wet tropical climates.

Throughout the Eocene the climate gradually cooled from that thermal maximum. Interestingly enough, one hypothesis suggests that a bloom, a massive population explosion of a plant something like the mosquito fern *Azolla* (Fig. A31 B), took place in the ancestral Arctic

Ocean, then a landlocked, freshwater sea. This much plant growth absorbed and held enough carbon to reverse the greenhouse climate. As the poles cooled, they started becoming the frozen ice boxes of today. The release of carbon dioxide stored in plant biomass over a very short period of time is one of the reasons biologists are so concerned about the impact of present-day deforestation on a worldwide basis. Trees represent huge reservoirs of carbon, and normally everything except the energy gets recycled in a regular pattern of growth, reproduction, death, and decay. Trees get older and larger, and eventually they encounter an environmental mishap or simply begin to fall apart as they get too large for the strength of the material they are constructed from, wood. When old trees fall, the gaps they leave in the forest canopy provide nurseries for young trees, a cycle that can take 100 to 300 years. So when forests are deliberately burned to remove the vegetation, or the trees are cut for lumber or paper, all of that stored-up carbon is being released in a relatively short time frame and forest regrowth, deprived of so many nutrients, remains slowed or even problematic if much soil erosion has taken place, so a new carbon reservoir may not replace the one destroyed.

The continental movements continued to break up former supercontinents Laurasia and Gondwana as the major land masses moved toward their present positions. North America, Greenland, and Eurasia, which had formed Laurasia, drifted apart as the north Atlantic opened up. India made contact with Asia, and the Himalayas began forming as the collision folded the continental masses into each other. Australia continued moving northward toward Asia, gradually raising the volcanic archipelago of islands that now form biological stepping stones between these two long separate floras and faunas. Organisms that were good at island hopping began moving both directions. Organisms that were poor at dispersing across water did not, so a biological disjunction still exists between floras and faunas of Laurasian and Gondwanan origins. This latter divide in the distributions of organisms became known as Wallace's line after the great naturalist Alfred Russell Wallace who discovered it, the same fellow whose letter to Darwin about natural selection spurred Darwin to finish his outline on the same topic, the *Origin of Species*. Wallace's line runs more or less north-south between Borneo and Sulawesi to the North and curving a bit between Bali and Lombok to the south.

The breakup of Gondwana had a major impact on the climate of the Northern Hemisphere. The separation of South America and Australia from Antarctica opened seaways that allowed cold water currents to flow northward, producing more high-pressure systems in the equatorial regions, which resulted in drier climates in the Northern Hemisphere. The uplift of the Rocky Mountains further altered the climate of North America, just as did the uplift of the Himalayas, by creating rain shadows that produced drier climates to their east and north, respectively. In the high latitudes polar ice became a permanent fixture, and as the tropical and then temperate floras retreated, coniferous forests spread, and when the climate became too cold for coniferous trees, tundra appeared. The later uplift of the Cascade and Sierra Nevada ranges in North America during the Miocene and Pliocene strengthened the rain shadow effect, and arid-adapted vegetation spread across the great basin between these mountain ranges and the Rocky Mountains.

THE RISE OF GRASSES AND SUNFLOWERS

Biologically the big events of the late Tertiary (the last 34 my) are the appearance of grasses (Eocene), the first appearance (Oligocene) and diversification (Miocene) of sunflowers, and the rise of grassland communities, all accompanied by a changing climate (Potts and Behrensmeyer 1992). The world's major grasslands include the North American prairie, the African veldt, the Asian steppes, and the South American pampas. Grasses (~10,000 species) and sunflowers (~23,000 species) are two of the biggest, most diverse, and most important families of flowering plants, and they are two of the major biotic components of grassland communities. Grasses are the most important primary producers of grassland communities, the trees of the prairie "forest," generating the bulk of grassland biomass, and so grasses represent a great opportunity for herbivores. In turn, adaptations for grass grazing greatly influenced the evolution of these herbivores, and that in turn affected the carnivores that preyed upon them. This geological era also featured the rise of the very large brown algal seaweeds and the appearance of kelp "forests" in the deeper portions of the coastal zone, and of course, this afforded ever more opportunities for animal diversification.

The climatic circumstances that resulted in grassland communities involved a long-term climatic cooling trend that resulted in increased seasonality and increasing aridity, producing more distinct and severe differences between winter-summer in temperate regions and between wet-dry in subtropical/tropical regions. Both types of increased seasonality favor grasses over woody plants, which allowed the expansion of grasslands into areas where forests once grew. Inclement cold and/or dry seasons cause the aerial shoots of herbaceous perennials like grasses and sunflowers to die. Large stands of dry grassland vegetation lead to a fire ecology that regularly burns off the standing biomass of grassland communities. Far from being terribly destructive, fire returns mineral nutrients to the soil and stimulates regrowth of new vegetation of higher nutritional value, a boon to herbivores. Fires help maintain grasslands by keeping invasive woody plants at bay, often preventing reestablishment of a forest even if a milder and wetter climate would permit it.

Neither fire nor grazing damages the growing tips (meristems) of most grasses because they are at or near ground level. The aerial foliage of most herbaceous perennials in grasslands can be grazed, or burned, without any damage to the meristems, especially after the aerial foliage has died. Undamaged meristems and intact root systems where food reserves are stored allow grasses and other herbaceous perennials to renew foliage quickly, to regrow annually or seasonally, and in fact, grazing may actually stimulate growth.[8] Herbivory may in fact reduce the reproductive potential of grasses and herbs, although it depends upon when in their growth cycle they are eaten, but since these long-lived perennials survive and regrow, many more opportunities for producing offspring will be forthcoming. Some studies have even found reproductive overcompensation following herbivory, meaning grazed plants actually produce more offspring than ungrazed ones! But fire and browsing do damage many woody perennials, so fire ecology eliminates many trees and shrubs and prevents their reinvasion into grassland communities where their size would give them a competitive advantage.

Grasses produce diffuse masses of roots that deeply penetrate and bind soil. Grasses in general do not use toxic chemicals to defend themselves against herbivory because their mature foliage is tough, fibrous, and of low nutritional value. Herbivores must eat large volumes of such vegetation to get adequate nutrition. Big mouthfuls and a large, long gastrointestinal tract are needed to consume enough grass and retain it long enough for digestion[9] to take place. Thus grassland grazing selects for larger herbivores, and this in turn selects for larger carnivores. Small woodland browsers turn into taller grassland grazers and runners—for example, bovids of all sorts, bison, buffalo, gaur, and bentang, the wildebeest and other antelopes, and equids, horses, donkeys, and zebras. Grazers' teeth become modified into cutting incisors and huge flat, ridged molars for grinding the tough grass leaves. Moving up the food chain, carnivores develop the size, speed, and hunting behaviors to capture such large prey in a variety of ways, including getting bigger too, although pack hunting behavior allows smaller predators to succeed in capturing larger prey through numbers and cooperation. This was not the first time such an evolutionary cycle took place. In general, gymnosperm foliage is high in bulk and low in nutrition, and since lots of them are trees with crowns of vegetation held aloft, herbivores got bigger, a lot bigger, sauropod big, which had quite an influence on the predators, producing carnivores like *Tyrannosaurus rex*, but their story ended where this chapter began.

The most recent event in the rise of grasses is the evolution of a new photosynthetic pathway. Tropical grasses evolved a new autotrophic pathway called C_4 (see-four) photosynthesis. Without going into the specific details, carbon is fixed by producing a 4 carbon molecule (thus the C_4) called malate (MAY-late). The malate is translocated to specialized mesophyll cells that surround vascular bundles (bundle sheath cells) where the malate is partially respired, releasing carbon dioxide and a 3-carbon molecule that is cycled back to regular mesophyll cells. Here the carbon dioxide is fixed by the standard photosynthetic pathway. However, an enhanced carbon dioxide concentration in the bundle sheath cells makes rubisco[10] more efficient because there is less competition from oxygen. This increased efficiency makes the biggest difference at higher temperatures (and lower carbon dioxide concentrations), so C_4 grasses are at a great advantage in subtropical and tropical climates. As a result such grasses dominate warm climate grasslands. The ecological expansion of C_4 grasses is only the latest event in the rise of grasses, having taken place in the late Miocene and early Pliocene 4 to 8 mya (Osborne 2008), whether the primary driving factor was climatic change or declining carbon dioxide concentration, both of which favor the more efficient handling of carbon dioxide provided by C_4 photosynthesis. C_4 photosynthesis has apparently evolved numerous times because it is found in several lineages unrelated by recent common ancestry; in total it occurs in over 400 genera of monocots and more than 90 genera of "dicots."

The best known tropical C_4 grass is maize (*Zea mays*), better known in North America as "corn."[11] C_4 photosynthesis is why maize grows so well in the hot summers of the Midwestern "corn" belt. How maize came to be a crop in this location, displacing the native prairie grassland, is an example of the most recent interaction between flowering plants and animals, in this case, with that animal called *Homo sapiens*. No question about it, this interaction can be viewed as mutualistic since it has enhanced the success of both parties, but it occurred at the expense of prairie species. Although humans cannot consume grass foliage, the one-seeded fruits of grasses have become our species' nutritional mainstay, the foundation of our agriculture, a necessity, whether flakes, bread, or beer.

DESERTS

The increased aridity and rain shadows caused by mountain building during the late Tertiary produced some even drier habitats in some locations, giving rise to deserts. In North America weather moves generally from west to east. The movement of the North American continent has been largely westward where, as it rides up over the descending oceanic plates, mountains and volcanoes rise. These mountains force air masses upward and produce rain. As the air passes the coastal ranges it loses much of this moisture, which is why Seattle is so wet and Salt Lake City so dry. Then more moisture is lost as the air passes up over the Sierra Nevada range, and by the time it gets to the great basin, not much moisture is left. In such areas less than 25 cm of precipitation can fall in a year and such areas are called deserts.

Mention desert and people immediately get an image of cacti, which is a family of flowering plants largely adapted to desert habitats. Quite a number of plants have similar appearances because they have similar adaptations, an evolutionary convergence in form and function. In general many such plants are termed succulents, a term that means juicy, a reference to soft water-storage tissues. Plants increase their capacity to store water by increasing the volume of leaves, stems, or roots; they get thick and soft because parenchyma has thin cell walls. In general the leaves of desert plants get smaller because sunlight is not in short supply and the reduced surface area limits water loss. The general similarities among succulents are such that many people mistakenly think all such plants are related, often using the term "cactus" to mean succulent, but a number of plants in the lily,[12] stonecrop, living stone, spurge, and milkweed families are succulents. Some succulents develop massively thick stems for their size, pachycauls, meaning thick stemmed.

Not all deserts look like that image because some are cold rather than hot. In western North America cold deserts have sagebrush-dominated communities that intergrade with the dry grasslands. Not all cacti look like desert cacti, either. The basal members of the cactus family are spiny shrubs with thick leathery leaves, and they live in short-statured woodlands usually with distinct wet and dry seasons. In such areas woody plants often drop their leaves in the dry season. Prickly pear cacti have short-lived vestigial leaves, but photosynthesis is primarily conducted by thick, but flattened "cactus pads," modified branches. Thus succulent stems serve to store water and conduct photosynthesis. These stems range from thick stout columns to almost globose stems, shapes that conserve water loss by presenting a limited surface area. The rest of the cacti have no leafy organs at all except for those modified into spines. But still some other cacti have readapted to wetter tropical habitats and these have slender, softer stems or stems flattened into almost leaf-like shapes. These so-called orchid cacti mostly live as epiphytes, growing upon tree trunks and sometimes living high in the canopy. Whatever genetic changes suppressed leaf development, the benefit that could be gained (the "need") for broad leaves did not result in a reversal of those changes, so natural selection acted upon stem development, resulting in leafy stems. Many epiphytes have some succulent features in spite of ample rainfall because they grow in exposed locations upon substrates that hold limited amounts of water.

The main idea here is that increasing aridity produced new habitats and the very adaptable flowering plants took advantage of these areas. Both grasslands and deserts continue to occupy a major proportion of the terrestrial environment to the present day. And our familiar groups of animals evolved in reaction. Deserts intergrade with grasslands, which intergrade with savannas, open grassy forests, and then finally with forests, a range of habitats along a climatic gradient largely based upon precipitation.

FLOWERING PLANTS CLASSIFICATION AND PHYLOGENY

Our present-day biotic world is dominated by flowering plants, some with ancient heritages, many more with histories no older than the mid-Tertiary. Naturally these are the plants with which we are most familiar, and yet, still the diversity of flowering plants remains unknown to most people. Taxonomy is the oldest component of the biological sciences, and often the most maligned, probably because so many users of taxonomy are not practitioners of taxonomy. Linnaeus became the father of plant taxonomy by consistently using binomial nomenclature, species names, those consisting of a genus and a specific epithet[13]—for example, *Helianthus annuus*. Similar species were grouped into genera, a collective taxonomic category. And as more general similarities were observed and organized, higher collective categories were used and eventually codified into a hierarchy of species, genus, family, order, class, phylum, and kingdom, and as discussed in the very beginning, concepts about these collective categories have changed.

After Charles Darwin many taxonomists decided that classifications should attempt to reflect the actual evolutionary history of the organisms, which is to say that all the members of the plant kingdom, the magnolia family, or the genus of magnolias, each taxonomic grouping, should have a common ancestry. This has proven harder than anyone imagined, but it has made taxonomy a lively science, which has undergone a renaissance since the application of molecular data began. And so it behooves us to consider the broad outlines of angiosperm classification before considering the most recent chapter in plant animal interactions.

In the course of this narrative so far, several well-known and familiar taxonomic categories have been falsified as classification hypotheses of common descent. There is no taxonomic category for all nonvascular plants (bryophytes), for all non–seed-producing vascular plants (the former pteridophytes, herein redefined), nor for all gymnosperms. And presently our phylogenetic understanding of flowering plants, which improved so vastly in just the years since my career began, is making a great big mess out of traditional classifications. A pragmatic approach is needed until things get sorted out. What this means is that to some extent flowering plant classification will be used in a "business as usual manner" because this is what is most familiar to readers such as yourselves. Many of these classification categories remain useful as a means of labeling even if phylogenetic hypotheses have falsified some of these groupings. Another problem exists for which there is no good solution. A bit of explanation will make this plain enough.

The 250,000 or so named species of flowering plants are organized into 400 to 500 families. The common names of many families are familiar: lily, orchid, ginger, rose, bean, grass, gourd, mint, nightshade, parsley, aster/sunflower, but not all of their family members would look like those with which you are familiar. Some big, familiar families are now known to consist of an assemblage of lineages—for example, lily and snapdragon, and sometimes bits of other families too, such as "verbs," mints and verbenes. Gradually a number of smaller and better-defined families are taking their place, but as yet most of these names and groupings remain unfamiliar to everyone but the taxonomists who specialize in these groups. This is going to take some adjustment because many traditional family concepts and the recognition of family-level characteristics remain important and useful in identification.

Other higher classification groupings are also being greatly modified. Indeed, the entire classification hierarchy may find itself in jeopardy because this codified hierarchy is no longer adequate to describe the phylogenetic pattern of diversity known to exist. In most traditional classifications angiosperms are treated as Magnoliophyta, a Division (= phylum) of the Plant Kingdom. Such a classification does not accord well with current phylogenetic hypotheses that have angiosperms nested in a spermatophyte clade, which is nested in a ligniophyte clade, which is nested in a megaphyllous plant clade, which is nested in a tracheophyte clade, which is nested in a polysporangiate clade, which is nested in an embryophyte clade, which is nested in a streptophyte clade, which is nested in a chlorophyte clade, which is nested in a bikont clade, which is nested in a eukaryote clade. That's quite a hierarchy to exist between phylum and kingdom. At what level should the label kingdom be applied? Wherever it is placed, the problem remains of what to label everything else. Various authors at various times have accorded all green photosynthetic organisms, all chlorophytes,[14] all embryophytes, and all tracheophytes phylum/division status, but the same problem remains. The pragmatic solution is to continue treating angiosperms as if a phylum and using the traditional hierarchy of classes, orders, and families of flowering plants in spite of this phylogenetic problem. Sorry, but it is not going to be resolved here.

The hypothesis that angiosperms form a single lineage, a monophyletic group, is well supported by shared characters (chapter 10) and virtually all molecular phylogenies. The woody ranalean hypothesis has had a tremendous influence upon flowering plant classification of the past 50 years primarily via the Cronquist tradition (1981, 1988). Plant taxonomy in the years BC (Before Cronquist) consisted of other classifications that reflected earlier hypotheses. Art Cronquist was not the only angiosperm taxonomist of the past 50 years, but his classification hypotheses were always delivered with great conviction in a stentorian voice. That plus his encyclopedic knowledge and detailed taxonomic publications assured him an influential position. As a result, up until recently much of the taxonomy used, particularly in its broad outlines, has had his signature on it. Cronquist illustrated his ideas about the broad relationships among flowering plants using a lobed diagram (Fig. 11–3) where the biggest subdivision is between monocots (bottom) and dicots (top).

This idea dates back to the very, very early days of plant taxonomy and the work of Albertus Magnus in the early 1200s who recognized that angiosperms consisted of plants with two embryonic leaves (cotyledons) on their embryos (dicotyledons = dicots) and those with one embryonic leaf on their embryos (monocotyledons = monocots). Virtually every biology textbook has a table listing the differences between monocots and dicots (e.g., flower parts in 3s versus 4s or 5s, etc.) (Table 11–1). Cronquist continued this tradition, treating angiosperms as having two large classes, Magnoliopsida[15] (dicots) and Liliopsida (monocots). When so delineated, both classes have impressive diversity. Dicots have more

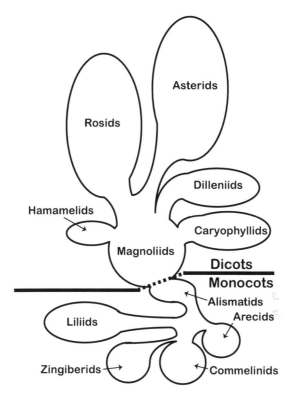

Fig. 11–3 Cronquist's phylogenetic concepts as he illustrated them with a lobed diagram that implies common ancestries. Dicots (Class: Magnoliopsida) are above the bold horizontal line; monocots (Class: Liliopsida) are below. Each of the balloons is a subclass (-ids). Plants that are now basal lineages, the ANA grade, members of the magnolialean clade, and Ranunculales clade were all placed in the Magnolids that represented a morphological nexus from which both the dicots and monocots were derived. The monocot transitional group is the Alismatids, which are largely aquatic plants like pickerel weed, some of which have flowers similar to some Magnoliids (3-parted or many parts in multiples of 3). Arecids includes the palms, screwpines, and aroids; Commelinids includes the grasses, sedges, rushes, and cattails; Zingiberids includes the pineapple family and ginger-banana complex of families; Liliids includes the lilies, now divided into many families and the orchids. Hamamelids includes most of the families that would have formed the old amentiferae, largely wind-pollinated trees with cone-like inflorescences. Along with the Hamamelids, the Rosids, Asterids, Dilleniids, and Caryophyllids compose the rest of the Eudicot clade. (Image source: After Cronquist 1981, 1988)

than 165,000 species organized into 64 orders and 320 families. Monocots have some 50,000[16] species organized into 19 orders and 66 families. What could be simpler? But things are not quite so simple. As pointed out in the last chapter, an embryo with a pair of cotyledons is a character shared with gymnosperms and presumably, via common ancestry with gymnosperms, retained by the majority of flowering plants. However, this means the dicot character cannot be used to define dicots as a taxon. This also means the monocot embryo must be derived from a dicot

Table 11-1. Differences between monocots and dicots

Dicots	Monocots
Flower parts in 4s and 5s	Flower parts in 3s
Leaves with netted venation	Leaves with parallel venation
Leaves broad of diverse shapes	Leaves linear
Stem vascular bundles in a ring	Stem vascular bundles scattered
Two cotyledons	One cotyledon

ancestry, either independently from a gymnosperm ancestor, or from a dicot flowering plant ancestor. Indeed, studies of embryo development suggest that the cotyledon of monocots is not homologous to the cotyledons of dicots (Burger 1998).

For some time now botanists have known that dicots have two fundamental types of pollen, that which has a single pore, actually a lateral furrow, through which the pollen tube emerges and that which has three pores. Those dicots with single-pore pollen are largely trees whose wood has more primitive features and either very simple flowers or flowers with numerous parts in multiples of threes, which identifies them as the woody ranales, now members of the magnolialean clade or the ANA grade. Monocots share this character with the magnolialean and ANA grade plants. Single-pored pollen is the oldest evidence of angiosperms, dating to 132 mya, the early Cretaceous (Friis et al. 2000). Three-pored pollen appears in the fossil record at about 100 mya, so pollen with a single pore is the ancestral condition. This suggests monocots acquired this type of pollen by means of a common ancestry with magnolialean clade or ANA grade plants. Although this was long known, it was not until molecular data produced a similar pattern that botanists began calling dicots with three-pored pollen eudicots, the "true dicots." This means plants with dicotyledonous embryos are found in the ANA grade, the magnolialean clade, and eudicots, except for the waterlilies (N of ANA), which have a single cotyledon but are dicots in all other respects. One way or another class Magnoliopsida has been falsified as a taxonomic hypothesis, but the dicot-monocot division of flowering plants into two classes remains ingrained[17] in everyone's thinking.

Presently the monocots are sister group to all dicots except for those that compose the ANA grade (Fig. 11–4). The non-ANA dicots form two major clades, the magnolialean clade and the eudicot clade. In some phylogenies the *Chloranthus* family (Fig. 10–6) is the basal lineage to both the monocot clade and magnolialean clade; in others *Chloranthus* is basal to just the magnolialean clade. Monocots remain a single lineage, a legiti-

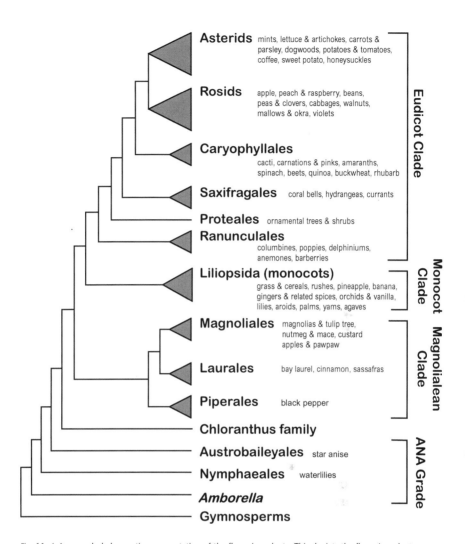

Asterids — mints, lettuce & artichokes, carrots & parsley, dogwoods, potatoes & tomatoes, coffee, sweet potato, honeysuckles

Rosids — apple, peach & raspberry, beans, peas & clovers, cabbages, walnuts, mallows & okra, violets

Caryophyllales — cacti, carnations & pinks, amaranths, spinach, beets, quinoa, buckwheat, rhubarb

Saxifragales — coral bells, hydrangeas, currants

Proteales — ornamental trees & shrubs

Ranunculales — columbines, poppies, delphiniums, anemones, barberries

Liliopsida (monocots) — grass & cereals, rushes, pineapple, banana, gingers & related spices, orchids & vanilla, lilies, aroids, palms, yams, agaves

Magnoliales — magnolias & tulip tree, nutmeg & mace, custard apples & pawpaw

Laurales — bay laurel, cinnamon, sassafras

Piperales — black pepper

Chloranthus family

Austrobaileyales — star anise

Nymphaeales — waterlilies

Amborella

Gymnosperms

Eudicot Clade

Monocot Clade

Magnolialean Clade

ANA Grade

Fig. 11–4 A general phylogenetic representation of the flowering plants. This depicts the flowering plants as consisting of the ANA grade with three lineages and three major clades: Monocot, Magnolialean, and Eudicot. The Chloranthus family is a separate lineage basal to Magnolialeans and Eudicots. The Eudicots consist of several basal lineages and two large clades: the Rosids and Asterids (cf. Fig. 11–3). The names following the clades are some familiar members of these lineages to provide some orientation. (Image source: The Author)

mate taxon, but not the dicots. The eudicot clade consists of a number of basal lineages and two major clades: the rosids and asterids.

The profound differences between Cronquist's classification and more recent phylogenetic hypotheses demonstrate how drastically these concepts have changed in just 25 years. Cronquist's Magnoliidae includes most members of the magnolialean clade, the paleoherbs, and most members of the ANA grade. The Magnoliidae are shown as basal to both the monocots and the rest of dicots, and in all recent phylogenetic studies they are neither. Many of the other subgroupings fare no better. So, no matter how the issue is examined, major revisions to angiosperm taxonomy are in the offing, but as yet many issues remain unsettled and it will take time to change the practices of decades, once enough people can decide what to do. Within the still large clades many other phylogenetic studies are finding results that further challenge, or sometimes support, traditional classifications of many families and even genera. Many well-accepted relationships of just a decade ago have been falsified. Some family concepts are being confirmed as pretty sound—for example, nightshades, beans, roses, orchids, but others will be reconfigured—for example, mints and verbenes, and others (lilies, snapdragons) have been completely dismantled. So stay tuned, folks, much more is to come.

The means by which monocots evolved remains uncertain, but one hypothesis is that they had an aquatic origin. This would mean monocots arose when their ancestors adapted to a freshwater aquatic environment, diversified, and then readapted to the terrestrial environment. No adaptive connection is known between aquatic life and a monocot embryo, but most basal monocots are aquatic plants, a condition also found in members of the ANA grade (N equals water lilies and *Hydatella*). An aquatic origin hypothesis explains a lot about why monocots are the way they are except for the embryo. An aquatic origin explains the lack of woodiness because aquatic plants have little need for support and conduction growing in an abundance water and in the buoyancy of water. In adapting to an aquatic environment the ancestors of monocots lost the ability for secondary growth and woodiness. Remember how natural selection works; any organism that expends energy upon unneeded structures has less energy for reproduction, so any plant not wasting as much energy on unneeded wood will produce more offspring carrying the genes for being less woody. Broad leaves can float (lily pads) or be held aloft as emergent vegetation (pickerel weed) in shallow still water. But if they are submerged in moving water, their resistance would be a tremendous disadvantage. Submerged aquatic plants have very reduced leaves as a result and develop thin, linear leaves without blades or with much-reduced blades and venation. The simple linear leaves of terrestrial monocots with their simple parallel venation may have originated via reexpansion as monocots diversified back onto land from an aquatic ancestry.

The hypothesis of the aquatic origin of monocots fits with the idea that early angiosperms were marginal plants because one place gymnosperm forest margins are found is adjacent to lakes and streams. Thus early flowering plants were presented with two equally difficult environments to invade, gymnospermous forests and freshwater streams or lakes. The vast majority of truly aquatic flowering plants are monocots. Most basal monocots are characterized by flowers with perianths in whorls of 3s, many spirally arranged sporophylls, and pollen with a single germination furrow, essentially the same as ANA grade or magnolialean plants with which they share a common ancestry.

Some general trends in flowering plant evolution and specialization are worth noting. Herbaceous plants, both perennials and annuals, have evolved in many lineages, as have vines, epiphytes, and parasites. Flowers have shifted toward those with fewer parts and more fusion between parts, producing tubular perianths and pistils with two or more carpels. Flowers further have shifted from radial symmetry to bilateral symmetry many times. Plants with small flowers have produced large attractive displays by clustering flowers into inflorescences, some of which have so much the appearance of a single flower that people are surprised to learn such inflorescences are composed of many small, individual flowers—for example, Queen Anne's lace or sunflowers. Weedy annuals have evolved in many lineages too, and they have really prospered owing to human activities. And some have even been domesticated and become part of our food, which brings us to the last and most recent development in the history of green organisms, the human-angiosperm interaction.

TENDING TO HUMAN NEEDS

Without flowering plants, we humans would be naked (or in an animal skin in addition to our own), miserable (no morning cup of coffee), and very hungry. For the longest part of human existence, humans lived as gatherers of plant foods and hunters of animals, and while meat provided much-needed protein, 70% to 80% of human calories came from plants.[18] Plant foods consisted of fleshy fruits, large seeds including nuts, and fleshy tubers, corms, roots, and bulbs, which with only a few exceptions are angiosperms. A few gymnosperm seeds are edible (pinyon pine), and one cycad provides a starchy food ("sago palm"), but that's about it. Flowering plants are also a source of medicines, fibers, building materials (framework and thatch), and even hunting tools and poisons, all the basic necessities of nomadic life.

But about 8,000 to 10,000 years ago in at least three different places, and maybe as many as six to eight different places, humans began shifting to a sedentary lifestyle of pastoral agriculturalists for one of a number of reasons that did not include hunger (Heiser 1981). Whatever the reasons, humans began interacting with certain plants, those that provided for our needs, in a new way, by actively assisting and promoting their growth and reproduction, and it goes without saying, at the expense of other native vegetation and of the animals that depend upon those plants. Human needs, preferences, and choices generated an artificial selection upon plants that resulted in domestication of many species. Selection for better-tasting (i.e., less bitter) vegetables resulted in less toxic foods, and as a consequence plants are more easily attacked by herbivores because they are less defended chemically. Vegetable foods were selected for larger size and softer, less fibrous, and more succulent tissues. Fruits were selected for larger size, more sweetness, better flavor, and sometimes fewer seeds than the wild types. These might prove maladaptive in nature, but human activities, our artificial selections, keep natural selection at bay.

Why humans find many ripe, fleshy fruits attractive is not such a mystery. Humans, like other animals, have sensory adaptations, instincts, for selecting safe, nutritious foods. We owe our arboreal, fruit-foraging ancestors for some of our instinctual food likes and dislikes, as well as our visual acuity, depth perception, opposable thumb, manual dexterity, and intellectual abilities. Fruits whose taste, color, and odor appeal to our likes are those fruits that generally offer safe and nutritious food, but do not place too much trust on your instincts. Humans are long removed from our natural environment that selected for these instincts, and our instincts must be augmented by learning, accumulated cultural knowledge about plant identities and safe foods. When people who lack such knowledge experiment with unknown fruits, it is a trial-and-error process where each error could be your last.[19]

Our instincts can be fooled in another way too. When our daughter was young she did not get candy, but all of our "good parenting" went out with the trash when a well-meaning store clerk gave her an orange lollipop. After only a brief look (fruity color, check), a quick smell (fruity odor, check), a tentative lick with the tip of her tongue[20] (sweet taste, check), and having used her instincts to confirm that this was clearly intended for human consumption, in her mush it went. A smile of delight spread across her face as the full sensory pleasure of this fruit-mimicking confection was realized. Our food preference instincts are fooled by such fruity confections, human artifacts added to our environment for exactly that purpose.[21] By contrast we quickly reject items that taste too sour or too bitter, or that look rotten or slimy, or that smell rank or spoiled.[22]

Human curiosity and the ability to learn led to many experiments with many potential plant foods, including ways to render the inedible edible. Such knowledge allowed humans to exploit the hard, dry, one-seeded fruits of grasses, cereal grains,[23] a food resource that now provides well over 50% of all human calories (Heiser 1981). Cereal grains are fruits that are not even slightly attractive to us, no pretty colors (except maize), no sweet tastes, no fruity odors, although roasted grains smell pretty nice. They are unappetizing and inedible without considerable process-ing: rolling, cracking, parching, grinding, and cooking. However, in one instance a wild cereal yielded an edible food rather easily and readily. Teosinte (tea-oh-SIN-tea), the wild ancestor of maize, has extremely hard and virtually inedible grains, but they pop! People would have rather eas-ily discovered this surprising property of teosinte because one means of rendering hard seeds edible and better tasting is to roast or parch them by placing them on a rock or stirring them into hot sand next to a fire.[24] Without question popcorn predates movies by millennia. In fact, humans like cereal grains best when converted to beer and bread, both resulting from an interaction with a microorganism unconsciously domesticated by humans, yeast (see chapter 2). The partial[25] mastery of fire learned some 1.5 mya provided a means of making the inedible edible, by mak-ing hard, unpalatable foods like roots and tubers more easily consumed and more nutritious, or tough meat easier to consume. Humans may also have used fire to alter, or maintain, certain communities for the benefit of herbivores, viewed by humans as prey organisms.

Only about 300 flowering plant species have been domesticated to any degree, less than 0.1% of the total, and only a little over 100 species account for 95% of the world human food budget in terms of total calo-ries consumed (Prescott-Allen and Prescott-Allen 1990). Of those, fewer than one in five, a total of about 20 species, are of major importance. The important ones are four cereal grains (grass)[26]: wheat, rice, maize, sugar cane; five legumes (bean): common bean, soybean, lentil, peanut, pea; five starchy staples: potato (nightshade), sweet potato (morning glory), yam[27] (yam), cassava (euphorb), banana/plantain, a starchy fruit (banana); and a few oil seeds, fruits, and vegetables: coconut (palm), sunflower (aster), tomato (nightshade), orange (rue), apple (rose), cabbage et al.[28] (mustard), onions (lily). These species are distributed pretty widely among angio-sperms phylogenetically (Fig. 11–4), but nearly all of this diversity greatly predated the appearance of humans.

In a few cases some other patterns emerge. Many spices come from plants in the magnolialean clade. Many herbs come from members of the mint and parsley families. Lots of fruits come from plants in the rosid clade. Some food plants are of major importance in certain countries, but of little importance elsewhere—for example, quinoa in Bolivia, taro in Samoa, and karite nuts in Ghana, Mali, and Nigeria (Prescott-Allen and Prescott-Allen 1990). Some foods become very important in new places far from their native land—for example, the "Irish" potato is a mountain native of Peru or Bolivia.

Beans (eudicots) and cereals (monocots) together form the basis of good human nutrition and sound agricultural practice, as they are complementary in both respects. So it comes as no surprise that a combination of cereals and beans were the foundations of more than one agricultural society.[29] Beans (legumes) have a symbiotic interaction with the bacterium *Rhizobacter* that results in the ability to fix atmospheric nitrogen, the most common nutrient limiting plant growth, so rotating nitrogen-hungry cereal crops with legume crops is a good agricultural practice. Bean seeds are the highest-protein plant food in human diets, having on average over twice the protein of whole-grain cereals. The soybean, the appropriately named *Glycine*[30] *max*, tops out at a protein content of 40% of their dry weight.

Many plants produce chemicals to protect themselves from animal herbivores, derivatives of their complex synthetic pathways. In a search for food, human trial-and-error sampling has discovered that many of these compounds have curative value for human diseases and disorders.[31] All such compounds are toxic[32] and whether the intoxications produced are curative or deleterious often depends upon dosage. Those that stimulate our central nervous system occupy a particularly exalted place in our lives, providing our morning coffee, our afternoon tea, a good cigar, or a sumptuous treat of chocolate. Smoking tobacco was one of the most-widespread cultural attributes of Native Americans, and unknown elsewhere prior to 1492, but the highly addictive qualities of nicotine virtually assured tobacco of continued success abroad.

Plant fibers from leaves, stems, and wood have proven very useful things for making cordage, fabric, and paper, itself named after the pith tissue (papyrus) from a wetland sedge native to the Nile River valley. And the human love affair for making our marks upon smooth surfaces, a romance that started with cave walls, moved to clay tablets, then onto hides (parchment) and papyrus, finally came into full bloom, and with other technical advances our cultural heritages, imaginations, aspirations, and knowledge were recorded and shared in books. How this need will be filled in the future is rather uncertain because the new media, acting more like chalk marks on sidewalks, are in no way as stable, satisfying, or durable as cellulose. And as this is written in what may be the twilight era of paper books, it is quite possible that you may be reading this on an electronic device.

Cotton ranks as the most important plant fiber, one of the most important nonfood plant commodities (Lewington 1990). This comes from cotton's versatility, its combination of lightness and durability, its ability to take dyes, and after the invention of the cotton gin, its cheapness. Cotton feels so much better in warm weather than wools and hides, not to mention its softness, so even when wearing heavier clothing cotton provides a buffer between them and our skin. Although they almost go unnoticed, has any aspect of recent culture other than Coca-Cola had any bigger impact around the world than that nearly ubiquitous cotton garment called blue jeans?[33] So whether burka or bikini, fabric from plant fibers is what so often provides our modesty, although clearly the cultural scale for modesty is broad. *Cannabis* provides the fiber hemp but also the euphoric marijuana, quite the two diverse uses, and as a result two distinct races have evolved at the hands of humans, and in both ways the plant has been remarkably successful. Escapees of the fiber race still can be frequently spotted along ditches in parts of Illinois where hemp was grown to replace Manila hemp (*Musa textilis*—a fibrous banana) during World War II, now almost 70 years ago.

All these domesticated plants have become tremendously successful because of their interactions with humans. However, the more that domesticated plants have become dependent on humans, the frailer their existence has become. Such is the nature of the evolutionary dance we call obligate mutualism where both interacting species absolutely require each other to survive. If something should happen to one partner, then the other may be doomed. Our primary means of safety is to depend upon multiple species, and to find and protect the wild relatives of our pet plants wherein their evolutionary heritage of genetic diversity resides. At certain times when too many people have depended upon too few plant species, disaster has struck—for example, the potato murrain in Ireland destroyed the only staple food available, leading to widespread starvation and the emigration of my ancestors to North America.

Many domesticated plants are unable to survive in the wild on their own, and extremely few modern humans could make it in the wild if enough wild still existed. But the wild relatives of many important plants remain unknown, and in only a few cases has action been taken to protect those wild relatives and their genetic diversity. Very few facilities exist for the long-term storage of seeds, and most of these are in the industrialized countries of the temperate zone that have the resources and infrastructure to support them. Yet most of the world's crops originated in subtropical and tropical countries. And then the question arises of who "owns" these genetic resources. Are they the property of indigenous peoples whose ancestors domesticated them? Are they the property of the countries in which they grow? Are they the property of institutes or corporations that identified particular genes or varieties or relationships? Are they a world heritage for all? What compensation if any does the industrialized world owe developing countries whose plant resources provide their food? These are difficult questions and a consensus on answers is nearly impossible to get, and no one seems to like the concept of a commonwealth.

In terms of human population numbers, agriculture has been a very successful human lifestyle, but we have reached the point when we can begin to wonder how long this success can continue, or even if an optimal population has been reached already. Agriculture does not necessarily provide more food or better food, but it does provide more food per unit area and per unit of effort. This means more humans can be supported in the same area by agriculture than by gathering and hunting. Human population growth over the past few thousand years certainly attests to this, and again we wonder how long it can continue. As humans became more numerous, and as humans have populated virtually all inhabitable terrestrial regions of the Earth, seeds of domesticated plants came with us. During the age of discovery and European colonial empires, plant resources (and animals, and people) were moved around the world for both good and bad reasons. In some manner of speaking, domesticated plants became more successful too. Sometimes crops became even more successful in foreign places than in their native regions. Coffee has been quite successful in Central and South America; it is an African native. Cacao (chocolate) and rubber are quite successful in the Old World tropics, and both are from the Neotropics. People today arguably have access to more diverse plant resources at a cheaper price than at any time in human history, and very few people appreciate this botanical heritage because it is just taken for granted. In just a few generations we have removed ourselves from self-sufficiency and primary productivity to the point that many people do not know what kind of plant a tomato grows upon, what vinegar is or how to make it, or that potatoes and peanuts grow underground. How many people know how to process flax to obtain linen fibers? Or which bark provides an aspirin-like painkiller for aching joints?

Human-altered communities have replaced natural communities across wide expanses of the Earth as our population has grown, sometimes almost to the point of eliminating the native vegetation. Nowhere is this truer than in Illinois where less than 1% of the original tall grass prairie community survives, having been largely supplanted by cultivation of a tropical Central American grass, maize, and an Asian legume, soybean. In other areas of the world similar destruction and fragmentation of large tracts of natural communities may have begun a major human-caused extinction event. The primary cause of species extinction is habitat destruction as our insatiable hunger for wood and agricultural land continues to grow. A toad here, a butterfly there, maybe a bird or an orchid somewhere else; what does it matter if a few species go missing as humans quest for our necessities? What we do not know is what long-term impact human activities will have on green organisms, and then upon everything else. We do not know how many strands of biological diversity can be severed before the fabric of life really starts to unravel. Our human societies and certainly our governmental bodies act largely in ignorance of, and even in contempt of, science, in part because some answers are not known—for example, the extent and impact of global warming, and in part because the answers that science provides are not to their liking. But those who ignore what is known about genetics, ecology, and evolution are acting with considerably more ignorance than others, and although some dispute this too, that knowledge does not diminish our ability to appreciate and enjoy that which is nature.

Of course, flowering plants also make our lives more enjoyable: a bottle of wine, a loaf of bread, and a pretty flower. We adorn ourselves and our abodes with plants that please our aesthetic senses. Plants, growing as they do by following a genetic program (pattern), subtly varied by the environment, produce good artistic design, and so we decorate with plants and plant images. Sheltered and buffered as we are from the harshness and even the reality of our natural environs, we nonetheless enjoy seeing and living with tame bits of nature in the form of our pets and plants. Nothing brings more joy than the sight of a lovely blossom or lush vegetation especially during the dark, bleak days of winter.

And finally, you now know that this blossom you admire represents the present-day status of a very successful cyanobacterium, part of an unbroken lineage of green organisms stretching back over at least 3.5 billion years. Understanding this and many other features of living organisms requires a historical perspective of descent with modification and an appreciation of the immense time involved. Darwin would be quite pleased by what we now know and understand, even though he would be disappointed to learn that we still have not produced a clear answer to the origin of angiosperms. This does not mean we have given up and declared this event so inexplicable that we must invoke some unknown or supernatural cause. Nothing in this admission should give any comfort to critics of evolution who cannot provide any adequate explanation for what we do know, and it shows because they seldom even make the attempt.

As best we understand it, this is how the Earth turned green, and it is a fascinating history that explains a great deal about the way the natural world works, and why nature is the way nature is, but much more has yet to be done; our understanding remains incomplete. In a decade or two this story will certainly be different than at present. So I shall close by thanking plants for being so interesting and so important that some of us could make a living by studying and teaching about them. And pleasure has always come from imparting this knowledge to others, to you.

The Best Biological Classroom

More than a couple of years have passed since I began writing this book. It has taken longer than I expected for a variety of reasons. Like ecosystems, books are fairly complicated things, and whenever you make a small change someplace, a cascade of effects can follow. Writing a book was not the only demand upon my time, and although my employer expects me to be a professional scientist and scholar, reading and writing are actually done in my "spare" time. My teaching duties and obligations to students, and the other requirements of my job at Illinois State University, easily constitute a full-time job, which according to the state of Illinois is 37.5 hours a week for nine months of the year. Now (November 2013) is my thirty-fifth academic year, and I am teaching rain forest ecology and leading a two-week field trip to Costa Rica. This may sound like a cushy trip, but it is a great deal of work, worry, and stress, and I'm doing it when everyone else is off on Thanksgiving break, and for all this effort nothing extra will be in the paycheck, either. So why bother? Over the past 40-some years I have been in quite a few classrooms, but rain forest is the best biology classroom for the simple reason that nothing else generates an excitement that makes teaching well-motivated students easy. I began writing this book with rain forest in full view, and to be symmetrical, I shall end this book the same way.

The history of green organisms has not ended. It will continue, and I cannot say with certainty what will happen. This is the way it is with evolution. Once you know how it works you understand this. A very famous biochemist, an evolutionary scofflaw, challenged me to make one good solid prediction about the future on the basis of evolutionary theory. I responded that things will change, and organisms will adapt. And more than anything else, we need to understand that they can do this without humans.

Appendix Contents

Brown Algae and Tribophyceans   367

Clubmosses and Fossil Stem Groups   373

Conifers and Ginkgoes   387

Coniferophytes: Cordaitales and Voltziales   399

Cycads   401

Ferns   409

Gnetophytes   431

Green Algae   439

Green Bacteria   451

Hornworts   455

Horsetails   457

Liverworts   465

Mosses   473

Phytoplankton   481

Red Algae   489

Rhyniophytes and Trimerophytes   493

Seed Ferns   495

Whisk Ferns   501

Brown Algae and Tribophyceans

Brown Algae

(synonyms: Phaeophyte, Phaeophyta, phaeophyceans)

The brown algae are marine seaweeds whose members include the well-known and familiar kelps and rockweeds. Their common names, and those of inclusive groups, are based upon their appearance; they are generally brown, tan, or olive brown in color because their chlorophylls (*a* and *c*) are masked or partially masked by yellow and gold accessory pigments like fucoxanthin and beta-carotene. The brown algae also are called the phaeophytes (FAY-oh-fights) or phaeophyceans (fay-oh-feye-SEE-anz), *phaeo-* meaning dusky. The brown algae together with the tribophyceans (TRI-bow-feye-see-anz) (see below) have been called the ochrophytes (OH-crow-fights), ochre referring to a yellowish-brown color (Graham and Wilcox 2000). A still more inclusive group that includes related phytoplankton has been called the chromophytes (CROME-oh-fights) or Kingdom Chromista (crow-MEEZ-tah), *chromo* meaning colored (see Appendix: Phytoplankton, stramenopiles). Members of this kingdom were also called the heterokont algae because their motile cells have two different flagella (= konts), one smooth and one hairy. Such flagella are the hallmark and namesake of the stramenopile clade, which also includes some nonphotosynthetic, fungal-like organisms, the labyrinthulids or slime net amoebae. The chloroplasts of brown algae and all the

Fig. A1 Examples of large brown algal seaweeds: A. *Macrocystis*, giant kelp. B. *Postelsia palmae*, sea palm. C. *Laminaria* on the beach. All have large, multicellular holdfasts anchoring the seaweed to rocks, well-developed stem-like stipes, and broad leaf-like blades. *Macrocystis* has pnematocysts, gas-filled floats, located at the stipe-blade junction. The tough, flexible stipe acts as a tether for a plant that can grow over 50 m long. *Postelsia*, sea palm, grows in the intertidal zone and has a thick stipe and small, thick, and tough blades to withstand surging tides and breaking waves. This seaweed is about 0.5 m tall. The beached *Laminaria* is about 1.5 meters long, showing a form similar to, but a bit less robust than, *Postelsia*; it grows below the intertidal zone and shows what happens when coastal organisms are not anchored securely. (Image sources: A—Courtesy of EncycloPetey, Wikimedia Creative Commons; B—Courtesy of Brocken Inaglory, Wikimedia Creative Commons; C—Courtesy of Gabriele Kothe-Heinrich, Wikimedia Creative Commons)

photosynthetic stramenopiles are delimited by four membranes indicating a secondary endosymbiont origin (see chapter 3 for discussion).

Brown algae vary from small filamentous forms to the largest of seaweeds, the kelps, whose anatomical complexity is as great or greater than that of some land plants. Cell division in many brown algae produces plasmodesmata, cytoplasmic connections between cells, so the larger seaweeds are composed of parenchyma like land plants. Kelps and other large brown algal seaweeds have a very plant-like organization with well differentiated rootlike holdfasts, stem-like stipes, and leaf-like blades (Fig. A1). Blade flotation by inflated, gas-filled bladders (pneumatocysts [new-MAT-oh-systs]) allows kelps to inhabit deeper water while still harvesting light near the surface. Flotation devices on or just below the blades pull on the stipe, placing it under tension like a string attached to a helium-filled balloon. Although stem-like, the stipe is not under compression from bearing the weight of the crown above like the stems of land plants. The strength of a stipe is measured by its tensile strength, the force it takes to pull it apart.

Photosynthetic products produced in the blades of large seaweeds like the kelps must be distributed to cells in the stipe and holdfast shaded by the crown of blades, so strands of phloem-like conducting tissue, or phloem, depending upon your criteria, extend from the blades down the stipe to the holdfast. Rather than storing starch, brown algae store the polysaccharide laminarin, a glucose polymer terminated by a sugar alcohol, mannitol. The cell walls are two-layered with an inner cellulose layer and an outer mucilaginous layer. This gummy carbohydrate can compose over 10% to 20% of the dry weight of the plant. Mucilaginous gums hold water, which is important for keeping intertidal zone seaweeds hydrated. It also makes them slippery.

Classification of brown algae is changing because many traditional groupings based on morphological features have been falsified by molecular data. Presently brown algae consist of three main lineages: the fucaleans ("few-kay-LEE-ans"), ectocarpaleans ("ek-toe-car-PALE-ee-ans"), and laminarialeans ("lamb-in-air-ee-AYE-lee-ans"), each named after a characteristic genus: *Fucus*, *Ectocarpus*, and *Laminaria* (Fig. A2). Fucaleans, traditionally classified as the order Fucales, are the rockweeds, a well-defined group of seaweeds that grow largely in the upper intertidal and splash zone (Fig. A3). *Sargassum*, the namesake of the Sargasso Sea west of Africa, is a fucalean with both anchored and free-floating species, the

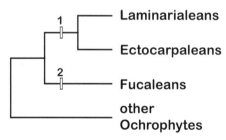

Fig. A2 Broad phylogenetic relationships among brown algae. Three main lineages have been identified with some confidence. The fucaleans, rockweeds, form a sister group to the rest of the brown algae and they share a life cycle similar to ours with diploid adults that produce gametes (2). All the rest have a life cycle with alternating diploid sporophytes and haploid gametophytes (1). Laminarialeans include the kelps and other large seaweeds. Ectocarpaleans are both filamentous and thalloid but generally smaller than laminarialeans. Brown algae have a common ancestry with other ochrophyte algae that share a similar chloroplast and motile cell. (Image source: After Graham and Wilcox 2000)

Fig. A3 Fucus occupying its typical rocky habitat near the high tide mark. At low tide like this they limply settle upon one another, but small air bladders will lift them when the tide returns. The ends of the thallus, called receptacles, look inflated and a bit like finger tips. The small dimples visible upon them are little cavities within which the gametes are produced. (Image source: The Author)

latter being an exception to the seaweed habit of being anchored. The rockweeds have apical growth and dichotomous branching of their thallus. Rockweeds are gamete-producing, diploid adults with no alternative haploid phase (Fig. 3–9 B). Gamete release is often timed to coincide with favorable tidal periods because motile zygotes must find an appropriate habitat and anchor in place quickly before the next retreating tide pulls them out to sea. Anchored rockweeds and free-floating *Sargassum* each occupy a different specific and particular habitat, and presumably an asexual (zoospore) dispersal stage and a free-living haploid phase was lost as part of their adaptation to their habitat.

The laminarialeans are large seaweeds, including the kelps, which can grow up to 50 m long in a season, literally forming seaweed "forests" just offshore. Other members of this group occur in the middle to upper intertidal zone where they are subjected to the considerable forces of moving water. Seaweeds like *Postelsia* have substantial holdfasts and a short, thick, flexible stipe topped by a moppish whorl of blades (Fig. A1 B). Like other forests, laminarialeans provide habitat for a diverse community of organisms.

The ectocarpaleans are largely anchored, branched filamentous algae with intercalary growth (Fig. A4). Others are smaller (compared with the kelps), simpler seaweeds like *Dictyota*, which has an apical cell that divides in three planes and grows into a dichotomously branching thallus three cells thick. The ectocarpaleans and laminarialeans have life cycles exhibiting an alternation of haploid and diploid phases. In the ectocarpaleans, both haploid and diploid phases look identical and you can tell them apart only because the diploid phase produces a unicellular sporangium whose cell undergoes meiosis, producing zoospores (Fig. A4 E-G). Zoospores, both diploid and haploid, and gametes all look alike, although the latter fuse to form zygotes; all have flagella characteristic of the stramenopile lineage (Fig. A4 F). Laminarialeans possess a similar life cycle except the diploid phase is the familiar large seaweed, while the haploid phase is small and filamentous and similar in form to ectocarpaleans (Fig. A5). Kelps like *Laminaria* reproduce differently from ectocarpaleans in another respect: their gametes are differentiated into larger, nonmotile eggs and smaller, motile sperm.

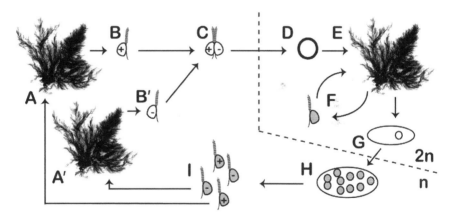

Fig. A4 Life cycle of *Ectocarpus*, a filamentous brown algae. This life cycle has an alternation of generations where both the haploid (A, A') and the diploid (E) generations are essentially identical-looking organisms; they differ only in reproductive cells. Sperm and zoospores look the same except they behave differently. Haploid individuals (A, A') produce motile gametes (B, B') that pair (C) when the find a complementary mating type (+/-) and fuse to produce a zygote (D), which grows into a new diploid individual (E). Diploids can produce zoospores (F, gray) for asexual reproduction. Single-celled sporangia (G) undergo meiosis (H) and then divide by mitosis to produce many haploid zoospores (I, gray) that disperse, anchor, and grow back into haploid individuals (A, A'). Dashed line divides haploid (n) and diploid (2n) generations. The scanned image of *Ectocarpus* used in this diagram comes from a specimen collected in 1893 (Courtesy of the Vasey Herbarium, Laboratory for Plant Identification and Conservation, Illinois State University). (Image source: The Author)

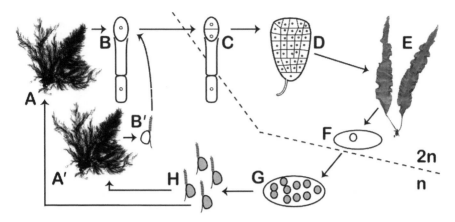

Fig. A5 Life cycle of *Laminaria*, kelp. This life cycle is an alternation of generations where the familiar large kelp seaweed is the diploid generation (E) and the haploid generation is small and filamentous, similar to *Ectocarpus* (A, A'). Haploid individuals produce gametes, either eggs (B) or sperm (B'). The egg exits its cell wall and "perches" at the end of the cell, and if fertilized it forms a zygote that then drifts away. Mitotic division of the zygote (C) produces a young kelp plant (D) that anchors itself with rhizoids and grows into the familiar kelp (E). Single-celled sporangia undergo meiosis (F), returning to the haploid condition (G). Mitotic divisions produce many zoospores (H, gray) that disperse, anchor themselves, and grow into female or male haploid organisms (A, A'). (Image source: *Laminaria* after A. Whittick, Ohio University; diagram by the Author)

Tribophyceans

(synonyms: Xanthophyceans, Xyanthophytes, yellow-green algae)

Tribophyceans have chlorophylls *a* and *c*, but they lack the fucoxanthins characteristic of the rest of the stramenopile alga, thus they appear yellowish-green as opposed to golden, brown, or olive. Many are hard to distinguish visually from green algae. Their cell walls generally consist of two halves, one on each end of the cell, overlapping in the middle. The walls are not composed of cellulose, although silica or chitin (like fungi) can be present. Their primary storage product is a lipid, not a starch. These algae are diverse in form, including flagellated unicells (similar to eustimatophyceans; see Appendix: Phytoplankton), nonmotile unicells, filaments with and without cross walls, and even amoeboid forms. Nonmotile unicells can be free floating; anchored by stalks and growing as epiphytes on larger algae; or rhizoidal multinucleate cells growing on damp soil. Filamentous *Tribonema* often forms pond scum.

Clubmosses and Fossil Stem Groups

(synonyms: lycopods, Lycophyta, Lycopodiophyta, Sphenophyta, sphenopods)

Clubmosses are the oldest and earliest diverging lineage of living vascular plants. Clubmosses appeared during the early Devonian and attained their maximum diversity, including many large arborescent forms, during the Carboniferous when they formed a dominant component of terrestrial floras, particularly the coal swamps. Living clubmosses are relicts of their past glory, significance, and diversity, but they still carry with them the aura of the primeval. In miniature, living clubmosses show us how the world used to look.

Ten genera are all that remain of clubmoss diversity, and even then they represent three distinct lineages (Øllgaard 1987) (Fig. A6). One genus, *Selaginella* (sell-ah-gin-ELLE-lah), spikemoss, first appears in the Carboniferous, which makes this the oldest living genus on Earth. *Selaginella* is the sole living representative in its lineage and consists of some 700 species worldwide. In some places *Selaginella* is quite common. *Isoetes* (eye-sow-EH-teaz), quillwort, with about 75 species, represents a second lineage, the last living relative of arborescent clubmosses, a surprising ancestry for a small aquatic or wetlands plant. *Lycopodium* (lye-coh-POH-dee-um) (15–25 species), ground pine or clubmoss; *Phylloglossum* (fill-oh-GLAH-sum) (1 species), which is found only in New Zealand, Tasmania, and southern Australia; and six more genera segregated out of *Lycopodium*: *Phlegmariurus* (~300 species), *Huperzia* (10–15 species), *Diphasiastrum* (15–20 species), *Phalhinhaea* (10–15 species), *Lycopodiella* (8–10 species), and

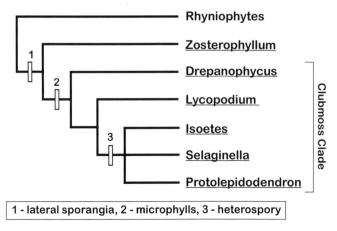

Rhyniophytes

Zosterophyllum

Drepanophycus

Lycopodium

Isoetes

Selaginella

Protolepidodendron

Clubmoss Clade

1 - lateral sporangia, 2 - microphylls, 3 - heterospory

Fig. A6 Phylogenetic relationships of the clubmoss clade. Rhyniophytes (see Appendix) are the outgroup and have naked stems and terminal sporangia. 1. *Zosterophyllum* and its relatives are a transitional stem group with lateral, dorsoventrally flattened sporangia but lack true leaves (microphylls). 2. Members of the clubmoss clade possess leaves called microphylls. 3. *Isoetes*, *Selaginella*, and all the arborescent fossil species are heterosporous. *Lycopodium* and associated genera, *Isoetes*, and *Selaginella* are the three lineages with living species. *Protolepidodendron* and *Drepanophycus* are lineages with only fossil members. (After Kenrick and Crane 1997)

Pseudolycopodiella (12 species) comprise a third lineage consisting of about 400 species in total. The genus *Lycopodium* also provides the generic root of formal taxonomic names (Lycopsida or Lycopodiophyta) and the common name for clubmosses used by most botanists, lycopods.

Phlegmariurus and *Huperzia*, the former largely tropical and subtropical, the latter largely temperate to arctic, lack horizontal shoot systems (rhizomes). They do not form discrete strobili because leaves associated with sporangia are not different in size or shape, and/or the sporangia are not localized near the terminal end of the aerial shoots. Many members of *Phlegmariurus* are epiphytes with long, dangling aerial shoots; thus the common name hanging fir-moss.

The remaining genera have horizontal shoot systems, which in some species can run along the ground for some distance. *Lycopodium* and *Diphasiastrum* are both largely temperate genera of dry uplands and their gametophytes are nonphotosynthetic and subterranean. The leaves of *Lycopodium* are arranged helically on the aerial shoots projecting laterally all around the stems. The aerial shoots of *Diphasiastrum* are flattened or four-sided branches bearing leaves in four rows or ranks. *Palhinhaea, Lycopodiella*, and *Pseudolycopodiella* largely live in tropical and subtropical wetland areas. *Palhinhaea cernua* (formerly *Lycopodium cernua*) is perhaps the most abundant clubmoss in the world. Their gametophytes form a discoid thallus, grow on the surface of the substrate, and are photosynthetic. Two additional clubmoss lineages, represented by *Drepanophycus* (druh-PAN-oh-feye-cuss) and *Protolepidodendron* (pro-toe-lep-pee-doe-DEN-dron), are known only from their fossils; they contain no living species (Stewart and Rothwell 1993).

All living clubmosses have leaves arranged helically or in four rows upon the stem. Clubmosses produce either determinate aerial shoots, arising from a clump or a rhizome, or a pendent branch system. The branching patterns range from equal Y-branching (Fig. A7 A) to one main stem or axis with distinctly lateral branches.[1] Aerial branches may include a combination of both types of branching, starting with a main axis and shifting to Y-branching on lateral axes. *Isoetes* and *Phylloglossum* have short, erect stems (corms) bearing a helix of leaves on the upper end and a helix of roots on the lower end. Sporangia are borne laterally on the stem in the angle between a stem and leaf or upon the upper surface of leaves. Leaves associated with sporangia are called sporophylls (spore leaves). Sporangia may be borne along any portion of an aerial axis and associated with regular-sized vegetative leaves, or sporangia may be aggregated terminally on an axis and associated with leaves of a modified shape, size (usually reduced), and orientation, and this terminal aggregation of differentiated sporophylls and sporangia forms a strobilus or cone (Fig. A7 B). The common name clubmoss refers to these cones, some of

Fig. A7 Tropical species of the clubmoss *Phlegmariurus*. A. Y-branching of the young shoot (white arrows 1–3) and helically arranged, awn-like leaves (microphylls). B. Pair of terminal cones on a pendent branch showing reduced scale-like sporophylls, each clasping a sporangium. (Image source: The Author)

which look like clubs held aloft upon modified aerial branches bearing reduced, scale-like leaves.

Members of the *Lycopodium* lineage are homosporous—that is, producing only one-sized spore that develops into a hermaphroditic gametophyte. In form these gametophytes can be a simple discoid or turnip-shaped thallus usually growing on the soil surface, and therefore photosynthetic, or a subterranean gametophyte that obtains nutrition from an association with fungi. While the subterranean gametophytes may grow more slowly, they can persist for a considerable period of time in this environment. Archegonia and antheridia are embedded in the surface of the gametophytes. As with other land plants, the early development of the sporophyte embryo results in a foot and an axis. An early branching of this axis produces a lateral and an aerial shoot system.

The helical arrangement of numerous leaves on erect stems causes some species of clubmosses to resemble little evergreen trees, resulting in another common name, ground pine. These leaves have a simple organization with stomates on both surfaces, upper and lower, and ample intercellular spaces within, but no organized internal tissue as in the leaves of seed plants. The leaves of clubmosses have only a single vascular strand running down their middle so while generally small, the leaves are also narrow. Vascular tissue occupies a central cylinder, and the organization of the xylem and phloem vary. In some species the xylem forms a star-shaped pattern with phloem occupying the area between the arms of xylem. In other species the xylem is organized into broad bands surrounded by phloem. Roots arise from outer layers of tissue of the vascular cylinder near branch junctions.

The Heterosporous Lineage

Selaginella, *Isoetes*, and extinct members of that clade are all heterosporous—that is, producing two sizes of spores, big ones (megaspores) and small ones (microspores) that develop into female and male gametophytes, respectively. Furthermore, the development takes place within the spore, relying on food reserves within. Such tiny gametophytes can develop quickly in equally tiny places. The size of the spores affects the number of spores produced (fewer megaspores) and how far they can disperse (tiny microspores can disperse the furthest).

Selaginella

In comparison with *Lycopodium* and its relatives, *Selaginella* is a smaller, more delicate plant bearing a superficial resemblance to prostrate mosses

Fig. A8 Selaginella. A. *S. delicatula* showing leaves arranged in four rows, two lateral and two rows of smaller leaves along the axis. This species does not have well-differentiated cones at the ends of axes where sporangia are clustered (arrows). B. *S. martensii* showing two rows of lateral leaves from the back and well-differentiated terminal cones (bracket). (Image sources: A-B. With permission and courtesy of D. Nickrent, www.plantsystematics.org)

or leafy liverworts. However, they can be distinguished by their leaf arrangement and central vascular strand in the leaves, which the leafy bryophytes lack. Most have a creeping rhizome and produce prostrate lateral or erect branches; a few grow in more upright tufts of erect branches. The leaves tend to be smaller than those of *Lycopodium* and its relatives, often broadly triangular or nearly ovate, sometimes even scale-like and clasping the axis. The leaves are arranged in alternating pairs, especially the scale-like ones, or they occur in four rows, two lower rows extending out laterally at right angles to the axis, and two upper rows of smaller leaves lying flat upon upper surface of the axis pointing toward the apex (Fig. A8 A). Each leaf possesses a small tongue-like appendage upon its upper surface, the ligule, an apparently functionless vestigial organ that is nonetheless a feature shared among all members of this clade.

The axes of *Selaginella* have a well-developed epidermis, cortex, and vascular cylinder, but the epidermis lacks stomates. Whether the lack of stomates is an ancestral condition, meaning they never had them, or a loss of character remains uncertain. The vascular cylinder consists of either one or several cylindrical or flat bands of vascular tissue suspended within a tubular air space. The innermost layer of the cortex, the endodermis, produces spoke-like projections of cells that suspend the vascular cylinder within the hollow central region of the stem. The axes of *Selaginella* produce roots or rhizophores from branch junctions. Rhizophores appear rootlike but they are modified branches, sometimes bearing small leaves, and they can Y-branch as well. Rhizophores are rather robust and can often grow downward several centimeters to reach the ground. After contacting the ground, the rhizophore produces a number of short roots. Roots arising from stems or leaves are termed adventitious roots, as opposed to an embryonically derived primary root or its branches, secondary roots. Rootlike branches occur in early fossil members of this lineage, which suggests roots arose from modified stems.

Selaginella is heterosporous, producing both megaspores and microspores in separate sporangia. Although the sporangia are always terminally aggregated on apical portions of axes, the associated sporophylls are similar to sterile leaves, and thus cones are small and poorly differentiated in most species although clearly differentiated in other species (Fig. A8 A-B). Megasporangia produce just four functional megaspores, the result of one large megaspore mother cell undergoing meiosis, while microsporangia produce thousands of microspores (Fig. A9). Megaspores have at least a ten-fold larger diameter, which means they have about 1,000 times the volume (10^3) of microspores, clearly demonstrating the tradeoff between spore size and number.

Megaspores develop into female gametophytes, while microspores develop into male gametophytes (Fig. A9 B-E). The development of the male gametophyte is very truncated. The spore divides unevenly, producing a vegetative cell, which is all that is left of the gametophyte thallus, and the remaining cells produce an antheridium with one layer of jacket cells surrounding sperm-generating cells. Thus the male gametophyte consists of one vegetative cell plus one antheridium all contained within its spore wall. The sperm are biflagellated.

The development of the female gametophyte is a bit more complex. The portion of the megaspore adjacent to the scar remaining from the spore tetrad becomes multicellular, and such growth can commence even before the megaspore is dispersed. After dispersal, the megaspore or the very young female gametophyte absorbs water and grows enough to split open the megaspore wall along the trilete scar, allowing the gametophyte tissue to bulge outward. The lower portion of the megaspore remains an undifferentiated mass of megaspore cytoplasm that will become cellular only as an embryo develops. Rhizoids are produced extending upward and outward from the exposed cellular surface. They can anchor the female gametophyte to a substrate, function in water uptake, and also trap windblown microspores, thus assuring that males are in the immediate

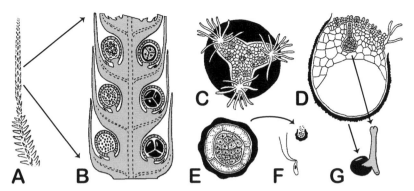

Fig. A9 Selaginella *cones, sporangia, spores, and gametophytes. A. Diagram of a cone. B. Short portion of cone in longitudinal section showing microsporangia (left) and megasporangia (right), each revealing three of the four megaspores composing each tetrad. Each sporangium is associated with a modified leaf (a sporophyll). C-D. Mature female gametophyte. C. Apical view showing the gametophyte barely expanding outside of the megaspore wall (black) through the scar left by where the spores were packed into a tetrad, a stack of four. Filamentous rhizoids extend out to capture microspores and absorb water. Several sets of four cells mark the apex of archegonia. D. Longitudinal section showing development of the megaspore cytoplasm into cells of the female gametophyte. At the upper surface are several archegonia each with an egg, and one containing a young sporophyte embryo (gray). E. Maturing male gametophyte (within the microspore wall) consisting of a layer of jacket cells around cells that will each produce a sperm. F. Microspore wall rupturing and releasing biflagellated sperm. Microspore shown on size scale similar to C and D. G. Young sporophyte (gray) emerging from within megaspore wall and female gametophyte (arrows). (Image source: After Corner 1964)*

vicinity of a female gametophyte. In the central portion of the "thallus" small archegonia differentiate, consisting of four rows of neck cells surrounding two neck canal cells above an egg. Following fertilization, the zygote divides transversely. The upper cell divides transversely several times, producing a row of cells that pushes the lower cell down into the female gametophyte's tissue. The embryo develops from the lower cell in a transverse orientation, producing a shoot axis in one direction and a root axis in the other. From the time the megaspore is shed from its sporangium to the time the new sporophyte sprouts forth from the female gametophyte takes about a month, provided ample males are present to supply the sperm (Shultz et al. 2010).

Isoetes—Quillworts

Quillworts are fairly small plants, and they look like a small rush or reed growing in shallow water and wet places, often temporary or seasonal aquatic habitats, and this explains why so few people recognize this plant or know of its existence.[2] Only real gung ho naturalists will muck about enough to find quillworts, and as a result, they are much more

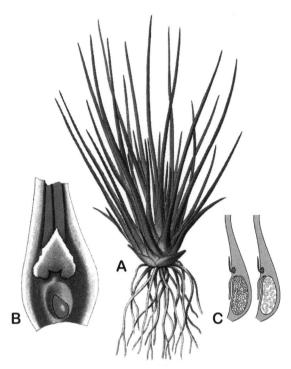

Fig. A10 Isoetes, quillwort. A. Classic illustration of a quillwort showing a tight whorl of long narrow leaves and a tuft of roots. The leaves are helically borne on a short, erect stem, a corm. The entire plant would be about 20–25 cm tall and growing in shallow water. Roots show Y-branching. B-C. A large oval sporangium is embedded and partially covered by the broad base of a fertile leaf and can contain either many microspores (C—left) or many, but fewer, megaspores (C—right). The flap of tissue above the sporangium (dark gray in C) is a vestigial appendage called a ligule (tongue) that is characteristic of the heterosporous clubmosses. (Image sources: A-B—After illustration from Thomé 1885; C—The Author)

common than most people realize. The 50 species of quillwort are widely distributed in the temperate zone and in alpine areas of the tropics. Quillworts grow from a short, stocky upright stem, a corm, with two or four lobes—that is, one or two dichotomies, and although it is not very woody, some secondary xylem is produced. Apical meristems at the top and bottom of the vertical axis produce either a helix of leaves or a helix of dichotomously branched roots (Fig. A10). The leaves have a broad clasping base tapering to cylindrical apices up to 25–30 cm long, the largest microphylls found among living plants. Befitting an aquatic plant, the leaves have air chambers running vertically around a central vascular strand. Sporophylls look the same as sterile leaves, so no strobilus is formed. Sporangia are borne within the expanded leaf base and a prominent ligule occurs embedded in the upper/inner surface just above the sporangium

(Fig. A10 B-C). The apical meristems occupy deep apical grooves between lobes of the corm. New tissues and appendages continue to grow up and out in a spiral. With their spirally arranged scars left from leaves and roots, the corms of quillworts resemble *Stigmaria*, the fossil rhizomes of their much larger antecedents, arborescent clubmosses of the Carboniferous.

Usually subject to seasonal environments, a whorl of microphylls is produced each year. The outermost leaves in the whorl will bear megasporangia. Continuing inward, the leaves will bear microsporangia, and the innermost leaves, usually bearing abortive sporangia, are sterile. The sporangia are quite large, often several millimeters across, and in section have sterile partitions. Spores are liberated by the eventual shedding of the sporophylls and rupturing of the sporangial walls. The development of the gametophytes is endosporic and similar to that of *Selaginella*. Dispersal is aided by the aquatic environment, and relatively more megaspores are produced. Each male gametophyte produces only four multiflagellated sperm. Young sporophytes can be quickly established in culture in a pond no larger than a petri dish.

Fossil Clubmoss Stem Groups: *Zosterophyllum, Nothia, Asteroxylon, Baragwanathia,* and *Drepanophycus*

In the Lower Devonian a number of fossil genera existed that have been interpreted as the earliest ancestors of the clubmosses, organisms intermediate between the rhyniophytes and the clubmosses proper and treated as stem groups, possessing some but not all of the characters defining the clubmoss lineage.

Zosterophyllum (Fig. 8–3; Fig. 8–4; Fig. 8–13) is reconstructed as producing a cluster of naked, dichotomous aerial axes up to 15 cm tall. The globose or flattened, kidney-shaped sporangia are borne clustered near the apices of aerial shoot branches. Their axes have a simple, solid core of vascular tissue like the rhyniophytes, which is interpreted as the primitive condition. One interesting feature is that a cuticle and stomates are found on upper branches but not on lower portions of the aerial stems. This together with slightly flattened axes suggests *Zosterophyllum* grew as emergent vegetation in a shallow-water habitat. Also consistent with this interpretation was their H-branching growth form, which produced tufts, clusters of aerial shoots (Fig. 8–4; Fig. 8–13). Another feature of zosterophylls is a coiled development of shoot apices, a feature that becomes common in ferns and other early vascular plants but a feature absent from other clubmosses. Other zosterophyll genera—for example, *Nothia* and *Sawdonia*—had aerial axes covered with multicellular outgrowths of

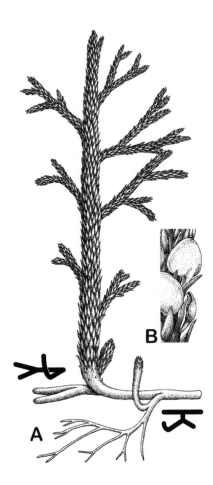

Fig. A11 *Asteroxylon*, representing a clubmoss stem group. A. This reconstruction shows a rhizome and a "leafy" aerial shoot with main axis and lateral branches. The rhizome Y-branches at its apex (left) where one branch will continue the rhizome and the other will produce an aerial shoot or a K-branch. A K- branch (lower right) produces an aerial shoot and a rootlike branch with more Y-branching. B. Flat, kidney-shaped sporangia as they would appear on the aerial shoot. (Image sources: A—After Kidston and Lang 1920; B—After Stewart and Rothwell 1993)

various sizes and shapes (Fig. 8–2), and some axes show the beginning of a shift toward monopodial branching via unequal dichotomies. The fact that all zosterophylls bear sporangia terminally on short lateral stalks is consistent with their origin as terminally positioned sporangia of rhyniophytes. Zosterophylls lack the microphylls of true clubmosses.

Genera such as *Asteroxylon, Baragwanathia,* and *Drepanophycus* are considered intermediates, stem groups, between the zosterophylls and clubmosses. These genera have a general resemblance to some modern clubmosses—for example, especially *Lycopodium* and *Palhinhaea*. The aerial axes of *Asteroxylon* (Fig. A11) consisted of a main stem and lateral branches. Their axes were covered with prominent enations. While not vascularized, a vascular strand extended out from the central cylinder toward the base of each enation, meaning that the growth of the enations influenced the development of the vascular tissue, which makes these leafy append-ages microphylls according to a broader interpretation of the microphyll concept (Kenrick and Crane 1997). The prostrate rhizomes of *Asteroxylon* and *Drepanophycus* produced K-branches (Fig. A11 A; cf. Fig. 8–13) that each resulted in a pair of axes, an aerial stem and a rootlike branch that grew into the substrate. That roots evolved from such modified stems is widely accepted, many fossils of early vascular plants offer examples of intermediates, and roots retain the solid core of vascular tissue like that of primitive stems. As the generic name suggests, *Asteroxylon* has a vascular cylinder with a solid core of xylem that is star shaped in cross-section.

Fossil Arborescent Clubmosses

A lineage of clubmosses characterized by the development of arbores-cence, woody trees, and heterospory arose in the early Devonian. The best known genus is *Lepidodendron*, a name applied to the entire organ-ism and the fossil trunk bearing the characteristic diamond-shaped leaf scars. As is often the case with large organisms, no single intact specimen has ever been found, and different parts were found as separate fossils and named separately as form genera. Thus roots, rhizomes, leaves, stems, cones all were given separate generic names. When physical connections are found and where structures show features that are shared or just fit together, the entire plant can be reconstructed, which is then referred to by the oldest form genus' name. *Lepidodendron* is reconstructed as a tree of the Carboniferous coal swamps (Fig. A12), wetlands that surrounded a shallow inland sea covering the lower Midwestern part of North America some 345 mya. Reconstructions suggest these trees could reach a height of 30 m and have a trunk whose basal diameter reached 1 m. To keep such a large tree erect, massive dichotomously branching rhizomes, named *Stig-maria*, radiated outward from the base of the stem (Fig. A12 A). *Stigmaria* are characterized by a helix of circular root scars (Fig. A12 B). The crown of such trees produced what were probably pendent branch systems that become more dichotomous branches as they approached their terminal portions. The leaves were long and awn-shaped, triangular- to diamond-shaped at the base, flattening into a blade in more distal portions, and of variable length in different species and genera. Some could be in excess of 40 cm long, although most were less than 1 cm wide. Imagine the impres-sive weeping form of this huge clubmoss tree! What a beauty!

Cones composed of sporophylls and sporangia could be over 30 cm in length and over 5 cm in diameter. Cones on some genera could bear both megasporangia and microsporangia. Megasporangia produced only

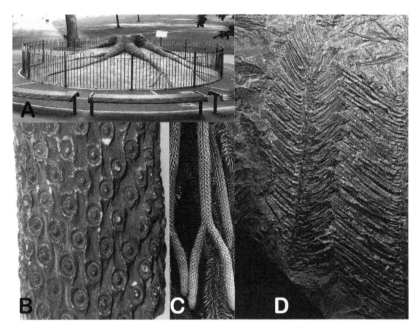

Fig. A12 Fossil remains of the arborescent Carboniferous clubmoss *Lepidodendron*. A. Base of the trunk and spreading Y-branched rhizomes (*Stigmaria*) some 5–6 m across. Such fossil trunks were up to a meter in diameter, indicating the tree could be over 30 m tall (more than 100 feet). B. Fossil of *Stigmaria*, the rhizome, bearing a helix of circular root scars. C. In real life the leafy branches would have looked something like the pendant shoots of this living, tropical relative of *Lycopodium*. D. Fossil axis bearing helix of long, narrow leaves called *Lepidophyllum*. (Image sources: A—Courtesy of D. Spencer, Wikimedia Creative Commons; B—Courtesy of Verisimilus, Wikimedia Creative Commons; C—The Author; D—Courtesy of Smith609, Wikimedia Creative Commons)

a single megaspore, which was not shed. Therefore, development of the endosporic female gametophyte took place while still attached to the sporophyte. Fertilization would result when microspores containing endosporic male gametophytes would disperse into the vicinity of female gametophytes still residing on the sporophyte tree. Following fertilization, a sporophyte embryo would begin developing within the female gametophyte's tissue. This unit, a megaspore containing a female gametophyte and its developing sporophyte embryo, fits the general definition of a seed. The entire unit was dispersed only after successful fertilization. This delay in spore dispersal is the critical event allowing the large, durable sporophyte to protect the megaspores, female gametophytes, and developing embryo from the environment, while providing greater resources for the female gametophyte and embryo of the next sporophyte generation. The concept of a seed is considered in more detail in chapter 9, but these arborescent clubmosses were the first land plants to evolve a "seed" habit, even if technically not a seed.

A second lineage of arborescent clubmosses is represented by *Protolepidodendron*, and as the name suggests, members of this lineage were once regarded as predecessors to *Lepidodendron*-type clubmosses; now they are regarded as having a separate origin from within earlier-appearing homosporous clubmosses. Not all members of this group were arborescent, and the herbaceous genus *Selaginella*, which first appears in the Carboniferous, perhaps had relatives of *Protolepidodendron* as an ancestor.

As the Carboniferous came to a close, a change in world climate marked the end of the warm, moist, swampy habitats and coincided with the extinction of the arborescent clubmosses. *Isoetes*, quillwort, appears to be the only living relative of these clubmoss trees. Although *Isoetes* bears little resemblance to a tree, its microphylls, sporangia, and corm-like axis are structurally similar to a series of more recent fossil genera representing smaller and more compact plants—for example, *Pleuromeia* (Triassic) and *Nathorstiana* (Lower Cretaceous), forms interpreted as increasingly reduced from larger, more arborescent ancestors leading to *Isoetes*.

Conifers and Ginkgoes

Conifers

(synonyms: evergreens, Coniferophytes [-ales, -opsida])

Conifers are better known to most people as "evergreens" and many do retain their tough scaly or needle leaves throughout the winter, but since many conifers have soft flexible needles that drop each fall[3]—for example, larch, bald cypress,[4] and dawn redwood, evergreen is not a good common name for conifers. Conversely, many flowering plants are evergreen and their prevalence increases as you move into the subtropical and tropical areas. Similarly many people call all evergreen conifers "pine" trees or Christmas trees because most people cannot distinguish the major genera—for example, firs, pines, spruces, and douglas-firs, four of the most common ornamental genera used as Christmas trees.[5] Conifer actually means cone-bearing and most have both pollen and seed cones, which are really cones producing microspores and cones producing megaspores. However, conifers are neither the only gymnosperm nor the only vascular plants to have cones, so even conifer is not a particular suitable common name.

Conifer leaves are generally rather simple, ranging from small scales that clasp the stem, to needles, to broad simple blades. All have either a single, central vascular strand or, if broad, parallel veins. Needle leaves range from a few millimeters long to 20–30 centimeters long. Some species shift between needle leaves and scale-like leaves, depending on their state of growth—for example, junipers. Some needles are hard and stiff, others are flexible and lax; needles can taper, be four-sided, or be flat. Needle leaves are borne helically on the stem, although reorientation sometimes produces the appearance of a flattened, two-ranked arrangement along the branch—for example, hemlock (*Tsuga*). Scale leaves are generally borne on stems in alternating pairs or whorls of three. They can be flat and appressed to the stem or folded (keeled) around the stem, sometimes both in alternating pairs, producing flat twigs. Some genera have both long shoots with typical internode elongation and short spur shoots with essentially no internode elongation, producing a tight helical whorl of leaves—for example, true cedar[6] (*Cedrus*) and larch or tamarack, *Larix*. Not all conifers have needle-like or scale-like leaves. Broad-leafed conifers are rather unfamiliar in the northern temperate zone, so often conifers from the Southern Hemisphere (*Podocarpus*, *Agathis*, and *Araucaria*) seem strange looking (Fig. A13).

The leaves of temperate evergreen conifers must endure low temperatures and very drying conditions during the winter (or all year long at high altitudes). Such tough leaves have very thick cuticles, stomates sunken into pits, and a thick-walled layer of sclerenchyma (sklair-INK-eh-mah—thick-walled ground tissue for support or protection) below the epidermis, which makes the needle tough and hard. In pines the cells of the photosynthetic ground tissue have their cell walls reinforced with ridges projecting into the cell lumen, a further toughening of what are usually thin-walled cells. The central vascular strand of the needles includes some sclerenchyma and is surrounded by radially oriented cells forming transfusion tissue to promote cell-to-cell movement of photosynthates and water. Resin canals are common, usually positioned just beneath the epidermis, a position that suggests they function to prevent herbivory. Some scale leaves have an external gland secreting aromatic oil for a similar purpose. Conifers also inhabit some fairly hot, dry climates, or places with easily drained sandy soils, and the same leaf adaptations work well under these conditions too.

All conifers are woody gymnosperms (trees or shrubs). The wood is simple and homogeneous, composed exclusively of tracheids in long, straight radial files and single-cell-wide rays (Fig. 8–8 A; Fig. 8–10). The tracheids have circular bordered pits (Fig. 8–8 A) and sometimes reach several millimeters in length, which is why conifer wood is so good for making paper. While the grain is not particularly attractive, conifer wood is valuable for rough lumber and plywood. Conifer wood is called softwood because it is generally of lower density than the hardwoods, such as oak and maple, and it lacks thick-walled fibers, making it easier to cut (i.e., soft). Although usually composed of boxy parenchyma cells, ray cells

Fig. A13 Broad-leafed conifers from the southern hemisphere. A. Podocarp (*Podocarpus henkelii*). B. Kauri pine (*Agathis robusta*). C. Monkey-puzzle tree (*Araucaria araucana*). Leaves are leathery to quite stiff in C. (Image sources: A—Courtesy of Stan Shebs, Wikimedia Creative Commons; B—Courtesy of Kahruoa, Wikimedia Creative Commons; C—The Author)

can differentiate as short boxy tracheids, thus enhancing radial transport. Resin canals originate in the rays and all other parenchyma tissue by the death of central canal cells, which leaves a sheath of secretory cells surrounding the canal. When tissues are damaged, the oozing resin helps protect the tissues and repels by being both sticky and smelly. Amber is fossilized resin, and occasionally amber includes trapped insects and other biological specimens.

Most conifers produce both pollen and seed cones. Conifers produce their ovules in pairs at the ends of modified lateral stems, which in most genera are aggregated into seed cones. Yews and podocarps bear seeds individually on the ends of modified lateral branches. Some conifers are dioecious (dye-EE-see-us—"two houses"), producing either pollen or seed cones but not both. Many more are monoecious (moh-KNEE-see-us—"one house"), meaning a single individual bears both pollen cones and seed cones. Generally seed cones are produced at the ends of branches, mostly toward the top of the tree, and pollen cones are borne on short lateral branches, mostly toward the bottom of the tree. This arrangement reduces the incidence of self-pollination, but a lot of variation exists, particularly in bushier forms.

Pollen cones are composed of a helix of modified leaves, sporophylls, each bearing two microsporangia or two clusters (sori?) of two to six microsporangia on their lower surface[7] (Fig. A14). Some fossil genera have even more numerous sporangia, so modern conifers are regarded as having a reduced number, especially when just two. Pollen cones are mostly fairly small, only a few mm long, although a few can be up to several centimeters long, and short-lived, dropping after shedding their pollen. In the yews (Taxales) several microsporangia are clustered in a whorl on modified axes at the ends of short lateral branches that bear several sterile bracts below, producing a cone-like structure (Fig. A14 F).

Seed cones are composed of a helical whorl of seed-bearing scales (Fig. A15). When newly formed, seed cones are often only 4–10 mm long. When mature, they can remain fairly small, 10–12 mm long, or they can become 40–50 cm long. Two ovules are found on the upper surface of each seed-bearing scale. The apex of these ovules is oriented inward toward the cone axis. In most conifers the seed-bearing scale, which is hard and dry at maturity, is of a composite nature consisting of a modified branch and a modified subtending leaf, a bract (Fig. A15 B-C). In most conifers the branch and bract fuse into one unit during development, but they remain separate and distinct in the seed cones of douglas-fir (*Pseudotsuga*), which clearly demonstrates that interpreting a cone scale as a modified branch and bract is correct (Fig. A15 C-D). In young seed cones of other genera, at the stage of pollination, the relationship between the seed-bearing scale and its subtending bract is clear, and in some species, precociously developed bracts function to assist in capturing pollen—for example, fir and douglas-fir. Fossil conifers assist in the interpretation of conifer seed cones; many fossil conifers have distinct seed-bearing branches subtended by bracts, and of course, the entire group has its common ancestry among pteridophytes and trimerophytes that bear sporangia either on the margins of megaphylls or terminally on Y-branched lateral branches. In the tropical *Podocarpus* and the yews, ovules are borne terminally and singly on short lateral branches. In yews a fleshy, colorful aril grows up and around the ovule, functioning like a fleshy fruit[8] to attract birds for seed dispersal (Fig. 10-2).

Fig. A14 Pollen cones of conifers. A. Larch. Each modified leaf (microsporophyll) bears two microsporangia (arrows) on its lower surface. B. Pine. Cluster of pollen cones showing large number of helically arranged fertile leaves composing the cone. C. Fir. Pollen cones with similar organization as pine but with fewer fertile leaves. D. Cedar (*Cedrus*). Long, slender and erect pollen cones of similar organization. E. Monkey-puzzle tree (*Araucaria*). Central portion of a mature pollen cone showing the large, leafy sterile apex of each fertile leaf (microsporophyll), each of which bears two clusters of microsporangia (sori), now empty, on their lower surface. F. Yew. Pollen cones showing cone-like bracts at the base (portion closet to twig) and several clusters of microsporangia at the apex of the "cone" axis. (Image sources: A—Courtesy of Yummifruitbat, Wikimedia Creative Commons; B—Courtesy of Riverbanks Outdoor Store, Wikimedia Creative Commons; C—Courtesy of Cruiser, Wikimedia Creative Commons; D—Courtesy of dontworry, Wikimedia Creative Commons; E—Courtesy of Forest and Kim Starr, Wikimedia Creative Commons; F—Courtesy of Lamiot, Wikimedia Creative Commons)

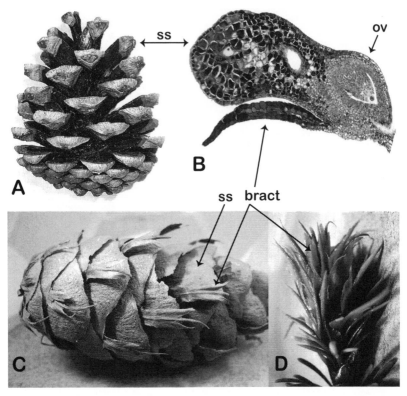

Fig. A15 Seed cones of pine and douglas-fir. A-B. Pine. A. Mature seed cone of pine showing helically arranged seed-bearing scales (ss). B. Longitudinal microscopic view of a young seed-bearing scale (ss) and its associated bract beneath it. The ovule (ov) is positioned with its apex oriented toward the cone axis, to the right. The bract is as long as the seed-bearing scale at this juvenile stage. At maturity the bract is much smaller, has fused to the seed-bearing scale, and is not discernible. The clear oval structure in the seed-bearing scale is a resin duct. C-D. Douglas-fir. C. Mature seed cone of douglas-fir showing stiff, papery bracts extending out beyond the seed-bearing scales (ss). D. Immature seed cone at or just after pollination and the still very small seed-bearing scales are hidden among the precociously enlarged and overlapping bracts. The bracts do not enlarge much after this stage, whereas the seed-bearing scales do enlarge, but the two don't fuse as in pine. (Image sources: A—Courtesy of Menchi 2004, Wikimedia Creative Commons; B—Courtesy of M. Clayton, U. Wisconsin Plant Teaching Collection; C-D—The Author)

Many conifers grow in low-diversity forests, a few species covering huge expanses, so wind pollination works efficiently, although copious amounts of pollen are still produced. Pollination occurs early in the growth season when new seed and pollen cones are produced. The young seed cones expand slightly, opening spaces between the seed-bearing scales. Pollen grains fall into the seed cones, which then reclose. Pollen drops are exuded from the apex of the ovule to capture or mop up the pollen grains and with reabsorption pull pollen grains into the ovule's pollen chamber. In *Ginkgo* and yew, wind-borne pollen grains are captured directly on pollen drops. Pollen grains of many conifers have asymmetrical bladders (Fig. A16 B′), which for years were described as aiding in wind pollination. A study of the aerodynamics of conifer pollen indicates the bladders slow the pollen grain's settling rate, which would increase dispersal (Schwendemann et al. 2007), but a careful analysis of ovule and pollen orientation suggest these bladders also function as water-wings, flotation devices to orient the pollen grain in the pollen drop so that the germination of the pollen tube occurs toward the apex of the megasporangium (Doyle 1945). In conifers the pollen grains all germinate from their distal end (from that portion of the spore that faced away from the center of the meiotic tetrad). Nothing in biology prevents such adaptations from having dual functions.

At the time of dispersal, pollen may still be at the microspore stage (uninucleate), or it may contain a partially or wholly developed male gametophyte, which in any case involves only a few mitotic divisions (Fig. A16). Two, four, eight, or sixteen vegetative cells can be formed,

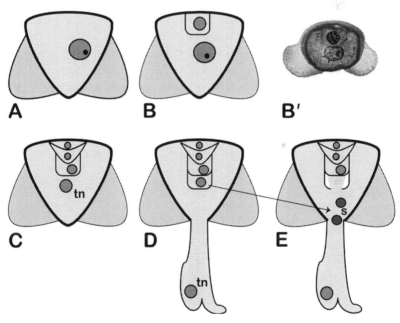

Fig. A16 Conifer pollen grains and male gametophyte development. A. Pollen—microspore stage. B. After first division. B′. Pollen grain at two-cell stage as in B showing two sack-like wings; dispersal often takes place at this stage. C. One or two more divisions produce one or two vegetative cells (top). The next division produces a generative cell and a tube nucleus (tn), which directs growth of the pollen tube (D-E). D. The generative cell divides to produce another sterile cell and a spermatogenous cell. E. Division of the spermatogenous cell produces two sperm nuclei (s). (Image source: A-E—The Author; B′—Courtesy of and with permission of M. Clayton, U. Wisconsin)

depending on the genus or family. A tube cell, which comprises most of the microspore cytoplasm, and a generative cell are formed. The generative cell will divide to make a "stalk" and "body" cell, like in cycads (Fig. A21). The body cell will divide to produce two sperm cells or two sperm nuclei. The pollen tube must grow from the pollen chamber a very short distance through megasporangium wall tissue toward the apex of the female gametophyte.

The temporal delay between pollination and fertilization can be readily observed in seed cone development. In pines, seed cones of three ages can be observed in the spring: mature cones that have shed or are ready to shed their seed, medium-sized one-year-old seed cones pollinated the previous spring, and the brand-new seed cones and pollen cones. The temporal separation between pollination and fertilization is about 13 to 15 months in pine. This means that in the one-year-old seed cones, those pollinated the previous spring, fertilization has still not quite happened; those males are still waiting! The mature seed will not be shed until that fall or the following spring, two years after pollination. In other conifer genera, pollination to seed dispersal takes one year, and fertilization follows pollination by only a couple of months.

The embryos of conifers usually have several cotyledons, and the embryo will be embedded centrally in the fleshy, nutritive tissue of the female gametophyte. The edible seeds of *Pinus edulis*, pinyon pine, are called pine nuts or pignolis and they form an important food resource for birds and mammals in their communities. If you have these around, the female gametophyte can be readily dissected to observe the embryonic pine tree within.[9]

Gingkoes

(synonyms: Ginkgophytes [-ales, -opsida])

Presently this lineage is represented by only a single species, *Ginkgo biloba*, but the fossil record shows many extinct species bearing the distinctive bi-lobed leaves (Fig. A17 A-B). Fossil ginkgo-like trees were fairly common, widespread, and important components of forests over vast portions of the supercontinent Laurasia (North America plus Europe plus Asia). The other supercontinent constituted by the remaining land masses was Gondwana (Fig. 9–5). *Ginkgo* is sometimes called a living fossil because western botanists knew of the fossils before living trees from Japan and China were known, and because all known Asian populations of *Ginkgo* are cultivated groves associated with monasteries. *Ginkgo biloba* used to be considered extinct in the wild, but now some wild forest populations

Fig. A17 Leaves and reproductive structures of *Ginkgo*. A. Fan-shaped, bilobed leaf of *Ginkgo biloba*. B. 49-million-year-old fossil leaf impression of an extinct ginkgo. C. Spur shoots bearing several ovuliferous branches or sporophylls, each with a pair of apical ovules. D. Spur shoot bearing several pollen cones composed of a loose helix of reduced sporophylls, each bearing two microsporangia. E. Spur shoot bearing two ovuliferous branches, each with one mature (ovule) and one abortive seed. F. Mature embryo surrounded by tissue of female gametophyte. (Image sources: A—Courtesy of J. L. Staub, Wikimedia Creative Commons; B—Courtesy of Kevmin, Wikimedia Creative Commons; C and E—Courtesy of H. Zell, Wikimedia Creative Commons; D—Courtesy of Marcin Kolasinski, Wikimedia Creative Commons; F—Courtesy of Curtis Clark, Wikimedia Creative Commons)

in China are known (Tang et al. 2012). Even this species might have become extinct without human intervention in the form of cultivation. Now a widely spread temperate zone ornamental, *Ginkgo* has proven very hardy and very tolerant of diverse climates and urban conditions. The tree suffers from no specialized herbivores, pests, or diseases, an evolutionary consequence of its near brush with extinction. When populations of the species were so reduced in numbers and very fragmented, just a few cultivated groves of trees, all the organisms that depended on *Ginkgo* (specialized consumers and disease organisms) became extinct. Models of extinction suggest such an extinction cascade can happen when a dominant plant species diminishes to dangerously low numbers. This phenomenon explains why conservation efforts must concentrate on protecting intact habitats and communities. No one understands what brought such a hardy tree to near extinction, but *Ginkgo biloba* almost suffered the fate of all the other species in this lineage.

Because of its characteristic broad leaves, classic examples of megaphylls, and deciduous habit, the obvious similarities between *Ginkgo* and conifers are often overlooked. The trees are conical or pyramidal when young, but the lateral branches are oriented upwards at an angle rather than held horizontally as in many conifers. When mature the pyramidal shape becomes obscured because the continued apical growth of lateral branches produces a rounded crown. *Ginkgo* trees are dioecious, producing either pollen or seed. Pollen trees produce lax, pendulous pollen cones, bearing a helix of loosely arranged sporophylls each bearing two microsporangia (Fig. A17 D). The organization is similar to pollen cones of conifers, but the sporophylls are not so tightly packed into a cone. On seed trees, ovuliferous branches are borne on spur shoots subtended by a vegetative leaf or a bud scale (a modified leaf). Each ovuliferous branch bears a pair of terminal ovules (Fig. A17 C, E), and because vegetative leaves and ovuliferous branches are positionally and developmentally interchangeable, the ovuliferous branch can even be interpreted as a modified fertile leaf, a sporophyll, bearing two ovules on its margin (Rothwell 1987). This interpretation is supported by developmental anomalies where only one half is fertile, bearing one ovule, and the other half is sterile, and more or less a typical half leaf blade. Sometimes small leaves will bear marginal ovules, and some fossils species have more branched ovuliferous shoots. Considering the megaphyll hypothesis this homology is not unexpected, and indeed, its explanatory power contributes to our confidence.

Like other gymnosperms, *Ginkgo* ovules have a small apical opening, and pollen is captured via an extruded and reabsorbed pollen drop. As the developing seed enlarges, the seed coat differentiates into an outer fleshy layer and an inner stony layer, which is similar to cycads. When mature, the entire seed looks something like a small, overripe apricot with a yellowish-orange color. However, the resemblance is quite superficial because the outer fleshy layer produces a foul (putrid) odor whose function remains uncertain because the one surviving species of ginkgo has been long divorced from its natural community. Was this a device to attract seed dispersers or discourage seed predators? An experimental study suggests the fleshy coating may prevent premature germination (Holt and Rothwell 1997), but this does not preclude another function. No matter, the fleshy seed coat is why people want only pollen-producing, not seed-producing, ginkgo trees in their yards or along sidewalks.[10]

The development of the large female gametophyte is similar to that of cycads. In the fall when the seeds are shed, the embryos are 4–5 mm long (Fig. A17 F) and ready to germinate (Holt and Rothwell 1997), a finding contrary to earlier reports of immature embryos in newly dispersed seeds. Cold weather induces dormancy so germination occurs in the spring.

Coniferophytes: Cordaitales and Voltziales

These two groups of fossil gymnosperms are the first coniferophytes, the first gymnosperms to display a coniferous organization. As such they are considered ancestral to conifers. The Cordaitales are the oldest, appearing in the upper middle Carboniferous and continuing into the upper Permian, broadly overlapping in time with the seed ferns, pteridosperms (other early fossil gymnosperms). The Voltziales appear in the late Carboniferous and persist until modern families of conifers (pines, cypresses, podocarps, araucarias) appear in the late Jurassic. Gingkoes are an exception; the lineage appears much earlier, in the Permian.

Cordaites (core-DYE-teaz) is one of the best known genera, and it may have been one of the largest trees of the Carboniferous. However, some reconstructions (Cridland 1964) present cordaitalean gymnosperms as mangroves, shrubby or sprawling trees with prop roots adapted to muddy areas of marine coastal areas and estuaries. Their leaves were flat and long (up to 1 m), strap-like, up to 12–15 cm wide with parallel veins running base to apex. The reproductive organs were cones composed of overlapping bracts or scales below and a helical whorl of more elongate sporophylls within and above. Each microsporophyll usually had six terminal microsporangia. Ovules were borne on similar structures, but like *Ginkgo*, in place of a megasporophyll, one or more ovules were produced terminally on a modified shoot extending out beyond sterile bracts.

Voltziales were studied extensively by the famous paleobotanist Rudolf Florin in the 1940s and 1950s. Vegetatively these plants most closely resembled Norfolk Island pines (*Araucaria*) because they had whorls of limbs and small needle-like leaves arranged helically around stems like a bottle brush. The reproductive structures are rather complex and difficult to describe because they are of a compound organization, fertile ovule-bearing shoots and associated subtending bracts. Seed cones of modern conifers also have a compound organization, being composed of units consisting of a modified, ovule-bearing shoot and a modified leaf or bract (see Fig. A15).

(synonyms: Cycadophytes [-ales, -opsida, -ophyta])

All living and extinct cycads, and cycad-like plants, used to be classified as one taxonomic group with two lineages: one lineage with about a dozen living genera and one of fossil plants called the cycadeoids. Now, however, the cycadeoids are treated as a group of pteridosperms similar in appearance to cycads but with very different reproductive structures (Fig. 9–10). Sometimes cycads are called "living fossils," leftovers from the age of dinosaurs, but according to recent molecular studies, the modern genera of cycads all arose from a common ancestry about 12 million years ago (Nagalingum et al. 2011), a surprisingly recent origin considering their worldwide tropical and subtropical distribution where genera are restricted to particular continents, suggesting an older origin. Cycads were widespread and diverse during the Jurassic and Cretaceous, some 200 to 65 mya, but they suffered considerable extinction after the K-T event associated with the demise of dinosaurs and the rise of flowering plants. A climate change producing more grassland savannas is associated with the recent diversification of cycads to about 300 species, so our living species are not the living fossils they are sometimes called.

To the casual observer, cycads look palm like because they generally have short, stout, mostly unbranched or sparsely branched trunks bearing an apical whorl of long, pinnately compound leaves[11] (Fig. A18). Most cycads grow no more than a couple of meters tall; some are very short with low-growing or subterranean stems. A few cycads grow to several

Fig. A18 Cycads. A. A specimen of *Encephalartos ferox* showing a whorl of pinnately compound leaves and two large (pineapple-sized) seed cones. B. Specimen of *Encephalartos princeps* showing a stout, unbranched trunk with leaf scars covering the stem and a terminal whorl of pinnately compound leaves. (Image sources: A—Courtesy of Wendy Cutler, Wikimedia Creative Commons; B—Courtesy of JMK, Wikimedia Creative Commons)

meters, and one species grows to 20 meters, *Lepidozamia hopei* of far northern Queensland.[12] Many cycads have stout, unbranched or monopodial trunks (Fig. A18 B); some produce basal or small lateral branches. A few low-growing forms—for example, species of *Zamia*, may Y-branch one to several times much like some ferns with short upright stems—for example, cinnamon fern. Cycads are woody, but the accumulation of wood remains limited and the stems are composed largely of cortex. While palm like in appearance,[13] cycad leaves do not encircle the stem as the leaves do in palms, which produces the characteristic leaf scars that encircle palm trunks. However, like palms and other monocots (and a few dicots), the vascular supply for a cycad leaf arises from all around the vascular cylinder, so some vascular strands from the far side must circle halfway around the stem to enter the leaf stalk.

Cycads are rather slow-growing, long-living plants.[14] Although the stems have no growth rings, some genera produce an annual whorl of leaves, and the scars left from these annual whorls can be counted, yielding estimates that some cycads only 1 or 2 meters tall can be several hundred to one thousand years old. The leaves are generally tough and leathery, and many cycads grow in drier climates where tough, hard, leathery leaves are adaptive. Grassland savannas and tropical or subtropical shrublands are typical habitats. A few are native to wetter tropics. Cycads growing in moister habitats have somewhat softer, more flexible foliage. Foliage toughened by a lot of sclerenchyma and tough edges can serve to make long-lasting, desiccation-resistant leaves, and leaves too tough for many herbivores, and in the case of cycads, tough foliage probably functions to some degree in all three ways. Some cycad foliage is so tough, so stiff and hard, and so well armed with spines—for example, the aptly named *Encephalartos horridus*—that the plants are used to form living livestock fences in southern Africa to keep predators at bay. Cycad leaf development is very fern-like; fronds uncoil from fiddleheads, and in *Cycas*, so do the leaflets (Fig. A19). In fact, it is perfectly correct to refer to cycad leaves as fronds and their leaflets as pinnae.

The leaflets of cycads are generally simple, long, and narrow, and generally entire smooth edged, although toothed and lobed margins are common in some genera, as are spine-tipped lobes and apices, and a few with broader leaflets. In some species basal leaflets are reduced in size, forming two lateral rows of spines down the leaf stalk. In still other cycads the leaf stalk bears spines whose size and distribution show no leaflet homology. The leaf organization and venation of leaflets is used to delimit three or four families of extant cycads (Jones 1993). Most cycads have pinnately compound leaves, meaning two rows of leaflets are aligned on either side of a central axis. *Bowenia*, from the wet tropics of Queensland, has a twice pinnately compound leaf (pinnate leaflets on secondary leaf stalks), a distinctive enough feature that some taxonomists place this genus in its own family (Boweniaceae). Otherwise, *Bowenia* is placed in a family together with the very fern-like genus *Stangeria* from southeastern Africa. Leaflet venation consists of either a single vascular strand in the midrib (*Cycas*, Cycadaceae), a single-stranded midrib with Y-branched secondary veins (*Stangeria*, Stangeriaceae), a very ferny venation, or Y-branched venation at the base of the leaflet making many parallel and similar-sized veins without a distinct midrib vein, which is found in all other cycad genera (placed in Zamiaceae) and *Bowenia*.

Like ferns, cycads bear their sporangia on modified leaves (sporophylls) aggregated into cones (strobili). Cycads are dioecious (literally "two houses"), meaning that each plant will bear either pollen cones or seed cones but never both. In most cycads the sporophylls are highly modified, hardly looking like leaves, and tightly aggregated into very discrete

Fig. A19 A whorl of uncoiling pinnately compound "fronds," fiddleheads, of a specimen of *Cycas revoluta*. The leaflets are uncoiling from along the inside of the curved leaf stalk. (Image source: Courtesy of Esculapio, Wikimedia Creative Commons)

cones (Fig. A18 A; Fig. A20 B, D). In *Cycas*, the megasporophylls are large, leaf-like, and only loosely aggregated into cones (Fig. A20 A). These large megasporophylls bear several ovules in two rows, one on either side of the leaf stalk. The terminal portion of the megasporophyll bears a number of reduced, basally fused leaflets. The most reduced megasporophylls bear a pair of ovules on the bottom margin of a shield-like structure (Fig. A20 B-C). It would be difficult to consider this a fertile leaf if not for intermediate forms in both living and fossil plants.

The microsporophylls are all modified through reduction, although the microsporophylls of a couple of genera still bear a pair of vestigial leaflets. The microsporangia are similar in development, organization, and

Fig. A20 Cycad cones and sperm. A. Seed cone of *Cycas*. Several ovules are borne on the margins of very leafy cone scales that are in fact modified leaves (sporophylls) with a substantial sterile apex bearing vestigial leaflets. The seed cones in this genus are large and are rather loosely arranged. B-C. Seed cones of *Zamia*. The modified leaves composing these cones are highly modified and tightly packed together in the cone. C. One sporophyll (circle and arrow in B) bearing two ovules and no leafy vestiges. D-E. Pollen cone of *Zamia*. The pollen cone is about half the diameter and somewhat taller than the seed cone (same species), but here it is somewhat lax as the cone expands to release pollen. E. Pollen cone broken open to show several sporophylls. Each individual sporophyll (oval in D) bears two clusters of some 12–15 microsporangia (sori; circle). Cones this size can produce copious amounts of pollen. F. Illustration of cycad sperm, which is the largest sperm of any organism. It has a large number of flagella in a spiral. (Image sources: A—Courtesy of Dipu tr, Wikimedia Creative Commons; B—Courtesy of Tanetahi, Wikimedia Creative Commons; C—The Author; D—Courtesy of Forest and Kim Starr, Wikimedia Creative Commons; E—Courtesy of David Webb and the Botanical Society of America; F—From Bonnier 1907)

size to those of basal fern lineages[15] and are borne in two clusters, sori, on the lower surface of each pollen cone scale (Fig. A20 D-E). Most of the pollen cones are a few centimeters tall (Fig. A20 D), some are 50–60 centimeters tall, a couple even larger, all capable of producing copious pollen. Many references say cycads are wind pollinated primarily because they are gymnosperms that produce copious amounts of pollen, but this is more legendary than factual. Evidence suggests cycads are basically insect pollinated (Jones 1993), particularly small, understory species of *Zamia* and *Bowenia* where wind dispersal would be inefficient (Norstag 1987; Wilson 2002). Insect pollination in gymnosperms is significant because it indicates that the involvement of animals in spore dispersal greatly predates the flowering plants. In *Bowenia* weevils feed upon, lay eggs in, and have larval development in the soft ground tissue of the pollen cones, activities that do not harm the microsporangia or pollen. The seed cones apparently offer no reward and presumably attract beetles by mimicking the odor of rewarding pollen cones, a situation not unlike the beetle pollination of some magnolialean flowering plants (Armstrong and Irvine 1989; Armstrong 1997) that grow in the same rain forest community.

A microspore becomes an endosporic male gametophyte with just three mitotic divisions (Fig. A21 a-d). The first division produces a vegetative cell. The remaining cell divides again, producing a generative cell and a tube cell, which occupies most of the microspore volume. Later, at the time of fertilization, the generative cell will divide once more, producing two cells that have been called stalk and body cells as if homologous to the stalk and gametangium of an antheridium. The body cell will divide once more, producing in most genera two large sperm cells bearing a spiral helix of many cilia/flagella at one end (Fig. A20 F; Fig. A21 d). At nearly 300 micrometers in diameter, these are the largest sperm known in any organism! In some genera, up to three additional mitotic divisions take place after meiosis, resulting in up to 16 sperm, which as a consequence of being divided in half once, twice, or three times, are one-half, one-fourth, or one-eighth smaller.

At the time of pollen dispersal, the ovulate cones expand slightly, which opens spaces between the megasporophylls, and the cones produce a noticeable odor. Once pollen has had an opportunity to arrive within the cone, either via insect visits or wind, the cone closes and does not reopen until the seeds are ready to be dispersed. In the case of *Cycas* (Fig. A20 A), the cone is more lax during pollination and then closes up more tightly, but ovules on the outer sporophylls are still a bit exposed. Each ovule secretes a sticky pollen drop that reaches out and grabs any males (pollen grains) in the vicinity and, via the reabsorption of the pol-

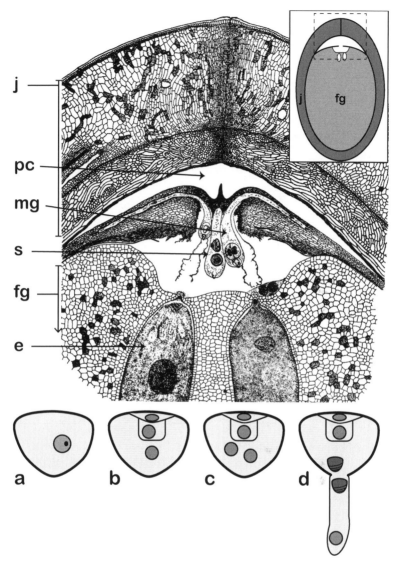

Fig. A21 Diagram of a cycad ovule at the time of fertilization and male gametophyte development. Diagram of the apical region of a cycad ovule (inset—detail from area outlined by dashed box) showing a jacket layer (j) surrounding the female gametophyte (fg). Apical space between the jacket and the megasporangium is the pollen chamber (pc), where pollen resides after being pulled into the ovule by a pollen drop. Pollination occurred some time ago; now fertilization is about to occur as the several male gametophytes (mg) have nearly grown to the female gametophyte. The female gametophyte (fg) has two large eggs (e) and one of the male gametophytes has produced two sperm (s) that will be released "above" the female. Diagrams showing the development of a microspore into a male gametophyte (a-d). a. Pollen—microspore stage. b. First two divisions of the microspore nucleus produce one or two vegetative cells. c. The next division produces a generative nucleus and a tube nucleus, which directs growth of the pollen tube. d. The generative nucleus divides to produce two motile sperm (s) that swim down the pollen tube following the tube nucleus. (Image source: After Chamberlain 1919; line drawings—The Author)

len drop, draws them into the apical pollen chamber of the ovule (between the jacket and megasporangium) (Fig. A21). Once inside the ovule, the pollen grain germinates and the tube cell grows into the megasporangium wall tissue, forming a haustorium from which it derives nutrition.

Within the ovule, a single functional megaspore develops within its megasporangium into an oval-shaped female gametophyte. The mature female gametophyte produces archegonia at the apical end, each with an egg (Fig. A21). Although the archegonia are reduced, they have a whorl of neck cells and one neck canal cell. As the female gametophyte matures, the sporangial wall tissue just below the pollen chamber undergoes an enzymatic dissolution, producing a syrupy liquid into which the sperm are released. A very short swim accomplishes fertilization, but these huge sperm must swim through a thick syrupy liquid.

Pollination is the act of tiny male gametophytes dispersing into the immediate vicinity of jacketed female gametophytes, and although swimming sperm are released, it is only after the males have developed within the ovules, so free water is not needed for fertilization. This clearly demonstrates how the seed habit internalizes the gametophyte generation with the result that the sporophyte is no longer limited in range to habitats where free-living gametophytes are adapted.

Embryonic development commences with production of a suspensor, a short filament of cells that serves to position the embryo in a central portion of the female gametophyte. The embryo develops such that the tip of its root axis is oriented toward the apex. Two large cotyledons[16] develop and remain in contact with the gametophyte tissue, absorbing food during early seedling growth. Cycads have quite large seeds, and since such seeds represent a considerable food resource, they are quite well protected chemically. The downside of large, well-provisioned seeds is that size is often inversely related to their ability to disperse. At the time of seed dispersal, seed cones expand and expose the seeds. The outer layer of the seed coat is often shiny and colored red, pink, orange, or yellow to attract, and presumably reward, animal seed dispersers, mostly fairly large mammals (baboons, bears, hyraxes, monkeys, peccaries).[17] A hard, tough inner seed coat protects the female gametophyte and embryo within. At the time of seed dispersal the embryo may not be mature, and a period of seed dormancy may be needed before the embryo is mature and the seed ready to germinate. At germination, elongation of the embryonic axis below the cotyledons pushes the elongating root through the seed coat at the apex of the ovule.

(synonyms: Filophytes, Filopsida, Filicales, Polypodiopsida)

Ferns have constituted a nearly constant component of vascular plants from the late Devonian to the present, but whether they represent a single lineage or not will require some examination. As a name, fern evokes a mental image of a particular type of plant growing in a cool, moist, shady glen or a rocky grotto or a quiet forest understory. Describing fern fronds, large, highly dissected, compound leaves, without calling them "ferny" is difficult. Yet, many ferns have most "unfern"-like leaves because ferns grow in a surprising (to many people) variety of habitats from terrestrial, some of which are pretty dry, to epiphytic to aquatic. As a result ferns display a number of leaf adaptations: tough, leathery leaves, very reduced leaves, very small and simple leaves, submerged rootlike leaves, and of course, specialized fertile leaves.[18] Some ferns are tiny, some are fairly large, but even large "tree" ferns lack wood, and so, like palms, tree ferns are arborescent in form only. Some living ferns are relicts of ancient groups, but with about 12,000 known species, some groups have diversified along with flowering plants. Fern history and their fossil record date to the late Devonian. The fossil history of two living groups of ferns can be traced back to the Carboniferous; "modern" groups of ferns appear in the Triassic.

Ferns have been variously treated taxonomically, often as an inclusive class called Polypodiopsida (on the basis of the genus *Polypodium*) or an irregular name, Filicopsida, based on *filix*, the Latin for fern. A major

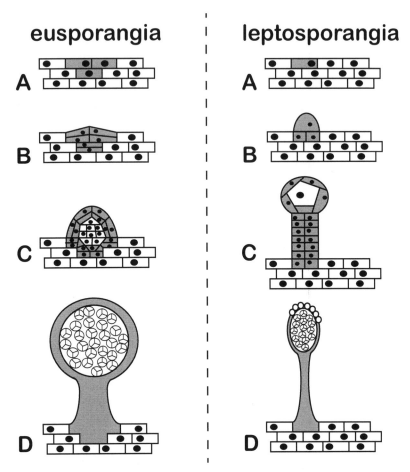

eusporangia | leptosporangia

Fig. A22 Developmental and size differences in eusporangia and leptosporangia. Eusporangia (left) develop from a group of cells in the epidermis and tissue below (A-B, gray cells), have two or more layers of jacket cells (B-C), and have many sporogenous cells (C) making spore mother cells, resulting in large sporangia capable of making thousands of spores (D). Leptosporangia (right) develop from a single initial cell (A, gray cell). The sporangium forms from a single apical cell (B) making a single jacket layer and one sporogenous cell (C). The resulting sporangium is small and delicate, with a limited number of mother cells undergoing meiosis to produce 64, 128, or 256 spores (D). The jacket may have a row or patch of specialized cells, an annulus to open the sporangium (see Fig. A26). The two illustrations are not to scale, which makes the leptosporangium look similar in size to the eusporangium. (Image source: The Author)

lineage of ferns shares a novel type of sporangium called a leptosporangium, which is small, delicate, and produced in great numbers. These sporangia arise from a single epidermal initial cell, have a single layer of jacket cells, and produce a limited number of spores (64, 128, or 256) depending on the number of spore mother cells undergoing meiosis (16, 32, or 64) (Fig. A22). Most leptosporangia have a well-developed annulus, a row or patch of specialized cells with unevenly thick walls that serve to open the sporangium, and in some cases the annulus can actively fling the spores (see Fig. A26).

Mosses, liverworts, hornworts, clubmosses, horsetails, whisk ferns, and basal groups of ferns retain eusporangia (Fig. A22), three living fern families, the marattioid ferns, the adder's tongue ferns, and the cinnamon ferns, plus two fossil groups, cladoxylopsid (klah-DOX-ee-LOP-sid) ferns, a basal sister group to all other ferns, and the coenopteridopsid (syn-OP-terr-eh-DOP-sid) ferns, a lineage basal to marattioid ferns, all the leptosporangiate ferns, and maybe the horsetails too (Fig. A23). Recent molecular studies have placed the whisk ferns as a sister group to the adder's tongue ferns and the horsetails as a sister group to the marattioid ferns. The cinnamon ferns have long been considered transitional between the eusporangiate ferns and the leptosporangiate ferns because the ancient cinnamon ferns actually produce both types of sporangia, a perfect intermediate. The leptosporangiate ferns consist of several distinct lineages, two largely tropical lineages containing the filmy ferns and forking ferns,[19] lineages of climbing ferns, water ferns, and tree ferns, and lastly, a large lineage of eupolypod ferns, which contains by far the majority of living fern species, about 80%. According to this phylogeny of the pteridophytes, ferns do not form a single lineage, but rather than making sense of things, this hypothesis does not explain the distribution of a number of characters and is not well supported by the fossil record either.

Fossil ferns, the cladoxylopsids and coenopteridopsids, date to the late Devonian, and so does the horsetail lineage. A strong relationship is thought to exist between the cladoxylopsids and the Hyeniales, the purported intermediates between the trimerophytes and horsetails, and indeed if these ferns and horsetail progenitors prove to be a single group then the origin of horsetails and coenopteridopsid ferns may trace back to a single group (Stewart and Rothwell 1993). The marattioid ferns have a fossil record back to the Carboniferous, and extant members are clearly relicts. The cinnamon ferns date to the Late Carboniferous. Some of the gleichenioid ferns (*Matonia*, *Dipteris*) date back to the Mesozoic (Triassic, Jurassic, Cretaceous), which makes them contemporaries of dinosaurs. Polypody ferns appear in the Triassic (Collinson 1996). The earliest living genera of polypody ferns, a sensitive fern (*Onoclea*) and deer and dwarf tree ferns (*Blechnum*), did not appear until the Paleoceae-Eocene (Fig. 11–2), lower subdivisions of the Tertiary, some 34–65 mya. Without any recognizable fossil records, no one knows how old the adder's tongue ferns or whisk ferns might be.

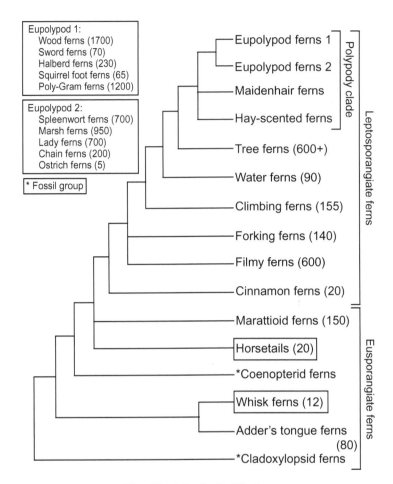

The Pteridophyte Clade

Fig. A23 Phylogeny of pteridophytes: ferns, horsetails, and whisk ferns. Pteridophytes are monophyletic, but ferns are not because recent molecular data embeds horsetails and whisk ferns within this clade. The adder's tongue ferns, whisk ferns, marattioid ferns, and horsetails compose the basal lineages, and all are eusporangiate (see Fig. A22). Although not shown, trimerophytes would be the sister group to the pteridophyte clade. Cinnamon ferns are basal and intermediate to the leptosporangiate fern clade. The polypody clade is characterized by having naked sori (no covering tissue), and in addition to the maidenhair ferns and hay-scented ferns, two large clades (the eupolypod ferns 1 and 2) each contain several groups; major ones are shown in the boxes (upper left). The approximate numbers of species found in each lineage is indicated by the numbers in parentheses. The common name used for each lineage is that of a major family or for their general features (e.g., tree, water) or by well-known genera (e.g., cinnamon, climbing). Two fossil groups are among the basal lineages. (Image source: After Smith et al. 2006)

Fern diversity decreased in the early Cretaceous as the diversity of flowering plants increased (Lidgard and Crane 1988), and then in the late Cretaceous the diversity of polypody ferns began increasing, presumably because they adapted to the shady habitats provided by canopies of flowering plants (Schneider et al. 2004). Once again a changing of the guard, the rise of flowering plants, and a shift of dominance from terrestrial communities dominated by gymnosperms and ferns reduced some fern lineages to relicts but provided other fern lineages with new opportunities for diversification. More detailed descriptions of the members of these groups follows, but to introduce the basics, a review of basic fern biology will be useful based upon the common polypody ferns.

Basic Fern Biology

The most common axis form in ferns is a rhizome, which can vary from thick and stout to thin and wiry. Rather than producing aerial branches, fern rhizomes most commonly produce helically arranged leaves, which through reorientation (bending of the leaf stalk base) are held vertically aloft. Like most clubmosses and horsetails, ferns have retained this growth pattern from early vascular plants. The rhizomes can Y-branch or be strongly monopodial. A few ferns have short, stout, erect stems, and a very few, so-called tree ferns, produce tall, hard, unbranched stems. However, since no wood is produced, sclerenchyma associated with the vascular cylinder and the vascular supply to leaves provides a supporting network of some considerable strength and flexibility. Tree ferns do not branch, so their "crown" simply consists of a helical whorl of leaves, so the nonwoody stem does not have to support a large crown composed of branches and leaves. Without a crown of branches, the small "crown" of fronds offers minimal wind resistance, and in high winds the fronds are quickly shredded, reducing wind resistance even more. Most of the large arborescent species are native to tropical areas where they sometimes compose a significant portion of the forest understory.

A common organization of fern fronds is the pinnately or bipinnately compound leaf. Both consist of a central leaf stalk upon which leaflets are arranged in two rows, or upon which there are two rows of secondary leaf stalks and the leaflets, called pinnae ("PEN-nay"), are then arranged pinnately upon the second order leaf stalks. In a number of species, the blade can be deeply lobed and so highly dissected that it becomes difficult to decide whether it is a simple leaf composed of one blade or a compound leaf composed of discrete leaflets.

Fern fronds vary tremendously in size. Some fronds are quite impressively large, ranging from one to several meters long; others are very small and simple, only a few millimeters in length, which are generally interpreted as reduced. In the climbing fern, the frond is indeterminate and grows apically in a vine-like fashion to several meters in length, bearing numerous compound pinnae along its length. The blade of fern fronds can be variously lobed or even simple and entire (unlobed). Some very specialized leaf forms are adaptations for modified functions. Some leaves are reduced in size for reproduction or an aquatic habitat, and there are even leaves that form ant domatia.[20] Some ferns produce two different forms of leaves for two different functions. Several epiphytic ferns like staghorn fern (*Platycerium*) produce broad, plate-like fronds that overlap each other and function as detritus[21] traps, producing their own hanging basket within which they root. These leaves have a heavy, waxy cuticle on their surface, and at maturity these leaves die, leaving their wax-papery skeletons forming the plant's basket. As they inner ones decay, they are overlapped by newer leaves to the outside. Large, elongate, broadly Y-lobed fronds extend out from the "basket." As adaptations to drier environments some ferns have thicker, more leathery "fronds." In aquatic environments fronds may have larger internal air spaces and have a dense covering of hairs, and in very moist dimly lit environments, fronds may have blades only a single cell layer thick. Such thin fern fronds are described as "filmy" and are characteristic of an entire family of mostly tropical ferns, the filmy ferns (Hymenophyllaceae; hymen = thin membrane).

Ferns bear sporangia upon the lower side or margins of their fronds. Many ferns produce only one type of leaf, which may be fertile or not; some species have fertile fronds that differ in form from the sterile fronds, or fertile pinnae that differ in form from sterile pinnae on the same frond. Usually the fertile fronds or fertile pinnae have little or no blade at all. Sporangia may be spread across a portion of a frond or pinnae, or more commonly they are found in clusters of a few to many sporangia called sori (SORE-eye; sing. sorus) (Fig. A24). Sori may be linear, oval, or round; they can be naked or covered by a protective flap of tissue, an indusium. Sori may be scattered, positioned parallel to veins, between veins, or parallel to margins. Some sori occur under marginal flaps (false indusium) or in marginal pockets of the leaf pinnae. The shape, size, and location of sori are important diagnostic features for identifying ferns.

The telome theory (hypothesis) predicts that when a lateral branch system bearing terminal sporangia becomes planated and webbed thus producing a leaf (a megaphyll), the terminal sporangia will now be on

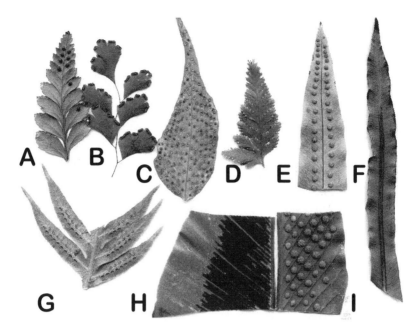

Fig. A24 Diversity of fern sori in shape and position. A. Round sori, submarginal. B. Oblong, marginal sori covered by flap of leaf tissue, a false indusium. C. Round sori each with a symmetrical indusium; scattered on the leaf. D. Round sori on leaf margins with an asymmetrical indusium. E. Round sori positioned midblade; "naked," no indusium. F. Linear sori positioned parallel to midvein of a leaflet; asymmetrical indusium. G. Oblong sori positioned parallel along lateral veins; asymmetrical indusium. H. Linear sori positioned parallel to secondary veins; naked. I. Round sori positioned between secondary veins; naked. A-G. Pinnae. H-I. One-half of frond with simple blade cut along midveins (center). (Image source: The Author)

such a leaf's margin (Fig. 8–15 E-H). While some sori are clearly marginal (Fig. A24 B, D), the many diverse positions of sori must be explained in light of this hypothesis of leaf origin. The apparent contradiction of nonmarginal positions of sporangia was resolved by studies showing sporangial development is always initiated upon the developing leaf margin. If sporangia are initiated marginally early in frond development, the continued marginal growth of the leaf blade leaves the sporangia in a nonmarginal position.

Leaf and pinnae primordia grow apically by means of a meristematic cell, adding length to the developing leaf or leaflet. This apical development occurs in such a manner that it produces a most characteristic feature, fiddleheads, where fern fronds and their leaflets unfold from a helical coil that evokes the head of a violin (Fig. 6–3). This helical coiling of aerial shoots shows up first in the fossil record among early vascular plants (Fig. 8–2). Among living plants this ancestral feature is retained only in ferns and cycads.

A few ferns retain residual meristems upon their fronds that can give rise to subsequent growth and discrete plantlets, buds which can detach and disperse as a means of vegetative reproduction (e.g., *Asplenium dauci-folium*). This feature should not come as a surprise since as modified branch systems, residual or bud meristems are expected. In most ferns, the fronds are clearly stem appendages and their junction at each node is quite well demarcated. However, in some basal ferns like *Stromatop-teris* (Schizaeaceae), the stem-frond junction is rather continuous with no clear demarcation of where a leaf begins and a stem ends, rather like the situation in the whisk ferns, *Psilotum* and *Tmesipteris*, where determinate aerial shoots are continuous with subterranean shoots and determinate in development like leaves. Like the whisk ferns, *Stromatopteris* has a sub-terranean gametophyte (Bierhorst 1971). Whether this is an ancestral condition or an evolutionary convergence cannot be determined.

Ferns have a life cycle typical of a free-sporing vascular plant (Fig. A25). The diploid sporophyte is the conspicuous and familiar organ-ism, what you think of as a fern. The free-living haploid gametophyte is seldom observed, but it is no less a fern (Fig. 6–3). Most ferns are ho-mosporous, producing free-living, photosynthetic, and hermaphroditic gametophytes resembling a small liverwort or hornwort thallus. The wa-ter ferns are heterosporous and have fast-developing endosporic gameto-phytes. Some of the basal ferns, as well as the whisk ferns, have subter-ranean gametophytes that obtain nutrients from a symbiotic interaction with fungi. Haploid spores produced by meiosis in sporangia are the dis-persal phase, and as described in chapter 9 (Hell's Half Acre), fern spores can disperse long distances, but they need to land in a habitat favorable for the growth and reproduction of the gametophyte, a plant that grows like a bryophyte because it is haploid and has a bryophyte-like organiza-tion. Under good conditions, a spore can germinate and grow into a sexu-ally mature gametophyte thallus in just a few weeks. Fertilization of an egg produces an embryo that is dependent on the maternal gametophyte at first until it outgrows its mother and establishes independent growth. No dispersal phase exists at this stage and the new sporophyte must grow in that location (Fig. A25 K).

Dispersal is a critical part of any terrestrial plant's life cycle, and spores are the dispersal phase for ferns. The eusporangia of basal ferns, horse-tails, and whisk ferns produce thousands of spores, and this might seem an advantage over the few-spored leptosporangia, but that does not take into account the number of sporangia and timing of development. Some-times the adaptive advantage of features is not immediately apparent so biologists must consider all aspects of their biology. Leptosporangia cost

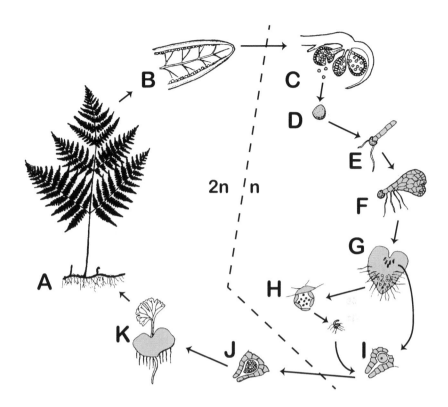

Fig. A25 Fern life cycle. The familiar fern (A) is a sporophyte bearing sporangia in clusters (sori) upon its leaves (B-C). Meiosis takes place within the sporangia producing haploid spores (D, gray). After dispersing, the spores develop into a gametophyte initially growing as a filament (E) and then spreading out into a flat thallus (F). The mature gametophyte (G) in ferns is a hermaphrodite producing both antheridia (H) and archegonia (I). Asymmetrical sperm fertilize an egg, producing a zygote and restoring the diploid (black and white) chromosome number. The resulting sporophyte embryo (J) outgrows the gametophyte (K) to ultimately establish a new diploid fern. (Image source: After Hoshizaki and Moran 2001)

very little to produce because they are so very small, with little stalks and thin walls, so they can be produced quickly and at low cost. Lots of spores are released all at once from eusporangia, so all are subject to similar dispersal opportunities ranging from ideal to poor depending upon the environmental conditions, and the number of chances is more limited because fewer sporangia are produced. Releasing spores in small batches over long periods of time ensures that some will be released when dispersal opportunities are optimal. And spores released under ideal conditions can readily disperse because the resulting gametophytes of leptosporangiate ferns have been found hundreds of miles from any adult

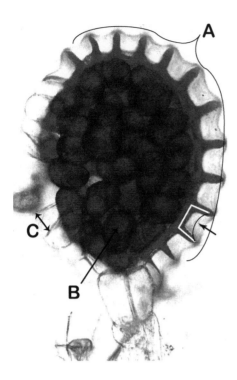

Fig. A26 Fern leptosporangium showing (A) annulus, a row of cells with thick side and bottom walls (outlined, lower right), (B) suture by which the sporangium opens, and (C) spores. As the cells of the annulus dehydrate, the thinner top wall pulls inward, in turn pulling on the side walls and generating tension like pulling on a bow. "Cocking" the annulus also pulls the top half of the sporangium back, and when it springs back, the spores within are actively dispersed. (Image source: Courtesy of and with permission of G. Shepherd)

sporophyte. Both types of sporangia clearly function, and the spore packaging and dispersal may seem equal in terms of the total number of spores produced, but having spores in lots of small packages allows for more complex dispersal strategies.

Leptosporangia also have an annulus, a row or patch of thick-walled cells, that functions to open each sporangium along an anatomical seam or suture (Fig. A26 A). If the suture is apical the annulus consists of a lateral patch of cells at each end of the suture or a diagonal row of cells opposite the suture. If the suture is lateral on the sporangium, the annulus consists of a single row of cells arching over the apex of the sporangium. The annular cells have thick side walls and a thin top wall. As the annular cells mature, die, and dehydrate, they pull the sporangium open along the suture. The most sophisticated form of annulus acts like a sling-shot. After pulling the sporangium open, the tension in the annular cell walls increases until it becomes great enough to spring back, flinging any enclosed spores out of the sporangium.

Most ferns are homosporous (producing a single type of spore) and the resulting gametophytes are hermaphroditic, developing both male and female sex organs, antheridia and archegonia, respectively. Archegonia usually mature first and are located centrally upon the upper surface of the thallus. Antheridia mature later and are borne on the lower surface among the rhizoids. Such a developmental pattern provides an opportunity for cross-fertilization and then for self-fertilization. In some species, gametophytes with mature archegonia exert a hormonal control over nearby developing gametophytes causing them to arrest their growth and produce antheridia, effectively becoming dwarf males, an adaptation to increase the chances of cross-fertilization when gametophyte populations are dense.

In some of the basal lineages of ferns, the gametophytes are subterranean and persist through an association with fungi; others have a small, stout erect thallus resembling a clubmoss gametophyte. Such gametophytes generally have a cylindrical form, sometimes similar to a rhizome except for the lack of vascular tissues. Are such gametophytes the ancestral condition in ferns or a derived condition? Are subterranean gametophytes similar through common ancestry or because they are adapted to a highly derived life style? These questions have not been resolved although the groups with such gametophytes occupy basal lineages. Subterranean gametophytes appear to be slow developing, but they are in a protected environment and may persist for years. They can even reproduce vegetatively via gemmae. Fertilization via swimming sperm is possible every time rainwater fills the spaces between soil particles. However, cross-fertilization remains a problem unless a population of gametophytes occurs together.

Major Groups of Ferns

Fossil Ferns: Cladoxylopsid Ferns and Coenopterid Ferns

Cladoxylopsid ferns appear in the late Devonian and persist into the Lower Carboniferous, and they are hypothesized to have had a common ancestry with trimerophytes. Reconstructions of *Pseudosporochnous* (sue-dough-SPOR-ak-nus) (Fig. A27) and the even older *Eospermatopteris* (Fig. 8–7) resemble modern tree-ferns with a helical whorl of lateral branches arising from single unbranched stem. The lateral branches were clearly photosynthetic organs that were once or twice Y-branched, and they bore second order branches arranged in a flat plane upon which were pairs of Y-branches (Stewart and Rothwell 1993) (Fig. A27 B-C). No webbing is evident so the axes themselves must have functioned in photosynthesis. On fertile appendages, sporangia terminated these ultimate branches. *Cladoxylon* had branched main axes that were covered

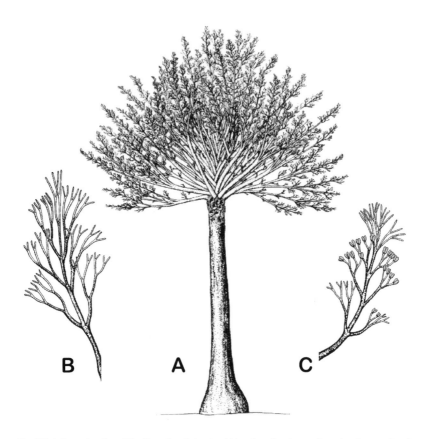

Fig. A27 A. Reconstruction of the Devonian cladoxylopsid tree fern *Pseudosporochnus* showing a main axis bearing a helical whorl of Y-branched lateral "fronds" bearing Y-branched appendages that were arranged in a single plane. B. Sterile branch system. C. Fertile branch system with terminal sporangia (gray). (Image source: A—After Berry and Fairon-Demaret 2002; B-C—After Leclercq and Banks 1962)

with small lateral branches that were Y-branched. Fertile shoots had fan-shaped, terminal portions (i.e., they were leafy) with Y-branched veins and marginal sporangia. The cladoxylopsid ferns are usually interpreted as a fern stem group intermediate between trimerophytes and ferns (Fig. A23). While they lacked fully developed fern fronds, their vascular system was constructed like ferns. This group is of particular significance because their leafy branches provide perfect intermediates in the evolution of megaphylls.

The coenopterid ferns also appear in the late Devonian and persist through the Carboniferous into the Lower Permian. Many think this assemblage of ferns represents several lineages, and some of the families are treated as separate orders. Certainly a diversity of forms existed, ranging from tree ferns of several meters to small, probably epiphytic ferns. No clear demarcation exists between the shoot system and the fronds. The ultimate appendages were very "ferny," very "leafy," with leaflets, whose veins were Y-branched, arranged in two rows. But these in turn were appendages of another branch and sometimes another until a main stem was reached. In such a transitional form, nothing allows biologists to determine whether such branch-leaf systems are one large compound leaf or a branch system bearing many smaller, simpler leaves.

The xylem in the leafy branches of these fossil ferns had a characteristic omega shape (ω) in cross-section. In modern ferns the xylem in leaf stalks is c-shaped, with the open side oriented to the upper (inner) surface, a form perhaps related to these early ferns. Vascular cylinders of fern stems have the appearance of a woven basket, not a solid mass of xylem. Leaf gaps, filled with ground tissue, are places immediately above the location where the vascular supply to a leaf attaches to the vascular supply of the stem.[22] The leaves of clubmosses do not have this effect on the vascular tissue of the main axis, which shows one of the differences that result from the different origins of leaves in these two lineages.

Just as the outer portions of the stem in clubmosses and horsetails may be considered a network of leaf bases running down the stem (Kaplan 2001), so too the vascular system of ferns may be considered as consisting of the interconnected basal portions of the vascular supply to leaves. Developmentally, vascular tissue does not grow out from the stem into the leaf. Vascular tissue differentiates at the base of a developing leaf and then it "grows" (differentiates) in two directions: basally, where it interconnects with the vascular network of the axis below; and outwardly into the leaf to form its vascular network. In many ferns the Y-branched veins form an open network that ends at the leaf margins, a feature retained from ancient common ancestors. In others, the veins form a closed network of cross branches.

Over all, in terms of their large fronds, their leaflet form, their fertile branchlets, and their sporangia, coenopterid ferns most closely resemble cinnamon ferns (see below), a small family whose fossil record dates to the Carboniferous. For these reasons, cinnamon ferns are thought to have a common ancestry with members of the coenopterid ferns.

Adder's Tongue Ferns

The adder's tongue ferns are a small family (Ophioglossaceae) consisting of only three genera, of which only two bear mentioning: *Ophioglossum* (oh-fee-oh-GLAH-some), adder's tongue ferns; and *Botrychium*

Fig. A28 Rattlesnake or grape ferns and adder's tongue ferns (Ophioglossaceae). A. *Ophioglossum vulgatum* showing a simple tongue-shaped sterile leaf and a fertile spike or fertile frond in front. Basally the two fronds are fused. B. *Botrychium virginianum* showing a finely divided sterile frond with a fertile frond rising above it. C. Two *Botrychium lunaria* ferns each consisting of a pinnate sterile frond with rounded pinnae and a fertile frond with rounded sporangia rising above the sterile frond. D. Portion of the reduced fertile frond of *Ophioglossum* showing that it consists of little more than two rows of sporangia. Both ferns have short, erect stems just below ground and fleshy roots. (Image sources: A—Courtesy of Aorg1961, Wikimedia Creative Commons; B—Courtesy of Jaknouse, Wikimedia Creative Commons; C—Courtesy of Abalg, Wikimedia Creative Commons; D—Courtesy of Przykuta, Wikimedia Creative Commons)

(bow-TRICK-ee-uhm), rattlesnake ferns or moonworts (Fig. A28). The adder's tongue ferns are largely tropical, but species occur in the northern temperate zone where they are rather common but inconspicuous and frequently overlooked. *Botrychium* is a common woodland fern of the northern temperate zone. All are small ferns with ephemeral aerial parts usually consisting of a single sterile frond plus a fertile frond if the plant is reproductive. All have perennial subterranean portions composed of a short, erect stem and fleshy, contractile roots. Contractile roots are found in many herbaceous perennials, and their contraction in length functions to keep the apex of the slowly elongating stem at or below the soil surface. Adder's tongue ferns may also form elongate lateral rhizomes, stolons or runners, that produce new plantlets at their end.

Both the sterile and fertile fronds of *Botrychium* are pinnate or pinnately dissected, although the fertile frond is of reduced size (Fig. A28 B-C). *Ophioglossum* has a simple, broadly oval, spatulate blade with reticulate venation, and the fertile frond is reduced to a spike-like, pinnate row of sporangia embedded in a reduced frond (Fig. A28 A, D). Some species bear a single fertile spike, others bear several in a pinnate arrangement suggesting the sterile frond might represent a terminal leaflet of a pinnate leaf. The resemblance of the fertile fronds to an erect rattle is the origin of another common name, the rattlesnake ferns. The sporangia are large. The fertile frond and sterile frond have their leaf stalks fused together.

These ferns differ from other ferns in that their leaf development does not show a helical coiling and their tracheids have circular bordered pits (see Fig. 8–8 A), a feature generally considered characteristic of conifers. Their subterranean stems, which can persist for years, were thought to possess a limited amount of cambial growth, but this has recently been shown incorrect; they lack any woody tissue (Rothwell and Karrfalt 2008). These features were used to suggest current members of this lineage might have been reduced from larger, more arborescent plants, but without secondary growth, they fit within a fern lineage without any serious problems. If these small ferns are reduced in size, it may explain why these ferns seem to lack a fossil record; fossil ancestors are there, but we don't see the resemblance.[23]

Their gametophytes grow in association with fungi, either at or below the soil surface. Chlorophyll does not develop and if laboratory cultured, they require a carbon source,[24] which would be provided by a symbiotic fungus in nature. The young sporophyte may grow below ground for a considerable time, obtaining nourishment from symbiotic fungi before producing its first aerial frond. In these features the ferns are reminiscent of whisk ferns, which molecular data places as their sister group (Fig. A23). Another interesting feature of these ferns is the prevalence of very high numbers of chromosomes ($2n = 500$–$1,000$). Whether this is some extreme expression of polyploidy, chromosome duplication, or whether it serves some other function remains unknown. Individually the chromosomes are tiny rods, and obtaining good counts is tedious and difficult even for experts or their graduate students.

Marattioid Ferns

The marattioid ferns have existed since the Carboniferous. Although of limited occurrence today, this tropical/subtropical family has some 200 species. One of my favorite places is Wright's Creek in far northern Queensland, Australia. In its cool, wet, rain forest grottos, *Marattia* grows in the shade of *Angiopteris*, whose fronds are so tall you can walk under them without stooping (Fig. A29 A).[25] Both of these genera have short,

Fig. A29 Marattioid ferns. A. A specimen of *Angiopteris evecta* showing its large, bipinnate fronds, one of which is being held by a normal-sized botanist. B. Small portion of a frond showing the leaflets (pinnae). C. Portion of a leaflet from *Marattia* showing two marginal rows of oblong sori (clusters of sporangia). D. Portion of a sterile frond of Carboniferous fossil *Pecopteris*. (Image sources: A—Courtesy of and with permission of R. Moran; B—Courtesy of Poyt448, Wikimedia Creative Commons; C—Courtesy of Ton Rulkens, Wikimedia Creative Commons; D—Courtesy of Haplochromis, Wikimedia Creative Commons)

stout erect stems (another genus is rhizomatous), which develops not from a single meristematic cell but from a broad, multicellular meristem, a unique feature among living ferns. *Marattia* and *Angiopteris* have large pinnate or bipinnate fronds, which in *Angiopteris* can reach 5 meters in length (over 15 feet) and have leaf stalks several centimeters in diameter at the base. The leaves develop typical fern fiddleheads. The leaflets are simple with a single midrib and secondary veins at right angles, some showing Y-branching near their origin (Fig. A29 B). Other members of this family have either palmately compound leaves or simple leaves with an ovate blade.

The large sporangia are clustered in sori near the leaflet margins and parallel to the veins. In *Angiopteris* 8 to 12 large, closely appressed sporangia form each oblong sorus (Fig. A29 C). In *Marattia* a fused pod-like synangium takes the place of a sorus. Like other eusporangia, each sporangium produces a thousand or more spores. The gametophytes begin growth with a brief filamentous phase before developing into a flat thallus. Since these ferns usually occupy moist tropical habitats, the gametophytes can be rather long-lived, growing to a couple of centimeters in length and quite resembling a liverwort.

Fossil ferns beginning in the Carboniferous and throughout the Mesozoic age are attributed to the marattioid ferns. The fossil fern foliage *Pecopteris* (Fig. A29 D) and the tree-fern trunks *Psaronius* have the anatomical and morphological characteristics of this family. Fertile pinnae of carboniferous ferns show a small group of large sporangia fused laterally into sori or fused into synangia. This group of ferns, including large tree ferns, reached its highest diversity during the late Carboniferous, then from the Permian on the diversity gradually dwindled to its present-day few. However, these relicts are very similar anatomically and morphologically to their ancient Carboniferous ancestors, remarkably unchanged over 325 million years.

Cinnamon Ferns

Cinnamon ferns (*Osmunda*, Osmundaceae) are a family consisting of three genera totaling only about 20 species worldwide. In North America, *Osmunda* is a common genus in moist environments with three species, which bear their sporangia either on separate fertile fronds in cinnamon fern (*O. cinnamomea*), on terminally positioned fertile pinnae in royal fern (*O. regalis*), or on fertile pinnae situated in the central region of the frond, with sterile pinnae above and below, in the aptly named interrupted fern (*O. claytoniana*). All three species are easily grown and happily reside in my garden. Osmundas have short, stout, erect stems that with age can Y-branch a few times. The stems produce a dense mantle of roots, all of which makes for a short, stout hummock-like axis bearing several apical whorls of leaves. The spores of these ferns are green because they contain developed chloroplasts; such spores germinate shortly after being shed. The gametophyte grows into a rather robust thallus and can persist for a considerable time. Cinnamon ferns have a fossil record dating to late Carboniferous. The family reached its maximum diversity during the age of dinosaurs—the Triassic, Jurassic, and Cretaceous—after which their diversity dwindled. Fossils from western North America identical to *Osmunda cinnamomea* have been dated to the late Cretaceous, some 70 million years ago (Serbet and Rothwell 1999), making cinnamon fern the oldest known species on Earth, a champion of evolutionary longevity.[26]

Filmy Ferns and Forking Ferns

The next two lineages are largely tropical ferns forming a small clade that is the sister group to all the remaining ferns. Filmy ferns are quite small ferns that often grow in quite moist, often low-light, habitats within tropical forests.[27] The leaf blades are only a single cell thick with a delicate dichotomous web of veins. Sporangia are borne in marginal pockets upon the leaves. Forking ferns, which may form a single family, consist of about 10 genera and some 140 species. Their fossil record dates to the Cretaceous or even older.

Climbing Ferns and Their Nonclimbing Relatives

The climbing fern (*Lygodium*) has a wide tropical and subtropical distribution; one species reaches southern New Hampshire in the east and southern Ohio in the Midwest. The rest of this lineage consists of three other nonclimbing, tropical genera that are generally unfamiliar to most people. The four genera have some 140 species in total. Sometimes all four genera are included in a single family; other treatments place them into three families. As a group these ferns are considered relicts retaining primitive characters.

The climbing fern has short, subterranean stems bearing indeterminate, vine-like fronds. The coiling, vine-like leaf stalk can reach 20–30 m in length. The leaf stalk of tropical species is used to weave lightweight mats. Each secondary branch is compound, bearing either fertile or sterile pinnae. *Anemia* is a small terrestrial fern that bears its sporangia on a pair of basal pinnae held vertically aloft above the rest of the pinnately compound frond. Superficially *Anemia* resembles *Botrychium*, but in *Botrychium* the sporangia are borne on a modified fertile frond whose leaf stalk is fused basally to an adjacent sterile frond. The sporangia are fairly large and occur in two rows. Sporangia have an apical annulus composed of radially arranged cells that open the sporangium along a vertical slit. Gametophytes are either photosynthetic and filamentous or mycotrophic, cylindrical, and subterranean like whisk ferns (see A53 F).

Water Ferns

The water ferns form a lineage consisting of two families, the mosquito ferns (Salviniaceae) and the clover ferns (Marsiliaceae) (Fig. A30). Discovery of a fossil water fern, *Hydropteris*, that is intermediate between the rhizomatous clover ferns and the free-floating mosquito ferns supports treating them as sister groups (Rothwell and Stockey 1994). Both families are aquatic and both are heterosporous, a derived character related to rapid gametophyte development, an adaptation to their aquatic environment. The highly modified features of the water ferns make it difficult to assess their closest relatives on morphological grounds. Molecular studies have hypothesized a common ancestry with the climbing ferns (Pryer et al. 1996); Eames (1936) long ago suggested the same relationship on the basis of scanty morphological similarities of the sporangia (Fig. A23).

Fig. A30 Clover fern family. The three genera in this family often grow in shallow water with their horizontal stems, rhizomes, growing along and rooted in muddy soil. A. *Pilularia* has bladeless fronds and grows submerged or emergent in shallow water. Arrow marks a fiddlehead. B. *Marsilea* has fronds with two pair of leaflets (pinnae). C. *Regnellidium*—Frond consisting of a pair of pinnae. D. Pair of sporocarps of *Pilularia* consisting of a pair of tough, desiccation-resistant, fused pinnae. E. Germinating sporocarp of *Marsilea* showing a gelatinous fertile pinnate axis that bears two rows of pinnae (2 marked by * at their base), each bearing a sorus containing megasporangia with one megaspore (arrow) and microsporangia with many microspores. (Image sources: A, D—Courtesy of Christian Fischer, Wikimedia Creative Commons; B—Courtesy of and with permission of Forest and Kim Starr, Wikimedia Creative Commons; C—Courtesy of and with permission of D. W. Stevenson; E—Courtesy of Curtis Clark, Wikimedia Creative Commons)

Marsiliaceae has three genera of tropical, rhizomatous ferns typically growing in shallow still water (Fig. A30). The rhizome grows along the bottom, producing roots into the muddy substrate and long-stalked fronds bearing small, modified blades. Water clover, *Marsilea* (mar-SILL-ee-ah), has only two pair of pinnae, and except for the Y-branching veins, the frond resembles a 4-leafed clover, and in fact these fern fronds are commonly used in constructing novelty good-luck charms (Fig. A30 B). *Regnellidium* (reg-nell-ID-ee-um) has only a single pair of pinnae (Fig. A30 C), and *Pilularia* (pill-you-LAIR-ee-ah), has no pinnae at all, its fronds reduced to a slender whip (Fig. A30 A). However, all three exhibit the characteristic fiddlehead development. Fronds of *Pilularia* reach only a few centimeters tall and with no obstructing blade, this fern often grows completely submerged; the fronds of the other two genera are usually emergent.

All three genera produce highly modified fertile pinnae, called sporocarps, at the base of leaf stalks (Fig. A30 D-E). The sporocarps are interpreted as being composed of a pair of fused pinnae within which a fertile pinnate axis bears two rows of sori. The outer covering is exceedingly hard, resistant to drying, and capable of withstanding considerable drought. Thus the sporocarp represents a seed-like stage only in that it can persist whenever their shallow-water habitat dries out. However, unlike a seed, the sporocarp is a means of delaying spore dispersal until favorable conditions return, then the sporocarp breaks open because the outer covering either deteriorates or is physically damaged, aided by the swelling of the water-absorbing gelatinous axis within. A fertile, sori-bearing gelatinous axis can expand to several centimeters long, bearing pairs of sori or pairs of fertile secondary axes, those proximal bearing megasporangia and those distal bearing microsporangia (Fig. A30 E). Spores are shed shortly after the sporocarp germinates and the gelatinous axis expands. The gametophytes quickly develop endosporically followed by fertilization. The development of the sporophyte begins very quickly, establishing new plants rapidly, clearly an advantage for an organism adapted to shallow water and therefore often intermittent aquatic habitats.

The mosquito ferns consist of two genera, *Salvinia* (sal-VIN-ee-ah) and *Azolla* (aye-ZOH-lah) (Fig. A31). These free-floating, aquatic ferns bear little resemblance to typical ferns. Both genera have rhizomes that branch frequently, and vegetative growth with fragmentation is the prevalent form of asexual reproduction. Both genera can have impressive growth rates under proper conditions, and both may become ecological problems where they have been introduced. *Salvinia* bears a whorl of three simple leaves at each node, two oval floating leaves, and a dissected, rootlike leaf hanging below the rhizome functioning as a keel and a fertile frond (Fig. A31 C). The floating leaves are very hairy, an adaptation for trapping air and increasing their buoyancy. The rootlike leaves absorb whatever water and nutrients are needed. *Azolla* bears small, simple leaves alternately along the rhizome, each consisting of two pinnae. The upper pinnae is green and photosynthetic, the lower pinnae is nongreen and membranous. True roots are produced from branch junctions. Fragmentation of clones is frequent, and growth can be rapid. Both can grow upon a muddy substrate if water levels subside. *Azolla* has a symbiotic relationship with the

Fig. A31 Mosquito ferns family. A. Free-floating plant of *Salvinia*. Rhizomatous axis bears whorls of three leaves at each node (C): an upper pair that float on the surface and a rootlike submerged leaf. The hairy surface holds air and prevents the surface from wetting. B. Free-floating plants of *Azolla* showing very reduced leaves. C. One node of *Salvinia* showing the pair of floating leaves on the top and a rootlike submerged leaf. The submerged leaves do triple duty functioning as a keel, an absorbing organ, and the fertile leaf. Each ball-like structure is a sorus (arrow) on a fertile leaflet. (Image sources: A—Courtesy of Le.Loup.Gris, Wikimedia Creative Commons; B—Courtesy of Kurt Stüber, Wikimedia Creative Commons; C—The Author)

nitrogen-fixing cyanobacterium *Anabaena*, which allows it to be used in aquaculture to increase rice production.

Both genera produce spore-bearing structures that are interpreted as a sorus surrounded by a modified indusium, which in *Salvinia* are borne on a submerged leaf (Fig. A31 C). Usually both microsporangia and megasporangia begin development within each sorus, but abortion of one type or the other results in sori containing either microsporangia or megasporangia. Megasporangia produce a single functional megaspore, which develops into a small, protruding female gametophyte bearing at least one archegonium. A unique net-like mesh surrounds the megaspore in which developing male gametophytes, armed with hook-like counterparts, become ensnared, thus producing a bisexual "raft" consisting of a female gametophyte and a number of male gametophytes, a description sounding a bit like a spring break pool party. Each male gametophyte produces a single antheridium and eight sperm are released in the immediate vicinity of the female gametophyte. Embryo development is rapid; sporophytes 1 cm long can be observed in as little as 5 to 6 days following spore dispersal.

Tree Ferns

Ferns with an arborescent habit (a single unbranched trunk and a terminal whorl of fronds) generally are placed in a single family. Depending upon the classification there are about five genera totaling over

600 species. With a fossil record dating to the early Cretaceous (Smith et al. 2006), tree ferns are about the same age as flowering plants.

Polypody Ferns

This clade includes thousands of species and many diverse ferns, and a great many familiar native and cultivated ferns belong in this lineage. In many cases generic and family delimitations remain very uncertain, and such detail goes well beyond the scope of this book. The sori in this clade lack a true indusium. Sporangia have an apical annulus, and the general fern characteristics described above generally refer to polypody[28] ferns, named after a prominent genus, *Polypodium*, a reference to its branched rhizome ("many feet"). This clade consists of four lineages, the dennstaedtoid ferns, the pteridoid ferns, which includes the maidenhair ferns, and two big clades denoted eupolypod ferns I and eupolypod ferns II (*eu-* means true). Each of these last two clades consists of numerous genera placed into many families.

Maidenhair ferns used to be placed near the climbing ferns, so their inclusion in the polypody clade represents a major phylogenetic rearrangement. While like other members of this clade they lack a true indusium, they have a false indusium formed by a margin of the fertile pinnae folded over (Fig. A24 B). The pinnae of the common maidenhair fern, *Adiantum*, have a fine dichotomous venation that was thought to resemble a young woman's pubic hair. Such rather bawdy common names are typical of an earlier era. Later when western botanists rediscovered the *Gingko*, it became known as the maidenhair tree because its leaves are similar in shape and venation to the maidenhair ferns (Fig. A17 A). In this era of political correctness, even mentioning such a historical note is certain to offend the delicate sensibilities of someone, but hopefully none of you, my dear readers.

In comparison with seed plants, ferns remain quite understudied, considering their prevalence and diversity. Their ecology has been little studied, particularly in the gametophyte stage, although they can be important elements in many communities. In an age when botany departments are fast becoming extinct, and when the human biomedical tail is wagging the biological dog, the prospects for botanists specializing in nonseed plant groups are not promising, even though much fascinating work remains to be done.

(synonyms: Gnetophyta [-ales, -opsida])

The gnetophytes (KNEE-toe-fights) consist of three strange and quite un-usual genera of gymnosperms, *Ephedra* (ee-FED-rah), *Gnetum* (KNEE-tum), and *Welwitschia* (well-WHITZ-ee-ah). Their relationship to other gym-nosperms is obscure, and their unusual appearances preclude any really close relationship to other gymnosperms or even each other. As a result each genus is usually placed in its own family, and the group is treated as a separate order, class, or division. None of these genera even resemble each other, perhaps because the fossil records indicates these three gen-era are relicts of a formerly much more diverse and common group of plants. Some gnetophyte features suggest a common ancestry with flow-ering plants, which is why such a strange group of gymnosperms gets our attention.

Ephedra

Ephedra is a relatively common shrub in semiarid regions of western North America like the sagebrush deserts and pinyon pine/ juniper forests. This is where it became known by the common name Mormon tea because of its use as a stimulating herbal beverage by Mormon settlers of Utah. *Ephedra*, called *ma-huang* in China, has long been used as an herbal treat-ment for asthma and is the source of the alkaloid ephedrine. The genus consists of about 40 species. The plants are much-branched shrubs with

Fig. A32 Ephedra. A. Shrubby habit. B. Detail of stems showing whorls of vestigial leaves (arrows). C. Pollen cones composed of overlapping pairs of bracts, showing branches terminated by several microsporangia. D. Female cone showing terminal ovule surrounded by pairs of bracts. A pollen drop at the tip of the ovule awaits pollen. E. Seeds surrounded by red, fleshy bracts ready for dispersal. (Image sources: A-B—The Author; C—Courtesy of Curtis Clark, Wikimedia Creative Commons; D—Courtesy of N. Friedman, with permission of the *American Journal of Botany*; E—Courtesy of Le.Loup.Gris 2010, Wikimedia Creative Commons)

reduced scale-like leaves in pairs or a whorl of three (Fig. A32 A-B), which is similar to scale-leafed conifers like junipers. While green in youth, the leaves as they mature dry into a papery sheath around the ridged stems at the nodes, leaving the younger green stems to perform photosynthesis. Ultimately, the production of wood and bark ends the photosynthetic function of older stems.

Pollen cones are short lateral branches composed of several pairs or whorls of bracts. Modified branches arise from above the upper bracts, each bearing a pair of bracts that surround 2 to 8 terminal clusters of fused microsporangia (Fig. A32 C). The modified branch is interpreted as a fused pair of sporophylls, modified fertile leaves (Hufford 1996), but its position above a leafy organ means it can equally well be considered a modified shoot. Pollen grains are football shaped with long ridges; similar fossil pollen is known from the Triassic and Jurassic. Seed cones are similar to pollen cones, consisting of an axis bearing several pairs or whorls of bracts and terminating with one or two short-stalked ovules surrounded by a pair of bracts (Fig. A32 D). At the time of the wind-vectored pollination, a tubular extension above the ovule's jacket layer exudes a pollen drop to capture pollen. By the time the seed is mature, the bracts surrounding the seed have become fleshy and colored to attract seed dispersers (Fig. A32 E).

The female gametophyte develops by repeated nuclear divisions until hundreds of nuclei are present. Subsequent cell wall formation results in the production of two or three archegonia. Fertilization occurs shortly after pollination, which is more like flowering plants than other gymnosperms where there is often a considerable delay between pollination and fertilization. Two sperm nuclei are produced and engage in a fertilization event, but the second fertilization produces a second embryo. Similar double fertilizations occur in other gymnosperms, but double fertilization takes on special significance in flowering plants where the second fertilization results in endosperm (chapter 11). Each zygote may divide into several cells each capable of becoming an embryo, but at maturity the seed will have only a single embryo. The bracts of the seed cone become fleshy and red colored, an attractive display and reward for an avian seed disperser.

Gnetum

All but two of the 40 species of *Gnetum* are tropical lianas, vines, bearing opposite pairs of broad, reticulate-veined leaves; the other two species are shrubs. Rather than having the parallel-veined leaves of other gymnosperms, the leaves of this gymnosperm have a network of leaves more like that of many dicotyledonous flowering plants (Fig. A33 A-B). Strobili are lateral branches bearing whorls of short shoots on swollen nodes (Fig. A33 C-D). Each whorl consists of shoots bearing microsporangia and shoots bearing ovules, but one or the other is abortive, producing either pollen or seed plants. At the time of pollination, a pollen drop exudes from an apical tube formed by the inner jacket layer (Fig. A33 C). Similar to *Ephedra*, the anther-like pollen-producing unit consists of a modified branch apically bearing a pair of microsporangia, surrounded by a pair of bracts (Fig. A33 D). The megaspore develops into a female gametophyte via a

Fig. A33 Gnetum. A. Leafy shoot showing pairs of broad, net-veined leaves. B. Leaf from back side showing looping veins. C. Seed cone showing whorls of ovules, many of which bear pollen drops. D. Portion of a pollen cone showing abortive ovules but yet bearing pollen drops; branches below each have a terminal pair of microsporangia. (Image sources: A-C—The Author; D—Courtesy of K. Nixon, Wikimedia Creative Commons)

free nuclear stage, where mitotic divisions are not followed immediately by cytokinesis. This is similar to the development of all nonflowering plant female gametophytes, but in all these other groups cellularization commences and archegonia are formed prior to fertilization. In *Gnetum* the pollen tube penetrates the female gametophyte while it is still in the free nuclear stage; any of the nuclei apparently can function as a gamete, and no archegonium or specific egg is produced. The oblong seed has a leathery outer layer that is a reddish color and a stony inner layer.

Welwitschia

Welwitschia mirabilis, the only species of *Welwitschia,* is a strange plant native to the coastal areas of the Namib Desert of southwestern Africa. *Welwitschia* is a slow-growing desert plant consisting of a short, stocky, woody stem that is continuous with a woody root system. Specimens have ages in excess of 1,000 years, and the plant produces only two pairs of leaves during that entire time. The first leaves are a pair of cotyledons followed a few weeks after germination by one pair of true leaves, which grow throughout the life of the plant from a basal meristem while they die distally at the tips (Fig. A34 A-B). The parallel-veined leaves can be

Fig. A34 *Welwitschia*. A. Habit of a greenhouse-grown specimen showing a single pair of broad, parallel-veined leaves. Note dead apical portions near the base of the figure. A branch bearing pollen cones is growing up from between the leaf and stem. B. Seedling showing a pair of cotyledons (oriented vertically in this image) and the first and only pair of true leaves (oriented horizontally). C. Several branches each bearing several seed cones, each composed of overlapping bracts. Long, thin tubes bearing pollen drops extend out beyond the bracts covering the ovules. D. Pollen cone showing branches extending out beyond the bracts and bearing terminal microsporangia. (Image sources: The Author)

40–50 cm broad at the base tapering toward the apex, which when dead looks like twisted, old rope. Small lateral branches growing from the leaf axils produce pollen and seed cones. The cones consist of overlapping and alternating pairs of bracts and can be 10–15 cm tall; sporangial units are produced in the axils of these bracts. On pollen-producing plants, each unit of a cone consists of a subtending bract, two pairs of bracts at right angles, two branched sporangial axes bearing 3 microsporangia each, and between these two axes, a terminal and abortive ovule (Fig. A34 D). Seed cones are larger but similar in construction; however, they lack sprorangial axes completely and the inner pair of bracts is completely fused around the ovule. The inner jacket layer elongates into a slender tube that emerges from within the cone's bracts (Fig A34 C). Early in the mornings the ovule produces a pollen drop, but since the pollen is a bit clumpy, whether the pollen is vectored by wind or animal is uncertain.

The two pair of bracts enclosing anther-like sporangial axes and an ovule is a very flower-like organization, and, this, along with double fertilizations, both resulting in zygotes, makes gnetophytes attractive candidates as common ancestors of flowering plants. No question angiosperms have a common ancestry somewhere within the gymnosperms, and when gnetophytes have been placed as a sister group to the flowering plants, the resulting clade has been called the anthophytes, literally flowering plants (*antho-* means flower). Now additional molecular data is placing the gnetophytes in the same clade with conifers. This means no living gymnosperms have a close ancestral relationship with angiosperms.

The distinction between a cone or strobilus consisting of sporophylls or modified sporangia-bearing branches arranged in a helix or series of whorls versus a flower consisting of a helix or whorls of sporangia-bearing modified leaves surrounded by a sterile perianth, hinges on the presence of a sterile perianth. The presence of sterile bracts surrounding "sporophylls" in *Ephedra*, *Gnetum*, and *Welwitschia* is why some people refer to the subunits of the cones as flowers or floral units.

All three gnetophyte genera are dioecious, although bisporangiate strobili have been observed in some specimens (Endress 1996), and since pollen cones have abortive ovules it suggests a monoecious ancestry with bisporangiate cones, another similarity to angiosperms. The gnetophytes have compound reproductive structures embodying features of both strobili and flowers, in particular sterile bracts associated with sporophylls. In both conifers and ginkgoes seed-bearing shoots are associated with sterile bracts. Gnetophyte microsporangia are borne on branched axes that superficially look like anthers. Yews have similar pollen structures too.

All three genera have a double jacket layer around their megasporangium, a feature usually associated with flowering plants, whereas other gymnosperms have only a single jacket. The inner jacket in gnetophytes forms the long tubular apex rising above the ovule. All three gnetophytes lack resin canals that are so common in conifers. All three gnetophytes have embryos with two cotyledons, like cycads and some flowering plants, and unlike conifers. All three gnetophytes have vessels in their wood, like most flowering plants, but vessels are also found in some ferns and clubmosses. However, in gnetophytes the vessels appear to be derived from conifer-type tracheids. All three gnetophytes have opposite or whorled (some species of *Ephedra*) leaves, which in *Gnetum* are broad, net-veined leaves, a character that fools many people into thinking *Gnetum* is a flowering plant. The gnetophytes have a fertilization mechanism that often involves both sperm in two separate fusion events, which makes a perfect intermediate between other gymnosperms and the double fertilization of flowering plants, which results in an embryo and endosperm (Friedman and Carmichael 1996).

Green Algae

(synonyms: Chlorophyta, Chlorophytes)

clade nesting: green algae/plant/bikonts/eukaryotes

Green algae include both freshwater and marine species in a diversity of forms ranging from unicellular phytoplankton to seaweeds. Since they form one lineage, the green algae and land plant clade might be considered a "plant kingdom," the chlorophytes. However, the name "plant" and the concept of a plant kingdom in many places still refer just to land plants. Traditional classifications of green algae grouped organisms on the basis of similar morphology—for example, biflagellated cells, nonmotile unicells, unbranched filaments, motile and nonmotile colonies, branched filaments— however, phylogenetic studies have demonstrated that morphology was generally a pretty poor indicator of relatedness, although a few familiar groupings remain. Another way of saying this is that new data in the form of ultrastructural studies of mitosis, cytokinesis, and flagellar structure, and then more recently macromolecular sequence data, have falsified many of the traditional taxonomic hypotheses.

Charophyceans are a small but diverse group of green algae whose members share a common ancestry with the land plants. Together charophyceans and land plants form a clade called the streptophytes because they share the same type of mitosis and cytoplasmic division involving a phragmoplast (Fig. 6–5; Fig. A35 D). The remaining green algae are a sister group to the streptophytes and they consist of two primary lineages, the

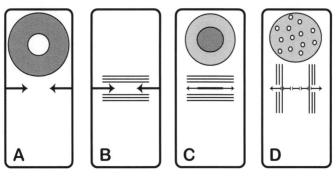

Fig. A35 Different mechanisms of cytoplasmic division at the end of mitosis. A-C. Cytoplasmic divisions that produce a solid cell wall. D. Cytoplasmic division that produces pores between the two daughter cells. A. Cytoplasmic furrowing where the cell membrane furrows inward (arrows) producing new wall from the outside inward (diagrammed in top of cell, C and D also). B. Cytoplasmic furrowing where cell membrane furrows inward (arrows) guided by parallel plates of microtubules; new cell wall forms as in A. C. Cell plate formation (double-headed arrow) starts in the center and grows and thickens from the center outward (diagram at top) guided by parallel plates of microtubules. D. Cell plate formation guided by a phragmoplast, which is a proliferation of microtubules, the remnants of the spindle. As in C, the cell plate and new cell wall material start from the center growing outward toward the cell periphery. Where the spindle fibers interrupt the cell plate they leave pores called plasmodesmata in the new cell wall. (Image source: The Author)

prasinophyceans (unicellular flagellates) and a large clade, itself consisting of three lineages, that includes the ulvophyceans, chlorophyceans, and trebouxiophyceans. Prasinophyceans alternatively may form a basal lineage to both the streptophytes and the green algae. Traditionally algal lineages are named using the suffix -phycean.

The chloroplasts of green algae and land plants contain both chlorophyll *a* and *b*. Some cells have a single chloroplast per cell of many diverse shapes; others have two or more chloroplasts including many small, discoidal chloroplasts like most land plants (Fig. A36). Chloroplasts may or may not exhibit a pyrenoid, a central visible mass of the enzyme rubisco. Flagella occur in pairs or multiples of two; both are similar in length and form. The primary storage product of photosynthesis is starch formed inside the chloroplast.

Prasinophyceans

(synonyms: scaly green monads, micromonads)

Prasinophyceans are primarily unicellular, marine flagellates although some occur in freshwater, some are nonmotile, and some form sessile colonies. Their cell surface is covered with small, individual, plate-like scales. The flagella, varying in number from one to 16, are inserted apically at the base of a depression. Prasinophyceans may exhibit phagocytotic feeding along with autotrophy, a mixotrophy (mixed feeding). The single chloroplast may be lobed and may contain an eyespot. At least one prasinophycean genus (*Mesostigma*)—that is, possessing typical scales and flagella, has been demonstrated to be a basal lineage of the charophyceans (Fig. 6–5 B), so perhaps other genera will be found to be basal to other lineages, demonstrating that the prasinophyceans are not a monophyletic group. Generally prasinophyceans are considered to possess the ancestral characters of the whole green algae–land plant clade, and whether they are actually their sister group remains to be determined.

Chlorophyceans
(synonyms: green algae, chlorophytes)

Chlorophycean algae have some 7,000 described species, and they are found largely in fresh water and certain terrestrial habitats like soil, rocks, and bark. They range in form from flagellated unicells to both motile and nonmotile colonies, both branched and unbranched filaments, and multinucleate filaments that lack cross-walls. Cell division in chlorophyceans is characterized by a "closed" mitosis, meaning the nuclear membrane remains intact during mitosis, together with a unique cytoplasmic division (Fig. A35 C). At the end of mitosis, a phycoplast is formed, bands of microtubules parallel to the plane of cell division. This band of microtubules appears to function by keeping the daughter nuclei separated after the spindle ceases functioning (Graham and Wilcox 2000). Cytoplasmic division is accomplished either by cytoplasmic furrowing (outside inward) or by cell plate formation (inside outward) within the phycoplast (Fig. A35 C-D). Cytoplasmic division in prokaryotic cells and some eukaryotic cells is accomplished by furrowing without any spindle fibers involved; in some green algae microtubules guide the furrowing (Fig. A35 A-B). In the streptophytes, remnants of the spindle form a phragmoplast, a cylinder of microtubules that proliferates and grows outward as a cell plate coalesces around a small cluster of microtubules, producing pores, plasmodesmata (Fig. A35 D).

Chlorophyceans form two major clades: the DO clade, consisting of motile and nonmotile unicells and colonies; and the CW clade, consisting of motile unicells and colonies and diverse filamentous forms. DO (directly opposite) and CW (clockwise) describe differences in the arrangement of flagellar basal bodies (centrioles) in motile cells (Inouye 1993). Molecular data supports the validity of using these ultrastructural characters to define lineages.

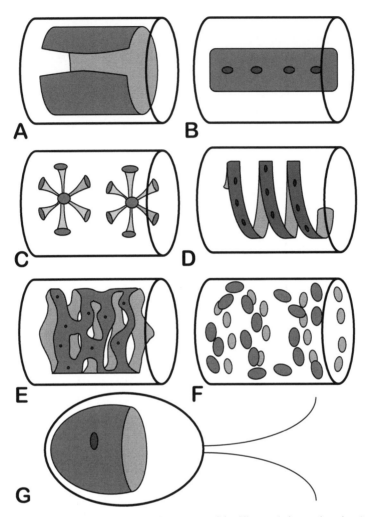

Fig. A36 Chloroplast diversity in green algae showing some of the differences in form and number. A. One peripheral band-shaped chloroplast. B. One flat ribbon-shaped chloroplast. C. Two chloroplasts with arms radiating from center (see Fig. A37 D). D. One long flat chloroplast in a helical spiral around cell periphery (see Fig. A37 C). E. One irregular, reticulate chloroplast around cell periphery. F. Many disk-shaped chloroplasts arranged around cell periphery. G. One cup-shaped chloroplast. Pyrenoids shown in B, D, E, G as darker bodies. (Image source: The Author)

Three familiar orders compose the CW clade: Volvocales; Chaetophorales, first proposed by Mattox and Stewart (1984); and Oedogoniales, long identified as a singularly unique group. *Chlamydomonas* represents the characteristic biflagellated unicell of the Volvocales, possessing a single cup-shaped chloroplast and an eyespot (Fig. A36 G). Spherical and flat 2-dimensional raft colonies are composed of various numbers of cells, each one resembling a unicellular *Chlamydomonas*. Lengthwise cellular divisions and continued association of the daughter cells produce a colony from a unicell. In smaller colonies all cells are identical, and depending on the number of divisions, colonies are composed of 4 to 512 cells. In many cases the cells are encased in an extracellular gelatinous matrix.

Volvox arguably represents the largest and most specialized phytoplanktonic organism with hollow, spherical colonies measuring nearly 1 mm in diameter and containing "as many as 20,000 cells," although 16,384 (2^{14}) would seem an appropriate number (Fig. A37 A). Such colonies display coordinated movement, both rotational and linear, meaning the colony has anterior and posterior poles relative to their direction of travel. Some specialized cells function in asexual reproduction (daughter colony formation) and sexual reproduction. Daughter colonies are initially released into the interior of the parent colony and are only freed upon its loss of structural integrity, a rather difficult birthing. *Volvox* is motile enough to occupy the light-rich upper layer of an aquatic habitat by day and to migrate several meters vertically to nutrient-rich depths by night.

Chaetophorales (key-toff-or-AYE-leez) was an order containing branched filaments in contrast to the unbranched filaments of the Ulotrichales (you-low-trick-AYE-leez), but this latter grouping has been falsified by cytological and molecular data. Chaetophorales now consists of both branched and unbranched filaments. Their motile cells have two pairs of flagella, and cell division is accomplished by a cell plate that forms plasmodesmata but of a simpler type than that found in the charophyceans and land plants. Many members of this order are simple "seaweeds" anchored by one or more basal holdfast cells. Most members have a single bracelet-shaped chloroplast (Fig. A36 A).

Oedogoniales (ee-doh-gon-ee-AYE-leez) consists of only three genera but perhaps well over 500 species. *Oedogonium* (ee-doh-GON-ee-uhm), the best-known genus, is a simple, unbranched filamentous "seaweed," meaning anchored. A single reticulate chloroplast with numerous pyrenoids (Fig. A36 E) wraps around the periphery of each cylindrical cell. Members of this group share a unique cytoplasmic division that involves intercalation of new wall material and a new daughter cell at one end of the parental cell, rather than in the middle of a cell, which leaves highly distinctive rings ("cell caps") of parental wall material at the end of the daughter cell (Fig. A38).

Oedogonium has oogamous (= gametes differentiated as egg and sperm) sexual reproduction, which has evolved several times in the green algae. Vegetative cells enlarge into a spherical shape and differentiate into an egg

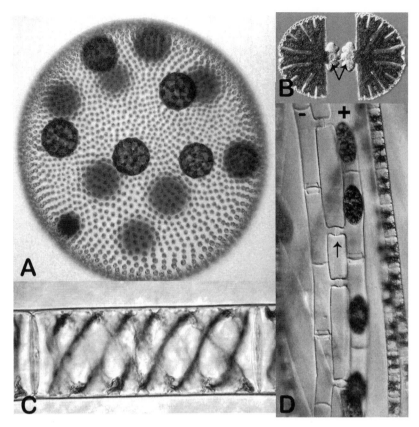

Fig. A37 Green algae and conjugation. A. A spherical colony of *Volvox* composed of hundreds of identical biflagellated cells embedded in a gelatinous matrix. Specialized cells have divided to produce daughter colonies that appear as the darker spheres within the "mother" colony, a type of asexual reproduction. For them to escape and become free-living, the maternal colony must be disrupted. B. Cell division of the desmid *Micrasterias* (me-kraw-STARE-ee-as, "little star"). After nuclear division, the two hemicells have each started to grow a new half, which at this stage looks like a slightly lobed bulge (arrows). The chloroplast will form a similar bulge that will later be partitioned into a separate chloroplast. The nucleus in each daughter cell, looking like a bubble, sits at the junction of the two hemicells (above arrows). 300–400 diameters magnification. C. Vegetative cell of the filamentous *Spirogyra* (spy-row-JEYE-rah) showing the typical spiral chloroplasts around the periphery of the cell. D. Two adjacent filaments of *Spirogyra* have recently reproduced sexually by conjugation. The filaments were opposite mating types (sexes; + and -) where the minus cells act as motile gametes and migrate through a connecting tube (arrow) formed by the two cells and fuse with a plus cell/gamete to form a dense oval zygote that will form a thick cell wall of a zygospore. (Image sources: A—Courtesy of A. Nedelcu, Volvocales Information Project; B—Courtesy of and with permission of Spike Walker; C—Courtesy of Spicywalnut, Wikimedia Creative Commons; D—Courtesy of and with permission of M. E. Cook)

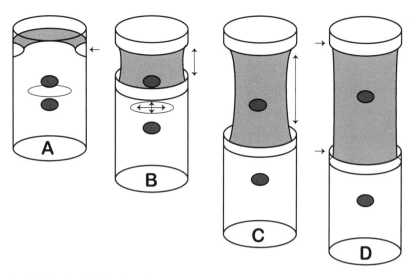

Fig. A38 Cytoplasmic division in *Oedogonium*. A. Telophase with circular cell plate growing outward in between the two daughter nuclei. A torus of new wall material (light gray) begins forming near one end of cell (arrow). B. Cell elongates and the torus of new wall material broadens (double-headed arrow). Cell plate migrates toward bottom of torus while continuing to expand outward (crossed arrows). C. Daughter cell continues to elongate and cell plate grows to periphery of the cell, completing cytokinesis. D. Cytoplasmic division completed. The new daughter cell with its newly intercalated cell wall (light gray) is marked by two rings of maternal cell wall material forming cell "caps" on daughter cell (arrows). (Image source: The Author)

(Fig. A39 B). When receptive, a pore opens in the cell wall to allow sperm entry. Multiflagellated sperm are produced in pairs from small single cells that are formed by a series of transverse divisions of a vegetative cell (Fig. A39 A). Some species are dioecious (two-housed), so filaments produce either sperm or eggs but not both; others are monoecious (one-housed) so filaments produce both eggs and sperm. In some species, male filaments are epiphytic dwarfs (Fig. A39 C). To produce such males, female filaments divide transversely as if making sperm, but the resulting zoospores disperse and are attracted to a developing egg by a chemotactic response. The zoospores anchor to female filaments and develop into dwarf males, and thus they are in a location to release sperm in the immediate vicinity of an egg.

The DO clade consists mostly of nonmotile unicellular and colonial phytoplankton, whose populations under favorable conditions can be dense enough to tint water green. Some genera like *Pediastrum* (pee-dee-AZ-trum) and *Scenedesmus* (skin-ee-DEZ-muhs) have a fixed number of cells composing the colony, 16 and 4 or 8 cells, respectively (Fig. 4–1). Others like *Hydrodictyon* (high-droh-DICK-tee-on; meaning "water net") have an indeterminate number of cells interconnected in a reticulate

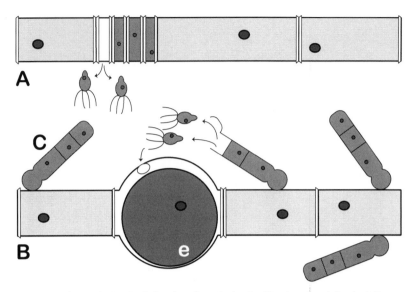

Fig. A39 Sexual reproduction in *Oedogonium*. Reproduction is either between similar-sized filaments (A, male and B, female) or between normal-sized female filaments (B) and dwarf male filaments (C) anchored to female filaments in the vicinity of an egg (e). A large spherical egg differentiates within an enlarged cell. An access pore forms in the cell wall through which sperm can enter. Pairs of multiflagellated sperm are formed in small cells produced by mitotic subdivisions of a vegetative cell in both normal-sized and dwarf males (A, C). Dwarf males (C) consist of a basal, anchoring vegetative cell and two or more sperm-producing cells. (Image source: The Author)

pattern forming large cylindrical colonies sometimes up to several cm long. Members of this clade produce biflagellated isogametes and similar zoospores.

Ulvophyceans (uhl-voh-FIE-see-ans) include organisms with considerable morphological diversity ranging from unicellular forms to filamentous, both simple and branched, to the largest seaweeds found among the green algae. Most interesting in terms of form are the unicellular seaweeds, which are large, relatively speaking, and display complex forms with leaf-like aerial branches arising from a horizontal axis and with anchoring rhizoidal branches. Precambrian fossils (late Proterozoic) similar to members of this clade date to 700–800 mya (Briggs et al. 1994). The large marine seaweeds of the Ulvales are the most recognizable and familiar members of this group. The genus *Ulva*, sea lettuce, has a broad blade consisting of two layers of cells and is anchored by a holdfast (Fig. 5–8). Members of this group often form a zone near the high-water mark in cooler waters. Detached blades looking like limp bright green lettuce leaves are commonly found among tidal debris.

Ulva features an alternation of haploid and diploid generations where both phases look identical. Haploid organisms produce biflagellated iso-gametes (+ and -) (see Fig. 6–6 A'). After fertilization, the zygote becomes sessile and develops into a diploid seaweed. The diploid seaweeds produce zoospores by meiosis. Following dispersal and anchoring, the zoospores develop into haploid seaweeds. This is very different from the basic life cycle of green algae, which is characterized by haploid adults that produce gametes. The zygote is the only diploid phase, and its division by meiosis produces four or more zoospores. This suggests the life cycle of *Ulva* began with the developmental delay of meiosis. However, since both haploid and diploid seaweeds are identical, the ecological and evolutionary advantages of this life cycle are rather uncertain. Are the haploid seaweeds more variable as a result, better able to adapt to many tiny differences in the patchy coastal environment? Are the diploid seaweeds genetically buffered against changes in their environment? Haploid and diploid *Ulva* have not been experimentally studied, to my knowledge. Perhaps the primary advantage of having a large multicellular diploid phase in a life cycle is that it allows many more cells to undergo meiosis, thus producing many more sexually variable and dispersible offspring while avoiding a tradeoff in the number and size of offspring.

Other seaweeds are essentially single celled although they exhibit impressively large and complex forms usually associated only with multicellular organisms. Some have a single nucleus (*Acetabularia*); most are multinucleate filaments of complex form, meaning they consist of elongate tubular cells without cross walls, or they are regular filaments (with cross walls), although the cells are quite large and remain multinucleate. *Acetabularia* is one of the largest, most complex, and best known of these unicellular seaweeds because of its experimental utility in cell biology. *Acetabularia* has a basal rhizoidal holdfast, a long, slender, erect, tubular cell body, and a terminal whorl of branches where reproductive cells form (Fig. 5–4).

The largest of these seaweeds with the most complex forms consist of a branching rhizomatous body anchored to its substrate by rhizoidal branches—for example, *Caulerpa*. Aerial branches ("fronds") arise at intervals and grow 10–20 cm above the rhizome, which can reach lengths up to a meter (Fig. 5–4). Yet, remember, this organism consists of only a single cell. Other seaweeds in this group consist of multiple erect axes "woven" together, forming what appears as a substantial stem-like axis holding aloft a "broom" of branched axes—for example, *Pencillium*, a generic name similar to *Penicillium*, the green bread mold of antibiotic fame, which has similar-looking asexual reproductive structures. All these seaweeds have numerous discoidal chloroplasts and cell walls composed of carbohydrate polymers constructed of sugars other than glucose. Sexual reproduction in this group is like in *Ulva*, except that the haploid and diploid phases have different forms (some of which were described as different genera before their life cycle was understood), so haploid and diploid phases of these seaweeds could each be adapted to a different habitat. *Caulerpa* and similar genera are proving to be invasive "weeds" in exotic environments, transported and released by aquarium owners.

The last group, Cladophorales (klad-ah-for-AYE-leez) or Siphonocladales (seye-fon-oh-klad-AYE-leez), is largely marine but with at least one genus that is well known in freshwater habitats, *Cladophora*. Most of these seaweeds are branched filaments with large, multinucleate cells containing either many discoid chloroplasts or one large reticulate chloroplast. Marine members generally have rhizoidal holdfasts. Intercalary branching occurs by a budding outgrowth near the distal end of cells, rather like a rhizoidal outgrowth. After the branch is formed, the cytoplasm divides by furrowing across the branch junction, which suggests they have derived from a filamentous ancestry that lacked cross walls. Reproduction in this group is generally like that of *Ulothrix* or *Ulva*. Some of these seaweeds are ecological problems in human-altered habitats. *Cladophora* forms dense mats of filaments in warm, nutrient-rich water associated with human-generated water pollution. Following a population explosion, their death and decomposition can deplete dissolved oxygen, creating an anoxic environment leading to fish kills.

Trebouxiophyceans

Trebouxiophyceans (trey-bucks-ee-oh-FEYE-see-ans) also represent a morphologically diverse array of organisms. *Trebouxia* is best known as the most common algal symbiont of lichens (Brodo et al. 2001). Many are nonmotile, unicellular, freshwater phytoplankton or algae found living in soil or on bark or leaves, especially in the tropics. Common forms are branched filaments, packets of cells, or sheets of cells resembling some marine seaweeds. Trebouxiophyceans are characterized by mitosis featuring centrioles positioned laterally near the metaphase plane. Why they separate from the polar MTOCs that form the spindle is not clear. In telophase the spindle is persistent and cytoplasmic division occurs by furrowing, which produces no plasmodesmata.

Charophyceans

Some quite diverse green algae compose the charophyceans, a sister group to the land plants (Bremer 1985; Mishler and Churchill 1985; Melkonian and Surek 1995; Waters and Chapman 1996). The charophycean–land plant clade, called the streptophytes (Fig. 6–5 B), is defined by a number of derived characteristics including mitosis and cytoplasm division using a phragmoplast and producing plasmodesmata (Fig. A35 D), along with other shared biochemical and ultrastructural features including asymmetrical motile cells. Botanists have long hypothesized a green algal ancestry of land plants, but the identification of the exact sister group has been a relatively recent development in the past 35 years. The last common ancestry between charophyceans and land plants is estimated to have taken place around 500 mya on the basis of the rate of molecular change, but unfortunately the fossil record offers no data, no possible intermediate forms, so scientists must make inferences about the origin of land plants on the basis of the features of living members of this clade.

Many charophyceans formerly were associated with other green algal lineages. On the basis of its general morphology *Mesostigma* would be a prasinophycean, but it possesses a flagellar basal structure like other streptophytes, a shared derived character that suggests common ancestry (Melkonian 1989), a hypothesis confirmed by molecular data (Lemieux et al. 2000). *Chlorokybus*, consisting of nonmotile cells in loose packets, looks like a chlorophycean. *Klebsormidium* is an unbranched filament whose bracelet-like chloroplasts formerly placed it with *Ulothrix* and the Ulotrichales. The Zygnematales, either the unicellular desmids or filamentous genera, is characterized by unusual chloroplast shapes and a sexual reproduction involving fusion of nonmotile (nonflagellated) isogametes called conjugation. The Charales include complex branched filamentous forms like the large, anchored, multicellular, freshwater "seaweeds" *Chara* and *Nitella*. The Coleochaetales (co-lee-oh-key-TAY-leez) are either highly branched filaments or a flat parenchymatous thallus that grows attached to substrates often at the water/atmosphere interface. Both the Charales and Coleochaetales have oogamy (i.e., eggs and sperm), and a life cycle where the zygote divides by meiosis. Both retain the egg until after fertilization, a feature with important implications for the origin of the land plant life cycle. Charales have many discoidal chloroplasts, and Coleochaetales have one peripheral chloroplast per cell.

Many Zygnematalean algae are rather handsome and easily recognized. *Spirogyra* has a thin band-like chloroplast forming a spiral around the cell periphery and *Microsterias* has two star-like chloroplasts (Fig. A36 C-D; Fig. A37 C-D). Zygnematales lack flagellated motile cells for either asexual or sexual reproduction. The fusion of nonmotile isogametes is called conjugation, which is accomplished when filaments in close proximity form conjugation tubes through which the protoplasm of an entire cell functioning as an isogamete moves (Fig. A37). The donor filament is designated the (-) mating strain, and the receiving filament is the (+) mating type—or, as in the less committal *Zygnema*, the two protoplasts move towards each other, forming the zygote in the conjugation tube between the filaments. The resulting zygote forms a thick-walled resting stage, a zygospore, whose walls contain sporopollenin, a very difficult substance to biodegrade. Thus charophyceans share the biosynthesis of sporopollenin with land plants where it is found in spore walls. Often sexual reproduction is instigated by deteriorating environmental conditions, shorter days, colder water, and nitrogen depletion, so the zygospore remains dormant until favorable growth conditions return.

Desmids are important components of freshwater phytoplankton. Some can grow on alpine snow fields or glaciers where their UV protective pigments produce a "pink" tint to the snow. Desmids have two basic forms, one of which consists of two uniquely shaped symmetrical hemicells connected by a narrow isthmus where the nucleus is located ("waist") (Fig. A37 B). When dividing, the cell separates at the waist and each hemicell grows itself a new half. Their cell walls are two-layered, a thin inner cellulosic layer and an outer, thicker mucilaginous layer composed of cellulose fibrils, pectins, and hemicelluloses. These cells are capable of gliding motility. Desmids of the other form do not have hemicells or mobility, and their cell wall is single layered. These differences were used to place desmids into two separate orders, but molecular studies have falsified this taxonomic hypothesis; the two cell forms do not reflect common ancestry. Presumably the two forms represent adaptations to similar environments.

The Charales have fossils dating to the Silurian. Some of these freshwater "seaweeds" have complex filamentous organizations and reach macroscopic size, often being mistaken for aquatic vascular plants. Others form much-branched filaments or flat discoidal sheets, a thallus. All have oogamy, apical growth, mitosis, and cytokinesis like land plants.

Green Bacteria

Only three groups of bacteria possess chlorophyll: the green nonsulfur bacteria, the green sulfur bacteria, and the cyanobacteria. No archaeans are photosynthetic. All other photosynthetic bacteria (heliobacteria, purple sulfur bacteria, and purple nonsulfur bacteria) solely use bacteriochlorophylls, and as their names suggest, they look brownish-purple.

Green Nonsulfur Bacteria

The phylogenetic position of green nonsulfur bacteria suggests they share a common ancestry with the cyanobacteria among the gram-positive bacteria (Fig. 2–8). They possess chlorophyll *a* but primarily use bacteriochlorophylls. Small organic molecules serve as a source of both carbon (instead of carbon dioxide) and hydrogen (instead of water), or they use hydrogen or hydrogen sulfide, in a very unusual photosynthetic pathway similar to the citric acid cycle (CAC) running in reverse. Such a photosynthetic pathway probably is derived from a heterotrophic fermenting metabolism, and indeed, green nonsulfur bacteria can function heterotrophically if provided with appropriate food molecules to ferment. Green nonsulfur bacteria are mostly flexible, gliding filaments sometimes called the green flexibacteria or chloroflexi bacteria. Many are thermophilic and residents of hot springs where they produce yellow to red carotenoid pigments and form films or bacterial mat communities, sometimes forming a layer below cyanobacteria. They may well have contributed to stromatolite formation on early Earth.

Green Sulfur Bacteria

The green sulfur bacteria have typical photosynthesis, like that of cyanobacteria and chloroplasts, except they have only photosystem II and acquire hydrogen from hydrogen sulfide (H_2S), which smells like rotten eggs, rather than from water. This results in the release of the unused sulfur as a photosynthetic byproduct. Hydrogen sulfide is found naturally in many anaerobic environments where decomposition is taking place, in volcanic gases, in natural gas (up to several percent), and in crude petroleum. As a result of their metabolism green sulfur bacteria are limited to very specialized aquatic environments, such as hot springs, where hydrogen sulfide occurs in a sufficiently bright place. Green sulfur bacteria have chlorophyll *a* and usually one to several bacteriochlorophylls, so depending upon the relative proportions of these pigments, the cells appear either green or brownish purple.

Cyanobacteria

(synonym: blue-green algae)

For two-thirds of Earth history, cyanobacteria were the predominant producers of Earth's surface-dwelling ecosystems, and they remain important photosynthetic producers and nitrogen-fixing symbionts. Free-floating cyanobacteria (phytoplankton) are estimated to be responsible for as much as 10% of ocean productivity (Waterbury and Stanier 1981). Presently they are common occupants of a wide variety of aquatic environments, both marine and freshwater, from hot springs to wet soil. Cyanobacteria probably achieved this success in part by switching their source of hydrogen from hydrogen sulfide to water, a much more common raw material.

All cyanobacteria possess chlorophyll *a*, and their bluish pigmentation (thus the name blue-green algae) comes from accessory pigments (phycocyanins) that help capture certain wavelengths of light. Three genera possess chlorophyll *b* as well, the same two chlorophylls found in the chloroplasts, yet surprisingly these genera do not appear related to chloroplasts. Cyanobacteria have no bacteriochlorophylls.

Cyanobacteria display one of four organizational forms: unicellular, colonial, unbranched filaments, and branched filaments (Fig. A40). These forms traditionally defined taxonomic groupings, but recent molecular studies have falsified most of these hypotheses. One filamentous lineage that includes several well-known genera, *Anabaena*, *Nostoc*, *Scytonema*, and *Fischerella*, is characterized by heterocysts, which are larger and some-

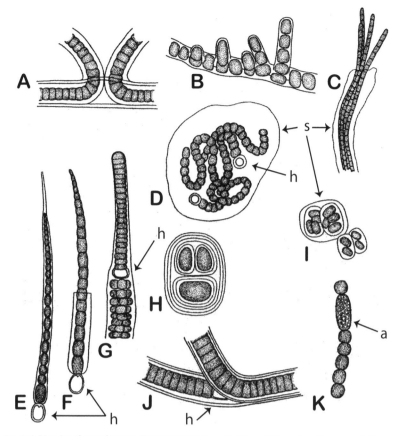

Fig. A40 Diversity of cyanobacteria showing unicellular, colonial, and filamentous forms. A. *Scytonema*, B. *Hapalosiphon*, C. *Schizothrix*, D. *Nostoc*, E. *Rivularia*, F. *Gloeotricha*, G. *Stigonema*, H. *Gloeocapsa*, I. *Chroococcus*, J. *Tolypothrix*, K. *Cylindrospermum*. A-J: With gelatinous sheaths (s); D-G, J: with heterocysts (h); H-I: unicellular colonial held together by gelatinous sheaths within sheaths; C-D: filamentous colonial within sheath or gelatinous matrix; A, J: false branching at dead cell (A) or heterocyst (J); B: true branching; K: filament with akinete (a), a spore-like stage for surviving poor environmental conditions. (Image source: After Scagel et al. 1984)

what empty-appearing specialized cells that function in nitrogen fixation. The formation of heterocysts can be stimulated by a low-nitrogen environment. Since nitrogen-fixing enzymes, nitrogenases, are inhibited by oxygen, heterocysts compartmentalize this function in cells separate from oxygen-liberating photosynthetic cells. Nitrogen-fixing genera that lack heterocysts are active in nitrogen fixation only at night when photosynthesis is not taking place. Since green organisms cannot use gaseous N_2 directly, a number of land plants have formed symbiotic relationships with nitrogen-fixing cyanobacteria (e.g., the mosquito ferns, cycads, and

Gunnera, a rain-forest flowering plant that lives on poor soil). A few genera exhibiting filaments with true branching (cell divisions in two perpendicular planes) may also form a phylogenetic lineage. Some filamentous forms, like *Oscillatoria*, exhibit gliding motility moving linearly, or rotating the end of the filament helically, although on a microscope slide this appears like bending rhythmically back and forth. Presumably the mechanism is based on excretion of hydrogels.

(synonyms: Anthocerophyta [-ales, -opsida])

clade nesting: hornworts/embryophyta/streptophytes/chlorophytes/plant/bikont/eukaryote

Hornworts are probably the least seen and least recognized group of land plants. Perhaps, like many small organisms, hornworts are more common than generally thought because they are so easy to overlook. Somewhere between five and eleven genera have been described; six genera are generally accepted as valid (Renzaglia and Vaughn 2000). Superficially hornworts bear a resemblance and similarity to liverworts, but hornworts have a smaller and more irregularly lobed thallus (Fig. 7–6 B). Unless the distinctive sporophytes are evident, a hornwort would be easy to mistake for a liverwort. Under a microscope, hornworts are easy to distinguish because each cell contains a single chloroplast with a pyrenoid, a cellular organization very similar to some green algae including *Coleochaete*, a charophycean. Liverworts and mosses have many oval chloroplasts in each cell like the rest of land plants. The hornwort gametophyte is simple in organization, just parenchyma with no specialized features and an organization no more complex than that of thalloid charophyceans.

Antheridia are embedded in pits in the thallus where an entire cluster of antheridia can develop. No organized archegonia exist. An egg cell differentiates within the thallus and neck canal cells differentiate above it. At maturity, the neck canal cells die, opening a channel in the dorsal surface of the thallus so sperm may access the egg within. The sporophyte begins development from within the gametophyte thallus, where the basal portions, the placental foot, remain embedded. As the sporophyte elongates, a sheath of maternal tissue grows upward around the sporophyte until finally the enlarging sporophyte pushes through, leaving this basal collar of gametophyte tissue to help support the tall (1–1.5 cm) columnar sporophyte.

The name hornwort is derived from the spike-like appearance of the sporophyte sticking up out the gametophyte thallus (Fig. 7–6 A-B) The hornwort sporophyte is quite unique among the bryophytes, although it shares common features too. The sporophyte is basically cylindrical with a well-developed foot. The sporophyte has a stem-like organization with a cuticle-coated epidermis, stomates and guard cells, and photosynthetic cells in its cortex region (Fig. 7–6 C), a much more complex internal organization than the gametophyte. A narrow column of sterile tissue occupies the center of the sporophyte, surrounded by a cylinder of spore-producing tissue. Just above the foot is an intercalary meristem providing for an indeterminate basal development, so new sporophyte tissues are added at the bottom as they mature at the top. New spore-generating tissue differentiates proximally around the column as the tissues mature distally. Thus this development allows apical spore dispersal to be continuous over a prolonged period of time rather than having all the spores mature and disperse at once. Since the sporophytes grow basally they can persist longer than sporophytes of mosses and liverworts. The basal regions of the sporophyte axis can form rhizoids and the sporophytes have been reported to persist on moist soil even after the surrounding gametophyte thallus has died. This suggests what an intermediate stage between a dependent and an independent sporophyte might be like.

Horsetails

(synonyms: Sphenopsida, Sphenophytes, Sphenopods, Sphenopsids, Equisetales)

clade nesting: horsetails/pteridophytes/megaphyllophytes/tracheophytes/embryophytes/ streptophytes/chlorophytes/plants/bikonts/eukaryotes

The generic name *Equisetum* (ek-why-SEE-tum) is constructed from two roots, the Latin for horses (*Equus*) and hair (*setae*), literally meaning horsehair or horsetail. Today the horsetail lineage consists of only some 20 species of *Equisetum*, and while this is a tiny fraction of former diversity, the relict genus is far from being in danger of extinction. *Equisetum* is a geographically widespread and successful genus. *Equisetum* can also be considered pretty successful in a geological sense because it, or a very similar genus, has a fossil record extending back to the Cretaceous, so this single genus has a history nearly as long as the entire history of flowering plants. Fossil evidence dates the earliest appearance of the genus *Equisetum* to the Jurassic (Channing et al. 2011). All living species of *Equisetum* are fairly small plants, even though *Equisetum giganteum* has 2.5 cm diameter stems that may reach close to 3 m in height, a glimmer of their past glory. The horsetail lineages appeared in the late Devonian, and like the clubmosses, horsetails left a rich, diverse fossil legacy and were once a major component of terrestrial floras. Arborescent horsetails existed during the Carboniferous, but following the climatic change of the Permian, the lineage began to decline in diversity and importance.

While presently all living species are placed in *Equisetum*, Hauke (1963, 1978) proposed segregating those species that have stiff, perennial stems into a different genus, *Hippochaete* (hip-poe-KEY-tee), which is constructed from the Greek roots for horse and hair. If this taxonomic revision becomes accepted by botanists, and so far it has not,[29] there would be two living genera, *Equisetum* and *Hippochaete*, both meaning horsehair, one in Latin and one in Greek. Botanists usually call members of the horsetail lineage sphenopods (SFEN-oh-pods) or sphenopsids, after the fossil genus *Sphenophyllum*, reserving the name horsetails just for living species.

Several species of *Equisetum* are widespread and so common that horsetails are a relatively familiar sight in the northern temperate zone around the globe. Given their antiquity, the absence of horsetails from Australia and New Zealand is an interesting biogeographical problem. Many species live in moist soils near fresh water, but some inhabit somewhat drier places. All grow rather aggressively at times, forming an extensive system of rhizomatous stems. Some species produce two types of aerial shoots: either green, branched, and sterile or nongreen, nonbranched, and fertile bearing a terminal cone. Fertile aerial shoots tend to be quite short-lived, dying after spores are shed. Growing as an herbaceous perennial, the sterile aerial shoots die after one season. Other species have tougher, scantly branched aerial shoots that persist for two or more seasons. The stems have a rough feeling produced by modified epidermal cells that house silica crystals. For this reason, horsetail stems can be used for an abrasive, a handy item for scrubbing a dirty pot, which results in another common name, scouring rushes.[30]

The sporophyte of *Equisetum* is an unmistakable and generally quite familiar plant and the eleven species found in North America are easily identified. The stems have prominent vertical ridging and nodal joints from which arise a whorl of leaves and branches (Fig. A41 A-B). In modern species the leaves are mostly vestigial, laterally fused into a sheathing whorl clasping the stem or reduced to nonphotosynthetic scales. Rhizomes produce whorls of roots at the nodes, and if the aerial stems branch, then whorls of branches, which break through the leaf sheath. The branches are just smaller versions of the main stem, and they also are noded and may be branched themselves. The vascular system consists of a ring of vascular bundles surrounding a hollow pith. Large air channels occupy the cortical region of the stems. The xylem has few heavily lignified cells, so the stem ridges are reinforced by thick-walled ground tissue. Because the leaves are small or vestigial and often nonphotosynthetic,

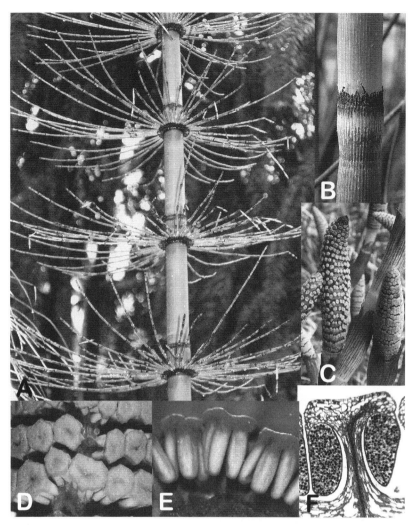

Fig. A41 Sporophyte of horsetail, *Equisetum*. A. Branching stem of *Equisetum giganteum* showing leaf sheaths and a whorl of branches at each node. B. Detail of unbranched stem showing fused leaf sheath above the node. C. Cones of *E. arvense* showing sporangiophores tightly packed together (right) or separated (left) as the cone elongates just prior to spore release. D-F. Sporangiophores. D. Sporangiophores showing angular tops shaped by packing and sporangia. E. Close-up side view showing whorl of sporangia attached to each sporangiophore. F. Longitudinal section through sporangiophore and two sporangia. Spores are visible within the sporangia. (Image sources: A—The Author; B—Courtesy of Frank Vincentz, Wikimedia Creative Commons; C—Courtesy of James Lindsey's Ecology of Commanster Site, Wikimedia Creative Commons; D-E—Courtesy of Curtis Clark, Wikimedia Creative Commons, F—Courtesy of and with permission of J. Jernstedt and the Botanical Society of America)

photosynthesis occurs in the cortical region of the stem, and the stomates occur primarily in the inter-ridge valleys.

Sporangia are borne terminally on modified lateral branches, sporangiophores that are aggregated terminally to produce strobili (Fig. A41 C-F). The sporangiophore, the multisporangiate subunit of *Equisetum* cones, is homologous to a lateral branch system bearing reflexed sporangia at their apices (Fig. A41 D-E). On fertile axes an abrupt developmental transition takes place between whorls of leaves and whorls of sporangiophores, accompanied by an abrupt cessation of internodal elongation. This results in whorls of fertile appendages being crowded together at the apex of a stem and forming a cone. On some aberrant shoots the transition from sterile to fertile is not so abrupt, and then the homology between leaf and sporangiophore, as predicted by the megaphyll hypothesis, becomes very apparent (Kaplan 2001). Some of these aberrant sterile leaves and sporophylls display an apical dichotomy, which is reminiscent of the leaves of *Sphenophyllum*. Modified, dichotomously branched foliar organs are associated with the sporangia of whisk ferns as well, although in most other respects, their appearance is different. This suggests more similarity between horsetails and whisk ferns than commonly recognized.

Equisetum is homosporous (producing only one kind of spore) (Fig. A41 F). The spore wall has an outer layer that unravels and flexes with changes in humidity, thus functioning as an elater in dispersal. The spores appear green because they contain functional chloroplasts, and thus they are constructed for rapid and nearly immediate growth. The gametophyte is terrestrial and photosynthetic, rather like a small, many-lobed, hornwort thallus. When growing at low density the gametophytes are hermaphrodites, producing both antheridia and archegonia. Although technically hermaphroditic, the gametophytes can respond to density in such a way as to improve chances of cross-fertilization. Gametophytes produce archegonia first, thus having an opportunity for cross-fertilization before producing antheridia, which almost assures self-fertilization. Functionally female gametophytes release a hormone that stimulates any nearby immature gametophytes to cease growing and quickly reach sexual maturity by producing antheridia. Thus a dominant female induces formation of a harem of dwarf males, a situation akin to heterospory but with only one type of spore. Similar hormonal interactions among populations of gametophytes also have been found in ferns.

Antheridia are embedded in the gametophyte thallus on lower lobes and constructed very much like those of *Lycopodium*, producing many sperm, but the sperm cells have a spiral shape and as many as 100 flagella. Archegonia are found among lobes on the upper surface of the gametophyte thallus. The neck cells elongate and protrude above the thallus. Fertilization requires free water. The zygote divides transversely, the lower cell producing a foot and the upper cell a primary aerial shoot. This shoot is very determinate in development, but it soon produces a basal branch from which a second shoot arises, which is bigger than the first. This continues until the new branch shoot is of a size characteristic of the species, then a basal branch establishes a rhizome from which many other aerial shoots arise.

Fossil Horsetails

Presently three fossil groups of horsetails are recognized: a group of arborescent species of the Carboniferous, a group of herbaceous species of the Carboniferous, and a transitional group (Stewart and Rothwell 1993). In many ways the development and decline of the horsetails parallels that of the clubmosses. The horsetail lineage arose from a trimerophyte ancestry in the mid-Devonian via an intermediate group now often placed in the cladoxylopsid "ferns," so as a whole the group is transitional between trimerophytes and pteridophytes, but this grouping may oversimplify things. The cladoxylopsids showing similarities to horsetails had stout dichotomously branched rhizomes and numerous aerial shoots. The somewhat monopodial axes had small, dichotomous lateral branch systems that functioned as photosynthetic organs. Sporangia were borne on the recurved ends of some of these lateral branches, often mixed with sterile segments. The fertile branches tended to be clustered terminally but not tightly enough to be considered cones. Thus this group possesses characters derived from trimerophytes that are appropriate for horsetails.

Both the herbaceous and arborescent groups of horsetails formed important components of Carboniferous coal swamp forests, occupying both its canopy and its understory. In many vicinities *Sphenophyllum* (Fig. A42) was one of the most common fossils, suggesting this was a plant adapted to scrambling or sprawling through the understory of a clubmoss-horsetail-fern forest. Slender axes bore whorls of 6 or 9 fan-shaped leaves that varied in the amount of fusion from species to species, or even whorl to whorl on the same plant (Fig. A42 A-B). The vascular strand entering the base of a leaf branched dichotomously one to several times to fan out across the blade, which was sometimes dichotomously highly dissected (Fig. A42 A). Thus *Sphenophyllum*'s leaves have the exact morphology appropriate for megaphylls. Some leaves are reduced to a slender, elongate blade with a single central vascular strand. With such leaves in the fossil record, the slender, single-veined leaves of *Equisetum*, often interpreted as microphylls, must be reinterpreted as megaphylls transformed into microphylls via reduction to vestigial structures.

About 30 years ago, horsetails and clubmosses were often grouped as microphyll-bearing plants, a hypothesis of common ancestry. But when the character suggesting common ancestry is at odds with other characters—for example, position of the sporangia (terminal versus lateral)—every effort must be made to determine whether the similar features are homologous (of one ancestral origin) or from two different origins. In this case the fossil record resolves the problem, placing horsetails firmly into the megaphyll-bearing lineage.

Axes of *Sphenophyllum* have a very distinctive solid, triangular core of xylem, a "primitive" feature retained from rhyniophytes, but they do produce some secondary xylem (Fig. A42 C). Straight files of secondary xylem arise along the sides of the three arms of primary xylem; the axes

Fig. A42 Sphenophyllum. A-B. Fossil compressions of leafy axes of *Sphenophyllum*, an herbaceous Carboniferous horsetail. Whorls of leaves are borne at each node. Each whorl here is 1–1.5 cm in diameter. C. Cross-section of a stem axis showing limited secondary growth. Primary xylem forms a concave triangle in the center (two vertices marked by arrows). Short rows of secondary xylem are between the two arrows. External to the xylem, the phloem region of this specimen has no cellular preservation, but the cortex is largely intact. (Image sources: A-B—Courtesy of and with the permission of Hans Steur, Hans' Paleobotany Page website; C. Courtesy of and with the permission of J. Jernstedt and the Botanical Society of America)

Fig. A43 *Calamites*, an arborescent horsetail of the Carboniferous. A. Reconstruction of *Calamites* showing rhizome and aerial shoot, both distinctly noded and ridged. B. *Palaeostachya*, a cone of a *Calamites*-type plant composed of whorls of sporophylls. C-D. Fossilized casts of hollow stems showing distinct nodes and ridges. C. Stem several centimeters in diameter. D. Stem apex showing several nodes close together, ridges, and bud scars. E. *Annularia*, fossil whorled leaves of *Calamites*. (Image sources: A—After Hirmer 1927; B—After Delevoryas 1955, with permission of the Botanical Society of America; C—Courtesy of Daderot, Wikimedia Creative Commons; D—Courtesy of Linda Spashett, Wikimedia Creative Commons; E—Courtesy of Ghedoghedo, *Museum Mensil und Natur*, Wikimedia Creative Commons)

never become large and woody, but *Sphenophyllum* could have been a perennial plant. The strobili of *Sphenophyllum*, which were named *Bowmanites*, consist of fused whorls of leaves alternating with whorls of short lateral branches whose recurved apices terminate in sporangia. Some phylogenetic studies have suggested that *Sphenophyllum* represents a separate lineage with its own origin from trimerophyte ancestry (Stein et al. 1984). The problem encountered here is the inability to determine homologies among foliar organs and reproductive organs.

Arborescent horsetails lived primarily during the Carboniferous, and they also possess most of the definitive features associated with modern horsetails. *Calamites* (kal-ah-MY-teez) is a common fossil axis, and the name also refers to the whole reconstructed organism (Fig. A43 A, C-D). Stems were rhizomatous or aerial, both with distinct ridges, nodes, whorled branches, roots, and leaves. Although organized like *Equisetum*, the vascular system developed a cambium between the xylem and phloem that produced considerable woody tissues to thicken the stem. The inner diameter of pith casts of *Calamites*, when compared with the hollow piths and heights of modern species of *Equisetum*, suggest that these plants reached 15–20 meters in height. Leafy stems consisting of simple whorled leaves on slender axes called *Annularia* and *Asterophyllites* are the foliage of these arborescent horsetails (Fig. A43 E). Cones of these arborescent horsetails were composed of lateral sporangia-bearing branches associated with subtending leaves (Fig. A43 B) similar to those of *Sphenophyllum*.

Both fossil groups appear at the beginning of the Carboniferous and diversify rapidly in the middle Carboniferous. Both groups decline toward the end of the Carboniferous and become extinct in the early Permian. The modern group of horsetails appears at the beginning of the Permian, reaching its maximum diversity during the Triassic and Jurassic. During the Cretaceous the diversity dwindles until a single genus remains.

(synonyms: Marchantiophyta, Hepatophyta, Hepaticophyta)

clade nesting: liverworts/embryophytes/streptophytes/chlorophytes/plant/bikont/eukaryote

Generally liverworts are small and grow in wet places, often at pretty low light levels; as such they frequently fall below the perceptional threshold of most people. The most familiar and common liverworts of the northern temperate zone are composed of a flat, ribbon-like thallus. The lobing of the thallus[31] caused people to think it looked like a liver, thus the common name and traditional use of the root hepato- for the taxonomic group. Liverworts may well be the most ancient lineage of land plants (Kenrick and Crane 1997; Crandall-Stotler and Stotler 2000). Traditionally liverworts were grouped with mosses and hornworts as the nonvascular land plants, phylum/division Bryophyta, but presently bryophytic plants are not treated as a single group. Liverworts possess several unique features that help delimit them as a group and differentiate them from other mosses and hornworts (Crandall-Stotler 1984; Mishler and Churchill 1984), although all three share a gametophyte/haploid phase dominant life cycle.

As in the hornworts and mosses, the common and familiar liverwort organism is the haploid phase, the gametophyte, which exhibits one of two general types of organization: either a flat, ribbon-like thallus or a flattened or three-sided axis with leafy enations arranged in two or three rows. Most North American biology and botany books illustrate liverworts using *Marchantia*, a temperate genus with a ribbon-like thallus (Class Marchantiopsida) (mar-SHANT-ee-op-sid-ah). While a great-looking organism, *Marchantia* is not at all typical of the majority of liverworts even though it has become a familiar poster child for the whole group. Ninety percent of all species are leafy liverworts, usually treated as Class Junger-manniopsida (youn-ger-man-ee-OP-sid-ah), but they remain less known because leafy liverworts are primarily tropical. Liverworts in this group are often mistaken for a prostrate moss, but the arrangement of enations makes them easy to distinguish with a careful look. Liverworts have a row of enations alternating along each side of the axis (Fig. A44 A), and if a third row exists, they are along the bottom of the axis and often vestigial or modified as anchoring structures. Prostrate mosses have one row of ena-tions along each side of the axis and usually two rows of enations prostrate along the top of the axis. At one extreme, liverworts have a basic complex-ity not much greater than the charophytes (Appendix: Green Algae) and at the other extreme, they possess a considerable number of specializa-tions. Although a leafy liverwort looks more complex, more differentiated, anatomically they are the simplest. In many cases the enations consist of a single layer of cells. Antheridia and archegonia are borne in terminal buds of lateral shoots surrounded by enations. The simple sporophytes have short stalk and a terminal sporangium (Fig. A44 B-C). Elongation of stalk cells pushes the sporangium up and out of the bud. The sporangium splits open apically by four valves and spring-like elater cells help eject the spores (Fig. A44 C).

Liverwort gametophytes commonly grow as a clone with an advanc-ing growth front, with each axis branching dichotomously, producing two equal branches. At some distance behind the growing apices the older portions of the thallus senesce and die. Liverworts have two general means of vegetative reproduction: fragmentation of a growing, branching thallus, and the production of gemmae (JEM-mee), which are tiny spe-cialized branches or outgrowths of the thallus capable of dispersing and growing into a new thallus (Fig A45 C). Gemmae vary from simple (one or two cells or a short filament of cells) to more complex multicellular types that are a miniature thallus with distinct apical cells and a bilateral organization. Gemmae develop from and are borne upon the thallus sur-face, sometimes in clusters or at the apex, or on margins of enations. In more complex forms gemmae can be associated with gemmae cups, spe-cialized outgrowths of the thallus with gemmae at the bottom. Gemmae cups cause falling drops of water to rebound, a process that can dislodge gemmae and disperse them a meter or more, a considerable distance for a

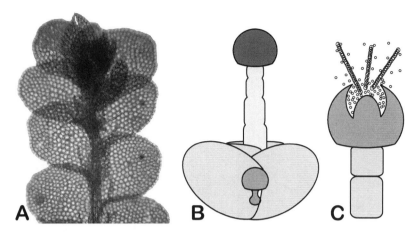

Fig. A44 Leafy liverworts. A. Small portion of a leafy thallus showing two rows of overlapping enations only one cell thick. B. Mature sporangium emerging from within apical enations by elongating stalk cells. Immature sporophyte consisting of sporangium, foot, and short stalk showing original position within terminal bud. C. Sporangium dehiscing by four valves releasing several long spring-like elators that assist in scattering the spores. (Image sources: A—Courtesy of B. Haynold, Wikimedia Creative Commons; B-C—The Author)

small organism. Such asexual reproductive mechanisms are used for colonization within a favorable patch (Fig. A45).

The gametangia of liverworts are typical for bryophytes in general (see Fig. 7-1). Antheridia are produced within pits or chambers sunk in the thallus or in the axils of enations. The antheridia have a stalk, an external layer of jacket cells. All the internal cells divide numerous times, making many small cells, each of which will produce one sperm. These sperm-producing cells often appear in packets that represent the boundaries of the original mother cell. The vase-shaped archegonia are produced in apical clusters, surrounded by enations in leafy forms, or on the dorsal surface of the thallus. In *Marchantia* and similar liverworts, both archegonia and antheridia are borne aloft upon specialized erect branches of the thallus (Fig. 7-7), a positioning that ultimately aids in spore dispersal because the sporophyte will develop at the location of the egg. Swimming sperm disperse in any film of water along the surface of the thallus or the leafy axis and along furrows and rhizoid-filled channels in the thallus. Splashing by rain drops can locally disperse sperm between nearby plants; still self-fertilization must be relatively frequent.

In contrast to the gametophytes, the sporophytes of liverworts are all similar in organization, small and simple in construction involving no meristematic growth. They consist of a sporangium, or "capsule," which

Fig. A45 Liverwort form and anatomy. A. *Riccia*. Simple thallus showing Y-branching. Internally the tissue is undifferentiated although cells on the upper side have more chloroplasts. B-D. *Marchantia*. B. Broad thallus showing numerous polygonal air chambers each with a central pore (circle and arrow). Penny serves as a size reference. C. Gemmae cups, one large, one small, on thallus each containing numerous oval gemmae, specialized branches for asexual reproduction. The cup serves to enhance the rebound of rain drops to detach and disperse the gemmae. D. Diagram of cross-sectional anatomy of the thallus through an air chamber (ac), circled in B. Partitions (vertical arrows) delimit the margins of the air chambers. The central pore (p) is formed by a cylinder of cells. The air chambers are filled with short, branched photosynthetic filaments (pf), cells filled with chloroplasts. Larger cells composing the lower thallus lack chloroplasts and function in storage. (Image sources: A—Courtesy of Bernd Haynold, Wikimedia Creative Commons; B—The Author; C—Courtesy of Luis Nunes Alberto, Wikimedia Creative Commons; D—After and with permission of the Florida Center for Instructional Technology)

contains the sporogenous tissue; a foot, which is a placental organ; and a short stalk between the sporangium and foot. This means the sporophytes are nutritionally completely dependent upon the gametophyte. In the simplest case the sporangium disintegrates at maturity, leaving the spores within the gametophyte tissue. Dispersal takes place after the thallus dies and disintegrates, but this does not mean such gametophytes are short lived. Apical portions continue to grow as the oldest portions of the thallus die.

The sporophytes of liverworts have a discrete epidermis, but no stomates or guard cells develop. Whether the lack of stomates (present in all other land plant sporophytes) represents the retention of an ancestral state, or whether it is nothing more than a consequence of being encased within gametophyte tissues where gas exchange is not needed (Crandall-Stotler and Stotler 2000), remains a question. If the former, then stomates and guard cells are a shared, novel character for all other land plants, which a recent study suggests is the case (Ligione et al. 2012). The sporangium of liverworts also lacks a central core of sterile tissue (a columella), which is present in mosses, hornworts, and one early-appearing vascular plant (*Horneophyton*).

In some liverworts the cells of the sporangial stalk can elongate 10–20 times their original length, elevating the sporangial capsule to aid in spore dispersal. Most liverworts produce elaters to function in dislodging the spore mass from the sporangium, but these elaters also may serve during their development to nourish developing spores (Crandall-Stotler and Stotler 2000), a function usually carried out by cells lining the sporangium. In most liverworts, each initial cell of the sporogenous tissue divides to produce one elater and one spore mother cell, resulting in one elater for every four spores (Kenrick and Crane 1997).

Leafy Liverworts

Over 300 genera of leafy liverworts have been described. Because of the higher rainfall in the Pacific Northwest, leafy liverwort species are both more numerous and more common than thalloid species. Leafy liverworts are composed of a basal grade including Metzgeriales and a more specialized clade including Jungermanniales and Porellales, wherein two-thirds of the genera are placed (Crandall-Stotler and Stotler 2000). The oldest known liverwort fossil, *Pallaviciniites* from the late Devonian (Schuster 1966), is similar to extant members of basal lineages. The five small basal clades are interpreted as relics of former diversity, a pattern that will be encountered in other ancient lineages as well.

The axis of leafy liverworts shows no internal organization and the enations are only a single cell-layer thick and arranged in two rows (Fig. A44 A). When present, the third row of enations is on the ventral (bottom) side of the thallus. Careful observation will reveal a couple of interesting features. Enations arise as outgrowths of an apex with a pyramidal apical cell, which has three proximal faces. Each division of the apical cell is parallel to one of these three proximal faces, so cell derivatives are produced in three ranks, resulting in three rows of enations. But this three-ranked organization becomes flattened into two dimensions. Each enation in the two lateral rows has a folded base where the bottom portion of the enation folds to the underside of the thallus. The ventral row of enations are interpreted as vestigial, meaning an organ of reduced structure of no apparent function. However, in tropical species growing upon the surfaces of leaves (epiphyllic), these bottom enations are modified and serve to anchor the axis to the leaf surface. A three-ranked organization has become bilateral presumably to increase light-absorbing surface area by growing prostrately upon a substrate. Does this mean liverworts have a radially symmetrical ancestry, something like *Chara*? Possibly, but no fossils exist to support or clarify such conjectures.

Thallose Liverworts

To many people the flat, ribbon-like thallus of these liverworts has a certain aesthetic appeal so that such simplicity represents the minimalist land plant. The thallus grows apically, Y-branching and fanning out over its substrate, the surface of rocks, sand or soil, burnt ground, tree bark or other plant stems (Fig. A45 A-B). The thallus may be very broad, 1 cm or more wide, or quite slender, no more than 1 mm wide. The lateral proliferation of tissues derived from a wedge-shaped apical cell results in an apical notch. A range of structural specializations are represented from simple to the rather complex thallus of *Marchantia* and its relatives.

Riccia represents a simple structural organization with essentially an undifferentiated thallus (Fig. A45 A), but within this group simplicity may be derived and associated with an adaptation to an aquatic habitat. Cells in upper portions of the thallus have more chloroplasts, and those in lower portions fewer. Rhizoids are solitary and scattered on the lower surface.

Ricciocarpus demonstrates an intermediate level of internal organization where the lower part of the thallus has a more compact unpigmented tissue functioning in food and water storage (Fig. 7–4). The upper portion of the thallus consists of partitions of chloroplast-bearing cells separated by air chambers. Some species in this genus are floating aquatics, and in these species larger air chambers and thin, one-cell-wide partitions can be viewed as specializations for flotation. Epidermal cells covering the air chambers lack chloroplasts and function as translucent windows allowing light to penetrate deeper tissues of the thallus. The thallus has a dorsal central furrow with an inverted Y-shape that functions as a capillary channel for the capture and uptake of water, and through which sperm swim to reach the archegonia embedded in the thallus at the base of the furrow.

In *Marchantia* and its close relatives the thallus has a dorsal/ventral differentiation similar to that described above, but the air chambers are open to the atmosphere through pores, allowing more rapid gas exchange (Fig. A45 B, D). In some species the pores are a simple hole; others have pores ringed by cells providing a border. The bordering cells usually excrete a waxy, water-repelling cuticle that helps prevent water from filling the chambers. Within the chambers very algal-like branched photosynthetic filaments proliferate, providing a mass of photosynthetic tissue and considerable surface area for gas exchange. This thallus organization is analogous to stomates and leaf mesophyll, a functional convergence in form.

On the ventral surface of the thallus, rhizoids are aggregated into a central strip and enclosed by overlapping, scaly outgrowths of the thallus parallel to the strip of rhizoids. The rhizoids run horizontally along the channel created by the scales, and then grow outwards into or onto the substrate. Rhizoids outside the channel grow vertically. Rhizoids adhere to hard substrates very tightly, and the thallus can break before rhizoids will pull away. The rhizoid-filled ventral channel functions in absorption, taking up water by a brush-like capillary action and providing a convenient route for swimming sperm.

In this group of liverworts, the gametangia are borne aloft on specialized branches of the thallus (Fig. 7–7). Although highly modified, gametangia-bearing aerial branches retain the rhizoid channel in a groove that is their morphological lower side and the photosynthetic chambers characteristic of the unmodified thallus. Most of these species are dioecious, meaning they produce antheridia or archegonia but not both. Branches bearing archegonia have radiating lobes while those bearing antheridia lack them. Placing archegonia aloft results in the small, short sporophytes developing in a location that greatly increases spore dispersal. As a consequence of their position, sporophyte stalk cells do not elongate in these species. Antheridia are produced in pits upon the upper surface of their aerial branch. A rhizoidal water channel is continuous down to the ventral surface of the thallus, but sperm still must get to a female thallus. This might be accomplished relatively easily if thalli of different sexes were intergrowing. Drops of rainwater or dew hitting a thallus can splash sperm a considerable distance, a meter or two, which is much further than sperm can swim.

Antheridia placed aloft on specialized branches would seem a hindrance to fertilization, so there is no obvious good reason to do so. A probable explanation is that the development of aerial branches and gametangia are linked. The thallus divides apically, an event that usually produces a simple Y-branching of the thallus, but under the proper stimulus, one branch develops aerially upon which gametangia then develop. If reproductive success was improved by elevating archegonia and egg and by developing the sporophyte and its sporangium aloft to enhance spore dispersal, then as a consequence of their developmental linkage, the antheridia were elevated upon a specialized branch as well even though of no particular value. Natural selection would not produce such features unless the overall result produced more offspring, so fertilization cannot be negatively affected to any significant degree no matter how awkwardly placed the antheridia appear to us.

Mosses

(synonyms: Bryophyta [-opsida, -ales], Musci [class])

The leafy gametophytes of mosses are readily recognized by most people. Many familiar mosses consist of a tussock or cushion of closely spaced aerial axes, each bearing a helix of enations (Fig. 7–5; Fig. A46 A, D). Other mosses have a branching prostrate form where the axis bears four rows of enations, one along each side and two along the top of the thallus; thus they are easily distinguished from leafy liverworts, which have two lateral rows of enations. Furthermore, the enations of liverworts tend to overlap, while those of mosses do not. Some moss enations are very simple, consisting of a single layer of cells and a small supporting midrib (Fig. A46 B-C). Others have more complex organization, 2 or more layers of cells and a larger midrib, and the most complex have on their upper surface single-celled sheets of photosynthetic tissue, an organization that increases surface area for gas exchange and light absorption (Fig. A46 D-E).

The familiar leafy moss develops from spores via an intermediate stage that takes the form of either an irregularly lobed thallus or a prolonged, branching filament, which can grow across a considerable area (100 cm^2 or more), producing numerous upright leafy axes that, after the disintegration of the filament, grow as individual plants (see Fig. 7–1 B, life cycle of moss). Thus a mound or tussock of moss is composed of many individual axes representing a clone arising from a single spore. However, if filaments from several spores intermingle, a tussock of axes may

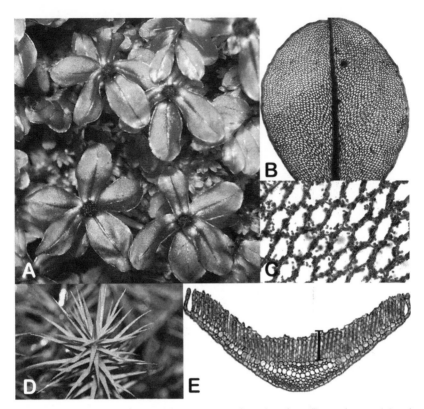

Fig. A46 Moss enations. A-C. Simple moss enation. A. A simple enation a few millimeters long consisting of a single layer of cells and a small midrib with a few central support cells. B. Enlargement showing a single layer of chloroplast-containing cells and midrib. C. Chloroplast-containing cells showing similar sizes and shapes. D-E. *Polytrichum*. D. Leafy axis bearing a helix of somewhat stiff, awn-shaped enations. E. Enation in cross-section showing a thicker, more complex structure. Upper surface consists of parallel sheets of chloroplast-containing cells (bracket) topped by cap cells with a heavy cuticle to keep water from filling the capillary spaces between layers. Thicker-walled cells below provide support and conduction. (Image sources: A—The Author; B-C—Courtesy of Kristian Peters, Wikimedia Creative Commons; D—Courtesy of Lindsey, Wikimedia Creative Commons; E—Courtesy of Thismia, Wikimedia Creative Commons)

represent two or more genetic individuals. Each fertile aerial axis produces gametangia terminally clustered among an apical whorl of enations. Prostrate mosses bear gametangia terminally on short lateral branches. Gametangia can be found fairly easily by anyone taking enough time and care to investigate for themselves, provided you have a microscope handy, some fine needles, and a steady hand.

Sperm must swim from the top of one axis bearing antheridia to another bearing archegonia. A heavy dew or light rain would easily provide adequate water. Growing closely in a cluster, these leafy axes can hold water between enations by capillary forces, and drops may even form on the surface. Sperm also get splashed around by falling droplets of water, an action that can disperse them a meter or more.

Moss sporophytes are fairly common and easily recognized, rising above the tufts of leafy gametophytes such that they appear as if the sporophyte were just a leafless extension of the gametophyte axis (Fig. 7–1 C, H). In most mosses a large round to oblong sporangium is held aloft on a long, thin stalk. After fertilization, the archegonium forms a sheath-like covering over the sporophyte early in its development. Near the end of development this sheath gets ripped loose at the base as the sporophyte stalk elongates, leaving the sheath as a detached hood covering the sporangium, a most violent birthing. A sporangium opens by means of one of three mechanisms, a single spiral slit down from the apex in the basal moss genus *Takakia* (Fig. A47 C, F), four vertical slits along the sides, or a circular line beneath an apical cap that is shed to release the spores. Some mosses have a row of hairs beneath the cap lining the opening of the sporangium that open and close the sporangium in response to changes in hydration, another function performed by uneven cell wall thickenings.

Considerable differences exist in the way various authors classify mosses. The problem is that six rather different genera require a taxonomic separation from the vast majority of mosses. This fact is reflected in a rather recent treatment of moss classification that consists of six classes (number of families/number of genera): Takakiopsida (1/1, *Takakia*), Sphagnopsida (2/2, *Sphagnum*), Andreaeopsida (1/2, *Andreaea*), Andreaeobryopsida (1/1, *Andreaeobryum*), Polytrichopsida (4/27), and Bryopsida (107/800+) (Buck and Goffinet 2000). The over whelming majority of families (107), genera (800+), and species resides in Bryopsida, one of the six classes. Two classes consist of a single genus; two classes have two genera, for a total of 6 genera in four classes. Of these six genera, only one is familiar, *Sphagnum*. These taxonomies suggest there are a few somewhat isolated or relic moss genera, perhaps representing very old lineages, and a great deal of diversity that has resulted from the multiplication of niches created by the diversification and spread of vascular plants. In particular mosses with a prostrate habit and an indeterminate thallus, as opposed to those with upright, determinate axes, diversified during the Cretaceous in the understory environment created by the rise to dominance of flowering/fruiting plants (Newton et al. 2007). A similar or parallel pattern exists in liverworts and ferns for much the same reason: changes in life, change life. Some opportunities may be removed as new opportunities arise. Some groups of former importance wane in diversity, other groups expand in diversity.

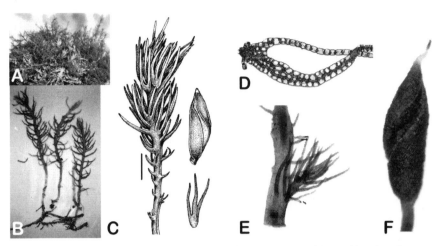

Fig. A47 *Takakia*, a basal moss. A. Habit of gametophyte on moist ground. B. Aerial branches of gameto-phyte attached to basal rhizome. C. Illustration of aerial shoot. Close-up of sporophyte capsule and leafy organs (enations) that are once or twice lobed, circular in cross-section, and borne in irregular clusters or whorls. D. Once-lobed enation. E. Detail of young aerial shoot. F. Mature sporophyte capsule opening via a spiral slit (compare with illustration C). (Image sources: A-B—Courtesy of the Digital Museum, University of Hiroshima, Wikimedia Creative Commons; C—After Spence and Schofield 2007; D-E—Courtesy of and with permission of K. Renzaglia; F—Courtesy of and with permission of J. M. Budke)

Many phylogenetic studies place *Sphagnum*, peat moss, as the earliest diverging lineage of mosses, although others accord that honor to the enigmatic genus *Takakia* (tah-KAH-kee-ah) (Fig. A47). *Takakia* is a very strange genus that was originally described as a leafy liverwort. The enations are cylindrical rather than flat, once or twice lobed, and borne in an irregular helix or in small clusters (Fig. A47 A-E). Placing *Takakia*, *Sphagnum*, and *Andreaea* as basal lineages leaves a clade of mosses that share the character of having cap-opening sporangia (Polytrichopsida and Bryopsida) (Magombo 2003). The basal genera all have capsules that dehisce via one, four, or more vertical slits, like liverworts. Furthermore, the basal mosses share the feature of having their short-stalked sporophytes elevated upon a gametophytic stalk rather than having sporophytes with elongating stalk cells.

Sphagnum is ecologically and economically the most important moss. *Sphagnum* alone is a significant component of widespread peat land communities, which in the high latitudes or at high altitudes can compose over 10% of all communities, particularly in bogs and tundra (Fig. 7–2; Fig. 7–3). The presence of *Sphagnum* contributes to the acidic conditions of bogs, and all organisms growing there are adapted to low pH conditions. *Sphagnum* moss axes grow apically and die proximally, accumulating considerable biomass as it compacts upon itself. In these low pH conditions, decomposition is slow and organic material accumulates, forming peat. Accumulations of the dead peat moss, both recently dead and partially decomposed, are gathered as peat moss for soil conditioning and potting mixtures. As a consequence of little microbial decomposition and the low pH, dead *Sphagnum* is nearly sterile, a feature, along with its ability to absorb moisture, that led to its former use as surgical and wound dressings. Ultimately the accumulation of decomposed *Sphagnum* gametophytes forms peat, a soil with high organic content. Peat soil contains so much carbon that it can be cut, dried, and burned as a fuel, albeit a rather smoky one; however, it is this smokiness that imparts the distinctive taste to drying barley malt in the making of scotch whiskeys. Now who would dare suggest *Sphagnum* is not economically important? Unfortunately, it appears that the harvest of peat for fuel is not sustainable.

Ecologically, *Sphagnum*-dominated communities are beginning to get more attention because their carbon turnover has been very slow. Like other low-turnover carbon reservoirs, the biomass of *Sphagnum* represents a lot of carbon dioxide that was formerly in the atmosphere. A very small increase in average temperature in the high latitudes could greatly accelerate the decomposition of peat moss, thus releasing carbon from this low-turnover reservoir back into the atmosphere, where the increased carbon dioxide could further contribute to a greenhouse effect and global warming, setting up a nice little positive feedback system. Although this ecosystem has been little studied outside the Scandinavian countries, understanding it may be critical to understanding the dynamics of global climatic change.

The particular feature that makes *Sphagnum* valuable in soil/potting mixtures is that this moss can hold twenty to thirty times its own weight in water, either dead or alive. How this is accomplished can be determined by close examination of the enations (Fig. A48). Like most moss enations, these are only 1-cell thick. In comparison with most green organisms, *Sphagnum* appears a very pale green, indicating in part a deficiency of nitrogen, which is not readily available in soluble forms under low pH conditions. But the pale green color also is due to the anatomy of the enations, which have only a network of long, narrow, living cells that bear chloroplasts (Fig. A48 B-D). In between the live "green" cells are larger, dead, thin-walled cells whose lateral walls are reinforced by thickenings that traverse the walls in thin spidery bands. Their lateral walls also have one or more circular pores allowing the dead cells to capture and hold water by capillary action. This means *Sphagnum* can hold almost as much water when dead as when alive. The arrangement of enations on axes

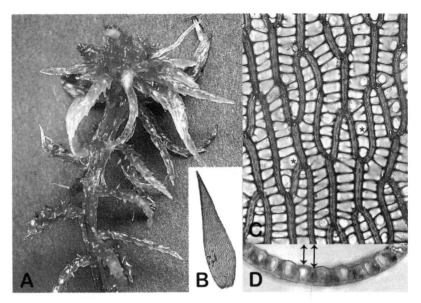

Fig. A48 Cellular organization of *Sphagnum* (peat) moss enation. A. Branched leafy axis of *Sphagnum*. B. Single enation showing a single layer of cells in a network. C. Enlargement showing a network of live chloroplast-containing cells (darker) surrounding large, dead, and empty water-holding cells (lighter). The water-holding cells have helical reinforcing bands and pores in the cell wall (*). These cells take up and hold by capillary action. D. Cross-section of enation showing smaller diameter, live, chloroplast-containing cells (right arrow) alternating with the larger diameter, water-holding cells (left arrow). (Image sources: A—Courtesy of Denis Barthel, Wikimedia Creative Commons; B-D—Courtesy of Kristian Peters, Wikimedia Creative Commons)

contributes to their function. Enations cover the branch axes in a closely overlapping pattern that produces many capillary spaces both between the enations and along the axis. The erect axes of *Sphagnum* are much branched, the branches occurring in clusters of three to eight drooping branches (Fig. A48 A; Fig. 7–2 A). The branch axes also have a single cortical layer of dead water-holding cells. This means the stems are not very robust, but larger species grow either upon water, supported by buoyancy, or emergent in a dense mat or mound of mutually supportive axes. Such mounds of *Sphagnum* are virtual living sponges, wicking up and holding massive amounts of water. This dense packing also means little light penetrates very far into such a mat or mound. As a result, only the distal few centimeters of each axis are alive and photosynthetic even though the axis may approach a meter in length in some species. In emergent species two types of branches are produced: divergent branches that are borne more or less perpendicularly from the main axis, and pendent branches that are longer and narrower and bear smaller enations. Under certain conditions *Sphagnum* produces considerable amounts of anthocyanins, red water-soluble pigments that mask the pale green. Such pigments may function in high altitudes and high latitudes by absorbing sunlight and raising the temperature to increase metabolic activity. They also may screen out harmful ultraviolet wavelengths.

Andreaea (ahn-dree-AYE-ah) is one of the few mosses I can identify reliably on sight. It grows as low, dark-brown to blackish clusters or tufts usually on bare silica-containing rocks. The genus has a worldwide distribution in alpine and high latitude regions and demonstrates that bryophytes can grow in extremely harsh conditions and are capable of inhabiting an otherwise bare continent. About 10 species are found in western North America; other members of this group are tropical and less known.

Polytrichopsida

Another group of mosses that possesses some unique specializations of the gametophyte is the genus *Polytrichum* and other members of its group. Some other distinctive mosses, *Tetraphis*, a genus commonly found growing on decaying coniferous logs in boreal forests, and *Buxbaumia*, a tiny plant usually found on acidic soil or decaying wood, are placed as members of this class in a different order. In some classifications all three groups are elevated to equal and separate taxa—for example, three subclasses, and all three may form a basal grade to the Bryopsida (Magombo 2003). They share at least some of a number of anatomical specializations that argues in favor of their relatedness, but unlike the three previous lineages, *Polytrichum* shows specializations that approach the vascular plant level of anatomical organization with three discrete tissue systems.

Polytrichum has unbranched aerial axes that can reach 10–20 cm in height, the tallest moss in North America. The enations are large, in some cases stiffly divergent from the axis, having a central midrib several cells thick. In some the leaves are arranged in five helical ranks. Cells within the midrib have thicker walls that look like vascular tissue but function more in support than conduction. In *Polytrichum* the dorsal surface of the enations bear sheets of photosynthetic cells (Fig. A46 D-E). Surface area is maximized and water loss minimized by the close spacing of the lamellae. Cutinized cap cells form the top row of the lamellae to prevent water from filling the spaces between the layers by capillary action and remaining there. The axes of this moss have a stem-like organization with a discrete epidermis, a cortex, and a central cylinder of conducting tissue. The central cylinder has specialized cells, hydroids and leptoids, which function in conduction like xylem and phloem, but the hydroids lack thick secondary cell walls and lignin characteristic of xylem. Strands of these conductive cells traverse the cortex to the enations much as in vascular plants, and this particular feature suggests the midrib has at least some conducting function.

The vast majority of mosses belong to the class Bryopsida consisting of 15 orders, over 100 families, over 800 genera, and over 14,000 species. It takes considerable patience and much experience to tell one from another. This group encompasses tremendous morphological diversity, growing in virtually all manner of habitats. The general description of mosses applies to members of this lineage. Virtually all of the mosses you regularly encounter will belong to this lineage.

Phytoplankton

In general phytoplankton are unicellular or small colonial, free-floating, photosynthetic organisms. They are the primary producers of open ocean ecosystems. Many are motile and have a light-sensing organ, an eyespot. The following groups of eukaryotic organisms (dinoflagellates; hapto-phytes; cryptophytes; glaucophytes; euglenozoa; chrysophytes, diatoms, eustigmophytes, and other stramenopiles) wholly or largely consist of phytoplankton. Some people would call them algae; others would not. Several groups include nongreen species. The red algae and green algae include some phytoplankton, and the brown algae, which are mostly seaweeds, have their own sections in the appendix.

Dinoflagellates

(synonyms: pyrrophytes, peridineans)

clade nesting: dinoflagellates/alveolates/bikonts/eukaryotes

Dinoflagellates are largely motile, biflagellated unicellular organisms whose cell wall consists of plates of cellulose (Fig. A49 B). Although the cell shapes are variable, the two laterally inserted flagella each occupy one of two grooves, one groove extending transversely around the perimeter of the cell forming a waist, and the other groove extending toward the cell's posterior (Fig. A49). The flagellum extended to the posterior pro-pels the cell forward, while the peripheral flagellum at the cell's waistline causes the cell to spin on its axis, providing their characteristic swimming

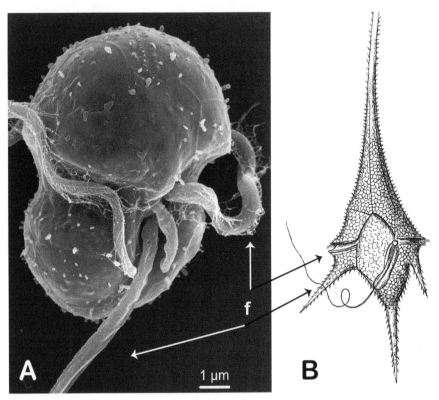

Fig. A49 Dinoflagellates. A-B. Although their cell shapes differ considerably these two genera both have a pair of flagella (f) oriented at right angles to each other, one in a groove around the "waist" of the cell (upper arrows), and the other extending to the posterior (lower arrows). Both have a cell wall composed of individual plates and are oriented similarly relative to each other. A. *Symbiodinium* (sim-bye-oh-DIN-ee-uhm). This genus lives both as a free-swimming organism and as an endosymbiont with coral. Scanning electron microscope image by Gert Hansen. Scale bar is one micron long, so about 100 cells this big would fit end to end in a space one millimeter long. B. *Ceratium* (sair-AYE-she-uhm). Classic illustration of a very distinctive dinoflagellate. (Image sources: A—Courtesy of Todd LaJeunesse, Wikimedia Creative Commons; B—After Oltmanns 1905)

motion and stabilizing their trajectory like a spinning bullet as it leaves a rifle barrel.

Dinoflagellates form one lineage within the alveolate clade, part of the more inclusive chromalveolate clade (Fig. 3–1). Most alveolates are predatory flagellates, including the well-known ciliates (*Paramecium, Stentor, Vorticella*) (Patterson 1999). Species of dinoflagellates lacking chloroplasts adopt an amoeboid form and they can be predatory or parasitic. Like other ciliated alveolates, some dinoflagellates possess a hair-like organ that can be expelled from the cell into a passing organism. Its function may be protective rather than predatory. Dinoflagellates are prominent components of the phytoplankton communities in coastal waters. Occasionally their populations undergo uncontrolled growth, producing the so-called toxic red tides. Dinoflagellate toxins can accumulate up food chains to the point where they can kill fish and other top consumers. Although too small to be directly consumed by filter feeders, dinoflagellates are consumed by zooplankton and larger ciliates, which in turn are consumed by filter feeders, like clams and oysters, or tiny, plankton-eating fish.

Haptophytes

(synonyms: coccolithophores)

clade nesting: haptophytes/bikonts/eukaryotes

Haptophytes are motile, biflagellated, unicellular organisms characterized by a haptonema, a flagella-like organ, often appearing coiled, that may function to anchor the cell. Their chloroplasts are basically the same as those found in the stramenopiles (chlorophylls *a* and *c*, fucoxanthin pigments, laminarin starch) to which they are a sister group (Fig. 3–1). Haptophyte cells bear finely sculptured plate-like scales composed of calcium

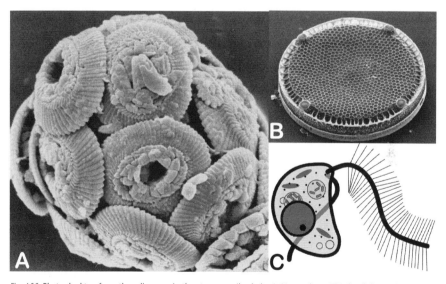

Fig. A50 Phytoplankton from three lineages in the stramenopile clade. A. Haptophyte. This fossil shows the overlapping calcium carbonate plates (coccoliths) characteristic of this group of organisms. Accumulations of these round plates form sediments that become chalks. B. Diatoms. This image shows the intricate detail of the pores in the opaline silicate, "glass," cell wall. This shape is centric (round), and the two halves of the cell wall overlap top to bottom (here upside down). C. Illustration of a motile cell of the Chrysophytes, the genus *Cafeteria*. This shows the asymmetrical position of the pair of flagella, one smooth and one hairy, which is characteristic of all motile cells in the stramenopile clade. (Image sources: A—Courtesy of Hannes Grobe/AWI, Wikimedia Creative Commons; B—Courtesy of Mary Ann Tiffany, Wikimedia Creative Commons; C—Courtesy of Dennis Barthel, Wikimedia Creative Commons)

carbonate, which when found as fossils were called coccoliths (Fig. A50 A). Deposits of coccoliths date to the Cambrian and are a principal component of Cretaceous age chalks—for example, the white cliffs of Dover, which in places are 500 meters thick. Since it takes hundreds of millions of coccoliths to make a cubic centimeter of sediment, imagine what countless numbers of cells rained down into the depths to produce such sediments, but even at the modest sediment accumulation rate of 0.5 mm (millimeter) per year, it takes only a million years to form 500 meters of sediments. Some studies have placed haptophytes in a very ancient common ancestry with the plant kingdom (Sogin 1994); however, recent studies have placed them as a sister group to the stramenopiles (Stechman and Cavalier-Smith 2002), a position that makes sense out of their possession of "brown algal" chloroplasts.

Cryptophytes
clade nesting: cryptophytes/bikonts/eukaryotes

Cryptophytes (KRIP-toe-fights) are very distinct, biflagellated, unicellular organisms with a flexible cell wall, a gullet-like furrow, and light-sensing pigments forming an eyespot. Like dinoflagellates, some are photosynthetic and others are heterotrophic. The gullet suggests cryptophytes can prey upon smaller cells and that photosynthetic species may be facultatively heterotrophic, either predatory or absorptive consumers. Their chloroplasts have chlorophylls *a* and *c*, as well as biliprotein pigments. Recent phylogenetic studies place cryptophytes as a sister group to a clade consisting of the haptophytes and stramenopiles (Fig. 3–1). Other studies have suggested a very ancient common ancestry with all other eukaryotes, in which case they perhaps acquired their chloroplasts thirdhand from haptophytes or some member of the stramenopiles. About 200 species have been described.

Glaucophytes
clade nesting: glaucophytes/plant/bikont/eukaryote

With only one known species, *Cyanophora paradoxa* is an aptly named organism. Its "chloroplast" is termed a "cyanelle" because it is not a fully integrated organelle, having retained at least one photosynthetic enzyme, which all other chloroplasts have lost to the nucleus. This suggests a separate and relatively recent endosymbiosis or an ancient divergence from the primary endosymbiosis event. Their phylogenetic position is problematic although some molecular studies have placed glaucophytes as a lineage basal to a red algae-green algae-land plant clade (Fig. 3–1).

Euglenozoa

(synonyms: euglenoids, Euglenophyta)

clade nesting: euglenoid/bikont/eukaryote

Euglenozoa have been placed in both the animal and plant kingdoms, but recent molecular data suggest they are a separate, very ancient lineage best treated as a stem group or a basal lineage of the bikont clade (Fig. 3–1). Both autotrophic and heterotrophic species of euglenozoa exist, and even autotrophic species can survive without light if provided with appropriate food molecules. Molecular studies suggest euglenozoa are related to parasitic flagellates, like trypanosomes (try-PAN-oh-sohms), which include the organism that causes sleeping sickness. Their chloroplasts are identical to those of green algae and land plants, and the third membrane envelope around their chloroplasts suggests they were obtained via a secondary endosymbiosis from green algae, a hypothesis supported by molecular studies of chloroplast phylogeny (Fig. 3–4).

Euglenozoa have two unequal flagella anchored at the base of an anterior gullet. One long flagellum emerges and provides locomotion while the second very short flagellum, associated with an adjacent eyespot, functions by transferring directional sensory data to the long locomotor flagellum. The cells have a flexible cell wall consisting of proteinaceous strips surrounding the cell like an overlapping helical ribbon, so the cells are capable of considerable shape variation and so-called euglenoid movements. Euglenozoa tend to be rounded bluntly on the anterior end and tapered to a point on the posterior end. They swim with a looping, somewhat helical, trajectory.

Stramenopile Phytoplankton

clade nesting: stramenopile/bikont/eukaryote

Based on differences in cell organization, the stramenopile clade includes several distinct groups of phytoplankton: chrysophytes, diatoms, eustigmatophytes, pedinellids, pelagophytes, phaeothamniophytes, raphidophytes, sarcinochrysidaleans, silicoflagellates, and synurophytes. Whew! All motile cells (unicellular phytoplankton, sperm, zoospores) in this clade share the characteristic features of biflagellated motile cells with one hairy and one smooth flagella inserted laterally, or some variation on this theme (Fig. A50 C). They also share similar chloroplasts with chlorophylls *a* and *c*, ß-carotene, xanthophylls, and thylakoids in stacks of three. All of the photosynthetic stramenopiles form a single lineage referred to as the ochrophytes (Cavalier-Smith and Chao 1996), so the secondary acquisition of chloroplasts had to take place only once in a common ancestor of this entire lineage, or in the common ancestor of the stramenopile-alveolate-cryptophyte-haptophyte clade. The heterotrophic fungal-like members of the stramenopile clade would have subsequently lost their chloroplasts, a testable hypothesis (look for nuclear genes of chloroplast origin). Only three groups will be described.

Chrysophytes
(synonyms: golden-brown algae, Chrysophyceans)

Most chrysophytes (CRY-sow-fights) are motile unicells; the rest are simple multicellular colonies or filaments. Unicellular forms can exhibit amoeboid behavior and ingest bacteria via phagocytosis. Some species bear spiked scales on their surface or an endocellular skeleton, both composed of silicon dioxide, glass. These siliceous structures are readily identified in fossil sediments from the Tertiary to present, but their phylogenetic position suggests chrysophytes are a more ancient lineage that developed these distinctive scales recently.

Diatoms
(synonym: baccillariophyceans)

Diatoms (DYE-ah-toms) are unicellular or colonial organisms surrounded by intricate opaline siliceous cell walls delicately sculptured with numerous pores in very distinctive patterns (Fig. A50 B). Accumulations of microscopic diatom walls form deep geological deposits known as diatomaceous earth, which when powdered is used as a fine abrasive in toothpastes and jewelry polishes. Such sediments indicate that diatoms were extraordinarily important producers of oceanic communities of the past. Diatoms probably remain the most numerous of eukaryotic planktonic organisms. In highly productive areas, their cells tint the water brownish.

Upwards of 10,000 species have been described in nearly 300 genera, and some experts think there may be several times more. Free-floating diatoms produce food reserves in the form of oils for buoyancy. Diatoms growing in shallow waters anchor to substrates or larger algae and plants or glide along surfaces by exuding mucilage through pores in the cell wall. Some can move by a yet undetermined mechanism. The cell wall consists of two halves that fit together such that one half overlaps the other. Radially symmetrical cells (centric) overlap top over bottom rather like a Petri plate (Fig. A50 B), and bilaterally symmetrical cells (pinnate) overlap side to side. When the cells divide, each daughter cell inherits one of the two cell wall halves and then synthesizes a new inner half. Glass cell walls are strong and durable, but they lack the ability to stretch or grow, so one daughter cell can grow as large as the parental cell, but the other is limited to a somewhat smaller size having inherited the inner half of the cell wall. Subsequent divisions continue to produce inner-half cell lineages of a more and more limited size (Fig. 4–2) until the cells reach some critical minimum size, or when environmental conditions deteriorate, then asexual or sexual reproduction is stimulated, resulting in an enlargement of the cell prior to synthesizing a whole new cell wall. Silicon dioxide occurs only in very low concentrations because of its relative insolubility, which is why we use glass containers for many liquids, so diatoms must accumulate silicon dioxide against huge concentration gradients, a feat that has yet to be explained.

Centric diatoms have many discoid chloroplasts, while pennate diatoms usually have only two large chloroplasts. While you might suspect otherwise, molecular studies indicate these two forms do not form distinct lineages, even though they also differ in characteristics of their sexual reproduction. Something about this seems odd because phylogenetic studies are confusing rather than sorting out characters.

Eustigmatophytes

Eustigmatophytes (you-stig-MAT-oh-fights) are biflagellated unicells possessing a prominent eyespot adjacent to the flagellar base or small, round nonmotile cells. The smooth flagellum is reduced in size and thought to function in phototropisms similar to the situation in euglenozoa. Eustigmatophytes look greener than other stramenopiles because their chloroplasts lack chlorophyll *c* and fucoxanthin. Eustigmatophytes are important components of oceanic food chains, but little else is known about these organisms.

Red Algae

(synonyms: Rhodophytes, Rhodophyta)

The red algae, some 800 genera and over 5,000 species, are mostly marine seaweeds, but the group includes some phytoplankton and a few freshwater species. In form some red algae are unicellular, but most are filamentous with simple to complex organizations. Red algae lack flagellated cells and have chloroplasts containing only chlorophyll *a* plus red to purple accessory pigments phycoerythrin and phycocyanin, which give them their characteristic red/purple colors. The accessory pigments have light absorption maxima near those wavelengths that penetrate deepest into water, allowing red algae to live deep in the coastal zone (Fig. 5–1). Red algae inhabiting shallow water or fresh water (a few small filamentous species) appear quite green because they produce little accessory pigments and have little need of them.

The unicellular and smaller, simpler filamentous algae form several basal lineages, which used to be grouped as the bangiophyceans. Phylogenetic studies have shown that they do not form a clade. The remaining red algae form a clade of some 12 to 18 orders, the florideophyceans, seaweeds with larger and more complex forms. This clade shares the novel characteristic of "pit connections" between adjacent cells, but these are very different from plasmodesmata. Bangiophyceans lack pit connections. This means red algae make no true parenchyma. Very complex filamentous forms produce a tissue-like organization called pseudoparenchyma, which is similar to the tissue-like filamentous organization of fruiting bodies of mushrooms and puffballs (basidiomycote fungi).

Cell walls have an inner layer of cellulose and an outer layer composed of pectin and other carbohydrates, mostly sulfated polysaccharides composed of the sugar galactose, which are colloidal and capable of binding water molecules up to two to three hundred times their weight. In addition to binding water these materials produce flexible cell walls of high tensile strength, important in coastal habitats for dealing with waves, tides, and exposure. These colloidal substances are economically important and valued for their thermal reversibility between sol (liquid) and gel phases. Irish moss, *Chondrus crispus*, is harvested for carrageenans (from *carrageen*, the Irish word for rock moss), a gelling, thickening agent used in many food and cosmetic applications. Agar, the basic gelatinous culture medium of biology, is obtained from the aptly named *Gelidium* and a few other genera. Most biologists do not know how much they owe to red algal seaweeds. Coralloid red algae, as the name suggests, produce a calcified shell that can contribute to reef construction.

Even large red algal seaweeds remain fundamentally filamentous (Fig. A51). A main axis often consists of multiple filaments and whorls of branch filaments, all surrounded by a gelatinous sheath giving them the appearance of tissue. The most complex forms have a distinct apical growth. When organized into a laminar shape these filaments encased in a continuous gelatinous sheath resemble a flat sheet of tissue. The largest of these seaweeds is but a few decimeters tall, but red algae represent the oldest (2.1 mya) lineage of large conspicuous organisms. Red algae have a very ancient common ancestry with other eukaryotes and their phylogenetic placement remains a question. Fossils of coralloid red algae appear 550 mya.

Red algae have rather complicated sexual life cycles that have spawned a great deal of unique terminology, which is a nice way of saying jargon. All involve an alternation between two free-living phases, a haploid phase that produces gametes—that is, a gametophyte, and a diploid phase that produces spores by meiosis—that is, a sporophyte (Fig. A52). Male gametes are called spermatia because they are not motile and therefore technically are not sperm. Spermatia and spores both disperse by drifting, the aquatic equivalent of wind dispersal, so large numbers must be produced. Repeated division of a vegetative cell produces a packet of small cells that become spermatia, nonmotile but drifting sperm. In filamentous forms the terminal cell of lateral branches can differentiate into an egg, which often grows a long, thin apical extension that functions to increase surface area for making contact with drifting spermatia. In the larger sea-

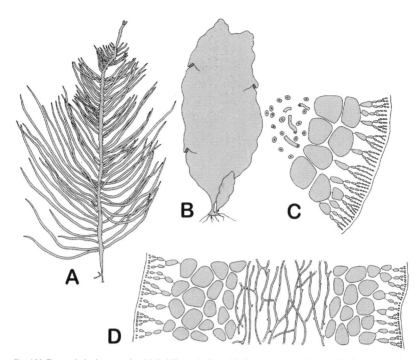

Fig. A51 Two red algal seaweeds with holdfasts. A. One with linear segments (*Neoagadhiella*; knee-oh-ah-GAHD-ee-el-ah). B. One with a solid blade (*Iridaea*). C. Cross- and D. longitudinal sections of a portion of the main axis in A. When examined in detail this seaweed shows that these forms are fundamentally filamentous with larger diameter filaments forming a longitudinal core and radiating filaments forming a cortex. An extracellular matrix holds the filaments together, making it appear as solid tissue. (Image source: After Scagel et al. 1984)

weeds, vegetative cells in the blade differentiate into eggs. In this case spermatia make contact with the flat gelatinous surface of the blade and fertilize an underlying egg.

The resulting zygote does one of two things depending upon the species. In the simplest case, as found in bangiophyceans like *Porphyra*, the zygote divides several times, producing a packet of spores that upon release each grow into a small filamentous sporophyte, but such divisions result in a halving of cell size in order to double the number of spores. Cells in sporangial branches undergo meiosis producing spores that disperse and develop into the familiar haploid seaweed. Both the sporophyte and the gametophyte also can reproduce asexually by spores.

In florideophyceans, the zygote develops into a small multicellular diploid organism whose purpose is to increase the number of dispersal units without sacrificing size (Fig. A52 K; Searles 1980). This diploid organism remains small and attached to the maternal parent, and it is

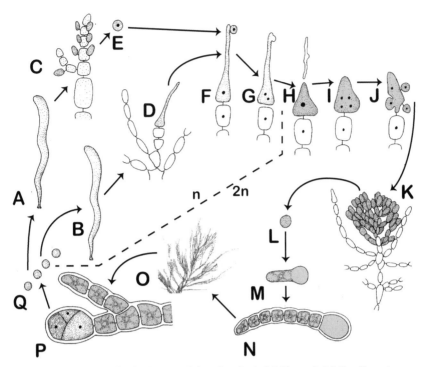

Fig. A52 Life cycle of the florideophycean red algae *Nemalion* (neh-MAL-ee-on). A-B. Two filamentous gametophytes, one male (left), one female (right). C. On males the ultimate cells become spermatia (E), nonswimming, drifting sperm. D. On females terminal cells become eggs with a long thin extended apex to increase surface area. F. A spematium makes contact with the egg. G. The gametes fuse and fertilization produces a zygote (H), which begins growing into a diploid organism (I-K), which makes many diploid spores that disperse and grow into a new diploid sporophyte (L-O), which is the large, conspicuous, and therefore familiar seaweed (O). P. Terminal cells of the mature seaweed divide by meiosis, producing haploid spores (Q) that disperse and grow back into filamentous gametophytes. Diagonal line separates haploid and diploid (shaded gray) phases of this life cycle. (Image source: After Scagel et al. 1984)

uncertain whether this multiplicative sporophyte phase derives any nutritional benefits from remaining in intimate association with the maternal plant. The spores released develop into free-living sporophytes that will produce tetrads of spores formed by meiosis. These spores develop back into gametophytes, completing the cycle. In filamentous forms like *Polysiphonia*, both free-living phases look similar. In larger seaweeds the familiar and conspicuous gametophyte alternates with a small filamentous sporophyte (Fig. A52). So much terminology is associated with these three-phase life cycles that it obscures the underlying functions of the life cycle's basic parts. Should you undertake a detailed study of algae, then you may have need of learning about carpogonia, carpospores, conchocelis, conchospores, conchosporangia, carposporophytes, tetraspores, tetrasporophytes, etc., but otherwise you have avoided much memorization.

Rhyniophytes and Trimerophytes

*clade nesting: rhyniophytes/tracheophytes/polysporangiophytes/embryophytes/
streptophytes/chlorophytes*

Rhyniophytes are the first-appearing vascular land plants, dating to about 425 mya late in the early Silurian. All of these early vascular plants used to be considered rhyniophytes, but now that their diversity is better understood, it appears that some genera represent pre-vascular plants—for example, *Cooksonia* (Fig. 6–7), and stem-groups of the clubmosses—for example, *Baragwanathia* and zosterophylls (Fig. 8–2; Fig. 8–3; Fig. 8–4). Presently, only *Rhynia* (Fig. 8–1), *Huvenia*, and *Stockmansella* remain as rhyniophytes.

Rhyniophytes were small plants whose sporophytes were no more than 15–20 cm tall. The axes were naked and they lacked true roots, but the very finely grained Rhynie chert from which the plants take their name show single-celled rhizoids. The aerial axes branched dichotomously and had terminal, spindle-shaped sporangia. Nothing is known of the haploid generation except for a specimen of *Cooksonia* where five axes arise from a basal amorphous structure that could be interpreted as a gametophyte thallus (Gerrianne et al. 2006).

Trimerophytes are introduced and discussed in chapter 8.

(synonym: pteridosperms, including glossopterids, cycadeoids, and Caytoniales)

clade nesting: pteridosperms/ ligniophytes/ megaphyllous plants/ tracheophytes/ polysporangiophytes/ embryophytes/ streptophytes/ chlorophytes

Seed fern or pteridosperm is a general name referring to a diverse assemblage of fossil gymnosperms, so basically any gymnosperms that isn't a cycad, gnetophyte, ginkgo, or conifer is a seed fern. The common name seed fern is derived from the fact that these woody plants bear ovules, and therefore seeds, upon very ferny foliage. They include the glossopterids, the cycadeoids (also called the Bennettitales), the Pentoxylales, the Caytoniales, and the Czekanowskiales. Four of these five groups of pteridosperms are important because their reproductive structures have some angiosperm-like features so they are candidates for common ancestry with angiosperms. All predate the angiosperms in the fossil record. Fossil seeds appear in the late Devonian and plants like *Medullosa* and other pteridosperms represent the oldest seed ferns (Fig. 9–4).

Glossopterids

Glossopterids (Fig. 9–9), the oldest of the five groups of seed ferns described here (Fig. 9–6), were arborescent gymnosperms that left an extensive fossil record during the Permian and Triassic. Glossopterid leaf fossils of have been found distributed across South America, Africa, Madagascar, India, Australia, and Antarctica, land masses which at that time in geological history formed the supercontinent Gondwana (Fig. 9–5). Masses of leaf fossils suggest these were deciduous trees with an annual leaf fall. Although the vegetative features were very similar, over 50 species have been described on the basis of a diversity of reproductive structures. The glossopterids are so named because they have a very distinctive tongue-shaped leaf (*glosso-* = tongue) with a well-defined midrib vein and reticulate secondary veins. The leaves were borne in helical whorls clustered upon the ends of branches, which is pretty typical of deciduous gymnosperms.

Glossopterids had unique pollen-bearing and ovule-bearing structures, and all were associated with vegetative leaves, bracts. This is important because, as with the gnetophytes, any gymnosperms with sterile, leafy structures associated with sporangia-bearing structures approaches the organization of a flower. In glossopterids, ovules in a cupule were associated with and borne upon bracts. In some cases the bract's enrolled margins partially enclosed the ovules, and such structures suggest some sophisticated pollination mechanisms.

Cycadeoids (Bennettitales)

Cycadeoids, as the name suggests, resemble cycads vegetatively, having similar stout, sparsely branched stems and helical whorls of pinnately compound leaves, but they differ from cycads and other gymnosperms in having complex bisporangiate reproductive structures (Fig. 9–10). Cycadeoids are younger than the glossopterids, appearing during the Triassic and having their greatest diversity in the Jurassic, and then declining in the first half of the Cretaceous as the flowering plants became dominant. Cycadeoids are certainly of a proper age to have had a common ancestry with flowering plants. The cones consisted of sterile bracts (a perianth?) surrounding whorls of microsporophylls with embedded microsporangia, which are the features most reminiscent of a flower. In some specimens the microsporophyll quite obviously has a pinnate organization and might even be considered a fertile frond. Stalked ovules were borne laterally on the central axis of the cone packed among sterile scales, which is not very much like a flower, except for having both microsporangia and megasporangia together. Several observations suggest the sterile scales and ovules are homologous structures (Stewart and Rothwell 1997), but it is uncertain if the stalked ovules and sterile scales represent leafy organs. Both are lateral appendages of a modified shoot, so interpreting them as leafy organs is certainly a logical conclusion, but ovules are usually found terminal on lateral shoots or on the margins of a leaf (a sporophyll). Neither seems to be evident here. Stalked ovules with an elongated tubular apical opening are very similar to those of gnetophytes. Clearly the sterile scales provide protection to the ovules/seeds (Stewart and Rothwell 1997), but they may have functioned primarily after pollination (i.e., after the microsporophylls and bracts were shed).

While it is fairly easy to imagine such a cone opening into a flower-like display, careful study of developing specimens led to the conclusion that the microsporophylls could not unfold in such a way (Delevoryas 1963, 1968). This is rather perplexing because most land plants have some mechanism to promote outcrossing, and a nonopening, bisporangiate cone would seem to result in obligate self-pollination. A permanently closed cone does not afford much opportunity for either wind or animal pollination unless, like the cones of modern cycads, the structure opens just enough to permit entry to small pollinators like beetles. Unfortunately the nature of fossilization dictates that the specimens are going to be largely developing buds or the postpollination developing seed stages because both of these stages last much longer than the relatively brief pollination stage, and both pre- and post-pollination stages were probably much tougher and more easily fossilized. The cone has been reconstructed in an open position since by others (Fig. 9–10 C).

This leads us to wonder whether our functional understanding of cycadeoid cones is accurate at all. Having worked on flowering plants my whole career, I know that figuring out how a flower looks and functions during pollen dispersal from examining only flower buds or developing fruits is nearly impossible. Floral organs greatly expand, changing shape and positions as the flower buds open, and such observations are virtually impossible with fossils, so it is not an overstatement to say the pollination biology of cycadeoids is not understood at all. On the basis of just vegetative features, a close relationship between cycads and cycadeoids might be justified, but the complex reproductive structures of cycadeoids indicate any relationship is rather distant (Stewart and Rothwell 1997).

Pentoxylales

The little-known group Pentoxylales (pen-tox-ee-LAY-lees) seems most similar to the cycadeoids. Their most distinctive feature is their woody stem composed of a ring of five segments. Simple laminar leaves some 7 cm long and 1 cm broad are associated with this group but have not been found attached to stems. As with cycads, the leaf scars are helical and closely packed. Pollen-bearing structures seem similar to those of the cycadeoids. Ovules were borne on terminal cones crowded together with the apex oriented outward, as in cycadeoids. This group appears and disappears during the Jurassic.

Caytoniales

The Caytoniales have leaves or leaflets with the reticulate venation of ferns, but no one knows exactly the phylogenetic affinities of these organisms (Harris 1964). Ovules were borne on a modified pinnate megasporophyll whose lateral appendages were recurved cupules enclosing several ovules. Pollen was produced in structures consisting of four fused microsporangia, a structure that rather resembles an anther. Pollen grains had the same type of sac-like wings as conifer pollen and such pollen grains have been found within the apex and pollen chamber of the ovules, which suggests the Caytoniales used a pollen drop mechanism. However, the organization of the cupule suggests the pollen drops of the enclosed ovules may have coalesced into a single pollen drop for the entire cupule. As such the margins of the cupule would function something like the stigma of an angiosperm where pollen is captured at a distance from the ovules. In angiosperms pollen tubes grow from the stigma to the ovules, and pollen drop reabsorption would pull the whole pollen grain to the ovules within. Unlike angiosperms, these cupules are folded or rolled top to bottom rather than side to side.

Czekanowskiales

Lastly, the Czekanowskiales (check-an-ow-ski-AYE-lees) is a little-known group of fossil gymnosperms that originally was thought to be related to gingkoes because of its Y-lobed leaves borne in a helical whorl on a short shoot (Harris 1951). However, detached fossils suggest *Czekanowskia* was a deciduous tree and the entire short shoot was shed as a unit, as in bald cypress (*Taxodium*) or dawn redwood (*Metasequoia*). The vegetative parts have been associated via anatomical similarities with ovule-bearing axes bearing spirally arranged scales at the base, similar to those at the base of the leafy shoots. The long slender axis has what can be described as clam shell–shaped cupules enclosing recurved ovules. If one half were sterile and one half fertile they might be similar to some glossopterid ovulate structures, but both halves were fertile, which suggests the structures were pairs of fertile cupules. Each pair of cupules completely enclosed two sets of ovules, and pollen grains have been found among epidermal hairs at the junction or suture between the two cupules (Harris 1976). Such an arrangement has considerable similarity to a primitive flowering plant ovary (megasporophyll) that has an unsealed suture functioning as a stigmatic surface. However, in these pteridosperms, the ovulate structure consists of a pair of cupules oriented top to bottom, not side to side.

While all these candidates have tantalizing features suggesting a relationship to angiosperms, none seem completely appropriate as the sole common ancestor with flowering plants. Rather, as a group the seed ferns demonstrate that gymnosperms predating angiosperms were experimenting with a number of features usually associated with flowering plants.

Whisk Ferns

(synonyms: Psilophyta, Psilophytes)

clade nesting: whisk ferns/ pteridophytes/ megaphyllous plants/ tracheophytes/ polysporangiophytes/ embryophytes/ streptophytes/ chlorophytes

Whisk ferns are wonderfully enigmatic plants consisting of only two genera, *Psilotum* (silent P: cy-LOW-tum) and *Tmesipteris* (silent T: mez-ip-TERR-iss). Both *Psilotum* and *Tmesipteris* are tropical to subtropical organisms, although the latter genus is confined to portions of Australia and New Zealand, a distribution suggesting their presence on Gondwana requiring an ancient history. Both are relatively small plants, with shoots seldom more than 20–30 cm tall or long. They certainly are not a conspicuous component of any flora and easily overlooked.

Psilotum has Y-branched, "naked" aerial axes that look rather similar to early vascular land plants, particularly *Renalia* (Gensel 1976) and other rhyniophytes (Fig. A53 A-C). Whisk ferns have no roots and their axes have a small, simple, solid core of xylem, another feature shared with rhyniophytes. For these reasons whisk ferns once were classified as rhyniophytes, living relics of those late Silurian-early Devonian plants. However, in the long interval since the rhyniophytes became extinct, no fossil whisk ferns have ever been identified. This stretches the credulity of skeptical botanists that whisk ferns managed to survive all that time without ever leaving a fossil.

Fig. A53 Psilotum. A. Habit of the whisk fern *Psilotum nudum.* B. Habit of *Psilotum complanatum,* a pendent epiphyte. C. Upper portion of a single aerial shoot of *P. nudum* showing repeated Y-branching and lateral sporangial structures. About one-third life size. Circle and arrows connect sporangia between magnifications. D. Detail of three fused sporangia composing each unit. E. Inset showing the bifid sporophyll associated with the sporangia. F. Subterranean gametophyte (g) of *Psilotum* (front) showing its rhizomatous nature measures some 4–5 mm long. Round structures on its surface are antheridia. Young Y-branched sporophyte (s—behind) is attached to this gametophyte. G. Scanning electron microscope view of apical portion of a gametophyte axis showing globose antheridia (an) on its surface and the upper portions of archegonia (ar) extending above the surface (groups of three or four cells). (Image sources: A—The Author; B—Courtesy of Forest and Kim Starr, Plants of Hawaii, Wikimedia Creative Commons; C—The Author; D—Courtesy of and with permission of J. Hayden; E—Courtesy of and with permission of D. Webb; F—Courtesy of and with permission of D. Whittier; G—Courtesy of K. Renzaglia and with the permission of the *American Journal of Botany*)

With some careful examination, the resemblance of whisk ferns to rhyniophytes proves superficial. Although *Psilotum* is commonly described as "leafless," both whisk fern genera have leafy organs. *Psilotum's* leafy organs are greatly reduced, appearing vestigial and often overlooked (Fig. A53), sometimes called enations because like the leafy organs of mosses and liverworts they lack a vascular supply. *Tmesipteris* has both vestigial leaves and larger (2 cm long) microphylls with a single central vascular bundle (Fig. A54). The best-known species, *Psilotum nudum,* is a double misnomer because it means "bare naked" (*psilos*—bare in Greek, *nudum*—naked in Latin), but considering it has no roots and barely has leaves, maybe this species name is appropriate. Both plants largely consist of stems, axes, a mass of more or less horizontally oriented rhizomes growing in a substrate and giving rise to cluster or tuft of aerial shoots. The resemblance of a tuft of aerial shoots to a whisk broom yields their common name.

How does a rootless terrestrial plant function? This question may have occurred to you following earlier descriptions of fossil organisms, but fossils can answer only certain questions. The extensive tangled network of subterranean axes, not exactly rhizomes because they lack a largely horizontal orientation, have a symbiotic association with fungi, termed mycorrhizae ("fungal roots"), whose hyphal filaments function as absorption organs. Since fungi were among the pioneers on land, fungal associations with land plants may be quite ancient.

Without any apparent pattern, a branch of a subterranean axis becomes negatively geotropic, growing straight up and becoming an aerial shoot. Upon reaching the surface, the apical meristem commences producing leafy appendages as well as branching once or twice (*Tmesipteris*)

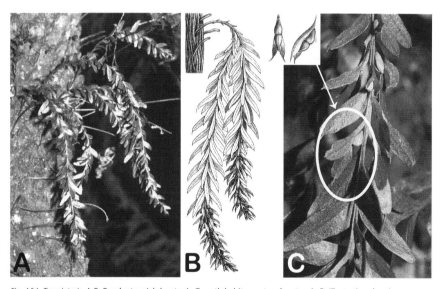

Fig. A54 Tmesipteris. A-B. Pendent aerial shoots. A. Growth habit on a tree fern trunk. B. Illustration showing one Y-branched pendent shoot with vestigial (at base) and full-sized (~2 cm long) sterile leaves and bifid fertile leaves (near apex). Inset (top) shows illustration of fertile leaf bearing a fused pair of sporangia in top and side views. Note similarity of growth habit to *Psilotum complanatum* and the bifid fertile leaf (Fig. A53 B, E). C. Detail showing vegetative and fertile leaves (oval). (Image sources: A, C—Courtesy of and with permission of D. W. Stevenson, www.plantsystematics.org; B and inset—After Bower 1908)

to several times (*Psilotum*). Aerial shoots of *Psilotum* grow such that the plane of each successive Y-branching is oriented more or less at right angles to the previous branching, so each aerial shoot occupies a spreading 3-dimensional space, and perhaps the regular simplicity of the design created by repeated Y-branching at right angles is part of this plant's appeal. The axes of aerial shoots seldom reach more than 3 mm in diameter. In cross section they have a well-defined epidermis and a small vascular cylinder with a solid core of xylem that provides little in the way of support. A band of thick-walled cells around the periphery of the axes contributes significantly to the support function. A broad cortex occupies the rest of the stem's volume; the outermost cells function in photosynthesis.

Psilotum nudum has angular stems with ridges upon which the leaves are borne, or from a developmental perspective, these ridges are the decurrent bases of the leafy organs (Kaplan 2001) and therefore the axes are of a composite structure (decurrent leaf bases surrounding a central stem). A leafless cultivar has more slender, unridged stems that are oriented more erectly. In *P. complanatum*, an epiphytic species, the aerial shoots are flattened and the axes hang pendently (Fig. A53 B), a habit and growth form it shares with *Tmesipteris* (Fig. A54).

The most prominent appendages on the aerial shoots of *Psilotum* are sporangia (Fig. A53 C-E). At first glance, the globose sporangia appear lateral on the main axis. This combined with the Y-branched stems, simple stem anatomy, "microphylls," and lateral sporangia would suggest a close relationship with clubmosses. However, the shape, organization, and positions of whisk fern sporangia are different from those in clubmosses. Anything more than a glance reveals that each globose "sporangium" consists of three laterally fused sporangia (Fig. A53 D). This fused sporangial structure of *Psilotum* has two different interpretations: first, as three sporangia terminal on a short lateral branch that is subtended by a small dichotomous "leafy" appendage, a bract (Bierhorst 1956; Rouffa 1978); or alternatively, the fused sporangia are borne on the upper surface of a bifid or dichotomous sporophyll (Kaplan 2001). The sporophyll interpretation is supported by developmental studies and comparisons with *Tmesipteris*. *Psilotum* is by far the more common and more easily grown genus, so *Psilotum* has received much more attention than *Tmesipteris*.

Tmesipteris shares so many similarities with *Psilotum* that no one doubts their relatedness. *Tmesipteris* is an epiphyte often growing upon tree-fern trunks (Fig. A54). The aerial shoots branch only once or twice dichotomously near their base. Like *Psilotum*, no clear point of demarcation separates an aerial shoot from a rhizome other than the rhizomes bear no leaves. The aerial shoots bear a helix of leafy appendages, which in some species are reduced in size such that they are similar to the small leafy organs of *Psilotum*. In other species the leaves are slightly asymmetrical microphylls 1–1.5 cm long and 4–5 mm broad. Even the leafy species bear small, reduced leafy organs on lower portions of their aerial shoots. One unusual and interesting feature is that the leaves of *Tmesipteris* are oriented perpendicular to the surface of the shoot,—o—, when viewed from end on, like wings on the fuselage of a plane. Most foliar organs are oriented such that their flattened blade is in a plane parallel to the surface of the shoot, |o|, when viewed from end on.

A spore-producing structure of *Tmesipteris* is composed of just two sporangia fused end to end (Fig. A54 B-C). They are borne upon the upper surface of a bifid leafy organ, one similar to, but considerably larger than, that of *Psilotum*. Again because of its insertion upon the shoot, the orientation of the blades, and its regular position in a helical series of leafy appendages, both sterile and fertile, Kaplan (2001) argues that this is a leafy organ, a bifid sporophyll. Many specimens bear an occasional sterile bifid leaf. The orientation of the blades on the sporophyll, which are similar to and only slightly smaller than the regular leaves, argues that this is not a short shoot bearing two leaves. If this were a lateral branch, then the leaf blades should be oriented perpendicular to the axis,—o—, but they are actually positioned parallel to the axis, |o| (Siegert 1967). This orientation argues that the entire structure is a leafy organ on the main axis rather than a specialized branch bearing two leaves and two sporangia.

The fused sporangia together with the bifid nature of the sporophyll suggests the leafy organs of whisk ferns are reductions from megaphylls or modified branch systems as are the microphylls of horsetails. This interpretation of the leafy organs agrees with the placement of whisk ferns with ferns and horsetails in a pteridophyte clade. Nonetheless, the whisk ferns remain odd-looking plants. Both whisk fern genera have subterranean gametophytes, and similarities to those of basal ferns including the adder's tongue ferns support their placement within ferns (Bierhorst 1971). Recent molecular phylogenetic studies have nested whisk ferns within a fern clade as sister group to the adder's tongue ferns (Pryer et al. 1996; Stevenson and Loconte 1996), a position supporting the Bierhorst hypothesis.

If it is really a fern, should a determinate aerial shoot of a whisk fern be called a frond? Yes, if you want to make whisk ferns sound more like a fern (e.g., Bierhorst 1971). The same interpretation could be applied to clubmosses. Some of the basal ferns also lack a clear demarcation between aerial shoots and subterranean axes, and this makes it rather difficult to label such aerial shoots fronds if a leaf is defined as an appendage of a stem. Further, whisk ferns have leafy organs that develop apically in a normal enough leafy manner. This helps further demonstrate the problem of applying seed plant concepts of leaves as appendages to noded stems to lower vascular plants where no such distinction exists.

Any greenhouse collection possessing *Psilotum* ultimately will have new sporophytes sprouting from numerous pots, yet the gametophytes are never seen without making a concerted effort to find them. The subterranean gametophytes (Fig. A53 F-G) live mycotrophically, which means they have a nutritional dependency upon their association with fungi. This is rather different than the symbiotic relationship between fungi and the photosynthetic sporophyte, which is presumed to be mutually beneficial, at least in the long haul. The mycotrophic gametophyte and the young sporophyte obtain nutrition at the expense of the fungus. Although the gametophyte is pretty small and represents only a limited cost, the young sporophyte establishes a considerable subterranean network of axes long before an aerial shoot appears and becomes photosynthetic, which represents a much larger cost to the fungus. The fungus in turn obtains its needed nutrition from either decomposition or another symbiotic relationship with another vascular plant. Presumably the fungus pays the price (energy given away) of raising a whisk fern gametophyte and nourishing the young sporophyte because the fungus benefits in the long run by establishing a relationship with a gametophyte early on, a relationship that has the potential to become an interaction with a large, long-lived, photosynthetic sporophyte that will pay long-term dividends that offset the initial small investment. Only recently has anyone been successful in growing *Psilotum* gametophytes to allow easy study (Renzaglia et al. 2001). Antheridia are superficial, arising upon the surface of the gametophyte (Fig. A53 G), a fern-like characteristic, and the sperm are multiflagellated, a feature they share with horsetails. The archegonia are embedded in the gametophyte. The gametophytes also can produce gemmae, small, multicellular buds that can detach and grow as a means of vegetative reproduction.

Preface

1. Tentative explanations, with an emphasis on tentative.

2. The other being natural selection.

3. For convenience the appendices are alphabetically organized by a common name for the group.

Chapter 1

1. By this I mean populations of organisms, not individuals; individual organisms do not evolve.

2. Curiously, this concept has been slow to catch on in medical research that studies the interactions between disease and parasitic organisms and their human host.

3. Here and there when new and/or unfamiliar terms are introduced a phonetic pronunciation will follow.

4. In total accord with the second law of thermodynamics, entropy increases.

5. A good example of the intricacies of diverse interactions was presented in an amusing dialogue by Stuart Altmann titled "The Monkey and the Fig," published in *American Scientist* 77 (1989): 256–63.

6. Purple photosynthetic bacteria exist as well, but they need not be considered presently.

7. 413 Maple Ave., Cincinnati, Ohio, USA, Sol 3. This addresses mail to all the people living in the abode numbered 413. A whole set of abodes are located on Maple Avenue, which along with many other streets are located in a city, which with a number of other cities are located in the political entity known as Ohio, a state, which is united with others into a country, and perhaps someday indicating its location on Sol 3 (Earth), Milky Way Galaxy.

8. The thick, spiny, succulent, green stems of the cactus and euphorb (or spurge) families are a favorite botanical example. They were arrived at independently as adaptions to desert conditions and not the result of common ancestry.

9. Mushrooms, molds, rusts, smuts, and such.

10. I say of course because in lots of instances the biased perspective of humans has generated an animal-centric biology.

11. Botanists use Division in place of Phylum, but since phylum is the more familiar term it will be used here.

12. The official names for the five kingdoms were Animalia, Plantae, Fungi, Protista, and Monera.

13. Recent studies have falsified several aspects of this traditional classification hypothesis. The specifics will be taken up again later.

14. I chose *Magnolia* as an example because I like them and the genus forms the root of all its higher classification categories which can be identified by their endings: -aceae, -ales, -opsida, -ophyta. Mostly these will be avoided.

15. The D stands for deoxyribose, a 5-carbon sugar; the NA stands for nucleic acid.

16. A-adenine, T-thymine, G-guanine, C-cytosine.

17. I joke that an affinity for new terms is inversely related to how frequently my colleagues teach undergraduates.

18. No rocks this old exist on Earth. The date is based upon rocks gathered on the Moon's surface.

19. The Archbishop Ussher used Biblical genealogy, the begats, to calculate the time from the creation to present as 6,000 years. Amazingly enough, a lot of people in the United States think this should be included in science courses!

20. As good as this biologist can provide.

21. Light demonstrates via its characteristics that it acts as both a wave and a particle.

22. Information about the famous Miller-Urey experiments can be readily found if you are interested in learning more.

23. Modern analytical techniques have found that the experiment actually produced 22 amino acids, not just five (Johnson et al. 2008).

24. To understand the early universe, astronomers examine much older light and energy.

Chapter 2

1. And if you believe advertising something to be rendered sterile at all costs!

2. The so-called biogenic law was formulated by the famous German biologist Ernst Haeckel in 1866.

3. Protoplasm literally means "first material formed."

4. Ribosomes are organelles that function in protein synthesis.

5. Ribonucleic acid is built on a backbone of ribose, a sugar with one more oxygen than deoxyribose, the sugar backbone of DNA.

6. Children schooled in creationism use this as a ploy to discredit evolution.

7. Reminder: mya = million years ago; bya = billion years ago.

8. As a paleontological colleague once said, "That's what graduate students are for."

9. Isotopes are atoms of the same element with different atomic masses from extra neutrons.

10. pH is a scale from 0 to 13 measuring how acidic or basic a solution is. Pure water is neutral, pH = 7, where the concentration of hydrogen ions just about equals the concentration of hydroxyl ions. An excess of hydrogen ions makes a solution acidic; an excess of hydroxyl ions makes a solution basic.

11. Although this is a common claim of creationists.

12. A note to would-be science teachers: Although I learned the details of metabolisms in biochemistry and cell biology, the forest was lost for all the trees. Too often in teaching science a conceptual understanding is lost in the details, so even though I could recite the facts, I did not understand what was really important. Most textbooks omit a conceptual framework upon which to arrange the details, and non-understanding memorization is the result.

13. Not to be confused with breathing.

14. The Dvorak keyboard puts aoeuidhtns (instead of asdfghjkl;) under your eight finger tips for the home row allowing you to type about 70% of commonly used words without reaching up or down in contrast to just 32% for qwerty.

15. Your PC probably has a defkey (define keys) program that allows you to reassign your keyboard characters if you want to learn to type faster.

16. This footnote was added by voice recognition software that came with my PC. It works pretty well as long as you speak clearly enough and slow enough. Slow instead of slowly; only one mistake, not bad after just 30 minutes of training the software to understand my speech. Now back to fingers.

17. The only alternative is a recent supernatural creation as is. But then what of all the evidence of Earth's geological past and the organisms that lived there? A recent supernatural creation requires either that you deny mountains of geological evidence, which is against the rules of science, or that the creator was deliberately deceitful, trying to fool us by creating a universe complete with evidence of histories that did not occur. Either way, this means creationism offers no explanation for why things are the way they are, other than "that's the way they were created," which is scientifically useless.

18. Indeed, some of this needless parts diversity contributed to the inefficiency and pointless change that has led to the near demise of the company that fed my family.

19. All six carbon atoms are thus accounted for. Keeping track of carbon atoms is useful in metabolic accounting.

20. I drew those two-headed arrows in pathways for years before their significance dawned on me. So I was either a slow learner or so busy memorizing that this concept escaped me. In reviewing old course notes, I discovered that I was a good note taker and no mention was ever made of pathway reversal or its significance.

21. To add to the confusion, ATP pumps or photosystems are also called reaction centers in some biological literature.

22. Critics of evolution are fond of arguing that the absence of intermediates is embarrassing to biology.

23. There are several different chlorophyll molecules that differ slightly and these are designated by letters, *a*, *b*, *c*, *d*, and *e*. Chlorophyll *a* is universal among green organisms.

24. Red, Orange, Yellow, Green, Blue, Indigo, Violet in case you have forgotten.

25. This is a problem of science education that focuses too much on facts and not enough on understandings.

26. Several were introduced earlier in this chapter.

Chapter 3

1. Twenty-five percent of all known species are beetles.

2. Natural selection operates on individuals but only changes populations. If larger organisms are more successful at something that results in more offspring, then the mean size of organisms in the population will increase through generations as more genes for larger size get passed along to offspring.

3. In the fall when chlorophyll production ceases, the yellow-gold pigments are unmasked and the leaves change color.

4. Spiderworts are flowering plants in the genus *Tradescantia*. Their stamens have hairy filaments that make the flowers look spidery.

5. Polyphyletic refers to a group found to consist of two or more independent lineages, a systematic no-no.

6. Critics of evolution attempt to exploit this situation to suggest the whole concept of evolution is in trouble, but they never offer a testable alternative explanation to account for the known data. As stated by Wolozin (2007), "Science's critics appear to be immune to the tremendous amount of data driving our conclusions and instead seem to rely on weak or dubious facts" to bolster their position.

7. Chromosome means "colored or pigmented body," referring to the staining that makes them visible for microscopic observation.

8. As a vestige of a misspent youth, I cannot think about live organisms within food vacuoles without recalling the security "balloons" that kept inmates from escaping the Village in the classic 1960s sci-fi series "The Prisoner," starring Patrick McGoohan. All the episodes can be viewed online at http://www.amctv.com/originals/the-prisoner-1960s-series/.

9. This book was meant to be used as a biology textbook.

10. Frair and Davis refer to vestigial structures as "organs of unknown function."

11. Unfamiliar or not, you might have met this one on your travels.

12. Critics of evolution often say, "No one was there to observe what happened," but as demonstrated, logically deduced predictions can be tested to determine whether the hypothesis has merit or not.

13. Cell division involves two mechanisms: mitosis (nuclear division) and cytokinesis (cytoplasmic division).

14. This includes some unicellular seaweeds of surprisingly large size.

15. Even a small culture tube could contain hundreds of millions to billions of individuals.

16. Eggs and sperm are the most familiar gametes, but among unicellular organisms, gametes are often similar in size and motility.

17. DNA repair requires such pairing and homologous chromosomes can swap portions of their DNA with each other during pairing, a phenomenon called "crossing over."

18. In many organisms no secondary traits indicate gender (♂,♀), but this does not mean they lack an equivalent function. Most such organisms have mating types such that a genetic positive (+) must mate with a genetic negative (-), so yes, literally opposites attract!

19. The odds of flipping 23 coins simultaneously and coming up either all heads or all tails.

Chapter 4

1. Cycads are palm-like seed plants that grow in the tropics or subtropics so they are unfamiliar to us temperate-zone folks.

2. The term periphyton is usually applied to small seaweeds ranging from unicellular to filaments that grow anchored to a substrate of some sort, but the dividing line between periphyton and seaweed is quite arbitrary.

3. Visible to the naked eye.

4. Photosynthetic bacteria have recently been discovered in the pitch-black ocean depths; these bacteria use infrared "light" generated by hot magma (Beatty et al. 2005).

5. This also accounts for large terrestrial herbivores like bison and elephants.

6. In this context, large is a colony visible to the naked eye, something about as big as the period at the end of this sentence.

7. *Volvox* forms large, hollow, spherical colonies composed of hundreds to thousands of flagellated cells. The largest of these colonies can just be seen by the naked eye as they tumble along.

Chapter 5

1. Sailors must pardon me because to them tides are the horizontal movement of water, as in the tide is going out or coming in, while we landlubbers think of tides as the vertical motion of water.

2. Morel mushrooms and cup fungi are familiar examples. Even these complex-looking reproductive structures are filamentous in organization.

3. A unicellular organism's life span commences as mitosis ends, and it lasts until the cell divides into two daughter cells. This may be minutes, hours, days, or weeks depending upon the growth conditions.

4. Polysaccharide is a general term for polymers constructed of sugars.

5. Red algal chloroplasts have only chlorophyll *a*; brown algal chloroplasts have chlorophylls *a* and *c*.

Chapter 6

1. Thallus is a general term for a simple plant-like body that lacks internal differentiation into tissues.

2. Tardegrades, commonly called "water bears," still live in such places. They can often be captured by collecting and breaking up tree bark lichens, then washing them free of the debris.

3. One alternative hypothesis no longer in favor suggests that two invasions took place, one for bryophytes and one for vascular plants, but this will be explained later.

4. Until the Army Corps of Engineers comes along and dredges out the channels.

5. The best known mechanism is a contractile vacuole, small vacuoles that fill with water and then expel their contents outside of the cell. Their operation can be seen with a light microscope in many unicellular organisms.

6. A dichotomous key is a table constructed as a series of two contrasting choices such that each choice leads you along a unique path to identify a particular organism or object out of the set being considered.

7. Removal of the waxy cuticle layer with a strong detergent or a wax solvent will increase the rate of dehydration. You can demonstrate for yourself how a cuticle and an epidermis reduce water loss with a simple experiment. Weigh a control fruit (apple or cucumber), a dewaxed fruit, and a peeled fruit, and then record the change in weight over several days. All three will lose water; you predict which will lose the least and the most.

8. In some herbaceous (nonwoody) flowering plants, long strands of thick-walled phloem fibers provide considerable support to the stem. Hemp and flax (linen) are among those of economic importance.

9. Phenols are organic compounds based upon six carbons in a ring structure which is what makes these molecules so tough. Polyphenols are composed of many phenolic subunits.

10. When thick-walled xylem cells are produced annually and accumulated in stems this tissue is called wood. Woody plants appear later in the Devonian.

11. In chapter 8 the fossil record will cause this list to be amended by placing vascular plants into a larger, more inclusive lineage on the basis of a novel shared character obtained before xylem, a branched sporophyte.

12. I used to joke in classes that serious biology students would tattoo the rules on their hands, but no longer. The resurgence of interest in tattoos means some of my little muffins might take my suggestion seriously.

13. Well, almost any. Some red algae still have confounding sexual life cycles.

14. Various kinds of cross-fertilization as well as self-fertilization are possible (see Appendix: Ferns).

15. Even though I try to practice what I preach, it by no means ensures that my teaching is appreciated. Many students will take no notes during such explanations, but they pounce on definitions like a cat on a mouse because they have been trained to listen only for such material. They get a shock come exam time when they are asked to demonstrate their understanding of the explanation. Some will blame me for being a poor teacher; others will learn a new way of learning. But one part of my job is to determine which students, on the basis of their learning skills, will be eligible for medical school and taking out your gall bladder.

16. Biology is so animal biased that many people think meiosis is a gamete-producing process.

17. There is considerable variation among eukaryotes in their means of nuclear and cytoplasmic division although textbooks tend to portray mitosis as similar throughout.

18. This is the consequence of oxygen competing with carbon dioxide for the active site of the enzyme rubisco.

19. This impregnable and decay-resistant biopolymer is a component of land plant spore walls.

20. In contrast to "antithetic," this hypothesis was called "homologous" because both generations started out alike.

21. This is another example of biologists using Occam's razor. When you have two competing explanations, the simplest is most likely to be correct.

Chapter 7

1. "The inability to see or notice the plants in one's own environment, leading to the inability to recognize the importance of plants in the biosphere and in human affairs" (Wandersee and Schussler 1999).

2. For example, *Mosses, Lichens, and Ferns of Northwest North America* by Vitt et al. (1988).

3. See Malcolm and Malcolm (2000).

4. And to make the odds even worse, the typical biology department is animal biased, making the case for *any* botany positions difficult to impossible.

5. The one exception is when the ancestral group has a character that is lost in a particular lineage.

6. Most commonly this argument arises from critics of evolution as something like "If humans are descended from monkeys, then why do monkeys still exist?" And the answer is the same; monkeys and humans live in very different ways, although as humans destroy habitat, monkeys may become extinct.

7. Either unicellular or multicellular filamentous rootlike outgrowths.

8. Epiphyte means upon a plant, and it refers to plants that grow upon the trunks, branches, and leaves of bigger plants, mostly trees. Epiphyte communities are most prevalent and most diverse in the wet tropics.

9. Alpine zones and tundras are above the tree line in altitude and latitude, respectively.

10. The most familiar microorganism that exhibits a tolerance to drying is the unicellular fungus called yeast. In a dehydrated and dormant state these organisms can survive months in a jar in your kitchen cupboard, and then they regain metabolic activity within 10 to 15 minutes when you rehydrate them and feed them a bit of sugar.

11. Some of the more robust liverworts do have pores associated with photosynthetic tissue.

12. Furrowing refers to an invagination of the cell membrane to accomplish cytoplasmic division. See Appendix: Green Algae, figure A35, showing different mechanisms of cytokinesis.

13. FYI, algal filaments have perpendicular cross walls, while those in a moss protonema can be slanted.

14. Irregular means names not formed according to *International Code of Botanical Nomenclature* although some irregular names are preserved because of long use.

15. At one time resemblances between plant parts and human body parts were thought to provide clues for medical usefulness, so such similarities were readily sought. Names like liverwort are relicts of this practice.

16. Also called by the irregular name Anthophyta (antho- means flower).

17. In a bit of equal opportunity taxonomy, tracheophyta, the vascular plants, gets redefined in the next chapter.

18. Epidermal, vascular, and ground tissue systems.

19. A pyrenoid is a small body that can be observed within a chloroplast; it functions to synthesize starch.

Chapter 8

1. Bag lunches are seldom memorable even under less exciting circumstances.

2. Dichotomous is the botanical term.

3. I love this phrase and wish that I had thought of it, but alas, no. However, I cannot for the life of me remember where or when I encountered it, so all I can do is acknowledge someone else's creative alliteration.

4. The first being the release of oxygen as a by-product of photosynthesis.

5. One of my favorite figs is much smaller, but it shelters a bar and grill under its spreading canopy supported by multiple trunks.

6. In the tropics some bats have diversified into fruit eating. The most famous of these is the Australian flying fox.

7. The current record holder is a Norway spruce growing in Sweden.

8. Since pollen is the haploid generation, it only has once copy of each gene, and if that copy has a mutation then the pollen's viability may be compromised. The more mutations, the more likely this is. Since the tree is the diploid generation, each cell has two copies of each gene, so a functional copy can mask the effect of a mutation.

9. Zygote to *Archaeopteris*, although now *Wattieza* has ruined my alphabetical reference.

10. The basal intercalary meristem of the hornwort sporophyte is an exception that also prolongs development and increases the number of spores produced.

11. Unfortunately trees can be cut them down much faster than they can grow, and although we know this simple fact it has not influenced our exploitation of this resource at all. The concept of sustainability is not complex; if you remove a resource faster than it is formed you will run out. And then there is the issue of cutting down hardwood forests and replacing this complex community with a monoculture plantation of pine.

12. Water molecules are very cohesive, they cling together, so pull on one and you pull on them all.

13. Designated primary xylem and primary phloem, respectively.

14. Growth rings are a temperate zone phenomenon. Tropical trees show such organization in their wood only in a climate with annual wet and dry seasons. In the wet tropics wood may show demarcations that coincide with onset and cessation of cambial activity, but growth periods may occur two or more times a year. Cambial activity is associated with a flush of twig and leaf growth, the activity of apical meristems.

15. Or at least it used to. Demand for cork has so outstripped supply and so many alternatives to cork are in use that real corks are becoming uncommon—and you may buy wine in screw-top bottles, or even in boxes!

16. Brown algal kelps and their relatives have parenchyma too, so it evolved in at least two lineages.

17. Stands of collenchymas form ridges on the convex side of the large leaf stalks of celery and they can be pulled free.

18. This transition zone was the subject of my first botanical research project as an undergraduate at State University of New York at Oswego, conducted under the supervision of my advisor James Seago.

19. Dichotomous.

20. About 12 inches.

21. The opposite is also true. Plants adapted to dry environments, succulents, often have thick or even rounded leaves that decrease the surface area–to–volume ratio to reduce water loss.

22. Think teeter-totter; as a mass moves further out from the fulcrum its moments of force increase, which is why a little kid seated at the very end can balance a much larger adult seated closer to the fulcrum.

23. This is basically the same principle that allows you to hold your arms vertically with less effort than you can hold them horizontally.

24. Epiphytes are plants that grow upon (epi- = upon) larger plants. In a manner of speaking epiphytes are parasites of stature. Epiphyte-laden branches of rain forest trees often fail, breaking and crashing down, when heavy rain quickly increases the weight load.

25. Take a plastic drinking straw and push down vertically. You can push quite hard before the straw buckles, but the slightest lateral pressure when the straw is under some compression will cause immediate buckling.

26. This clade is called the euphyllophytes (Kenrick and Crane 1997), the true-leafed plants, in reference to megaphylls, but also shares monopodial branching, terminal sporangia on recurved branches, and sporangia opening along a single lateral suture.

27. *Osmunda cinnamomea* has remained unchanged for over 70 million years.

Chapter 9

1. In the original draft it was April, but now as spring get earlier, it made sense to change it to March.

2. Even if you think of these as vegetables, botanically they are fruits.

3. For any real taxonomic sticklers, flowering plant genera range alphabetically from *Aa* to *Zygotritonia*, an orchid to an iris.

4. The only close contender are virus, which by most definitions do not qualify as alive.

5. Granted, the amount of free water needed for sperm is not an Olympic-sized swimming pool but just more of a film of moisture, a bit of dew.

6. One need not be too intrepid because a line-of-sight nature trail is laid out from the parking lot.

7. A specimen in the ISU Herbarium collected by Gertrude Stowell documents this horsetail's presence in 1897.

8. Horsetails are homosporous, so their gametophytes are hermaphroditic.

9. Gardeners seldom see this because radishes are biennials and take two growing seasons to flower and make seed. We usually eat the storage root during the first growing season.

10. "Plant a Radish" is a musical number from *The Fantasticks*, an off-Broadway production that ran for 17,162 performances from 1960 to 2002, a record.

11. Megaspore and microspore do not refer to size any more, but the sex of the gametophyte.

12. In the Victorian era, young ladies of good-breeding studied botany to avoid learning about the sordid details of animal reproduction, but now what is one to think? Plants turned out to be as sordid as animals, or maybe more so! Of all the sciences, botany still has the largest percentage of women, so perhaps botany can thank this Victorian tradition for establishing the beginnings of generations of female mentors to encourage other young women to study and investigate plants.

13. *Lepidocarpon* is considered not a true seed for the same reason.

14. These obstacles function like snow fence, a common feature of the upstate New York snow belt; they obstruct the flow of air, slowing it, and causing a deposition of any snow being blown along. If you place the fences correctly, the drifts will form over there rather than across your driveway or highway.

15. Haustoria are anatomical outgrowths into the tissue of another plant used by parasitic plants as a means of gaining nutrition. In a manner of speaking, these tiny males are nutritional parasites.

16. Not to be confused with *Archaeopteryx*, long the oldest fossil bird.

17. Who says historical events cannot be tested? An out-of-sequence appearance or a different set of characters could falsify an evolutionary hypothesis.

18. As recounted in the previous chapter, these were thought to have been the Earth's oldest forests until the discovery of older forests composed of pseudosporochnalean cladoxylopsid "ferns" (Fig. 8–4).

19. Most botanical gardens and plant conservatories have cycads on display. Even our little university greenhouse has nine different genera of cycads. The biggest collection of cycads in North America can be viewed at Fairchild Tropical Gardens in Florida. A very good collection exists in the Garfield Park Conservatory in Chicago.

20. This custom arose from pagan celebrations of the winter solstice when the days began to lengthen and new greenery would follow; this tradition was co-opted by Christianity.

21. Only one living species, *Ginkgo biloba*, exists, although the group has an extensive fossil history.

22. Tracheids come in two basic types, those with circular bordered pits and those with scalariform (long-narrow) pits.

23. Some molecular studies make no attempt to reconcile differences with the fossil record or even the morphology of living plants, as if molecular data "rest the case."

24. Flowering plant anthers are interpreted as modified leaves, microsporophylls.

25. Finding a living glossopterid is the botanical equivalent of searching for the

abominable snowman, Bigfoot, the Loch Ness monster, and so on, a field called crypto-zoology (crypto- means hidden). A comparable field does not exist for botany, probably because too many fossil plants have been found alive.

Chapter 10

1. The now famous K-T extinction event associated with the demise of dinosaurs occurred at the end of the Cretaceous, and by then angiosperms were already dominant in many areas.

2. Many people estimate that as-yet-unknown, undocumented species will take the total to 250,000 to 300,000.

3. Recently a species of *Spiranthes*, lady's tresses orchid, appeared in the author's personal arboretum-like yard, much to our great delight.

4. *Eupomatia* comes to mind, a family of one genus with two species, both native to the wet tropics of Northeast Australia and New Guinea.

5. The seed minus its coat is the spice nutmeg; the fleshy red aril is the spice mace. When the fruit opens, it's the aril that attracts and rewards seed-dispersing birds.

6. For a visual reference, think pea pod.

7. A cluster of flowers each with a single pistil can fuse during development into what appears to be a single fruit. Pineapples and mulberries are examples of such multiple fruits.

8. Some flowers cheat by advertising a reward they do not provide. Evolutionary theory predicts why such cheating occurs.

9. Yours truly demonstrated how variable displays on male and female nutmeg trees manipulated beetles into foraging errors that moved pollen to non-rewarding female flowers and caused female trees to produce flowers just for display purposes (Armstrong 1997).

10. Bright blue and white fruit colors are more common in the tropics than among temperate plants.

11. The endosperm-filled seed is the spice called nutmeg.

12. On my prairie research plot, wind-pollinated grasses make up 58% of the total above-ground plant mass but account for only 14% of species diversity, which means there are lots of individuals in a relatively high frequency. The other 86% of the species are animal pollinated, and they occur in lower densities and frequencies.

13. Some of this data suggests tree turnover will accelerate as the average climatic temperature increases. Deborah and David Clark (Clark et al. 2003) have data showing that tropical tree respiration and mortality is highest and reproductive success lowest in the hottest years. As a result, global warming may allow wind-pollinated grassland savannas to replace tropical forests in a relatively short period of time. Pollen record data suggest this may have happened previously in the Amazon because of drought during the Pleistocene ice ages when so much water was locked up in ice. If tropical forests decline rapidly with global warming, all that carbon bound up in wood will be released, further exacerbating the climate change, a very depressing scenario.

14. Because of their small, inconspicuous flowers some people are surprised to learn that grasses and deciduous trees are flowering plants.

15. In the fall no one notices the inconspicuous wind-pollinated flowers of rag-weeds, so the conspicuous insect-pollinated flowers of goldenrods that flower at the same time often get blamed by hay fever sufferers.

16. Pollen grains range from 10 to 200 micrometers (10 to 200 millionths of a meter), so a 100 micrometer male racing 10 cm to an ovule is equivalent to 6-foot human males racing about 2.5 miles to win a bride.

17. If you ever eat fresh coconut, you can find the small columnar embryo under one of the three eyes. All the rest of the edible tissue is cellular endosperm.

18. Next time you are snacking, check out that peanut. This embryo consists of two huge cotyledons attached to a short plant axis with a pair of folded leaves on one end and a root apex on the other. A common botanical error in pre-college science books is labeling the cotyledons of a bean embryo "endosperm." This error happens if you think all seeds have endosperm.

19. For example, embryos derived from out-crossing might be favored over em-bryos derived from self-pollination, which results in inbreeding.

20. Allyn tragically died in a plane crash in South America while conducting aerial surveys of forest communities.

21. The bignon family includes only a couple of familiar temperate representatives, the catalpa tree and trumpet creeper. A number of bignons are widespread and com-mon tropical ornamentals.

22. Perfect is the term used to describe flowers with both stamens and pistils.

23. All members of the Aster or Sunflower family.

24. Because they lacked a perianth.

25. Ranales refers to *Ranunculus*, the genus of buttercups, which are herbs, there-fore woody was added to distinguish the arborescent magnolias and their relatives.

26. Many years ago I thought wood would be the subject of my career's work. Prof. Charles Heimsch, one of my mentors and a great comparative anatomist, in-troduced me to the subject. Quite sadly I learned while writing this very section that Charlie was seriously ill, and he died before it was done. His active botanical career began in the late 1930s and spanned eight decades.

27. Bailey begat Jim Canright, who begat Tom Wilson, who begat me.

28. Many people think Darwin called the ancestry of angiosperms an "abominable mystery," but really he was referring to their "sudden" appearance in the fossil record (see Friedman 2009 for a historical discussion).

29. Nutmegs have this type of pistil also (Armstrong and Tucker 1986).

30. In typical stamen development, the filament is last formed as an intercalary structure. Intercalary zones of meristematic growth produce stamen filaments, "fused" perianth tubes, and stamens inserted upon the perianth and they are generally features of more derived groups.

31. A drip tip functions like the pouring spout on a beaker or tea pot.

32. As opposed to evergreens, deciduous trees drop their leaves every fall or dry season and grow a new crown of leaves when warm weather or the rains come back.

33. Less than 10% of the light striking the crown of a sugar maple reaches the ground beneath it.

34. Whether green parasitic plants are in the process of losing their photosynthetic

ability or whether they are using parasitism in conjunction with parasitism is a question some of us are trying to answer.

35. The European common name *lignum vitae* means wood of life and refers to the use of its heartwood as a medicinal.

Chapter 11

1. The Paluxy River "human" footprints are highly eroded dinosaur footprints.

2. K stands for Cretaceous because C stands for Cambrian.

3. An earlier and more massive one occurred at the Permian-Triassic boundary. Some biologists argue that we are in a human-caused extinction event currently.

4. A new hypothesis is based on evidence that the Chicxulub impact was maybe 300,000 years too early and that massive volcanic activity in Siberia is to blame for the climate change catastrophe (Keller et al. 2009).

5. Binocular vision provides good depth of field, which is important when jumping from tree to tree.

6. The Pleistocene began 2.6 million years ago.

7. In 1837 John Deere invented a self-cleaning plow that could cut and turn the prairie sod. He was so successful that only 1% of the original prairie survives in Illinois.

8. So it isn't just the imaginations of lawn mowers. Grass does grow faster after it has been cut.

9. Aided by populations of microorganisms that possess the enzyme cellulase.

10. Remember rubisco from back in chapter 2? This is the slow, inefficient enzyme that fixes carbon in photosynthesis.

11. Ask for "corn" in Scotland and you will get oats. Corn means only "common grain of the region."

12. The lily family turned out to be a catchall group and recent molecular data have divided the genera into many smaller families, but since this is not known by most people, it makes no sense to refer to or provide what remain unfamiliar labels. *Yucca*, *Agave*, and *Aloe* are some of the better known succulent genera.

13. In this example, *annuus* is often incorrectly called the "species name," but species names are binomials.

14. All organisms with chloroplasts having just two membranes and only chlorophylls *a* and *b* (Chapter 3).

15. -opsida is the ending that signifies the taxonomic category called a class.

16. This adds up to 235,000 species because I used data from a different source than Fig. 10–1, where species added up to 220,000.

17. Sorry, couldn't resist.

18. This is based upon the food budgets of modern gatherer-hunters.

19. A few times every year I make a quick trip to the local poison control center because I must identify a plant so the physicians can determine what treatment, if any, is necessary. But it scares me to see physicians, who usually have virtually no educational background in botany, looking through plant identification picture books. Yikes! What would a physician say if they caught a botanist ready to use a pictorial guide to appendectomies?

20. Your sugar molecule sensors are localized on the tip of your tongue, which is why so many sweet confections are made to be licked.

21. Our tendencies to overeat when presented with an abundance of food and our overconsumption of meat are other poor dietary choices arising from instincts shaped by natural selection for living a very different human life style (Shepard 1998).

22. These latter features are generally associated with food spoilage organisms (fungi and bacteria). By making food unappetizing they win the competition with us big organisms. Natural selection explains why because any microorganism that did not make food disgusting would get eaten and digested along with the food.

23. The food value of cereal grains derives from their stock of starchy endosperm, although the embryo (germ), and seed coat/fruit wall (bran) are also nutritious.

24. In India, peanuts are commonly dry roasted in clean, coarse sand, which is then separated by sieving.

25. Our mastery of fire cannot be considered complete because humans still burn things by mistake or for the sheer fascination of watching things burn.

26. Family name in parentheses.

27. Yams and sweet potatoes are not the same thing. Confusion is the result of people applying familiar common names to new plants. In addition it appears the native name for lowland Peruvian sweet potato ("batatas") was mistakenly transferred to a highland Peruvian tuber and they have been called "potatoes" ever since.

28. Broccoli, cauliflower, Brussels sprouts, kale, kohlrabi, and cabbage are all different varieties of the same species.

29. Common bean with maize in Mesoamerica; lentils and peas with wheat and barley in the Middle East; soybeans with rice in China.

30. Glycine is also the name of the most common amino acid in proteins.

31. Technically diseases are caused by an infective organism or virus, while disorders have a genetic basis, but the medical field mixes up these designations, commonly causing much confusion.

32. Toxic in this context includes poisonous but also nondeleterious intoxications of curative value, stimulants and depressants including painkillers, and even hallucinogens.

33. Jeans supposedly are named after its parent fabric that was originally shipped from the Italian port of Genoa.

Appendix

1. Y-branching, an even division of the apex into two equal-sized stems, occurs in only a few living lineages, but it was more common among ancient land plants.

2. Two colleagues, Jim Hickey and Carl Taylor, both of whom I have known since our graduate school days, are experts on *Isoetes* and can spot what everyone else passes by.

3. Deciduous (dee-SID-you-us) is the botanical term for a seasonal or annual leaf drop.

4. Bald cypress and dawn redwood bear their soft, slender needles along lateral branches, which drop as a unit in the fall.

5. Pines have needles in clusters or bundles of 2s, 3s, or 5s; firs have single needles with round bases attached to smooth stems and resinous buds; douglas-firs have single needles with oval bases attached to smooth stems and buds with dry, shiny scales; spruces have single needles on short pegs, each attached to a segment of the stem and dry buds with outward curled scales.

6. Cedar is a much used common name that has been applied to aromatic plants from eleven different genera of plants, some of which are not even conifers.

7. In ferns a cluster of sporangia on the lower surface of a leaf is called a sorus, but this term is seldom used in describing gymnosperms.

8. Some flowering plants—for example, nutmegs and magnolias, have arils to attract and reward dispersers too. At maturity the protective fruit opens to display the aril-covered seed.

9. Yes, that's right, eating a pine nut means you are consuming a pregnant female.

10. A seed-producing *Ginkgo* over one hundred years old graces our campus and welcomes members of the music department to their building. On a regular basis people suggest cutting down this huge tree because its seeds make a smelly mess for two weeks each year. Such intolerance is most distressing, but it demonstrates that *Ginkgo* is less adapted to western culture than eastern. Quite likely our music majors and faculty are unaware that the pistachio green female gametophyte is edible and highly prized in parts of Asia.

11. Leaf with central stalk bearing two rows of leaflets, one along each side.

12. On a memorable day I happened upon my first specimen of this huge cycad and saw my first cassowary, too.

13. Sago palm is a common name for *Cycas revoluta*, a common ornamental in subtropical to tropical climates.

14. Some specimens of *Cycas circinalis* in our university greenhouse have not changed appreciably in appearance in the 30-plus years I have been observing them.

15. Eusporangia.

16. A dicotyledonous embryo in flowering plants was presumably inherited from their gymnospermous ancestors, if only we knew which one specifically.

17. Ginkgo produces a similar seed coat, but no one knows how its stinky outer layer functions.

18. However, in comparison with flowering plants, ferns possess only a limited range of leaf forms.

19. Forking refers to their unique Y-branched fronds.

20. Domatia, from the same root as domestic, means homes. Plants that provide homes for ants or mites do so because these predatory borders protect the plant from herbivores. Such associations can be found in a number of flowering plants, but they are uncommon in ferns.

21. Falling leaves, twigs, and other humus-forming materials.

22. To visualize a vascular cylinder with leaf gaps, take a cardboard cylinder from a roll of paper to represent the vascular cylinder. Draw a helix around the cylinder and at regular intervals make a downward slice into the cardboard. If you then bend the slice outward, an elliptically shaped hole opens above the slice. The outwardly bent piece represents a U-shaped vascular supply to a leaf. This describes only how a leaf gap looks

anatomically, not how it develops. Do not feel bad if this example failed to generate a useful image; it has always produced a glazed look on the faces of my students.

23. Such an explanation is little more than a probable guess, but in science we always attempt to explain the observations, and this kills two birds with one explanation.

24. Photosynthetic gametophytes require only mineral nutrients.

25. And you might even see a platypus as a bonus.

26. For comparison, paleontologists estimate that the average duration of any species is about 1 million years.

27. One filmy fern occurs in southern Illinois and eastern Kentucky in narrow wet rocky crevasses of cliffs.

28. Do not confuse polypody, many feet, with polyploidy, many sets of chromosomes.

29. *Hippochaete* is now treated as a subsection of *Equisetum* in the Flora of North America.

30. True rushes are a monocotyledonous flowering plant, family Juncaceae.

31. Thallus generally refers to a simple plant body that is not differentiated into an axis or leafy organs. The term is also applied to lichens and seaweeds.

Abiotic: Without life.

Actin: Protein found in all eukaryotic cells; polymerizes to form microfilaments part of the cytoskeleton and part of muscle contractile apparatus.

ADP/ATP: Adenosine diphosphate/adenosine triphosphate, a molecule consisting of a nucleotide plus two or three phosphate groups; they are used as an energy exchange in cellular reactions.

Aerobic: Requiring or using oxygen.

Algae: General term for mostly aquatic green organisms ranging from unicellular organisms to complex seaweeds. Algae is neither an ecological nor a taxonomic term.

Alleles: Different forms of a gene—for example, A, B, O blood proteins.

Alternation of generations: A sexual life cycle that alternates between multicellular haploid and diploid organisms, a gametophyte and a sporophyte.

Anaerobic: Operating in the absence of oxygen.

Angiosperm: Flowering plants.

Antheridium (-ia): Sex organ or gametangium that produces sperm.

Apical growth: Growth at tip or apex from a single cell or meristem.

Arboreal: Pertaining to trees; living in or among trees.

Arborescent: A plant with a tree-like form, usually characterized by a large trunk supporting a crown of branches.

Archaea: Name referring to archaebacteria; a Domain, one of three proposed superkingdom taxonomic groups.

Archaebacteria: Thermophilic, halophilic, acidophilic, or methane-producing bacteria; literally, ancient bacteria.

Archean: Period of Earth history from the origin of life to the appearance of eukaryotes; 4 billion years ago to 2.5 billion years ago.

Archegonium (-ia): Sex organ or gametangium that produces an egg.

Asexual: Without sex.

ATP: Adenosine triphosphate, a molecule consisting of a nucleotide plus three phosphate groups that is used as an energy exchange in cellular reactions.

Autogenesis: Self-generated.

Autotroph (-ic): Self-feeder; organism capable of capturing energy from the environment and synthesizing organic molecules from very simple raw materials.

Axil: The angle between a stem and leaf.

Bacteria: Common name for many diverse prokaryotic organisms.

Bacteriochlorophyll: Purplish-brown pigment used in photosynthesis by non-green bacteria.

Baleen: Whale "bone," actually composed of keratin (like hair and nails) and arranged in rows on the upper jaw of whales that feed by sieving small organisms out of seawater.

Bikont: Having two flagella.

Binomial: Two-parted name: first and last, genus and specific epithet.

Biogenic: Refers to naturally occurring inorganic molecules that can combine to produce organic molecules.

Biogeochemicals: Elements that cycle through the biotic component of the environment including nitrogen, carbon, sulfur, and phosphorus.

Biotic: Living; something associated with the activities of organisms.

Blue-green algae: Former name of cyanobacteria.

Carbohydrates: Class of organic molecules composed of hydrogen, oxygen, and carbon; sugars, starches, cellulose.

Cellulose: Insoluble carbohydrate polymer of glucoses; common cell wall material among many algae and land plants.

Centriole: Basal body of a flagellum, once thought to play a role in mitosis.

Centromere: Constricted portion of a chromosome (see chromosome) that binds sister chromatids together after duplication; point of attachment for spindle fibers, microtubules (see MTOC) during mitosis or meiosis.

Chemoautotroph: Organism that captures energy from biogeochemical reactions.

Chlorophyll: Light-capturing green pigment.

Chloroplast: Photosynthetic organelle of eukaryotic cells.

Chromosome: A DNA molecule whose sequence consists of many genes; condenses or coils into a compact, densely staining body during cell division.

Cilia: Many short flagella; used for locomotion.

Clade: Lineage with a single common ancestry.

Cladistics: Approach to study of phylogeny by identifying clades and the novel shared characters that define them.

Clone: Genetically identical offspring produced by asexual reproduction.

Colony: Refers to two or more individuals or similar cells that live together in a loose to regularly patterned association. At one extreme a mass of offspring from repeated cell divisions—for example, bacteria; at the other extreme a multicellular organization with a fixed number of cells in a specific pattern—for example, green algae like *Pediastrum* and *Scenedesmus*.

Competition: When organisms interact in an attempt to acquire some limited resource.

Conjugation: Fusion of two similar, nonflagellated sex cells.

Consumer: Heterotrophic member of a food chain that obtains food by absorption or ingestion.

Convergent evolution: When organisms attain similar adaptations in response to similar environmental stimuli. For example, stem succulents in the cacti and euphorb families look similar: thick green stems, reduced or no leaves, and spines, all adaptations to desert environments.

Coralloid: Having the form of coral.

Custard apple: Tropical fruit of the genus *Annona*; member of this family (Annonaceae).

Cyanobacteria: A group of gram-negative photosynthetic bacteria; formerly called blue-green algae.

Cycads: Gymnosperms resembling palms with a thick, stocky trunk and a helical whorl of palm-like leaves.

Cytokinesis: Division of the cytoplasm; usually follows mitosis.

Diploid: Cell or organisms with two complements of chromosomes, one from each parent.

Division: Botanical taxon equivalent to a phylum.

Electron microscope: Microscope that uses electrons to image specimens rather than light.

Embryo: Young organism being nurtured by its maternal parent.

Embryophyte: Land plant; green organisms whose alternation-of-generations life cycle produces an embryo.

Endoparasite: Parasite that lives within the body of its host.

Endoplasmic reticulum (ER): Membrane network for packaging and distributing materials within a eukaryotic cell.

Endosymbiosis: Refers to one organism that lives within another. See symbiosi

Epiphyte: Plant that grows upon another plant.

Eukaryotes: Organisms with nucleated cells (plants, animals, fungi, protists).

Extremophiles: Organisms that live in extreme environments, very hot, very acid, very salty.

Facultative: Organisms with alternative metabolisms or means of reproduction.

Fertilization: Fusion of sex cells in reproduction; application of mineral nutrients.

Filament (-ous): A multicellular organization characterized by a linear chain or row of cells.

Flagellum (-a): Whiplike organ used for locomotion.

Food chain: Sequence of producers and consumers passing along energy and materials by eating and being eaten.

Gametangium (-ia): Reproductive organ that produces gametes.

Gametes: Sex cells that fuse to form a diploid zygote, often but not always sperm and egg.

Gametophyte: Gamete-producing, haploid, sexual generation of land plants.

Genome: An organism's whole set of genes.

Genus: Collective taxonomic group of similar species.

GI: Gastrointestinal; inner tube of animal body.

Gondwana: Former supercontinent including South America, Africa, India, Madagascar, Australia, New Zealand, and Antarctica.

Hadean: Pre-life period of Earth history.

Haploid: Cell or organism with only one complement of chromosomes.

Hermaphrodite (-ic): A bisexual organism, one with both antheridia and archegonia, producing both sperm and eggs.

Heterocysts: Clear, empty-appearing (nongreen) cells that function in nitrogen fixation in some filamentous cyanobacteria; literally, a different cell.

Heterogeneous: Diverse in texture or composition.

Heterospory (-ous): Spores of two different sizes, microspores and megaspores, leading to male and female gametophytes, respectively.

Heterotroph (ic): Other feeders; organisms that must consume or absorb organic molecules for energy and raw materials.

Holdfast: Root-like structure that serves to anchor seaweeds.

Homogeneous: Uniform in texture or composition.

Homologous: With reference to chromosomes, having the same set of genes.

Homospory (-ous): All spores are the same and the resulting gametophytes are hermaphroditic.

Hydrogel: A protein or carbohydrate that can bind and hold water.

Hypothesis: Explanation for some observed phenomenon.

Intercalary growth: Growing from within the middle.

Invertebrate: Animals lacking a backbone or internal skeleton.

Isotopes: Elements of different atomic mass due to additional neutrons.

K-T boundary: Geological discontinuity that marks the transition from the Cretaceous (K) to the Tertiary (T); end of the age of dinosaurs.

Leptosporangium (pl. -ia): Literally the "slender" sporangium, a spore-producing organ; the small, thin-walled, few-spored sporangium of the clade or lineage of advanced ferns—that is, leptosporangiate ferns.

Linnaeans: Pre-Darwinian taxonomists; refers to Linnaeus (also known as Carl von Linné), the father of taxonomy.

Macroscopic: Visible to the naked eye.

Meiosis: A nuclear division that reduces the diploid chromosome number to haploid by separating homologous pairs of chromosomes.

Meristem: Group of small, perpetually juvenile cells found in locations where cell division and growth take place—for example, apical meristems of root and shoot tips, lateral cork and vascular meristems.

Metabolism: Biochemical processes used by a cell or organism for breaking down or synthesizing new organic compounds.

Microorganism: Any organism so small that it requires a microscope to observe them.

Micropyle: Literally means small hole; apical opening through an ovule's jacket.

Microtubules: Tubular cytoskeletal elements composed of tubulin proteins.

Mitochondria: Cellular organelle where aerobic respiration takes place.

Mitosis: Nuclear division.

Monera: Kingdom that includes all prokaryotic organisms; bacteria et al.

Motile: Organism capable of locomotion.

MTOC: Microtubule organizing center where tubulin proteins are assembled or disassembled.

Multicellular: Composed of many cells.

Mutation: Change in the genetic material.

Mutualism: A symbiotic relationship that benefits both parties.

Mycelium: Mass of filaments composing a fungal body.

Mycorrhizae: Literally fungus-roots; a symbiotic interaction between fungi and plant roots.

Nucleus: Membrane-delimited location of genetic material, chromosomes.

Ontogeny: Developmental sequence or process.

Oogamy: Sex cells differentiated as large, nonmotile eggs and small, motile sperm.

Opisthokonts: Clade of organisms whose motile cells have one posterior flagellum; clade consisting of animals and fungi plus their unicellular ancestors.

Organelles: Subcellular structures—for example, chloroplasts, mitochondria, nucleus.

Osmosis: Diffusion of water through a semipermeable membrane.

Parenchyma: General cell type composing block-organized ground tissue (pith, cortex, mesophyll, thallus) of land plants.

pH: Measure of acidity or alkalinity of a solution where pure water is neutral, a pH of 7.

Phagocytosis (-ic): Process by which a cell can engulf prey or food particles by folding in its cell membrane to form a food vacuole.

Phanerozoic: Portion of the geological column from the Cambrian to present; literally, evident life.

Phloem: Food-conducting tissue of vascular plants.

Photoautotroph: Organism that captures light energy.

Photosynthesis: Metabolism that uses captured light energy to synthesize organic molecules.

Phototropism: Growth or movement toward light (positive) or away from light (negative).

Phragmoplast: Structure that divides the cell cytoplasm (cytokinesis) in land plants and related algae; a cell plate.

Phylogeny (phylogenetic): An evolutionary history of common ancestries.

Phytoplankton: Free-floating unicellular or small multicellular photosynthetic organisms that occupy open ocean habitats.

Plasmodesmata: Microscopic cytoplasmic pores between cells; characteristic of parenchyma.

Plasmodium: An amoeboid cell.

Plasmodium: Genus of malarial parasite.

Plate tectonics: Geological theory that proposes the Earth's surface is composed of crustal plates that move relative to each other.

Polymer: A large molecule composed of many similar subunits called monomers.

Polyphyletic: Group with more than one ancestry, therefore not a clade.

Primate: Order of mammals that includes lemurs, monkeys, great apes, and humans.

Producers: Autotrophs at base of a food chain.

Prokaryote: Organism that have naked, circular DNA attached to their cell membrane rather than having chromosomes in a nucleus (eukaryotes); term literally means "before the nucleus," e.g., bacteria.

Proterozoic: Portion of the geological column from about 2.5 billion years ago to 540 million years ago, from estimated appearance of eukaryotes to the appearance of large organisms.

Protist: Originally referred to unicellular or simple multicellular organism.

Protoplasm: "Cell sap"; cytoplasmic contents of a cell.

Pseudopod: False foot; refers to lobes of an amoeboid cell.

Rain forest: Evergreen forest characterized by high annual rainfall and warm equitable temperatures, usually tropical.

Reproduction: Production of offspring, either asexual or sexual.

Respiration: A metabolic pathway that breaks down an organic molecule to obtain the energy stored in its chemical bonds.

Rhizoid: Rootlike outgrowth.

Rhizome: A stout horizontal stem growing at or just below ground level.

Ribosomes: Organelles composed of RNA molecules that function in protein synthesis.

Rubisco: Nickname for ribulose-1,5-bisphosphate carboxylase, the primary carbon-fixing enzyme of most photosynthetic organisms.

Science: Study of nature conducted by observation, hypothesis formation, and testing via experimentation or comparison.

Sclerenchyma: Thick-walled ground tissue that is dead at maturity, often forming very stiff, hard parts like stony fruit walls.

Seaweed: Algae adapted to coastal environments by anchoring; unicellular to complex multicellular organisms.

Semipermeable: With reference to membranes, lets some molecules through while blocking or containing others.

Sessile: Organism rooted or anchored in place.

Sorus (pl. sori): Cluster of sporangia on fern leaves.

Species (name): A distinct group of organisms with the potential to interbreed; a binomial label consisting of a genus and specific epithet—for example, *Taraxicum officinale*, common dandelion.

Spontaneous generation: Origin of life from nonliving materials.

Sporangium (pl. sporangia): A reproductive organ that makes spores.

Spore: Dispersible cell/body usually within a thick wall that grows directly into a new organism.

Sporophyll: A leaf associated with a sporangium. It may be modified in form or like vegetative leaves. If aggregated at the end of a stem, sporophylls may form a cone or strobilus.

Sporophyte: Spore-producing diploid generation of land plants.

Sporopollenin: An impregnable, complex biopolymer considered to be the organic substance most resistant to decay and decomposition.

Starch: Insoluble carbohydrate polymer of sugars.

Stolon: Slender, elongate horizontal stem growing at or just below the soil surface and usually bearing plantlets at intervals (syn. runner).

Stomate: A pore in the epidermis of most land plants whose opening is regulated by a pair of guard cells.

Strata: Layers.

Strobilus (pl. strobili): A cone, generally composed of stacked whorls or a helix of sporangium-bearing branches or sporophylls (leaves) on a central axis.

Stromatolites: Minutely layered geological structures associated with bacterial biofilm communities.

Sugar: Soluble carbohydrate that generally stimulates your taste sensors to produce the sensation of sweet.

Symbiosis (-tic): Literally, living together; an intimate interaction between organisms that may benefit one or both parties.

Taxonomy: Science of classification.

Tetrad: Group of four cells where each is in contact with the other three.

Thallus: A simple plant body not differentiated into tissues or organs.

Thermophilic: Organism that grows in a hot environment; literally, heat-loving.

Trilete: Referring to a scar radiating out from a single point in three directions.

Trophic level: That point along a food chain where the organism obtains its food.

Tundra: A treeless plain characterized by frost-molded landscapes, extremely low temperatures, little precipitation, poor nutrients, short growing seasons, and low-growing vegetation including dwarf shrubs, grasses, sedges, and lichens.

Turgor: Cells with a positive osmotic pressure (see osmosis) whose protoplasts generate an outward pressure on the cell wall. Turgor pressure contributes to plant stature, and plants wilt because of the loss of turgor pressure.

Undulopodia: Little-used term for flagella and cilia.

Unicellular: Composed of one cell.

Unikont: Having one flagellum.

Vacuole: Membrane-bound sac within a cell.

Vascular tissue: Conducting tissue; xylem and phloem in vascular land plants.

Wood: Accumulation of xylem from annual cambial growth of trees and shrubs.

Xylem: Water-conducting and supporting tissue of vascular plants.

Zooplankton: Small, free-floating consumer organisms occupying open ocean habitats.

Zoospore: A motile spore.

Zygospore: Zygote that forms a thick-walled stage resistant to desiccation and cold.

Zygote: Diploid cell formed by fertilization, the fusion of two gametes.

Adams, D. 1980. *The Restaurant at the End of the Universe*. London: Ballantine Books.

Adams, D. 1992. *Mostly Harmless*. New York: Harmony Books.

Algeo, T. J., S. E. Scheckler, and J. B. Maynard. 2001. "Effects of the Middle to Late Devonian Spread of Vascular Land Plants on Weathering Regimes." In *Plants Invade the Land: Evolutionary and Environmental Perspectives*, edited by P. G. Gensel and D. Edwards. New York: Columbia University Press.

Ally, D., K. Ritland, and S. P. Otto. 2010. "Aging in a Long-lived Clonal Tree." *PLoS Biol* 8(8): e1000454. doi:10.1371/journal.pbi0.1000454.

Armstrong, J. E. 1997. "Pollination by Deceit in Nutmeg (*Myristica insipida*, Myristicaceae): Floral Displays and Beetle Activity at Male and Female Trees." *American Journal of Botany* 84: 1266–74.

Armstrong, J. E., and G. E. Collier. 1990. *Science in Biology: An Introduction*. Prospect Heights, IL: Waveland Press.

Armstrong, J. E., and A. K. Irvine. 1989. "Floral Biology of *Myristica insipida*, a Distinctive Beetle Pollination Syndrome." *American Journal of Botany* 76: 86–94.

Armstrong, J. E., and D. Marsh. 1997. "Floral Herbivory, Floral Phenology, Visitation Rate, and Fruit Set in *Anaxagorea crassipetala* (Annonaceae), a Lowland Rain Forest Tree of Costa Rica." *Journal of the Torrey Botanical Society* 124: 228–35.

Armstrong, J. E., and S. C. Tucker. 1986. "Floral Development in *Myristica* (Myristicaceae)." *American Journal of Botany* 73: 1131–43.

Bada, J. L., and A. Lazcano. 2002. "Some Like It Hot, But Not the First Molecules." *Science* 296: 1982–83.

Bailey, I. W. 1954. *Contributions to Plant Anatomy*. Waltham, MA: Chronica Botanica.

Bailey, I. W., and B. G. L. Swamy. 1948. "*Amborella trichopoda* Baill., a New Morphological Type of Vesselless Dicotyledon." *Journal of the Arnold Arboretum* 29: 245–54.

Bailey, I. W., and B. G. L. Swamy. 1949. "The Morphology and Relationships of *Austrobaileya*." *Journal of the Arnold Arboretum* 30: 211–36.

Bailey, I. W., and B. G. L. Swamy. 1951. "The Conduplicate Carpel of Dicotyledons and Its Initial Trends of Specialization." *American Journal of Botany* 38: 373–79.

Bailey, I. W., and W. W. Tupper. 1918. "Size Variation in Tracheary Cells, I: A Comparison between the Secondary Xylems of Vascular Cryptogams, Gymnosperms, and Angiosperms." *Proceedings of the American Academy of Arts and Sciences* 54: 149–204.

Baldauf, S. L., A. J. Roger, I. Wenk-Siefert, and W. F. Doolittle. 2000. "Kingdom-Level Phylogeny of Eukaryotes." *Science* 290: 972–77.

Banks, H. P. 1968. "The Early History of Land Plants." In *Evolution and Environment*, edited by E. T. Drake, 73–107. New Haven: Yale University Press.

Barghoorn, E. S., and J. W. Schopf. 1966. "Microorganisms Three Billion Years Old from Precambrian of South Africa." *Science* 152: 758–63.

Beatty, J. T., J. Overmann, M. T. Lince, A. K. Manske, A. S. Lang, R. E. Blankenship, C. L. Van Dover, T. A. Martinson, and F. G. Plumley. 2005. "An Obligately Photosynthetic Bacterial Anaerobe from a Deep-Sea Hydrothermal Vent." *Proceedings of the National Academy of Sciences* 102: 9306–10.

Beck, C. B. 1960. "The Identity of *Archaeopteris* and *Callixylon*." *Brittonia* 12: 351–68.

Beck, C. B. 1962. "Reconstruction of *Archaeopteris* and Further Consideration of its Phylogenetic Position." *American Journal of Botany* 49: 373–82.

Beck, C. B., and D. C. Wight. 1988. "Progymnosperms." In *Origin and Evolution of Gymnosperms*, edited by C. B. Beck, 1–84. New York: Columbia University Press.

Becker, B., and B. Marin. 2009. "Streptophyte Algae and the Origin of Embryophytes." *Annals of Botany* 103: 999–1004.

Bell, G. 1994. "The Comparative Biology of the Alternation of Generations." *Lectures on Mathematics in the Life Sciences* 25: 1–26.

Bell, R. A. 1993. "Cryptoendolithic Algae of Hot Semiarid Lands and Deserts." *Journal of Phycology* 29: 133–39.

Bern, M., D. Goldberg, and E. Lyashenko. 2006. "Data Mining for Proteins Characteristic of Clades." *Nucleic Acids Research* 34: 4342–53.

Berner, E. K., R. A. Berner, and K. L. Moulton. 2003. "Plants and Mineral Weathering: Present and Past." *Treatise on Geochemistry* 5: 169–88.

Berner, R. A. 2001. "The Effect of the Rise of Land Plants on Atmospheric CO_2 during the Paleozoic." In *Plants Invade the Land*, edited by P. G. Gensel and D. Edwards, 173–78. New York: Columbia University Press.

Berry, C. M., and M. Fairon-Dermaret. 2002. "The Architecture of *Pseudosporochnus nodosa* Leclercq & Banks: A Middle Devonian Cladoxylopsid from Belgium." *International Journal of Plant Sciences* 163: 699–713.

Bierhorst, D. W. 1956. "Observations on Aerial Appendages in Psilotaceae." *Phytomorphology* 6: 176–84.

Bierhorst, D. W. 1971. *Morphology of Vascular Plants*. New York: Macmillan.

Bilinski, T. 1991. "Oxygen Toxicity and Microbial Evolution." *BioSystems* 24: 305–12.

Blackwell, W. 2004. "Is It Kingdoms or Domains? Confusion and Solutions." *American Biology Teacher* 66: 268–76.

Bonnier, G. 1907. *Le Monde Végétal*. Paris: Ernest Flammarion.

Boraas, M. E., D. B. Seale, and J. E. Boxhorn. 1998. "Phagotrophy by a Flagellate Selects for Colonial Prey: A Possible Origin of Multicellularity." *Evolutionary Ecology* 12: 153–64.

Borucki, W. J., David G. Koch, Gibor Basri, Natalie Batalha, Alan Boss, Timothy M. Brown, Douglas Caldwell, et al. 2010. "Characteristics of Kepler Planetary Candidates Based on the First Data Set: The Majority Are Found to Be Neptune-Size and Smaller." arXiv.org, Cornell University. http://arxiv.org/pdf/1006.2799v2.pdf

Bowe, L. M., G. Coat, and C. W. dePamphilis. 2000. "Phylogeny of Seed Plants Based on All Three Genomic Compartments: Extant Gymnosperms Are Monophyletic and Gnetales' Closest Relatives Are Conifers." *Proceedings of the National Academy of Sciences* 97: 4092–97.

Bower, F. O. 1908. *The Origin of a Land Flora*. London: Macmillan.

Bower, F. O. 1935. *Primitive Land Plants*. London: Macmillan.

Boyce, C. K. 2008. "How Green Was *Cooksonia*? The Importance of Size in Understanding the Early Evolution of Physiology in the Vascular Plant Lineage." *Paleobiology* 34: 179–94.

Bremer, K. 1985. "Summary of Green Plant Phylogeny and Classification." *Cladistics* 1: 369–85.

Briggs, D. E. G., D. H. Erwin, and F. J. Collier. 1994. *The Fossils of the Burgess Shale*. Washington, DC: Smithsonian Institution Press.

Brodo, I. M., S. D. Sharnoff, and S. Sharnoff. 2001. *Lichens of North America*. New Haven: Yale University Press.

Buchanan, R. E., and N. E. Gibbons. 1974. *Bergey's Manual of Determinative Bacteriology*. Baltimore: Williams and Wilkins.

Buck, W. R., and B. Goffinet. 2000. "Morphology and Classification of Mosses." In *Bryophyte Biology*, edited by A. J. Shaw and B. Goffinet. Cambridge: Cambridge University Press.

Burger, W. C. 1998. "The Question of Cotyledon Homology in Angiosperms." *Botanical Review* 64: 356–71.

Cairns-Smith, A. G. 1985. *Seven Clues to the Origin of Life*. Cambridge: Cambridge University Press.

Canright, J. E. 1952. "The Comparative Morphology and Relationships of the Magnoliaceae, I: Trends of Specialization in the Stamens." *American Journal of Botany* 39: 484–97.

Carlquist, S. 1996. "Wood Anatomy of Primitive Angiosperms: New Perspectives and Syntheses." In *Flowering Plant Origin, Evolution, and Phylogeny*, edited by D. W. Taylor and L. J. Hickey, 68–90. New York: Chapman and Hall.

Carlquist, S., and E. L. Schneider. 1997. "SEM Studies on Vessels in Ferns, 2: *Pteridium*." *American Journal of Botany* 84: 581–87.

Carlquist, S., and E. L. Schneider. 1999. "SEM Studies on Vessels in Ferns, 12: Marrattiaceae, with Comments of Vessel Patterns in Eusporangiate Ferns." *American Journal of Botany* 86: 457–64.

Carlquist, S., and E. L. Schneider. 2002. "The Tracheid-Vessel Element Transition in Angiosperms Involves Multiple Independent Features: Cladistic Consequences." *American Journal of Botany* 89: 185–95.

Carluccio, L. M., F. M. Hueber, and H. P. Banks. 1966. "*Archaeopteris macilenta*, Anatomy and Morphology of Its Frond." *American Journal of Botany* 53: 719–30.

Carsaro, D., D. Venditti, M. Padula, and M. Valassina. 1999. "Intracellular Life." *Critical Reviews in Microbiology* 25: 39–79.

Cavalier-Smith, T. 1993. "Kingdom Protozoa and Its 18 Phyla." *Microbiological Review* 57: 953–94.

Cavalier-Smith, T. 1997. "Amoeboflagellates and Mitochondrial Cristae in Eukaryote Evolution—Megasystematics of the New Protozoan Subkingdoms Eozoa and Neozoa." *Archiv für Protistenkunde* 147: 237–58.

Cavalier-Smith, T., and E. E. Chao. 1996. "18S rRNA Sequence of *Heterosigma carterae* (Raphidophyceae) and the Phylogeny of Heterokont Algae (Ochrophyta)." *Phycologia* 35: 500–510.

Cedergren, R., M. W. Gray, and Y. Abel. 1989. "The Evolutionary Relationships among Known Life Forms." *Journal of Molecular Evolution* 28: 98–112.

Chamberlain, C. J. 1919. *The Living Cycads.* Chicago: University of Chicago Press.

Channing, A., A. Zamuner, D. Edwards, and D. Guido. 2011. "*Equisetum thermale* sp. nov. (Equisetales) from the Jurassic San Agustin Hot Spring Deposit, Patagonia: Anatomy, Paleoecology, and Inferred Paleoecophysiology. *American Journal of Botany* 98: 1–18.

Chaw, S.-M., C. L. Parkinson, Y. Cheng, T. M. Vincent, and J. D. Palmer. 2000. "Seed Plant Phylogeny Inferred from All Three Plant Genomes: Monophyly of Extant Gymnosperms and Origin of Gnetales from Conifers." *Proceedings of the National Academy of Sciences* 97: 4086–91.

Clark, D. A., S. C. Piper, C. D. Keeling, and D. B. Clark. 2003. "Tropical Rain Forest Tree Growth and Atmospheric Carbon Dynamics Linked to Interannual Temperature Variation during 1984–2000." *Proceedings of the National Academy of Sciences* 100: 5852–57.

Cloud, P. 1988. *Oasis in Space: Earth History from the Beginning.* New York: Norton.

Colinvaux, P. A. 1978. *Why Big Fierce Animals Are Rare.* Princeton: Princeton University Press.

Collinson, M. E. 1996. "'What Use Are Fossil Ferns?'—20 Years On: With a Review of Extant Pteridophyte Families and Genera." In *Pteridology in Perspective*, edited by J. M. Camus, M. Gibby, and R. J. Johns, 349–94. Kew (UK): Royal Botanic Gardens.

Cook, M. E., and L. E. Graham. 1999. "Evolution of Plasmodesmata." In *Plasmodesmata: Nanochannels with Megatasks*, edited by A. van Bel and C. Kesteren. Berlin: Springer-Verlag.

Copeland, H. F. 1956. *Classification of the Lower Organisms.* Palo Alto (CA): Pacific Books.

Corner, E. J. H. 1964. *The Life of Plants.* Chicago: University of Chicago Press.

Corsaro, D., D. Venditti, M. Padula, and M. Valassina. 1999. "Intracellular Life." *Critical Reviews in Microbiology* 25: 39–79.

Cox, P. A. 1985. "Noodles of the Tide." *Natural History* 94: 36–40.

Crandall-Stotler, B. 1980. "Morphogenetic Designs and a Theory of Bryophyte Origins and Divergence." *BioScience* 30: 580–85.

Crandall-Stotler, B. 1984. "Musci, Hepatics, and Anthocerotes: An Essay on Analogues." In *New Manual of Bryology*, edited by R. M. Schuster, 1093–129. Nichinan, Japan: Hattori Botanical Laboratory.

Crandall-Stotler, B., and R. E. Stotler. 2000. "Morphology and Classification of the

Marchantiophyta." In *Bryophyte Biology*, edited by A. J. Shaw and B. Goffinet. Cambridge: Cambridge University Press.

Crane, P. R. 1987. "Vegetational Consequences of Angiosperm Diversification." In *The Origins of Angiosperms and Their Biological Consequences*, edited by E. M. Friis, W. G. Chaloner, and P. R. Crane. New York: Cambridge University Press.

Crane, P. R., P. Herendeen, and E. M. Friis. 2004. "Fossils and Plant Phylogeny." *American Journal of Botany* 91: 1683–99.

Crepet, W. L., and T. Delevoryas. 1972. Investigations of North American Cycadeoids: Early Ovule Ontogeny. *American Journal of Botany* 59: 209–15.

Crepet, W. L., and K. J. Niklas. 2009. Darwin's Second "Abominable Mystery": Why Are There So Many Angiosperm Species? *American Journal of Botany* 96: 366–81.

Cridland, A. A. 1964. "*Amyelon* in American Coal-balls." *Palaeontology* 7: 186–209.

Cronquist, A. 1981. *An Integrated System of Classification of Flowering Plants*. New York: Columbia University Press.

Cronquist, A. 1988. *The Evolution and Classification of Flowering Plants*. Bronx: New York Botanical Garden.

Darwin, C. 1859. *The Origin of Species by Means of Natural Selection: Or, the Preservation of Favored Races in the Struggle for Life*. London: John Murray.

Davies, P. 1999. *The Fifth Miracle: The Search for the Origin and Meaning of Life*. New York: Simon and Schuster.

De Duve, C. 1995. *Vital Dust: Life as a Cosmic Imperative*. New York: Basic Books.

Delevoryas, T. 1955. "A *Palaeostachya* from the Pennsylvanian of Kansas." *American Journal of Botany* 42: 481–88.

Delevoryas, T. 1963. "Investigation of North American Cycadeoids: Cones of *Cycadeoidea*." *American Journal of Botany* 50: 45–52.

Delevoryas, T. 1968. "Investigations of North American Cycadeoids: Structure, Ontogeny, and Phylogenetic Considerations of Cones of *Cycadeoidea*." *Palaeontographica B* 121: 122–33.

Delsemme, A. H. 1998. *Our Cosmic Origins*. Cambridge: Cambridge University Press.

Delsemme, A. H. 2001. "An Argument for the Cometary Origin of the Biosphere." *American Scientist* 89: 432–42.

Delwiche, C. F. 1999. "Tracing the Thread of Plastid Diversity through the Tapestry of Life." *American Naturalist* 154: 164–77.

Delwiche, C. F., R. A. Anderson, D. Bhattacharya, B. D. Mishler, and R. M. McCourt. 2004. "Algal Evolution and the Early Radiation of Green Plants." In *Assembling the Tree of Life*, edited by J. Cracraft and M. J. Donoghue, 121–37. New York: Oxford University Press.

Dilcher, D. L., and P. R. Crane. 1984. "*Archaeanthus*: An Early Angiosperm from the Cenomanian of the Western Interior of North America." *Annals Missouri Botanical Garden* 71: 351–83.

Dismukes, G. C., V. V. Kimov, S. V. Baranov, Y. N. Kozlov, J. DasGupta, and A. Tyryshkin. 2001. "Hypothesis for the Origin of Oxygenic Photosynthesis." *Proceedings of the National Academy of Sciences* 98: 2170–75.

Doolittle, W. F. 1999. "Phylogenetic Classification and the Universal Tree." *Science* 284: 2124–28.

Doyle, J. 1945. "Developmental Lines in Pollination Mechanisms in the Coniferales." *Scientific Proceedings, Royal Dublin Society* 24: 43–62.

Doyle, J. A. 1996. "Seed Plant Phylogeny and the Relationships of Gnetales." *International Journal Plant Science* 157: 3–39.

Doyle, J. A., H. Eklund, and P. S. Herendeen. 2003. "Morphological Phylogenetic Analysis of Living and Fossil Chloranthaceae." *International Journal of Plant Sciences* 164 (suppl.): 365–82.

Doyle, J. A., and P. K. Endress. 2000. "Morphological Phylogenetic Analysis of Basal Angiosperms: Comparison and Combination with Molecular Data." *International Journal Plant Science* 161: 121–53.

Duff, R. J., and D. L. Nickrent. 1999. "Phylogenetic Relationships of Land Plants Using Mitochondrial Small-Subunit rDNA Sequences." *American Journal of Botany* 86: 372–86.

Dyer, B. D., and R. A. Obar. 1994. *Tracing the History of Eukaryotic Cells*. New York: Columbia University Press.

Dyson, F. 1999. *Origin of Life*. Cambridge: Cambridge University Press.

Eames, A. J. 1936. *Morphology of Vascular Plants: Lower Groups*. New York: McGraw-Hill.

Eames, A. J. 1961. *Morphology of the Angiosperms*. New York: McGraw-Hill.

Eames, A. J., and L. H. MacDaniels. 1925. *Introduction to Plant Anatomy*. New York: McGraw-Hill.

Edwards, D. S. 1986. "Aglaophyton Major, a Non-vascular Land-Plant from the Devonian Rhynie Chert." *Botanical Journal of the Linnean Society* 93: 173–204.

Edwards, D. S. 1993. "Cells and Tissues in the Vegetative Sporophytes of Early Land Plants." *New Phytologist* 125: 225–47.

Edwards, D. S. 1996. "New Insights into Early Land Ecosystems: A Glimpse of a Lilliputian World." *Review of Palaeobotany and Palynology* 90: 159–74.

Edwards, D. S., J. G. Duckett, and J. B. Richardson. 1995. "Hepatic Characters in the Earliest Land Plants." *Nature* 374: 635–36.

Edwards, D. S., and C. Wellman. 2001. "Embryophytes on Land: the Ordovician to Lachkovian (Lower Devonian) Record." In *Plants Invade the Land: Evolutionary and Environmental Perspectives*, edited by P. G. Gensel and D. Edwards, 3–28. New York: Columbia University Press.

Egelman, E. H. 2001. "Molecular Evolution: Actin's Long Lost Relative Found." *Current Biology* 11: 1022–24.

Egelman, E. H. 2003. "Actin's Prokaryotic Homologs." *Current Opinion in Structural Biology* 13: 244–48.

Elick, J. M., S. G. Driese, and C. I. Mora. 1998. "Very Large Plant and Root Traces from the Early to Middle Devonian: Implications for Early Terrestrial Ecosystems and Atmospheric CO_2." *Geology* 26: 143–46.

Embley, T. M., and W. Martin. 2006. Eukaryotic Evolution, Changes and Challenges. *Nature* 440: 623–30.

Endler, J. A. 1986. *Natural Selection in the Wild*. Princeton: Princeton University Press.

Endress, P. K. 1996. "Structure and Function of Female and Bisexual Organ Complexes in Gnetales." *International Journal of Plant Science* 157: 113–25.

Endress, P. K. 2001. "The Flowers in Extant Basal Angiosperms and Inferences on Ancestral Flowers." *International Journal of Plant Science* 162: 1111–40.

Endress, P. K., and J. A. Doyle. 2009. "Reconstructing the Ancestral Angiosperm Flower and Its Initial Specializations." *American Journal of Botany* 96: 22–66.

Ewald, P. W. 1997. *Evolution of Infectious Disease*. Oxford: Oxford University Press.

Foster, A. S., and E. M. Gifford Jr. 1959. *"Comparative Morphology of Vascular Plants."* San Francisco: W. H. Freeman.

Frair, W., and P. Davis. 1983. *A Case for Creation*. Kansas City (MO): CRS Books.

Friedman, W. E. 2009. "The Meaning of Darwin's 'Abominable Mystery.'" *American Journal of Botany* 96: 5–21.

Friedman, W. E., and J. S. Carmichael. 1996. "Double Fertilization in Gnetales: Implications for Understanding Reproductive Diversification among Seed Plants." *International Journal of Plant Science* 157: 77–94.

Friedman, W. E., and S. K. Floyd. 2001. "The Origin of Flowering Plants and Their Reproductive Biology: A Tale of Two Phylogenies." *Evolution* 55: 217–31.

Friedman, W. E., and J. H. Williams. 2003. "Modularity of the Angiosperm Female Gametophyte and Its Bearing on the Early Evolution of Endosperm in Flowering Plants." *Evolution* 57: 216–30.

Friedman, W. E., and J. H. Williams. 2004. "Developmental Evolution of the Sexual Process in Ancient Flowering Plant Lineage." *Plant Cell* 16: S119–32.

Friis, E. M., K. R. Pedersen, and P. R. Crane. 2000. "Reproductive Structure and Organization of Basal Angiosperms from the Early Cretaceous (Barremian or Aptian) of Portugal." *International Journal of Plant Science* 161: 169–82.

Friis, E. M., K. R. Pedersen, and P. R. Crane. 2006. "Cretaceous Angiosperm Flowers: Innovation and Evolution in Plant Reproduction." *Palaeogeography, Palaeoclimatology, Palaeoecology* 232: 251–93.

Fry, I. V. 2000. *The Emergence of Life on Earth*. New Brunswick: Rutgers University Press.

Gensel, P. G. 1976. *"Renalia hueberi*, a New Plant from the Lower Devonian of Eastern Canada with a Discussion of Morphological Variation within the Genus." *Palaeontographica B* 168: 81–99.

Gensel, P. G., M. E. Kotyk, and J. F. Basinger. 2001. "Morphology of Above- and Below-Ground Structures in Early Devonian (Pragian-Emsian) Plants." In *Plants Invade the Land: Evolutionary and Environmental Perspectives*, edited by P. G. Gensel and D. Edwards, 83–102. New York: Columbia University Press.

Gentry, A. H. 1974. "Flowering Phenology and Diversity in Tropical Bignoniaceae." *Biotropica* 6: 64–68.

Gerrienne, P. 1988. "Early Devonian Plant Remains from Marchin (North of Dinant Synclinorium, Belgium), I: *Zosterophyllum deciduum* sp. nov." *Review of Palaeobotany and Palynology* 55: 317–35.

Gerrienne, P., D. L. Dilcher, S. Bergamaschi, I. M. Miracles, E. Pear Tree, M. A. C. Rodrigues. 2006. "An Exceptional Specimen of the Early Land Plant *Cooksonia paranensis*, and a Hypothesis on the Life Cycle of Earliest Eutracheophytes." *Review of Palaeobotany and Palynology* 142: 123–30.

Goffinet, B. 2000. "Origin and Phylogenetic Relationships of Bryophytes." In *Bryophyte Biology*, edited by A. J. Shaw and B. Goffinet, 124–49. Cambridge: Cambridge University Press.

Gould, S. J. 1980. *The Panda's Thumb: More Reflections in Natural History*. New York: Norton.

Graham, A. 1993. "History of Vegetation—Cretaceous to Tertiary." *Flora of North America* 1: 57–70.

Graham, L. E. 1985. "The Origin of the Life Cycle of Land Plants." *American Scientist* 75: 178–86.

Graham, L. E. 1993. *Origin of Land Plants*. New York: Wiley.

Graham, L. E., P. Arancibia-Avila, W. A. Taylor, P. K. Strother, and M. E. Cook. 2012. "Aeroterrestrial *Coleochaete* (Streptophyta, Coleochaetales) Models Early Plant Adaptation to Land." *American Journal of Botany* 99: 130–44.

Graham, L. E., M. E. Cook, and J. S. Busse. 2000. "The Origin of Plants: Body Plan Changes Contributing to a Major Evolutionary Radiation." *Proceedings of the National Academy of Sciences* 97: 4535–40.

Graham, L. E., and J. Gray. 2001. "The Origin, Morphology, and Ecophysiology of Early Embryophytes: Neontological and Paleontological Perspectives." In *Plants Invade the Land: Evolutionary and Environmental Perspectives*, edited by P. G. Gensel and D. Edwards, 140–58. New York: Columbia University Press.

Graham, L. E., and L. Wilcox. 2000. *Algae*. Upper Saddle River (NJ): Prentice Hall.

Graham, L. E., L. W. Wilcox, M. E. Cook, and P. G. Gensel. 2004. "Resistant Tissues of Modern Marchantioid Liverworts Resemble Enigmatic Early Paleozoic Microfossils." *Proceedings of the National Academy of Sciences* 101: 11025–29.

Grout, A. J. (1903) 1972. "Mosses with Hand-Lens and Microscope." Reprint, Ashton (MD): Lundberg.

Guerrero, R., C. Pedros-Alio, I. Esteve, J. Mas, D. Chase, and L. Margulis. 1986. "Predatory Prokaryotes: Predation and Primary Consumption Evolved in Bacteria." *Proceedings of the National Academy of Sciences* 83: 2138–42.

Haeckel, E. 1904. *Kunstformen der Natur*. Leipzig: Verlag des Bibliographischen Instituts.

Hampl, V., L. Hug, J. W. Leigh, J. B. Dacks, B. F. Lang, A. G. B. Simpson, and A. J. Roger. 2009. "Phylogenomic Analyses Support the Monophyly of Excavata and Resolve Relationships among Eukaryotic 'Supergroups.'" *Proceedings of the National Academy of Sciences* 106(10): 3859–64.

Han, T. M., and B. Runnegar. 1992. "Megascopic Eukaryotic Algae from the 2.1-Billion-Year-Old Negaunee Iron-Formation, Michigan." *Science* 257: 232–35.

Hao, S.-G., J. Xue, D. Guo, and D. Wang. 2010. "Earliest Rooting System and Root:Shoot Ratio from a New *Zosterophyllum* Plant." *New Phytologist* 185: 217–25.

Harris, T. M. 1951. "The Fructification of *Czekanowskia* and Its Allies." *Philosophical Transactions of the Royal Society of London, B* 235: 483–508.

Harris, T. M. 1964. *The Yorkshire Jurassic Flora, II: Caytoniales, Cycadales, and Pteridosperms*. London: British Museum of Natural History.

Harris, T. M. 1976. "The Mesozoic Gymnosperms." *Review of Palaeobotany and Palynology* 21: 119–34.

Hartman, H. 1998. "Photosynthesis and the Origin of Life." *Origin of Life and Evolution of the Biosphere* 28: 515–21.

Hauke, R. L. 1963. "A Taxonomic Monograph of the Genus *Equisetum* Subgenus *Hippochaete*." *Nova Hedwigia* 8: 1–123.

Hauke, R. L. 1978. "A Taxonomic Monograph of the Genus *Equisetum* Subgenus *Equisetum*." *Nova Hedwigia* 30: 385–455.

Heiser, C. B., Jr. 1981. *Seed to Civilization: The Story of Food*. San Francisco: Freeman.

Hennig, W. 1966. *Phylogenetic Systematics*. Urbana: University of Illinois Press.

Hessen, D. O., and E. van Donk. 1993. "Morphological Changes in *Scenedesmus* Induced by Substances Released from *Daphnia*." *Archiv für Hydrobiologie* 127: 129–40.

Hickey, L. J., and J. A. Doyle. 1977. "Early Cretaceous Fossil Evidence for Angiosperm Evolution." *Botanical Review* 43: 3–104.

Hickey, L. J., and D. W. Taylor. 1996. "Origin of the Angiosperm Flower." In *Flowering Plant Origin, Evolution, and Phylogeny*, edited by D. W. Taylor and L. J. Hickey, 176–231. New York: Chapman and Hall.

Hirmer, M. 1927. *Handbuch der Paläbotanik: Band I–Thallophyta, Bryophyta, Pteridophyta*. Munich: R. Oldenburg.

Hoiczyk, E., and W. Baumeister. 1998. "The Junctional Pore Complex, a Prokaryote Excretory Organelle, Is the Molecular Motor Underlying Gliding Motility in Cyanobacteria." *Current Biology* 8: 1161–68.

Holt, F. G., and G. W. Rothwell. 1997. "Is *Ginkgo biloba* (Ginkgoaceae) Really an Oviparous Plant?" *American Journal of Botany* 84: 870–72.

Hooke, R. 1665. *Micrographia: Or Some Physiological Descriptions of Minute Bodies Made by Magnifying Glasses with Observations and Inquiries Thereupon*. London: J. Martyn and J. Allestry.

Hoshizaki, B. J., and R. C. Moran. 2001. *Fern Grower's Manual*. Portland: Timber Press.

Hotton, C. L., F. M. Hueber, D. H. Griffing, and J. S. Bridge. 2001. "Early Terrestrial Plant Environments: An Example from the Emsian of Gaspé, Canada." In *Plants Invade the Land: Evolutionary and Environmental Perspectives*, edited by P. G. Gensel and D. Edwards, 179–212. New York: Columbia University Press.

Hufford, L. 1996. "The Morphology and Evolution of Male Reproductive Structures of Gnetales." *International Journal of Plant Science* 157 (Suppl.): 95–112.

Hughes, N. F. 1976. *Palaeobiology of Angiosperm Origins*. Cambridge: Cambridge University Press.

Humphreys, C. P., P. J. Franks, M. Rees, M. I. Bidartondo, J. R. Leake, and D. J. Beerling. 2010. "Mutualistic Mycorrhiza-Like Symbiosis in the Most Ancient Group of Land Plants." *Nature Communications* 1: 103.

Inouye, I. 1993. "Flagella and Flagellar Apparatuses of Algae." In *Ultrastructure of Microalgae*, edited by T. Berner, 99–133. London: CRC Press.

Jeon, K. 1991. "Amoeba and X Bacteria: Symbiont Acquisition and Possible Species Change." In *Symbiosis as a Source of Evolutionary Change*, edited by L. Margulis and R. Fester, 118–31. Cambridge: MIT Press.

Johnson, A. P., H. J. Cleaves, J. P. Dworkin, D. P. Glavin, A. Lazcano, J. L. Bada. 2008. "The Miller Volcanic Spark Discharge Experiment." *Science* 322: 404.

Jones, D. L. 1993. *Cycads of the World: Ancient Plants in Today's Landscape*. Washington, DC: Smithsonian Institution Press.

Kaplan, D. R. 2001. "The Science of Plant Morphology: Definition, History, and Role in Modern Biology." *American Journal of Botany* 88: 1711–41.

Kaplan, D. R., and T. J. Cooke. 1996. "The Genius of Wilhelm Hofmeister: The Origin of Causal-Analytical Research in Plant Development." *American Journal of Botany* 83: 1647–60.

Kaplan, D. R., and W. Hagemann. 1991. "The Relationship of Cell and Organism in Vascular Plants." *BioScience* 41: 693–703.

Karol, K. G., R. M. McCourt, M. T. Cimino, and C. G. Delwiche. 2001. "The Closest Living Relatives of Plants." *Science* 294: 2351–53.

Kasper, A. E., Jr., and H. N. Andrews Jr. 1972. "*Pertica*, a New Genus of Devonian Plants from Northern Maine." *American Journal of Botany* 59: 897–911.

Kasting, J. F. 2001. "The Rise of Atmospheric Oxygen." *Science* 293: 819–20.

Kaveski, S., L. Margulis, and D. Mehor. 1983. "There's No Such Thing as a One-Celled Animal or Plant." *Science Teacher* 50: 34–36, 41–43.

Keddy, P. A. 1981. "Why Gametophytes and Sporophytes Are Different: Form and Function in a Terrestrial Environment." *American Naturalist* 118: 452–54.

Keeling, P. J., and W. F. Doolittle. 1995. "Archaea: Narrowing the Gap between Prokaryotes and Eukaryotes." *Proceedings of the National Academy of Sciences* 92: 5761–64.

Keller, G., S. Abramovich, Z. Berner, and T. Adatte. 2009. "Biotic Effects of the Chicxulub Impact, K-T Catastrophe and Sea-Level Changes in Texas." *Paleogeography, Paleoclimatology, Paleoecology* 271: 52–68.

Kenrick, P. 1994. "Alternation of Generations in Land Plants: New Phylogenetic and Morphological Evidence." *Biological Review* 69: 293–330.

Kenrick, P., and P. R. Crane. 1991. "Water-Conducting Cells in Early Fossil Land Plants: Implications for the Early Evolution of Tracheophytes." *Botanical Gazette* 152: 335–56.

Kenrick, P., and P. R. Crane. 1997. *The Origin and Early Diversification of Land Plants: A Cladistic Study*. Washington, DC: Smithsonian Institution Press.

Kerr, R. A. 1999. "Early Life Thrived despite Earthly Travails." *Science* 284: 2111–13.

Kidston, R., and W. H. Lang. 1920. "On Old Red Sandstone Plants Showing Structure, from the Rhynie Chert Bed, Aberdeenshire, Part III: *Asteroxylon mackiei*." *Royal Society of Edinburgh Transactions* 52: 643–80.

Kim, E., A. G. B. Simpson and L. E. Graham. 2006. "Evolutionary Relationships of Apusomonads Inferred from Taxon-Rich Analyses of 6 Nuclear Encoded Genes." *Molecular Biology and Evolution* 23: 2455–66.

Knoll, A. H. 1999. "A New Molecular Window on Early Life." *Science* 285: 1025–26.

Knoll, A. H. 2003. *Life on a Young Planet*. Princeton: Princeton University Press.

Knoll, A. H., E. J. Javaux, D. Hewitt, and P. Cohen. 2006. "Eukaryotic Organisms in Proterozoic Oceans." *Philosophical Transactions of the Royal Society, B* 361: 1023–38.

Kodner, R. B., and L. E. Graham. 2001. "High-Temperature, Acid-Hydrolyzed Remains of *Polytrichum* (Musci, Polytrichaceae) Resemble Enigmatic Silurean-Devonian Tubular Microfossils." *American Journal of Botany* 88: 462–66.

Korn, R. W. 1999. "Biological Organization—A New Look at an Old Problem." *BioScience* 49: 51–57.

Kotyk, M. E., J. F. Basinger, P. G. Gensel, and T. A. deFreitas. 2002. "Morphologically Complex Plant Macrofossils from the Late Silurian of Arctic Canada." *American Journal of Botany* 89: 1003–12.

Krings, M., H. Kerp, E. L. Taylor, and T. N. Taylor. 2001. "Reconstruction of *Pseudomariopteris busquetii*, a Vine-Like Late Carboniferous-Early Permian Pteridosperm." *American Journal of Botany* 88: 767–76.

Kronestedt, E. 1981. "Anatomy of *Ricciocarpus natans* (L.) Corda, Studied by Scanning Electron Microscope." *Annals of Botany* 47: 817–27.

Kumar, S., and A. Rzhetsky. 1996. "Evolutionary Relationships of Eukaryotic Kingdoms." *Journal of Molecular Evolution* 42: 183–93.

Labandeira, C. C., J. Kvaček, and M. B. Mostovski. 2007. "Pollination Drops, Pollen, and Insect Pollination of Mesozoic Gymnosperms." *Taxon* 56: 663–95.

Lappalainen, E., ed. 1996. *Global Peat Resources*. Jyskä, Finland: International Peat Society, 161.

Lazcano, A., and S. L. Miller. 1994. "How Long Did It Take for Life to Begin and Evolve to Cyanobacteria?" *Journal of Molecular Evolution* 39: 546–54.

Lazcano, A., and S. L. Miller. 1999. "On the Origin of Metabolic Pathways." *Journal of Molecular Evolution* 49: 424–31.

Leclercq, S., and H. P. Banks. 1962. "*Pseudosporochnus nodosus* sp. nov., a Middle Devonian Plant with Cladoxylalean Affinities." *Palaeontographica* 110B: 1–34.

Lemieux, C., C. Otis, and M. Turmel. 2000. "Ancestral Chloroplast Genome in *Mesostigma viride* Reveals an Early Branch of Green Plant Evolution." *Nature* 403: 649–52.

Lewington, A. 1990. *Plants for People*. New York: Oxford University Press.

Lewis, L. A., and R. M. McCourt. 2004. "Green Algae and the Origin of Land Plants." *American Journal of Botany* 91: 1535–56.

Lidgard, S. H., and P. R. Crane. 1988. "Quantitative Analyses of the Early Angiosperm Radiation." *Nature* 331: 344–46.

Ligione, R., J. C. Duckett, and K. S. Renzaglia. 2012. "Major Transitions in the Evolution of Early Land Plants: A Bryological Perspective." *Annals of Botany* 109: 851–71.

Liu, S. V., J. Zhou, C. Ahang, D. R. Cole, M. Gajdarziska-Josifovska, and T. J. Phelps. 1997. "Thermophilic Fe (III)-Reducing Bacteria from the Deep Subsurface: The Evolutionary Implications." *Science* 277: 1106–9.

Lockhart, P. J., A. W. D. Larkum, M. A. Steel, P. J. Waddell, and D. Penny. 1996. "Evolution of Chlorophyll and Bacteriochlorophyll: The Problem of Invariant Sites in Sequence Analysis." *Proceedings of the National Academy of Sciences* 93: 1930–34.

Lumine, J. I. 2001. "Cold Beginnings." *American Scientist* 89: 484.

Lüning, K. 1985. *Meeresbotanik. Verbreitung, Ökophysiologie, und wirtschaftliche Bedeutung der marinen Makroalgen*. Stuttgart: Thieme Verlag.

Madigan, M. J., and B. L. Marrs. 1997. "Extremophiles." *Scientific American* April: 82–87.

Magallón, S., and A. Castillo. 2009. "Angiosperm Diversification through Time." *American Journal of Botany* 96: 349–65.

Magombo, Z. L. K. 2003. "The Phylogeny of Basal Peristomate Mosses: Evidence from cpDNA, and Implications for Peristome Evolution." *Systematic Botany* 28: 24–38.

Malcolm, B., and N. Malcolm. 2000. *Mosses and Other Bryophytes: An Illustrated Glossary*. Nelson, New Zealand: Micro-Optics Press.

Margulis (Sagan), L. 1967. "On the Origin of Mitosing Cells." *Journal of Theoretical Biology* 14: 225–74.

Margulis, L. 1974. "On the Evolutionary Origin and Possible Mechanism of Colchicine-Sensitive Mitotic Movements." *BioSystems* 6: 16–36.

Margulis, L. 1981. *Symbiosis in Cell Evolution*. San Francisco: Freeman.

Margulis, L. 1982. *Early Life*. Boston: Science Books Institute.

Margulis, L. 1996. "Archaeal-Eubacterial Mergers in the Origin of Eukarya: Phylogenetic Classification of Life." *Proceedings of the National Academy of Sciences* 92: 1071–76.

Margulis, L., J. O. Corliss, M. Melkonian, and D. J. Chapman, eds. 1989. *Handbook of Protoctista*. Boston: Jones and Bartlett.

Margulis, L., M. F. Dolan, and R. Guerrero. 2000. "The Chimeric Eukaryotes: Origin of the Nucleus from the Karyomastigote in Amitochondrial Protists." In *Variation and Evolution in Plants and Microorganisms toward a New Synthesis 50 Years after Stebbins*, edited by F. J. Ayala, W. M. Fitch, and M.T. Clegg, 21–34. Washington, DC: National Academy of Sciences Press.

Margulis, L., and D. Sagan. 1986. *Origins of Sex: Three Billion Years of Genetic Recombination*. New Haven: Yale University Press.

Margulis, L., and K. Schwartz. 1988. *Five Kingdoms*. 2nd ed. New York: Freeman.

Martin, W. 2000. "A Powerhouse Divided." *Science* 287: 1219.

Martin, W., and M. J. Russell. 2003. "On the Origins of Cells: A Hypothesis for the Evolutionary Transitions from Abiotic Geochemistry to Chemoautotrophic Prokaryotes, and from Prokaryotes to Nucleated Cells." *Philosophical Transactions of the Royal Society of London, B* 358: 59–85.

Mattox, K. R., and K. D. Stewart. 1984. "A Classification of the Green Algae: A Concept Based on Comparative Cytology." In *Systematics of the Green Algae*, edited by D. E. G. Irvine and D. M. John. London: Academic Press.

Mayer, F. 2003. "Cytoskeletons in Prokaryotes." *Cell Biology International* 27: 429–38.

Mayr, E. 2004. *What Makes Biology Unique?* Cambridge: Cambridge University Press, 246.

Melkonian, M. 1989. "Flagellar Apparatus Ultrastructure in *Mesostigma viride* (Prasinophyceae)." *Plant Systematics and Evolution* 164: 93–122.

Melkonian, M., and B. Surek. 1995. "Phylogeny of the Chlorophyta: Congruence between Ultrastructural and Molecular Evidence." *Bulletin de la Société Zoologique de France* 120: 191–208.

Miller, N. G., and S. F. McDaniel. 2004. "Bryophyte Dispersal Inferred from Colonization of an Introduced Substratum on Whiteface Mountain, New York." *American Journal of Botany* 91: 1173–82.

Miller, S. L., and A. Lazcano. 2002. "Formation of the Building Blocks of Life." In *Life's Origin: The Beginnings of Biological Evolution*, edited by J. W. Schopf, 78–112. Berkeley: University of California Press.

Mishler, B. D., and S. P. Churchill. 1984. "A Cladistic Approach to the Phylogeny of the 'Bryophytes.'" *Brittonia* 36: 406–24.

Mishler, B. D., and S. P. Churchill. 1985. "Transition to a Land Flora: Phylogenetic Relationships of the Green Algae and Bryophytes." *Cladistics* 1: 305–28.

Morowitz, H. J. 1992. *Beginnings of Cellular Life: Metabolism Recapitulates Biogenesis*. New Haven: Yale University Press.

Müller, K. M., M. C. Oliveira, R. G. Sheath, and D. Bhattacharya. 2001. "Ribosomal DNA Phylogeny of the Bangiophycidae (Rhodophyta) and the Origin of Secondary Plastids." *American Journal of Botany* 88: 1390–1400.

Nagalingum, N. S., C. R. Marshall, T. B. Quental, H. S. Rai, D. P. Little, and S. Mathews. 2011. "Recent Synchronous Radiation of a Living Fossil." *Science* 334: 796–99.

Nei, M., P. Xu, and G. Glazko. 2001. "Estimation of Divergence Times from Multiprotein Sequences for a Few Mammalian Species and Several Distantly Related Organisms." *Proceedings of the National Academy of Sciences USA* 98: 2497–502.

Newton, A. E., N. Wikström, N. Bell, L. L. Forrest, and M. S. Ignatov. 2007. "Dating the Diversification of Pleurocarpous Mosses." In *Pleurocarpous Mosses: Systematics and Evolution*, edited by A. E. Newton and R. S. Tangney, 337–66. Boca Raton, FL: CRC Press.

Niklas, K. J. 1981. "Airflow Patterns around Some Early Seed Plant Ovules and Cupules: Implications Concerning Efficiency in Wind Pollination." *American Journal of Botany* 68: 635–50.

Niklas, K. J. 1992. *Plant Biomechanics: An Engineering Approach to Plant Form and Function.* Chicago: University of Chicago Press.

Niklas, K. J. 1994. *Plant Allometry: The Scaling of Form and Process.* Chicago: University of Chicago Press.

Niklas, K. J. 1997. *The Evolutionary Biology of Plants.* Chicago: University of Chicago Press.

Nisbet, E. G., J. R. Gann, and C. L. van Dover. 1995. "Origins of Photosynthesis." *Nature* 373: 479–80.

Nisbet, E. G., and N. H. Sleep. 2001. "The Habitat and Nature of Early Life." *Nature* 409: 1083–91.

Norstag, K. J. 1987. "Cycads and the Origin of Insect Pollination." *American Scientist* 75: 270–79.

Oliveira, M. C., and D. Bhattacharya. 2000. "Phylogeny of the Bangiophycidae (Rhodophyta) and the Secondary Endosymbiotic Origin of Algal Plastids." *American Journal of Botany* 87: 482–92.

Øllgaard, B. 1987. "A revised classification of the Lycopodiaceae *s. lat.*" *Opera Botanica* 92: 153–78.

Oltmanns, F. 1905. *Morphologie und biologie der Algen.* Jena, Germany: Gustav Fischer.

O'Neil, J., R. W. Carlson, D. Francis, and R. K. Stevenson. 2008. "Neodymium-142 Evidence for Hadean Mafic Crust." *Science* 321: 1828–31.

Oro, J. 1983. "Chemical Evolution and the Origin of Life." *Advances in Space Research* 3: 77–94.

Oro, J. 2002. "Historical Understanding of Life's Beginning." In *The Beginning of Biological Evolution*, edited by J. W. Schopf. Berkeley: University of California Press.

Osborne, C. P. 2008. "Atmosphere, Ecology and Evolution: What Drove the Miocene Expansion of C_4 Grasses?" *Journal of Ecology* 96: 35–45.

Palmer, J. D. 2000. "A Single Birth of All Plastids?" *Nature* 405: 32–33.

Patterson, D. J. 1999. "The Diversity of Eukaryotes." *American Naturalist* 154: 96–124.

Payne, J. L., A. G. Boyer, J. H. Brown, S. Finnegan, M. Kowalewski, R. A. Krause Jr, S. K. Lyons, et al. 2009. "Two-Phase Increase in the Maximum Size of Life over 3.5 Billion Years Reflects Biological Innovation and Environmental Opportunity." *Proceedings of the National Academy of Sciences* 106(1): 24–27.

Pettitt, J., and C. B. Beck. 1968. "*Archaeosperma arnoldii*—A Cupulate Seed from the Upper Devonian of North America." *Contributions from the Museum of Paleontology, University of Michigan* 22: 139–54.

Phillips, T. L., H. N. Andrews, and P. G. Gensel. 1972. "Two Heterosporous Species of

Archaeopteris from the Upper Devonian of West Virginia." *Palaeontographica B* 139: 47–71.

Poli, D., M. Jacobs, and T. J. Cooke. 2003. "Auxin Regulation of Axial Growth in Bryophyte Sporophytes: Its Potential Significance for the Evolution of Early Land Plants." *American Journal of Botany* 90: 1405–15.

Pollan, M. 2001. *The Botany of Desire: A Plant's-Eye View of the World.* New York: Random House.

Porter, C. L. 1967. *Taxonomy of Flowering Plants.* San Francisco: Freeman.

Potts, R., and A. K. Behrensmeyer. 1992. "Late Cenozoic Terrestrial Ecosystems." In *Terrestrial Ecosystems through Time,* edited by A. K. Behrensmeyer, J. D. Damuth, W. A. DiMichele, R. Potts, H.-D. Sues, and S. L. Wing, 419–541. Chicago: University of Chicago Press.

Prescott-Allen, R., and C. Prescott-Allen. 1990. "How Many Plants Feed the World?" *Conservation Biology* 4: 365–74.

Proctor, M. C. F. 2000. "Physiological Ecology." In *Bryophyte Biology,* edited by A. J. Shaw and B. Goffinet. Cambridge: Cambridge University Press.

Pryer, K. M., H. Schneider, A. R. Smith, R. Cranfill, P. G. Wolf, J. S. Hunt, and S. D. Sipes. 2001. "Horsetails and Ferns Are a Monophyletic Group and the Closest Living Relatives to Seed Plants." *Nature* 409: 618–21.

Pryer, K. M., E. Schuettpelz, P. G. Wolf, H. Schneider, A. R. Smith, and R. Cranfill. 2004. "Phylogeny and Evolution of Ferns (Monilophytes) with a Focus on the Early Leptosporangiate Divergences." *American Journal of Botany* 91: 1582–98.

Pryer, K. M., A. R. Smith, and J. E. Skog. 1995. "Phylogenetic Relationships of Extant Pteridophytes Based on Evidence from Morphology and rbcL Sequences." *American Fern Journal* 85: 205–82.

Qiu, Y.-L., J. Lee, F. Bernasconi-Quadroni, D. Soltis, P. Soltis, M. Zanis, E. Zimmer, Z. Chen, V. Savolainen, and M. Chase. 1999. "The Earliest Angiosperms: Evidence from Mitochondrial, Plastid and Nuclear Genomes.". *Nature* 402: 1403–11.

Qiu, Y.-L., L. Li, B. Wang, Z. Chen, V. Knoop, M. Groth-Malonek, O. Dombrouska, J. Lee, L. Kent, J. Rest, et al. 2006. "The Deepest Divergences in Land Plants Inferred from Phylogenetic Evidence." *Proceedings of the National Academy of Sciences USA* 103: 15511–16.

Rasmussen, B. 2000. "Filamentous Microfossils in a 3,235-Million-Year-Old Volcanogenic Massive Sulphide." *Nature* 405: 676–79.

Raven, J., and P. Crane. 2007. "Trees." *Current Biology* 17: 303–4.

Remy, W., P. G. Gensel, and H. Haas. 1993. "The Gametophyte Generation of Some Early Devonian Land Plants." *International Journal of Plant Science* 154: 35–58.

Remy, W., and R. Remy. 1980a. "Devonian Gametophytes with Anatomically Preserved Gametangia." *Science* 208: 295–96.

Remy, W., and R. Remy. 1980b. "*Lyonophyton rhyniensis* nov. gen. et nov. sp., ein Gametophyt aus dem Chert von Rhynie (Unterdevon, Schottland)." *Argumenta Palaeobotanica* 6: 37–72.

Renzaglia, K. S., T. H. Johnson, H. D. Gates, and D. P. Whittier. 2001. "Architecture of the Sperm Cell of *Psilotum.*" *American Journal of Botany* 88: 1151–63.

Renzaglia, K. S., and K. C. Vaughn. 2000. "Anatomy, Development, and Classification of Hornworts." In *Bryophyte Biology*, edited by A. J. Shaw and B. Goffinet. Cambridge: Cambridge University Press.

Ribeiro, S., and G. B. Golding. 1998. "The Mosaic Nature of the Eukaryotic Nucleus." *Molecular Biology and Evolution* 15: 779–88.

Riding, R. 1992. "The Algal Breath of Life." *Nature* 359: 13–14.

Roger, A. J. 1999. "Reconstructing the Early Events in Eukaryotic Evolution." *American Naturalist* 154: 146–63.

Rothschild, L. J., and R. L. Mancinelli. 2001. "Life in Extreme Environments." *Nature* 409: 1092–101.

Rothwell, G. W. 1979. "Evidence for a Pollination-Drop Mechanism in Paleozoic Pteridosperms." *Science* 198: 1251–52.

Rothwell, G. W. 1981. "The Callistophytales (Pteridospermopsida): Reproductively Sophisticated Gymnosperms." *Review of Palaeobotany and Palynology* 37: 7–28.

Rothwell, G. W. 1987. "The Role of Development in Plant Phylogeny: A Paleobotanical Perspective." *Review of Palaeobotany and Palynology* 50: 96–114.

Rothwell, G. W. 1999. "Fossils and Ferns in the Resolution of Land Plant Phylogeny." *Botanical Review* 65: 188–218.

Rothwell, G. W., W. Crepet, and R. A. Stockey. 2009. "Is the Anthophyte Hypothesis Alive and Well? New Evidence from the Reproductive Structures of Bennettitales." *American Journal of Botany* 96: 296–322.

Rothwell, G. W., and E. E. Karrfalt. 2008. "Growth, Development, and Systematics of Ferns: Does *Botrychium* s.1. (Ophioglossales) Really Produce Secondary Xylem?" *American Journal of Botany* 95: 414–23.

Rothwell, G. W., and S. E. Scheckler. 1988. "Biology of Ancestral Gymnosperms." In *Origin and Evolution of Gymnosperms*, edited by C. B. Beck, 85–134. New York: Columbia University Press.

Rothwell, G. W., and R. Serbet. 1994. "Lignophyte Phylogeny and the Evolution of Spermatophytes: A Numerical Cladistic Analysis." *Systematic Botany* 19: 443–82.

Rothwell, G. W., and R. A. Stockey. 1994. "The Role of *Hydropteris pinnata* gen. et sp. nov. in Reconstructing the Cladistics of Heterosporous Ferns." *American Journal of Botany* 81: 479–92.

Rouffa, A. S. 1978. "On Phenotypic Expression, Morphogenetic Pattern and Synangium Evolution in *Psilotum*." *American Journal of Botany* 65: 692–713.

Royer, D. L., I. M. Miller, D. J. Peppe, and L. J. Hickey. 2010. "Leaf Economic Traits from Fossils Support a Weedy Habitat for Early Angiosperms." *American Journal of Botany* 97: 438–45.

Rudall, P. J., M. V. Remizowa, G. Prenner, C. J. Prychid, R. E. Tuckett, and D. D. Sokoloff. 2009. "Nonflowers near the Base of Extant Angiosperms? Spatiotemporal Arrangement of Organs in Reproductive Units of Hydatellaceae and Its Bearing on the Origin of the Flower." *American Journal of Botany* 96: 67–82.

Saarela, J. M., H. S. Rai, J. A. Doyle, P. K. Endress, S. Mathews, A. D. Marchant, B. G. Briggs, and S. W. Graham. 2007. "Hydatellaceae Identified as a New Branch near the Base of the Angiosperm Phylogenetic Tree." *Nature* 446: 312–15.

Scagel, R. E., R. J. Bandoni, J. R. Maze, G. E. Rouse, W. B. Schofield, and J. R. Stein. 1984. *Plants: An Evolutionary Survey*. Belmont, CA: Wadsworth.

Scharaschkin, T., and J. A. Doyle. 2006. "Character Evolution in *Anaxagorea* (Annonaceae)." *American Journal of Botany* 93: 36–54.

Schidlowski, M. 1988. "A 3,800-Million-Year Isotopic Record of Life from Carbon in Sedimentary Rocks." *Nature* 333: 313–18.

Schneider, E. L., and S. Carlquist. 1999. "SEM Studies of Vessels in Ferns, 11: *Ophioglossum*." *Botanical Journal of the Linnaean Society* 129: 105–14.

Schneider, E. L., and S. Carlquist. 2000. "SEM Studies of Vessels in Ferns, 17: Psilotaceae." *American Journal of Botany* 87: 176–81.

Schneider, H., E. Schuettpelz, K. M. Pryer, R. Cranfill, S. Magallón, and R. Lupia. 2004. "Ferns Diversified in the Shadow of Angiosperms." *Nature* 428: 553–57.

Schopf, J. W. 1993. "Microfossils of the Early Archean Apex Chert: New Evidence of the Antiquity of Life." *Science* 260: 640–46.

Schopf, J. W. 1994. "The Oldest Known Records of Life: Early Archean Stromatolites, Microfossils, and Organic Matter." In *Early Life on Earth*, edited by S. Bengtson. New York: Columbia University Press.

Schopf, J. W. 1999. *Cradle of Life*. Princeton: Princeton University Press.

Schopf, J. W. 2000. "Solution to Darwin's Dilemma: Precambrian Record of Life." *Proceedings of the National Academy of Sciences* 97(13): 6947–53.

Schopf, J. W. 2002. "When Did Life Begin?" In *Life's Origin*, edited by J. W. Schopf, 158–79. Berkeley: University of California Press.

Schopf, J. W., and E. S. Barghoorn. 1967. "Algal-Like Fossils from the Early Precambrian of South Africa." *Science* 165: 508–12.

Schuster, R. 1966. *The Hepaticae and Anthocerotae of North America East of the Hundredth Meridian*. New York: Columbia University Press.

Schwendemann, A. B., G. Wang, M. L. Mertz, R. T. McWilliams, S. L. Thatcher, and J. M. Osborn. 2007. "Aerodynamics of Saccate Pollen and Its Implications for Wind Pollination." *American Journal of Botany* 94: 1371–81.

Searcy, D. G., and D. B. Stein. 1980. "Nucleoplasm Subunit Structure in an Unusual Prokaryotic Organism: *Thermoplasma acidophilum*." *Biochimica et Biophysica Acta* 609: 108–95.

Searcy, D. G., D. B. Stein, and G. R. Green. 1978. "Phylogenetic Affinities between Eukaryotic Cells and a Thermoplasmic Mycoplasma." *BioSystems* 10: 19–28.

Searles, R. B. 1980. "The Strategy of the Red Algal Life History." *American Naturalist* 115: 113–20.

Seilacher, A., P. K. Bose, and F. Pflüger. 1998. "Triploblastic Animals More than 1 Billion Years Ago: Trace Fossil Evidence from India." *Science* 282: 80–83.

Selosse, M.-A. 2005. "Are Liverworts Imitating Mycorrhizas?" *New Phytologist* 165: 345–49.

Selosse, M.-A., and F. Le Tacon. 1998. "The Land Flora: A Phototroph-Fungus Partnership?" *Trends in Ecology and Evolution* 13: 15–20.

Sephton, M. A. 2001. "Life's Sweet Beginnings." *Nature* 414: 857–58.

Serbet, R., and G. W. Rothwell. 1999. "*Osmunda cinnamomea* (Osmundaceae) in the Upper Cretaceous of Western North America: Additional Evidence for Exceptional Species Longevity among Filicalean Ferns." *International Journal of Plant Science* 160: 425–33.

Seward, A. C. 1933. *Plant Life through the Ages*. New York: Hafner.

Sharma, M. V., and J. E. Armstrong. 2013. "Pollination of *Myristica* and Other Nutmegs in Natural Populations." *Tropical Conservation Science* 6: 592–605.

Shaw, J., and K. Renzaglia. 2004. "Phylogeny and Diversification of Bryophytes." *American Journal of Botany* 91: 1557–81.

Shen, Y., R. Buick, and D. E. Canfield. 2001. "Isotopic Evidence for Microbial Sulphate Reduction in the Early Archaean era." *Nature* 410: 77–81.

Shephard, P. 1998. *Coming Home to the Pleistocene*. Washington, DC: Island.

Shou-Gang, H., and P. G. Gensel. 2001. "The Posongchong Floral Assemblage of Southeastern Yunnan, China—Diversity and Disparity in Early Devonian Plant Assemblages." In *Plants Invade the Land: Evolutionary and Environmental Perspectives*, edited by P. G. Gensel and D. Edwards, 103–19. New York: Columbia University Press.

Shultz, C., D. P. Little, D. W. Stevenson, A. Nowogrodzki, and D. Paquiot. 2010. "Growth and Care Instructions of a New Model Species of Lycophyte *Selaginella apoda*." *American Fern Journal* 100: 167–71.

Siegert, A. 1967. "Morphologische, enwicklungsgeschichtliche und systematische Studien an *Psilotum triquetrum* Sw. III. Das Blatt aus der Sicht der Homolgien." *Beitrage zur Biologie der Pflanzen* 43: 285–328.

Simpson, G. G., and W. S. Beck. 1965. *Life: An Introduction to Biology*. 2nd ed. New York: Harcourt, Brace and World.

Smith, A. R., K. M. Pryer, E. Schuettpelz, P. Korall, H. Schneider, and P. G. Wolf. 2006. "A Classification for Extant Ferns." *Taxon* 55: 705–31.

Sogin, M. L. 1994. "The Origin of Eukaryotes and Evolution into Major Kingdoms." In *Early Life on Earth*, edited by S. Bengtson, 181–92. New York: Columbia University Press.

Sogin, M. L. 1997. "Organelle Origins: Energy-Producing Symbionts in Early Eukaryotes." *Current Biology* 7: 315–17.

Sournia, A. 1982. "Form and Function in Marine Phytoplankton." *Biological Review* 57: 347–94.

Spence, J. R., and W. B. Schofield. 2007. "Takakiaceae." *Flora of North America* 27: 42–44.

Stechmann, A., and T. Cavalier-Smith. 2002. "Rooting the Eukaryote Tree by Using a Derived Gene Fusion." *Science* 297: 89–91.

Steemans, P., A. Le Hérissé, J. Melvin, M. A. Miller, F. Paris, J. Verniers, and C. H. Wellman. 2009. "Origin and Radiation of the Earliest Vascular Land Plants." *Science* 324: 353.

Stein, W. E., F. Mannolini, L. V. Hernick, E. Landing, and C. M. Berry. 2007. "Giant Cladoxylopsid Trees Resolve the Enigma of the Earth's Earliest Forest Stumps at Gilboa." *Nature* 446: 904–7.

Stein, W. E., D. C. Wight Jr., and C. B. Beck. 1984. "Possible Alternatives for the Origin of the Sphenopsida." *Systematic Botany* 9: 102–18.

Stevenson, D. W., and H. Loconte. 1996. "Ordinal and Familial Relationships of Pteridophyte Genera." In *Pteridology in Perspective*, edited by J. M. Carmus, M. Gibby, and R. J. Johns, 435–67. Kew (UK): Royal Botanic Gardens.

Stewart, W., and G. Rothwell. 1993. *Paleobotany and the Evolution of Plants*. 2nd ed. Cambridge: Cambridge University Press.

Stossel, T. P. 1990. "How Cells Crawl." *American Scientist* 78: 408–23.

Sun, G., Q. Ji, D. L. Dilcher, S. Zheng, K. C. Nixon, and X. Wang. 2002. "Archaefructa-ceae, a New Basal Angiosperm Family." *Science* 296: 899–904.

Tang, C. Q., Y. Yang, M. Ohsawa, S.-R. Yi, A. Momohara, W.-H. Su, H.-C. Wang, Z.-Y. Zhang, M.-C. Peng, and Z.-L. Wu. 2012. "Evidence for the Persistence of Wild *Ginkgo biloba* (Ginkgoaceae) Populations in the Dalou Mountains, Southwestern China." *American Journal of Botany* 99: 1408–14.

Taylor, D. W., and G. Kirchner. 1996. "The Origin and Evolution of the Angiosperm Carpel." In *Flowering Plant Origin, Evolution, and Phylogeny*, edited by D. W. Taylor and L. J. Hickey, 116–40. New York: Chapman and Hall.

Taylor, T. N., H. Kerp, and H. Hass. 2005. "Life History Biology of Early Land Plants: Deciphering the Gametophyte Phase." *Proceedings of the National Academy of Sciences* 102: 5892–97.

Terry, L. I., G. H. Walter, J. S. Donaldson, E. Snow, P. I. Forster, and P. J. Machin. 2005. "Pollination of Australian *Macrozamia* Cycads (Zamiaceae): Effectiveness and Behavior of Specialist Vectors in a Dependent Mutualism." *American Journal of Botany* 92: 931–40.

Thien, L. J. 1974. "Floral Biology of *Magnolia*." *American Journal of Botany* 61: 1037–45.

Thomé, O. W. 1885. *Flora von Deutschland, Österreich, und der Schweiz.* Gera, Germany: Untermhaus.

Tomescu, A. M. F., and G. W. Rothwell. 2006. "Wetlands before Tracheophytes: Thalloid Terrestrial Communities of the Early Silurian Passage Creek Biota (Virginia)." *Geological Society of America Special Paper* 399: 41–56.

Tomescu, A. M. F., G. W. Rothwell, and R. Honegger. 2006. "Cyanobacterial Macrophytes in an Early Silurian (Llandovery) Continental Biota: Passage Creek, Lower Massanutten Sandstone, Virginia, USA." *Lethaia* 39: 329–38.

Trainor, F. R., and P. F. Egan. 1988. "The Role of Bristles in the Distribution of a *Scenedesmus*." *British Phycological Journal* 23: 135–41.

Tucker, M. E., and V. P. Wright. 1990. *Carbonate Sedimentology.* Cambridge: Blackwell Scientific Books.

Tudge, C. 2000. *The Variety of Life: A Survey and a Celebration of All the Creatures That Have Ever Lived.* Oxford: Oxford University Press.

Tunnicliffe, V. 1992. "Hydrothermal-Vent Communities of the Deep Sea." *American Scientist* 80: 336–49.

Tyler, S. A., and E. S. Barghoorn. 1954. "Occurrence of Structurally Preserved Plants in Pre-Cambrian Rocks of the Canadian Shield." *Science* 119: 606–8.

Van den Ent, F., L. Amos, and J. Löwe. 2001. "Bacterial Ancestry of Actin and Tubulin." *Current Opinions in Microbiology* 4: 634–38.

Van der Peer, Y., and R. de Wachter. 1997. "Evolutionary Relationships among the Eukaryote Crown Taxa Taking into Account Site-to-Site Rate Variation in 18S rRNA." *Journal of Molecular Evolution* 45: 619–30.

Van der Peer, Y., S. L. Baldauf, W. F. Doolittle, and A. Myer. 2000. "An Updated and Comprehensive rRNA Phylogeny of (Crown) Eukaryotes Based on Rate-Calibrated Evolutionary Distances." *Journal of Molecular Evolution* 51: 565–76.

Van der Staay, S. Y. M., R. De Wachter, and D. Vaulot. 2001. "Oceanic 18S rDNA Sequences from Picoplankton Reveal Unsuspected Eukaryote Diversity." *Nature* 409: 607–10.

Vitt, D. H., J. E. Marsh, and R. B. Bovey. 1988. *Mosses, Lichens and Ferns of Northwest North America*. Edmonton (AB): Lone Pine.

Wacey, D., M. R. Kilburn, M. Saunders, J. Cliff, and M. D. Brasier. 2011. "Microfossils of Sulphur-Metabolizing Cells in 3.4-Billion-Year-Old Rocks of Western Australia." *Nature Geoscience* 4: 698–702.

Wandersee, J. H., and E. E. Schussler. 1999. "Preventing Plant Blindness." *American Biology Teacher* 61: 82–86.

Waterbury, J. B., and R. Y. Stanier. 1981. "Isolation and Growth of Cyanobacteria from Marine and Hypersaline Environments." In *The Prokaryotes*, edited by M. P. Starr, H. Steele, H. G. Truper, A. Balows, and H. G. Schlegel, 221–23. Berlin: Springer Verlag.

Waters, D. A., and R. L. Chapman. 1996. "Molecular Phylogenetics and the Evolution of Green Algae and Land Plants." In *Cytology, Genetics, and Molecular Biology of Algae*, edited by B. R. Chaudhary and S. B. Agrawal, 337–50. Amsterdam: SPB.

Wellman, C. H., and J. Gray. 2000. "The Microfossil Record of Early Land Plants." *Philosophical Transactions of the Royal Society B* 355: 717–32.

Wellman, C. H., P. L. Osterloff, and U. Mohluddin. 2003. "Fragments of the Earliest Land Plants." *Nature* 425: 282–85.

Whittaker, R. H. 1969. "New Concepts of Kingdoms of Organisms." *Science* 163: 150–60.

Whittet, D. C. B. 1997. "Is Extraterrestrial Organic Matter Relevant to the Origin of Life on Earth?" *Origin of Life and the Biosphere* 27: 249–62.

Wikimedia Creative Commons. Illustrations and images attributed to this source are used under the provisions of the Attribution-ShareAlike 3.0 license. http://creativecommons/licenses/by-sa/3.0/.

Williams, B. S. 1868. *Select Ferns and Lycopods: British and Exotic*. London: B. S. Williams.

Williams, J. H., and W. E. Friedman. 2002. "Identification of Diploid Endosperm in an Early Angiosperm Lineage." *Nature* 415: 522–26.

Williams, J. H., and W. E. Friedman. 2004. "The Four-Celled Female Gametophyte of *Illicium* (Illiciaceae; Arustrobaileyales): Implications for Understanding the Origin and Early Evolution of Monocots, Eumagnoliids, and Eudicots." *American Journal of Botany* 91: 332–51.

Wilson, E. B. 1900. *The Cell in Development and Inheritance*. 2nd ed. New York: Macmillan.

Wilson, G. W. 2002. "Insect Pollination in the Cycad Genus *Bowenia* Hook. *ex* Hook. f. (Stangeriaceae)." *Biotropica* 34: 438–41.

Winsor, J. A., S. Peretz, and A. G. Stephenson. 2000. "Pollen Competition in a Natural Population of *Cucurbita foetidissima* (Curcurbitaceae)." *American Journal of Botany* 87: 527–32.

Woese, C. R. 1998. "The Universal Common Ancestor." *Proceedings of the National Academy of Sciences* 95: 6854–59.

Woese, C. R., and G. E. Fox. 1977. "Phylogenetic Structure of the Prokaryotic Domain: The Primary Kingdoms." *Proceedings of the National Academy of Sciences* 74: 5088–90.

Woese, C. R., O. Kandler, and M. L. Wheelis. 1990. "Towards a Natural System of Organisms: Proposal for the Domains Archaea, Bacteria, and Eucarya." *Proceedings of the National Academy of Sciences* 87: 4575–79.

Wolozin, B. 2007. "The Art of Persuasion in Politics (and Science)." *Skeptical Inquirer* 31: 15–17.

Zhang, Z., B. R. Green, and T. Cavalier-Smith. 2000. "Phylogeny of Ultra-Rapidly Evolving Dinoflagellate Chloroplast Genes: A Possible Common Origin for Sporozoan and Dinoflagellate Plastids." *Journal of Molecular Evolution* 51: 26–40.

The letter *f* following a page number denotes a figure, and bold font denotes a glossary definition.

abiotic, **523**; dispersal, 7; origin, synthesis (of organic molecules), 26, 29, 45, 71–72

aerobic respiration, 47–48, 50–51, 53–56, 55f, 61, 63f, 71, 82–83, **523**, 527

alcohol, 54, 369; ethanol, 523–24. *See also* fungi; respiration: anaerobic

algae, 5, 10–11, chap. 4 (131–42). *See also* brown algae; green algae; phytoplankton; red algae; seaweed

allele, 121, 190, **523**; blood type, 121

alternation of generations, 125, 178–79, 179f, 183–84, 187, 192, 193, **523**; in Brown algae, 371f, 372f; in Red algae, 490–92, 492f; in seed plants, 270–71, 270f

amino acid(s), 9, 16, 26–27, 35, 508, 520

ammonia, 25, 27, 57, 62

ancient: earth, 26, 44, 48, 120; history, 4, 8, 36, 501; life, 31–32, 35–40, 67, 71, 106, 165

ancient rocks, 21–22, 37; Apex chert, 38, 42; Strelley Pool Formation, 38

angiosperm. *See* flowering plants

Antarctica, 4, 28, 288–89, 343, 495, **526**

antheridia/archegonia, 193, **523**; clubmoss, 376, 378; cycad, 406; development of, 194–95; fern, 80, 417f, 419, 429; hornwort, 455; horsetail, 460; liverwort, 209, 467, 471–72, 474; moss, 199f; whisk fern, 502, 502f, 506

antimatter, 23–24

apical growth. *See* meristem: apical

archaea, archaebacteria, 43, 51, 68–69, 85, 94, 98–100, 105, 112, 114, 128, **524**

Archaeanthus, 318–19, 319f

Archaefructus, 315f, 318–20

Archaeopteris, 229–31, 277, 514

Archaeoptryx, 516

archegonia. *See* antheridia/archegonia

archezoan hypothesis, 106–7

asexual reproduction, 13, 117–19, 140–41, 145, 147, 152–53, 428, 443–44, 467–68, **524**; liverwort, gemmae, 468f; in red algae, 491. *See also* clone

ATP/ADP (adenosine tri- [di-]phosphate), 53–63, 509, **523**, **524**

autogenesis, 104–5, 107, 110–12, 128, **524**

autotroph(y), 46–47, 50, 57, 59, 63f, 71, **524**; C_4 photosynthesis, 346; chemoautotroph, 45, 57, 69, 71, 129, **524**; *Cooksonia*, 186; fungi, 86, 165; green photosynthetic, 6, 8, 66, 73; origin of, 45, 57–58, 71–72, 94; photoautotroph, 45,

autotroph(y) (*cont.*)
60, 83, **528**; phytoplankton, 142, 441,
485; prevalence of green, 66; producer,
44, 72, **529**; purple photosynthetic
(bacteria), 62; thermoautotroph, 65. *See
also* lichen; photosynthesis; rubisco

bacteria: *Acetobacter*, 53; gram – (negative),
59, 67, 69–71, **525**; gram + (positive),
59, 61, 67, 69–71, 451; green non-sulfur,
8, 55, 59–61, 63, 69–71, 84, 451; green
sulfur, 8, 58–59, 63, 65, 71, 84, 451–52;
heliobacteria, 59, 61, 63, 69, 70–71, 451;
Lactobacillus, 32, 54; purple/proteobac-
teria, 59, 62, 70, 94, 105; spirochetes,
67, 70, 110–11; *Thermus*, 70, 70f. *See
also* cyanobacteria; Monera/monerans
beer, 31–32, 259, 346, 357
binary fission, 83, 90, 92, 100, 107, 113–13f,
115, 122
biochemical: signature of life, 28, 37, 39;
pathway, 58
biosphere, 42, 134, 339, 513
birds, aves, 13, 163, 227, 263, 295, 303–4,
310–11, 361, 390, 394, 516–17, 522
bread, 31–32, 337, 346, 357, 361; green
bread mold (*Penicillium*), 447
brown algae, 367–72; *Dictyota*, 371; *Ectocarpus*, 193, 369–72, 371f; *Ectocarpus* life
cycle, 371f; *Fucus*, 369–70, 370f; *Laminaria*, 368–72; *Laminaria* life cycle, 372f;
Macrocyctis, 368f; *Postelsia*, 368f, 370;
Sargassum, 133, 369–70
bryophyte, clade, 213. *See also* hornworts;
liverworts; moss

cambium: cork, 242; vascular, 239–42, 464.
See also meristem; wood
Cambrian period, 20–21, 36–37, 133, 146,
158–59, 225, 484, 519, **528**
Cretaceous, Gondwana, 283f
Devonian, 21f, 171f, 174–76, 187f, chap. 8
(217–64); clubmoss, 373, 381, 383; fern,
409, 411, 419–20, 420f; horsetail, 457,
461, 469; rhyniophyte, 501; seed, 266,
269, 274, 276, 278, 281, 284–85, 285f;
woody plants, 512
explosion, 158–59
K-T boundary, 21f, 339, 342, 401, 517, **527**
Tertiary period, 20–21, 280, 284–85,
chap. 11 (337–62), 411, 486, **527**
carbon-fixing. *See* enzyme(s): rubisco

Carboniferous period, 21, 217, 224–25, 285,
307, 385
cell division, 510; apical cell, 153, 155, 157f,
231–32, 254–55, 255f, 321, 470; asexual
reproduction, 117, 119, 141f; binary
fission, 83, 90, 92, 100, 107, 112–13,
115, 122; in Brown algae, 368; closed
mitosis, 441, 445f; cytokinesis, 91, 113,
115–16, 434, 441, 445f, 450, 510, 513,
525, 528; development, 34, 80, 153,
193, 226, 229, 231; land plants, 109f,
181, 204; in meristems, 237; microtubules, 109–10, 109f; multinucleate
cells, 116; protein ftsZ, 111–13. *See also*
mitosis
centrioles, 109–10, 441, 448. *See also* MTOC
cheese, 31
chloroplasts: evolution, 63, 78, 94–111, 115,
128, 142, 146, 368, 482, 484–85; illustration, 92f. *See also* endosymbiosis
clades, 19, **525**; animal, 19; anthophyte,
286–88, 436; bacteria, 69; bikont, 84–
86, 107, 350, 485; bryophyte, 213; chromalveolate (*see* stramenopiles); conifer,
286–88, 436; embryophyte/land plant,
5, 9–10, 17, 86, 176, 181–82, 199, 258,
350, 439, 441, 449; eudicot, 351–53;
eukaryote, 69, 84–85, 118, 350; fern,
412, 425, 430, 505; flowering plant,
332; fungi, 19; gram –/+ bacteria, 69–71;
green algae, 460–61, 446; gymnosperm,
284, 287; ligniophyte/woody plants,
278, 285; magnolialean, 241, 314, 325,
350, 352–54, 357; megaphyllous, 251,
253, 258–60, 350, 515; moniliophyte
(*see* pteridophytes); monocot, 352–53;
moss, 257; opisthokont, 84–85, **528**;
plant, 19, 85, 88–89, 102–3, 484; polysporangiophyte, 233, 300; prokaryote,
69; protist, 19; pteridophyte, 259–60,
331, 412, 505; spermatophyte / seed
plant, 284–87; stramenopile, 86–88,
142, 369, 482–85; streptophyte, 181–82,
350, 439. *See also* clubmosses; Monera/
monerans
cladistics, 18–19, **525**
classification: five kingdom, 10–12, 17–19,
508; phylum, 10, 14–15, 17, 87, 200,
210–11, 348–50, 465, 508, 525; plant
kingdom, 1, 10–12, 15, 88, 211, 349,
484–85; taxonomic hierarchy, 19, 89,
211

clone, 117–19, 141, 153, 205, 227–29, 425, 466, 473, **525**

clubmosses, extant, 258, 374–76; *Diphasiastrum*, 373–74; *Huperzia*, 373–74; *Isoetes*, 252, 328, 373–76, 379–80, 380f, 385, 520; *Lycopodiella*, 373–74; *Lycopodium*, 373–74, 376–77, 382, 384f, 460; *Phalhinhaea*, 373; *Phlegmariurus*, 373–74, 375f; *Phylloglossum*, 373, 375; *Pseudolycopodiella*, 374; *Selaginella*, 245, 258, 266, 275, 278–79, 376–79, 377f, 381, 385; *Selaginella* life cycle, 379f

clubmosses, fossil: *Lepidodendron*, 383–85, 384f; *Lepidophyllum*, 384; *Nathorstiana*, 385; *Pleuromeia*, 385; *Protolepidodendron*, 384–85; *Stigmaria*, 381, 383, 384f

clubmosses, stem group fossils: *Asteroxylon*, 219, 252, 259, 381–82, 382f, 383; *Baragwanathia*, 381; *Drepanophycus*, 381; *Nothia*, 381; *Renalia*, 501; *Sawdonia*, 220f, 381; *Zosterophyllum*, 222–25, 222f, 323f, 245f, 253, 259, 374, 381

COBE satellite, 24

comets, 25–26

common ancestry, 5, 8–9, 16, 29, 33, 36, 67, 81–82, 89; angiosperms and gymnosperms, 266, 276, 495–96 (*see also* seeds / seed plants); animals and fungi, 85f, 86; archaea and eukaryotes, 68, 106, 114; bacteria, 67–68; Brown algae and phytoplankton, 134, 369; bryophytes and vascular plants, 187, 199, 213–14, 257; chloroplasts and cyanobacteria, 94, 99; conifers, 390; cycads (extant), 401; embryophytes, 5, 9, 12, 49, 162, 167, 179–85, 255, 439, 449; eukaryotes, 77, 83, 84, 90, 93, 94, 100, 106, 484, 490; ferns, horsetails, whisk ferns (*see* pteridophytes); flowering plants, 20, 288, 316, 320, 331, 336, 351–52, 355, 436; flowering plants and gnetophytes, 286f, 280, 431 (*see also* seeds / seed plants); green algae, 182, 193, 450; kingdoms, 17–19, 76; life, 35, 69; megaphyll bearing plants, 251; mitochondria and purple bacteria, 94, 99; progymnosperms and seed plants, 277–78, 281, 284–87, 286f, 287f; pteridophytes, 258–61, 266, 331, 349, 390, 411–12, 412f, 419, 421, 426; pteridophytes and trimerophytes, 261; whisk ferns and adder's tongue ferns, 261. *See also* clades

community, 3, 121, 268, 304–5, 311, 327, 370, 396, 514; bog, 201; coastal, 163; forest, 227, 328, 406; microbial mat, 41; ocean, 136; prairie, 361; soil crust, 165–66, 166f

competition, **525**; CO_2/O_2, 346 (*see also* rubisco); coastal/patchy space, 41, 145, 149, 160, 269, 306; for food, 520; gamete, 192; light, 41, 146, 153, 154f, 306; pollen / floral displays, 302, 308

cones (strobilus), 374–75, 380, 403, 433, 436, **530**; cycad, 279, 403, 405f, 406; horsetail, 460, 463; megasporophyll, 291–92, 404, 406 (*see also* flowering plants: flower, carpel); microsporophyll, 284, 289–91, 390–91, 395–96, 399, 405 (*see also* flowering plants: flower, stamen); pollen cone, 27–280, 284, 299, 312, 390–96, 391f, 395f, 403–6, 405f, 432–36; seed cone, 274, 279–80, 284, 99, 303, 387, 389–90, 392f, 394, 400–408, 433–36, 434f, 435f

conifer, 5, 278–79, 285–87, 297, 387–94; broad-leafed foliage, 389f; clade, 287f; cones, pollen and seeds, 280, 284, 299, 389–92, 391f, 392f, 436; families, 279–80, 284, 392; female gametophyte, 333f; foliage/leaves, 277, 283, 289, 325, 389f, 432; forests, 304, 306, 344, 392; geological history, 284–85, 285f; male gametophyte, 276, 393f; pollen, 291, 305, 333, 393, 393f; wood, 240f, 433. *See also* seeds / seed plants: phylogeny

conifers, extant: *Agathis*, kauri pine, 388, 389f; *Araucaria*, 280, 284, 339, 388–91, 389f, 400; cedar (*Cedrus*), 388, 521; douglas-fir (*Pseudotsuga*), 390; hemlock (*Tsuga*), 279, 388; larch (*Larix*), 378, 388; *Metasequoia*, 292, 493; pine (*Pinus*), 279–80, 284, 355, 387–88, 394, 399; *Podocarpus*, 289f, 390; *Sequoia*, 75, 159f, 226, 280; *Taxodium*, 292

conjugation, 120–22, 444, 444f, 449–50, **525**; tube, 449–50

consumer, 99, 136–37, 140, 303, 395, **525**. *See also* heterotroph

Cooksonia, 174–75, 184, 186, 187f, 189, 252, 258, 493

cork (Bark), 34, 242, 515, 527

cosmology, 23, 29; dark energy, 23, 29; dark matter, 29; redshift, 24

Creationism, 49, 105–6, 337, 508–9. *See also* intelligent design

Cretaceous, 21f, chap. 10 (295–400), **527**; angiosperm pollen, oldest, 352; arborescent clubmoss, 385; cinnamon fern, oldest species, 425; cycads, 401; *Equisetum*, horsetails, 457, 464; fern diversity, 411, 413, 426, 430; field trip, 217; flowering plants, rise of, chap. 10 (295–336); gnetophytes, 280, 285f; mammals, 339. *See also* Cambrian period: K-T boundary; haptophytes: coccoliths

Cryptozoic, 21, 36, 38

cyanobacteria (blue-green algae), 5, 8, 38, 132, 451–54 (appendix), **524, 525**

cycads, 279, 285, 287, 299, 332, 355, 402–3, 521; *Bowenia*, 403; cones, 405f; *Cycas*, 279, 403, 404f, 405f, 406, 521; *Encephalartos*, 402f, 403; female gametophyte, 333f; geological history, 285f; *Lepidozamia*, 402; ovule, 407f; pollen, male gametophyte, 407f; sperm, 405f; *Stangeria*, 403; *Zamia*, 402–3, 405f, 405–6

Darwin, Charles: abominable mystery, 316, 518; common ancestry/descent, 8, 18–19, 107, 348–49, 421, 484, 490; descent with modification, xi, 5, 9–10, 35, 361; finches, 263; fitness, 118, 191; House, in Kew Gardens, 187; natural selection, 2, 343; quote, 1

descent with modification, xi, 5, 9–10, 35, 361

disease, 32–33, 98–99, 358, 395, 507, 520; infection, 32–33, 98

dispersal: abiotic, 4, 7; biotic, 3–4, 302, 304, 306, 321; early land plants, 166, 173, 186, 188; land plant life cycle, 178–80, 179f, 183, 190; pollen, spore, 86, 177, 190; seed, fruit, 3, 271–73, 300–303, 306, 321–22, 390, 394, 408, 432; wind, 301–2, 305, 307, 406, 490. *See also* zoospore

DNA, 508, 510; chloroplast, 103, 213; chromosomes, 34–35, 82–84, 91, 107, 112–14, 124–25, **525**; deoxyribose, 508; mitosis, 109–13; mutations, 16, 119; naked DNA, prokaryote, 90, 92, 105, 111, 127, **529**; repair, 120–22, 510; replication, 120–27; ribose, 53, 508; sequence data, 16, 90, 93, 99, 212–13; shared characters, 9, 35

domains, 68–69, 69f, 86, **524**

ecology, 6, 32, 80, 321, 361, 430; fire, 45; forest, 363; microbial, 32; ocean, chap. 4 (131–42)

electron transport, 54–63, 55f, 60f, 92

elements: atomic fusion, 29; helium, 24–25, 47, 369; hydrogen, 24–25, 45, 47, 50, 55–63, 65, 70, 73, 451–52, 509, **524**; nitrogen , 25–27, 47, 62, 152, 201, 322, 358, 450, 477, **524, 526**; sulfur, 25, 38–39, 43, 45, 56–57, 59, 62, 70, **524** (*see also* sulfur-based metabolisms). *See also* iron; nitrogen fixation; nitrogen reduction; oxygen

endosymbiosis, 63, 78, 94–111, 115, 128, 142, 146, 368, 482, 484–85. *See also* chloroplasts: evolution

entropy, 29, 46, 507

enzyme(s), 46–47, 95, 97, 181, 484; glycolysis, 58; nitrogen fixing, 453; rubisco, 40, 48–49, 60, 66, 346, 440, 513, 519, **529**

eons, 20, 21f, 32, 159, 338–39, 341; Archean, 69, 85, 146, 159, **524**; Hadean Earth, 26–27, 45, 72, **526**; Phanerozoic, 20–21, 36, 159, 217, **528**; Proterozoic, 93, 116, 133, 146, 159, 162, 446, **529**

Euphrates, 31

evergreen, 279–80, 305, 307, 322, 326–27, 329, 376, 387–88, 518, 529, 521, **525**; foliage, 295, 303; forest canopy, 227

extremophile, 42, 44, 68–71, 129, **526**; hot springs, 38, 60, 218, 451–52

fermentation. *See* respiration: anaerobic

ferns, 177, 253, 259–61, 342, 406, 409–12; alternation of generation, 175–76, 176f; *Anemia*, 426; *Angiopteris*, 423–24, 424f; *Asplenium*, 416; *Azolla*, 342, 428, 429f; basic biology, 413–19; *Blechnum*, 411; *Botrychium*, 421–22, 422f, 426; cinnamon fern (*Osmunda cinnamomea*), oldest species on Earth, 261, 340, 402, 411–12, 421, 425, 515; *Dipteris*, 411; fossil, 229–30, 230f, 251; haploid-diploid, 175–78, 180, 267; life cycle, 417f; *Lygodium*, 426; *Marattia*, 281, 423–24, 424f; *Marsilea*, 427, 427f; *Matonia*, 411; *Onoclea*, 411; *Ophioglossum*, 421–22, 422f; *Pilularia*, 427f, 428; *Platycerium*, 414; *Polypodium*, 409, 430; *Regnellidium*, 427, 427f; *Salvinia*, 426, 428–29, 429f; sorus (sori), 410f, 415f, **529**; sporangium (lepto-), 410f, 418f; spores, 268; *Stromatopteris*, 416; tree ferns, 230, 249,

263, 281, 409, 411–13, 412f, 419–21, 425, 429–30

ferns, fossil: *Cladoxylon*, 419, 422f; *Eospermatopteris*, 229, 230f, 419; *Hydropteris*, 426, *Pecopteris*, 424f, 425; *Psaronius*, 425; *Pseudosporochnous*, 419

filament/filamentous: apical growth, 154–57, 154f, 156f, 157f; brown algae, 368–72, 372f; development, 81, 153–54, 157–58, 256; fungi, 94, 151, 165, 204, 490, 491f, 503, 511 (*see also* lichen); green algae, 131, 133, 148, 151, 162, 183, 204, 255f, 439–50 (appendix); intercalary growth, 155–57, 156f; organisms, 11, 78–79, 81; photosynthetic in liverworts, 468f, 471; prokaryotes, 38, 73, 81, 83, 152, 451–54 (appendix), 453f; protonema in mosses, 199f, 473; red algae, 88, 156, 162, 489–92 (appendix), 492f; rhizoids, 206, 379, 513; stamen, 313–14, 320, 510, 518

flowering plants, 9, 13–14, 84, 200, 202, 266, 287, chap. 10 (295–336), chap. 11 (337–62); ancestry, 20, 280, 431, 436, 499 (*see also* seeds / seed plants); basal flowering plants, 298–300; biology, 309–11; classification, 348–53, 351f, 352 (table 11–1), 353f; common ancestor, 288–90, 292; cone-flower similarity, 436; double fertilization, 308, 332; endosperm, 308, 332, 433; female gametophyte, 333f; flower, carpel (pistil), 289, 292, 299–301, 301f, 307–8, 312–15, 317–20, 319f, 355, 517, 518; flower, stamen, 299–300, 312–15, 314f, 315f, 317–20, 319f, 518; fruit, 300, 303; geological history / fossils, 4, 21, 29–30, 217, 266, 315f, 319f, 320, 339, 344, 401, 430, 496; grasslands, 344, 347–48; growth forms, 325–30, 347; human interactions, 4, 265, 295–96, 308, 346, 355, 357, 361; leaves, modified, 322, 323f, 324f; life cycle, 270f, 335f; monocot origins, 168, 355; oldest flower, 312–21 (see also *Archaefructus* [315f], *Archaeanthus* [319f]); ornamentals, 265; ovules, double-jacketed, 331–32, 437; pollination, 276, 291, 295, 301–7, 355, 357, 361; seed plant phylogeny, 286–87, 286f, 287f; teaching botany, 30; wood, 236–39, 236f, 238f, 241f, 330–31

food crisis, 72

fossil record: bacteria, 38–39, 41–42, 71, 133; clubmoss, 252–53, 259, 365, 373–75,

378, 381–85; fern, 251, 261, 277, 409, 411–12, 419, 421–26; flowering plants, 305, 312–15, 317–18, 321, 340, 352; horsetail, 261, 457, 461–64; land plant, 163, 166–67, 170–75, 185–87, 195, 198, 263, 415; pteridosperms, 284–85, 289, 291; seaweed, 21, 36, 133, 146, 158–59; seed plants, gymnosperms, 266, 274, 276–77, 282f, 285, 344; vascular plant, 201, 215, 218–20, 223, 229–30, 241, 246, 249–51, 257–59. See also clubmosses, stem group fossils: *Asteroxylon*; clubmosses, stem group fossils: *Zosterophyllum*; *Cooksonia*; microfossils

fungi, 34, 76, 78, 86, 95, 105f, 116, 118, 490; age of higher fungi, 165; asexual reproduction, 118; common ancestry, 12, 18, 85f, 86; conjugation, 121; decomposers, 6; kingdom, 10–12, 11f, 18–19, 18f, 19f, 86, 508; opisthokont clade, 85–86, 85f, **528**; plant kingdom, 6, 10; symbiosis, 86, 132, 165, 201, 224, 329; symbiosis with subterranean gametophytes, 376, 416, 419, 423, 503, 506. *See also* lichen

gametangia, 179–80, 193–94, 199, 208, 209f, 474, **526**; liverwort, 208–9, 209f, 467, 471–72. *See also* antheridia/archegonia

genetic: code, 9, 16, 35; information, 53, 190–91, 360–61, 520; isolation, 13, 19; material, 16, 34, 82, 90, 107, 110, 114–15, 120, 122, 128, 527, 528; studies, 104, 128, 154, 192–93, 214, 359; variation, 16, 85, 89, 100, 118–20, 124–25, 127–28, 145, 149, 155, 191–92, 195, 227–29, 232, 257, 326, 332, 348, 511. *See also* sex

geochemicals, 44, 50, 57, 224; biogeochemicals, 6, 45, 524

geothermal, 26, 28, 42–45, 71

Ginkgo, 266, 276, 278–81, 284–87, 292, 304, 312, 392, 394–97, 395f, 399, 495; geological history, 285f

glaucophyte, 78, 85f, 89, 95, 102, 105, 142, 481, 484

gliding motility, 77, 83, 108, 450, 454

Goldberg, Rube, 46

green: bacteria, 5, 8, 35, 39, 55, 58–65, 69–73, 81, 84, 108, 451–54 (appendix) (*see also* cyanobacteria); house, 310, 506, 516; -house effect/gases, 224, 342–43, 477; light, 62, 66; pigment, 66, 83, 86, 524; plants, 5, 34, 50, 66, 161,166,

green (*cont.*)
195, 207, 325; shoots, stems, tissue, 208, 218, 222, 246, 432, 458, 508; spores, 425, 460. *See also* terrestrial plants: chlorophyll

green algae, 88, 132, 439–50 (appendix); chloroplast diversity, 442; chloroplast origin, 13, 85, 88, 92f, 95, 101–3, 103f (*see also* endosymbiosis); colonial, 81, 139, 139f; conjugation, 121, 444, 444f; cytokinesis, 440–41, 440f; endolithic, 164; land plant ancestry, 9, 12, 17, 49, 86, 161–62, 167, 171, 179–83, 179f, 182f, 195, 204, 439, 455; large unicellular organism, 116; lichen, 165; life cycle, 175, 179f, 330f, 447; oogamy, 443, 446f; phytoplankton, 81, 139f, 142, 444f (*see also under* green algae: *Chlamydomonas, Micrasterias, Scenedesmus, Volvox*); plant kingdom, 10, 12, 76, 79, 85f, 88, 211, 441; plasmodesmata, 154 (*see also* green algae: cytokinesis); seaweeds, 133, 147, 446 (*see also under* green algae: *Acetabularia, Caulerpa, Ulva); Caulerpa,* 151f, 447–48; *Chara,* 162, 181, 214, 255, 335, 470, 449; *Chlamydomonas,* 192–93, 442–43; *Chlorella,* 49, 140, 155; *Chlorokybus,* 182, 445; *Cladophora,* 448; *Coleochaete,* 455; *Hydrodictyon,* 445; *Klebsormidium,* 182–83, 449; *Mesostigma,* 182, 442, 449; *Micrasterias,* 444f, 494; *Nitella,* 162, 181, 449; *Oedogonium,* 443, 445f, 446f; *Pediastrum,* 81, 138, 139f, 152, 445, 525; *Pencillium,* 447; *Scenedesmus,* 139f, 140, 445, 525; *Spirogyra,* 121, 444f, 449; *Ulothrix,* 448–49; *Ulva,* 157–58, 158f, 183–84, 204, 446–48; *Volvox,* 81, 139, 154, 443, 444f, 511; *Zygnema,* 450

gymnosperms. *See also* conifer; *Ginkgo*; wood: gnetophyte; pteridosperms

Haeckel, Ernst, 35, 151, 508
haptophytes, 78, 85, 87–88, 103, 133, 142, 481–84, 483f, 486; coccoliths, 483–84, 483f; white cliffs of Dover, 484
heterocyst, 39, 81, 152, 452–53, **526**
heterospory. *See also* microspores
heterotroph, 6, 8, 44–47, 59, 62, 77–79, **526**; consumer, 6, 95, **525**; decomposers, 6; facultative autotroph, 142; food chain evolution, 44–45, 72; fungi and animals, 86; microorganism, 95; zooplankton, 133. *See also* autotroph(y): producer

histone proteins, 83, 112
Hooke, Robert, 34, 75–76, 226
hornworts, chap. 7 (197–216), appendix 455–56; bryophyte classification, 199–200, 210–11, 465; early land plants, 5, 174, 198; gametophyte (haploid generation) / thallus, 198, 204, 232, 416; land plant phylogeny, 174, 211–15, 213f, 233, 258; life cycle, 180; sporophyte, 189, 198–99, 207–8, 208f, 212, 233–34, 456, 469
horsetails, 457–64, 459f, 462f, 463f (appendix); aborescent, 263f (*see also* horsetails, fossil: *Calamites*); eusporangia, 411, 416, 459f; geological history, 5, 21, 307, 340, 411; land plant phylogeny, 258–61, 258f, 260f, 411; leaf origin, 247, 250–51, 253, 255, 421, 505; life cycle, 178, 180, 267–69, 335; pteridophyte phylogeny, 412, 412f, 506
horsetails, extant: *Equisetum,* 255, 261, 269, 340, 457–58, 459f, 460–61, 464, 522; *Hippochaete,* 458, 522
horsetails, fossil: *Annularia,* 463–64, 463f; *Bowmanites,* 463; *Calamites,* 463f, 464; *Palaeostachya,* 463f; *Sphenophyllum,* 458, 460–64, 462f
hydrogel, 108, 454, **527**. *See also* gliding motility
hydrogen sulfide, 25, 56–57, 59–61, 65, 70, 451–52

intelligent design, 49, 66, 105
Iraq, 31
iron, 45, 50, 57, 59, 79

kingdom, 508, **527**; animal-fungi lineage, 85f, 86; chromista, 87, 367; cladistics, 19f, 349; classification, 9–13, 11f, 15f, 17–18, 18f, 211, 348, 350; eukaryotes, 85f, 89, 485; protists, 87; super-kingdom, domains, 68–69, 69f

land plants, 76, 88; ancestry, 9, 12, 49, 66, 148, 150, 161–62, 166–67, 180–84, 182f, 193; apical growth, 155; cell division, 109f, 181, 204 (*see also* mitosis; plasmodesmata); chloroplasts, 83, 101–3, 103f; dispersal, 170, 176, 190; embryophyte clade, 5, 9–10, 17, 86, 176, 181–82, 199, 258, 350, 439, 441, 449; eukaryote phylogeny, 85f, 174; fossils, 164–66, 174–75, 185 (see also *Cooksonia*); free-sporing, 259–60, 331, 412, 505; gametangia, 193–95 (*see also* antheridia/archegonia); geological history, 5, 21f, 30, 164, 198; hallmarks, 161, 170–75, 171f, 177, 202; invasion of land, chap. 6 (163–96); life cycle, 122, 125, 153, 175–85, 179f, 184f, 191–92 (*see also* alternation of generations); plant kingdom, 10–11, 14, 79, 88; sporophyte, benefit of, 188–92, 195; sporophyte, origin of, 181, 184f, 188. *See also* parenchyma; seeds / seed plant: habit

Leaven, 31

lichen, 86, 94, 132, 165, 195, 204, 229, 328, 448, 511, 513, 522

light: electromagnetic spectrum, 65–66; green wavelengths, 62, 66, 510; infrared (heat), 44, 62–63, 65, 511; ROYGBIV, 65; visible spectrum, 62, 64–66; UV, 66, 120, 146, 164, 450

lateral gene transfer, 56, 71, 104, 114

leaves. *See* megaphylls; microphylls; moss: enations, leafy organs

Leewenhoek, Anton van, 33

Linnaeus, 9, 14–15, 348, **527**

liverworts: adaptations, 243; apical growth, 232; appendix, 465–72, 467f; enations, leafy organs, 206f, 247, 252–53, 465–67, 467f, 470; geological history, 5, 166, 174–75; invasion of land, 165, chap. 7 (197–216), 205f, 206f; land plant phylogeny, 173–74, 211–15, 213f, 258; life cycle, 180, 189; sporophyte, 233, 467f; taxonomy, 209–10; *Marchantia*, 466–67, 468f, 470–71; *Pallaviciniites*, 469; *Riccia*, 468f, 470; *Ricciocarpus*, 205f, 470

megaphylls, 17, 250–53, 250f, 255, 258, 258f, 259–61, 263, 278, 280, 350; carpel, 317; conifer, 390, 396; ferns, 414, 420; horsetails, 253, 460–62; whisk ferns, 253, 258–59, 505

megaspore. *See* clubmosses; clubmosses, extant: *Selaginella* life cycle; cones; microspores; seeds / seed plants;

meiosis, 118, 123f, 124, **527**; brown algae life cycle, 371f, 372f; evolution of, 122, 126–27; evolution of seeds, 273–74, 275f, 278; functions in life cycles, 125, 126f, 176, 512; genetic variation, 125; independent assortment, 124, 191; plant life cycle, 173, 177–80, 179f, 183, 184f, 188–90, 226, 335, 406; produc-

ing plant spores, 186, 189, 208, 371, 378, 410, 416, 417f; red algae life cycle, 490–92, 492f; *Ulva* life cycle, 184f, 447, 449

meristem: apical, 239, 242, 248, 253–54, 270, 345, 380–81, 416, 424, 467; apical cell, 153, 155, 157, 194, 214, 231–32, 254–55, 255f, 371, 466, 470; intercalary, 207, 434, 456, 514; leaf development, 253–54, 415, 503. *See also* cambium

metabolisms/metabolic, chap. 2 (31–74), 27, 100, 509, **527**; aerobic respiration, 47–48, 51–56, 55f, 61, 63, 71, 82, 522; anaerobic respiration, 43, 47–48, 50–56, 60–61, 71–73, 99, 101, 115, 128, 452, **523**; ancient, 44; ATP pump, 57–59, 509; autotrophic (*see* metabolisms/ metabolic: chemoautotroph; metabolisms/metabolic: photoautotroph; metabolisms/metabolic: thermoautotroph; photosynthesis); bacterial/microbiological, 32–33, 67, 76, 83, 451–52; chemoautotroph, 45, 57, 69, 71, 129, **524**; components of, 46, 51–58, 60f, 61–62, 67, 72, 127; electron transport, 54–58, 61–63, 92; fermentation, 45, 47, 52, 56, 58, 63; glycolysis, 53, 55f, 56, 58, 83; Krebs cycle, 54, 83; pathways, 8, 45–47, 51, 55, 58; photoautotroph, 45, 57–58, 60, 62, 66, 71, 83, 98, 165, **528**; photosynthesis, **528**; pyruvate / pyruvic acid, 53–55, 58; respiration, **529**; thermoautotroph, 65; waste products of, 6, 37, 53, 56, 63, 67–68, 95, 99, 101, 150

meteor (-ite), 26, 28, 42, 339

methane, 27, 47, 56–57, 62, 68, 342, 524

microbial community, 32, 40–41, 133, 164–65, 195

microbial decomposition, 477

microbiologist, 32, 44, 67, 91

microfossils, 39, 41–42, 93, 133, 165, 174, 185, 213

microhabitat, 177, 200, 268, 327

microorganism, 6, 17, 31–35, 40, 43, 76, 80, 95, 108, 128, 185, 203, 357, 513, 519–20, **527**

microphylls, 251–55, 258, 374–75, 380–83, 385, 462, 502, 504–5

microscope: early microscopists, 33–34; electron, **525**; flagella/cilia, motile cells, 108, 147; hornwort chloroplasts, 455; microfossil, 38, 219; microorganism, microscope (*cont.*)

    76, 90, 98, 205, 454, 512, **527**; moss
    gametangia, 474; resolution, 34, 109;
    seed plant female gametophyte, 333f;
    seed plant megaspore, 272–73
microsporangium. *See* cones; microspores;
    pollen
microspores: clubmoss, 266, 376–84, 379f;
    heterospory, 266, 275f, 376, 516, **526**;
    pollen, 271, 273, 281, 334, 387, 393f,
    406–7, 407f; progymnosperms, 277,
    277f; seed habit, 272, 275f, 303; sta-
    mens, 314f, 320, 516
microtubules, 83, 91, 107–13, 117, 127, 181,
    440–41, 524; cytoskeleton, 77–78, 117;
    MTOC, 108–13, 127, 448; spindle fibers,
    108
mitochondria; 34, 83, 91, 92f, 104, 106,
    108, 110, 114–15, 527; aerobic respira-
    tion / citric acid (Krebs) cycle, 55–56,
    55f, 82; autogenesis, 105; bacterial
    symbiosis, 101; electron transport, 61;
    features, 92, 100–101; origin, 51, 85f,
    92, 94–105, 96f, 105f, 111, 128, 146 (*see
    also* endosymbiosis). *See also* archezoan
    hypothesis
mitosis, 82–83, 91, 124–25, 127, 179, 191,
    439, 510, 512, 527; autogenesis, origin
    of, 107, 112, 113f, 116; centrioles,
    448, 524 (*see also* MTOC); centromere,
    524 (*see also* MTOC); closed, 112, 441;
    MTOC, 108–10, 112, 127; stages of,
    109f. *See also* green algae: cytokinesis;
    meiosis
Monera/monerans, 10–12, 11f, 17–19, 18f,
    19f, 35, 69, 508, 527. *See also* archaea,
    archaebacteria; bacteria
moss, chap. 7: 197–216; appendix, 473–80;
    bog communities, 201–2, 201f, 202f;
    enations, leafy organs, 206, 206f, 243,
    253, 255f, 473–79, 474f, 478f; fossil,
    166; geological history, 5, 174–75;
    growth and development, 153; land
    plant phylogeny, 173, 199–200, 211–15,
    233, 257; life cycle, 180, 189, 199f, 205,
    210, 335; poikilohydric, 203, 243; spo-
    rophyte, 186, 189, 198–99, 199f, 207–8,
    231–32, 234, 242, 425; taxonomy, 210,
    425–26; thallus, 205, 479
Moss, genera: *Andreaea*, 475, 479; *Bauxbau-
    mia*, 477; *Polytrichum*, 474f, 479; *Sphag-
    num* (peat moss), 176, 206, 214–15,
    255f, 335f, 475–79, 478f; *Sphagnum*

communities distribution, 201, 202f;
    Takakia, 475–78, 476f; *Tetraphis*, 477
MTOC (microtubule organizing center),
    108–13, 113f, 122, 448, 524, **527**
mutation, 16, 64, 119, 127, 229, 514, **527**

nitrogen-fixation, 152, 201, 429, 452–53. *See
    also* heterocyst
nitrogen reduction, 62
nucleotides, 16, 35, 53, 68, 523, **524**;
    adenine, 35, 53, 508; cytosine, 35, 508;
    guanine, 35, 508; thymine, 35, 508

ontogeny (development), 34, 213, **528**;
    recapitulates phylogeny, 34
Oparin, A. I., 45
oxygen: in the atmosphere, 26–27, 44,
    47–50, 55, 61, 99, 146, 159–60; crisis,
    49–51, 115, 146; in living matter, 25;
    ozone , 120, 146, 160; produced by
    photosynthesis, 50, 59–63, 83, 453, 513;
    toxicity of, 47, 50–51

paleobotanist, 174, 217–18, 223, 245, 259,
    276, 316, 318, 336, 399
paleontologist, 37, 508, 522
Paleozoic, 21
panda's thumb, 66–67
parenchyma, 157–58, 165, 174, 181, 195,
    239, 347, 368, 388–89, 449, 455, **528**;
    block-constructed or block-organized
    tissue, 153, 162, 172, 194, 203–5, 234,
    237, 242, 244; pseudoparenchyma, 79,
    489; turgor, 234–35
Pasteur, Louis, 31
Pelomyxa (giant amoeba), 101, 106, 113–16;
    nuclear division, 113f
phagocytosis, 95–100, 96f, 107, 486, **528**.
    See also endosymbiosis
photosynthesis, **528**; age of, 42; bacterio-
    chlorophyll, 64; C_4, 346; carbon-fixing,
    48, 519 (*see also* enzyme(s): rubisco);
    carbon isotopes, 39; *Cooksonia*, 186;
    gene in parasite, 104; hydrogen source,
    50, 59–60, 70, 452; infared "light," 44,
    65; metabolic components, 51, 57, 60–
    63, 60f, 63f, 67–70; moss enations, 206;
    oxygenic, 48–62; parenchyma, 204, 244;
    productivity, 33, 127, 134–35, 143, 242;
    stems, 186, 329, 347, 432, 460, 504; suc-
    culents, 322, 324f; sugar, sunlight, 6. *See
    also* chloroplasts; light: visible spectrum;
    photosystems; starch

photosystems, 57, 59–61, 63, 63f, 70–71, 73, 452, 509; heliobacteria, 59, 61, 63f, 69–71, 451; *Oscillatoria*, 60, 108, 454; purple bacteria, 59, 62, 70, 94, 105

phytoplankton: chrysophtes, 483–84, 483f; dinoflagellates, 10, 87, 132–33, 142, 481–84, 482f; euglenozoa, 485; red tides, 483–84. *See also* haptophytes; straminopiles

pickles, 31

pigments: biliproteins, 101, 484; cytochromes, 56, 58, 63, 92; photosynthetic, 1, 6, 34, 57, 62, 65–66, 72, 83, 86; thermo-sensing, 65. *See also* porphyrin; terrestrial plants: chlorophyll

plasmodesmata, 154, 158, 162, 204, 368, 440–49, 440f

plasmodium, 87, 104, 116, **528**; malaria, 87

pollen: chamber, 275–76, 392, 394, 407–8, 498; drop, 275–76, 291, 392–93, 396, 406–7, 432–36, 498; grain, 271–72, 274–75, 291–92, 300–302, 307–8, 310, 392–93, 406, 408, 433, 498, 518; tube, 272–73, 275–76, 301, 308, 331, 335, 352, 393–94, 407, 434, 398. *See also* conifer: male gametophyte; microspores

porphyrin, 63–64, 64f

Precambrian, 21, 36–37, 41, 133, 446. *See also* eons: Archean

progymnosperm: *Archaeopteris*, 229–31, 277–78, 277f, 284–85, 514; *Callixylon*, 276–77

protein, 16, 25–26, 35, 47, 53–54, 59, 68, 83, 111, 150, 181, 355, 358, 485, 508, 520, **529**; actin / actin-binding, 97–98, 107–8, 128, 523; biliproteins, 101, 484; blood types, 121, **523**; cell division protein ftsZ, 111–13; histones, 82–83, 90–91, 112; tubulins, 83, 91, 107–11, **527** (*see also* MTOC). *See also* enzyme(s)

pteridophytes, 258–61, 260f, 266, 269, 297f, 331, 335f, 349, 390, 412f, 461, 505

pteridosperm (seed fern), 266, 279, 281–82, 282f, 284, 292, 495–99 (appendix); Caytoniales, 281–82, 285f, 289, 291–92, 495, 498; Cordaitales and Voltziales, 399–400; Cycadeoids (Bennettitales), 286–89, 290f, 317, 401; Czekanowskiales, 281, 285f, 289, 292, 495, 498–99; geographical distribution, 283f, 288; geological history, 281, 284–85, 285f; glossopterids, 281, 283f, 285–89, 285f, 287f, 289f, 292–93, 317, 325, 516; *Me-*

dullosa, 281–82, 282f, 495; Pentoxylales, 495–98. *See also* seeds / seed plants: phylogeny

QWERTY, 49, 509

red algae, 12, 76, 79, 132, 161–62, 489–92 (appendix); antiquity/fossils, 36, 490; chloroplasts, 13, 95, 101, 103f; coralloid, 490; life cycles, 490, 492f, 512; plant clade/phylogeny, 85f, 86, 88, 484; phytoplankton, 142, 156, 481; seaweeds, 79, 88, 132–33; source of agar/carrageen, 160, 490

red algae, genera: *Chondrus crispus* (Irish moss), 490; *Gelidium*, 490; *Iridaea*, 491f; *Nemalion*, 492f; *Polysiphonia*, 292; *Porphyra*, 491

resolution limit, 76

respiration, 34, 45, 47, 51–54, 57–58, 61, 70, 83, 99, 517; aerobic, 47–48, 51–56, 61, 63, 71, 82, 527, **529**; anaerobic (fermentation), 45, 47–48, 51–52, 54, 56, 58, 63

rhyniophytes, 493; *Baragwanathia*, 381–82, 493; *Cooksonia*, 174–75, 184, 186, 187f, 189, 252, 258, 493

ribosomes, '70s/'80s, 68–69, 83–84, 90, 98, 100, 105–6

RNA, 35, 53, 68, 99, 213, **529**. *See also* ribosomes

rock: bands (*see* rock: strata); igneous, 22; metamorphic, 37; moon, 23; oldest, 21–22, 36–38, 42, 133; sedimentary, 21, 164, 170; strata, 20–22, 36

rock moss, 490

rockweeds. *See* seaweed

Rocky Mountains, 344

rubisco, 40, 48–49, 60, 66, 346, 440, 513, 519, **529**

Samarkand, 31

science, 9, 36, 40, 65, 67, 72, 90, 104, 106, 159, 185, 192–93, 212, 361, 508–10, 516, 522, **529**; fiction, 27–28, 90, 337–38; taxonomy/classification, 5, 9–15, 348–49, **530**; teaching, 52, 178, 508–10, 518

sea water, 33, 43, 142, 168, 524

seaweed, 5, 11–12, 116, 125, 131–33, 163–69, 172, 178, 510–11, 522–23, 526, **529**; coastal ecology, 160–65, 167, 307; evolution, 41, 133, 146, 154, 160, 226, 344; fossil/geological history, 21, 36, 133,

seaweed (*cont.*)
158–60; freshwater, 162, 166; gels/hydration, 160, 166, 169; growth, development, form, 152–58, 154f, 156f, 157f, 204, 206, 234–35, 244; holdfast, 147, 152; land plant ancestry, 166, 168–69, 172, 181, 183–85, 184f, 193. *See also* brown algae; green algae; red algae
sedimentary rocks, 21–22, 38, 164, 170
seeds / seed plants, 5, 9, 17, 20, 21f, 82, 88, 180, 211, 230, 247, 250–51, 254, 258–59, 260–61, 260f, 263, chap. 9 (265–94); clubmoss "seed," 384; common ancestor, 229, 250, 276–77 (*see also* progymnosperm); dispersal, 3, 7, 271–74, 278f, 300–306; evolution, 233, 270, 273, 275f; food, 346, 355–60; fossil, 274, 282f, 291, 298; geological history, 21f, 218, 263, 269, 274, 277, 283f, 285f, 307; habit, 190, 272–74, 276, 307, 385, 408; life cycle, 270f, 272; phylogeny, 284–87, 286f, 287f, 336; pollen, 273; pollen drop mechanism, 276; seed ferns (*see* pteridosperm). *See also* cones
sex, 83, 117–22, 125, 127–28, 141, 145, 443–44, 448–50, 516; brown algae life cycle, 371f, 372f; cells, 168, 193, **526**; life cycles, 125–26, 126f, 147, 176–78, 180, 278; moss life cycle, 199f, 335f; *Oedogonium*, 446f; organs (gametangia, antheridia, archegonia), 179–80, 193–94, 199f, 208–9, 419, 467–72, 474, **523**, **524**; red algae life cycle, 491–92, 492f. *See also* asexual reproduction; conjugation; genetic: variation; meiosis
slime mold, 10, 77–78, 116, 128
species names, 14–15, 211, 519, **524**, **529**; binomial nomenclature, 348. *See also* Linnaeus
spontaneous generation, 35, **529**
starch, 6, 58, 60, 84, 139, 150, 239, 355, 357, 369, 440, 483, 514, 520, **530**
stomata / guard cells, 171–74, 176, 195, 207, 219, 234, 243, 258f, 388, 460, **530**; bryophytes, 199, 202–3, 207, 208f, 214–15, 456, 469; clubmosses, 376–77, 381, 388
stramenopiles, 86–88, 142, 369, 482–85; chrysophytes, 112, 481, 483f, 485–86; diatoms, 79, 87, 108, 132, 141f, 161, 481, 483f, 485–87; eustigmatophytes, 485, 487. *See also* brown algae

stromatolite, 36, 40–41, 129, 146, 160, 164, 307, 451, **530**
Sumer, 31
sulfur-based metabolisms, 38, 45, 57, 62
symbiosis, 33, 51, 75, 77, 86, 94, 100, 111, 128, 132, 201, 224, 226, 304, 322, 358, 428, 452–53, 506, 527, **530**. *See also* endosymbiosis; lichen

teaching science, x, 6, 11, 52, 362–63, 509
terrestrial: animals, 283, 340, 511; birds, 263; ecosystem, 135; habitat (patchy), 4–5, 141, 143, 182, 267, 360, 441; life, 142, 158, 164, 166, 170, 175, 181, 269. *See also* terrestrial plants
terrestrial environment: challenges of, 169–70; colonization, 80, chap. 6 (163–95); geological history, 132; life and changes to, 164, 167
terrestrial plants: bryophytes, 125; carbon sequestering, 224–25, 225f; chlorophyll, 66; cuticle, 172; hallmarks, 170–75, 171f; life cycle adaptations, 188–95; mat/soil crust communities, 164; spore dispersal, 177, 416; swimming sperm, 190. See also *Cooksonia*
themophile, 38, 43, 45, 62–65, 69–71, 100, 112, 129, 451, 524, **530**
Thermodynamics, second Law of, 46, 507
Thermoplasma, 112
Tigris, 31
trees. *See also* woody plants / trees
trimerophytes, 229, 251, 259–63, 262f, 278, 390, 411–12, 419–20, 461, 464, 493; *Pertica*, 262f
tyranny of data, xiii, 40. *See also* science

unicellular organisms, chap. 3 (75–130), **530**; algae, 49, 79, 87–89, 112, 132, 155, 164, 183, 192–93, 523; algae, phytoplankton, 132, 134, 140, 142, 150, 153, 226, 439–40, 445–46, 448–49, 481–86, 489, 528; algae, seaweeds, 116, 133, 147, 149, 151, 446–47, 510–11, 529; ancestors, 34–35, 80–82, 91–93, 95, 158, 528; eukaryotes, 34–35, 159, 162; fungus, 47, 78, 86, 151, 513; motility, amoeboid movement, 77–78, 83, 86–87, 90, 95–100, 107–8, 113–14, 116, 128, 372, 482, 486, 528–29; motility, flagella/cilia, 108–9, 112; motility, gliding, 77, 83, 108, 450–51, 454; protists, 10–12,

17, 80, 87–88, 95, 101, 116, 170, 512, 529; rhizoid, 513; sex, reproduction, 117–19, 121–22, 125–26, 140, 153, 511; sex cells, sex organs, 178–79, 181, 193, 371, 443, 510
universe, 23–29, 45–46, 75, 129, 508–9; big bang, 23–24; big chill, 29; big crunch, 29; big rip, **29**
UV, ultraviolet radiation, 44, 65–66, 120, 146, 160, 169, 295, 450, 479

vinegar, 32, 53, 360

whisk ferns, 501–6 (appendix); determinate aerial axes, 233, 248, 502f, 503f, 505; distribution and haploid phase, 178, 267; free-sporing land plant, 180, 416; megaphyll clade, 251, 258–60, 260f; microphyll, 253, 505; primitive stem anatomy, 244, 246, 257, 416; Psilotum, 502f; pteridophyte clade, 260–61, 260f, 411–12, 412f; rootless, 246; sporophyll, 460, 502f, 503f; subterranean gameto-phyte, 416, 423, 426, 502f; Tmesipteris, 503f; Y-branching, 245–47, 247f
wine, 31–32, 34, 53, 361, 515
wood, 172, 239, 279, 306, 343, 358, 361, 512, **531**; anatomy and evolution, 312–14; angiosperm, 241, 241f, 285, 316, 321, 330–31, 336, 352; annual growth, 7, 218, 235–37, 236f (*see also* cambium); gnetophyte, 280, 285; growth rings, 239–40, 240f, 514; gymnosperm, 240, 240f, 388; progymnosperm, 276–77; sapwood/heartwood, 239; softwood/hardwood, 331, 514, springwood/sum-merwood, 239–41, 240f, 519; tracheids, 171, 235–40, 236f, 240f, 241, 276, 285,

330–31, 388–89, 423, 437, 516; vessel elements, 240–41, 241f, 280, 285, 437
woody plants / trees: annual growth, 7, 235–36, 236f, 239–40, 240f (*see also* cambium); arborescent clubmosses, 224, 383–84, 384f; arborescent horsetails, 463, 463f; biggest organisms, trees, 160, 162, 218, 226–28, 228f, 248, 280, 325; biology, 305–12, 325–27; clade, 17, 211, 263, 266, 278; environmental impact, 224–25, 224f, 344–45, 347; forests, oldest, 218, 223–24, 230; form, 200, 325; gymnosperms, 226, 228–29, 281, 284, 312, 388, 495; progymnosperms, 276–77; tropical evergreen, 307. *See also* ferns
Woody Ranales, 313–16, 318, 320, 350, 352, 518

zoospore, 108, 180, **531**; invasion of land, 170, 190; land plant / charophycean, 181; in life cycles, 125–26, 126f, 179, 179f, 194; seaweed dispersal phase, 147, 184f, 188, 370–72, 371f, 372f, 445–47; size-number tradeoff, 188. *See also* meiosis
zygospore. *See* zygote
zygote, 34, 121, 124–26, 514, **531**; brown algae life cycle, 370–71, 371f, 372f; fern life cycle, 417f; flowering plant life cycle, 332–34; land plant life cycle, 179, 179f, 181, 183–84, 184f, 186, 188–90, 198–99, 199f, 208, 226, 231, 335f, 379, 461; seed plant life cy-cle, 271, 275f, 433, 436; sexual life cycles, 126f, 176–78, 449; red algae life cycle, 491–92, 492f; Ulva's life cycle, 184f, 447; zygospore, 122, 444, 450